Distribution Packaging

Distribution Packaging

WALTER F. FRIEDMAN
President
Walter Frederick Friedman & Co., Inc.

JEROME J. KIPNEES
Executive Vice President
Kipnees & Falkner, Associates, Inc.

ROBERT E. KRIEGER PUBLISHING COMPANY
HUNTINGTON, NEW YORK
1977

First Edition 1977

Printed and Published by
ROBERT E. KRIEGER PUBLISHING CO., INC.
645 NEW YORK AVENUE
HUNTINGTON, NEW YORK 11743

Material from Original Edition
Copyright © 1960 by
JOHN WILEY & SONS
New Material Copyright © 1977 by
ROBERT E. KRIEGER PUBLISHING CO., INC.

All rights reserved. No reproduction in any form of this book, in whole or in part (except for brief quotation in critical articles or reviews), may be made without written authorization from the publisher.

Printed in the United States of America

Library of Congress Cataloging in Publication Data

Friedman, Walter F.
   Distribution packaging.

   First edition published in 1960 under title: Industrial packaging.
   Bibliography: p.
   Includes index.
   1. Packaging.   I. Kipnees, Jerome J., joint author.
II. Title.
[TS195.F74   1976]   621.7'57   75-22096
ISBN 0-88275-222-7

TO LILLY and PEARL,
OUR DEVOTED WIVES

Preface to Revised Edition

The original edition of this book entitled *Industrial Packaging* was issued in 1960. The motivation to write the original text was derived from an awareness of the lack of suitable analytical and interpretative material in the field. This paucity of material relating to principles was particularly evident to us at that time in our teaching of university courses dealing with packaging and material handling.

Since the first edition was published, there have been many major changes in technology which have had a profound influence on our life styles. These changes have served to advance the discipline of packaging and have provided a wider range of options for problem solving than heretofore. Some of the remarkable developments which have occurred in this period include the increased emphasis on convenience orientation due to greater purchasing power and leisure time, the rapid growth of single service or disposable products, expanding world markets demanding more consumer goods and an increasing sophistication in production, marketing and distribution. The sum total of this changing scene has dramatically increased the consumption of packaging materials and containers but has also magnified the problems which are commonplace in contemporary packaging. For example, while the use of single service units have many excellent benefits, the impact of this development has led to the twin problem of material shortages and solid waste disposal. This in part has contributed to the concern of environmentalists and has stimulated increasing governmental involvement and legislation.

Another example of our changing technology relates to the impact of plastics on all aspects of packaging. Materials in common use today were virtually unknown when the original text was compiled. Interestingly, since plastics are petrochemical derivatives, they are therefore seriously influenced by the availability and cost of crude oil. It remains to be seen what magnitude of impact this will have on the field of packaging in the future. Certainly, greater attention will be placed on those raw material sources which are more readily available or can be more easily replenished.

Still another major consideration has been the emergence of *physical distribution management* as a viable discipline which provides a systems approach to the flow, storage and control of product. It has become recognized that packaging greatly influences the cost, safety and convenience of material handling, storage and transportation. Because of this interface, we felt compelled to change the name of the text from *Industrial Packaging* to a more meaningful title of *Distribution Packaging*. In addition to this title change, greater emphasis was placed on the full interrelationship of packaging with the distribution function.

Unquestionably, the discipline of packaging will retain its importance in coming generations emphasizing the continued need for trained specialists to cope with ever increasing complexities.

The expansion, refinement and upgrading of packaging education is largely dependent upon the availability of suitable teaching aids. While professional societies, trade associations, carrier agencies, commercial suppliers and branches of the government continue to make voluminous technical data available, there is a continuing need to formulate basic principles to permit the proper interpretation of available data.

The approach selected in the presentation of the subject matter was designed as a concise and objective treatment of distribution packaging with major emphasis on the selection of materials, systems, and equipment required to perform the physical distribution function of commerce and industry economically. We did not attempt to prepare a listing of descriptive material such as would be contained in a manual or handbook. Furthermore, it was considered impractical to include all aspects of distribution packaging in one volume. The direction that we pursued was towards the development of an understanding of the basic areas of the distribution packaging function and packaging accouterments.

For a better understanding of the area embraced by distribution packaging, the following definition may be applied: *distribution packaging is the function fundamentally concerned with the economical preparation and protection of merchandise for shipment and distribution.* If sales acceptance is or should be influenced by packaging techniques, the activity towards the accomplishment of this objective is commonly referred to as *consumer packaging*. It can be readily appreciated that there is no sharp division between these two major segments of the packaging function. Therefore, when overlapping occurred in the selection of materials, systems, or equipment, a discussion of what normally may be termed consumer packaging was introduced.

PREFACE TO REVISED EDITION

The distribution packaging function and related activities consume important proportions of the total product and distribution expenditure. Therefore, many economies can be achieved by the proper selection and application of materials, systems and equipment. In addition, the need for an adequate level of product protection is emphasized by the staggering damage claims paid by transportation agencies.

The modern complexities of the packaging field and the many scientific advances attained in recent years often make the selection of a suitable package a very difficult task. It is the purpose of this textbook to establish updated criteria whereby the packaging of a given product can be effected on an engineering basis and systems can be designed to assure a satisfactory level of product protection at minimum distribution cost.

The dynamic growth of the packaging field has presented us with a dilemma with respect to the inclusion or deletion of subject matter in the book. Whenever possible, an attempt has been made to cover the latest technological advances despite the lack of current analytical and comparative data. Similarly, we felt the necessity to condense or merely refer to well-documented principles and interpretations readily obtainable in the literature of the field. The bibliography contains representative material of such published matter. It should be recognized that since the rate of technological change in this field is very rapid, important revisions and additions are constantly being made. The reader should be alerted to this fact and should supplement his reading with literature specifically reporting on new developments.

For the proper organization of the subject of distribution packaging, this book has been divided into three essential parts. One part introduces the packaging field with specific emphasis on distribution packaging aspects and includes the historical evolution and milieu of packaging. The second part, entitled "Packaging Materials and Containers," covers the most important media used to contain and protect products in distribution. Particular emphasis was directed to those areas in which modification of the media is possible. Some standard types of materials and containers were only briefly covered because well-documented data are currently available. The final part, entitled "Packaging Systems and Equipment," deals with the integrated aspects of packing and packaging. This area includes various types of equipment as well as other packaging materials basically designed for the performance of these functions.

References to specific materials, designs, or machines are intended to illustrate principles or attainable variations. This mention does not preclude the existence of similar tools and equipment of equal quality. Furthermore, we do not lend our endorsement to a specific product solely on the basis of inclusion in this book.

We gratefully acknowledge the splendid assistance furnished by the many individuals and organizations whose contributions have made the revised edition possible. We particularly wish to extend our sincere appreciation to the following individuals who provided valuable advice and guidance.

Edward E. Janda, Package Research Laboratory, Alfred W. Hoffman, Fibre Box Association, Norman T. Baldwin, National Paper Box Association, C. G. Peterman, Paper Shipping Sack Manufacturers Association, Inc., Dr. E. George Stern, Virginia Polytechnic Institute and State University, E. B. Edwards, Techs, Inc., George Gero, Gero Consulting Company.

In addition, we wish to thank the following organizations for valuable statistical and/or technical information which they supplied.

Packaging Education Foundation, Inc., Packaging Machinery Manufacturers Institute, U.S. Department of Commerce, Society of Packaging and Handling Engineers, The Packaging Institute, U.S.A., Technical Association of the Pulp and Paper Industry, The American Society for Testing and Materials, National Safe Transit Committee, Inc., Container Testing Laboratory, Inc., American Paper Institute, Paperboard Packaging Council, Steel Shipping Container Institute, American National Standards Institute, Inc., and Composite Can and Tube Institute.

It would be impossible to list all of the many other organizations who contributed to this book. However, every attempt has been made to credit all those directly or indirectly concerned, and to update the identity and current affiliation of those illustrations and materials retained from the original text.

November, 1976 *Walter F. Friedman*
Jerome J. Kipnees

Contents

Chapter 1	Scope of Packaging for Distribution	1
	PACKAGING MATERIALS AND CONTAINERS	
Chapter 2	Corrugated and Solid Fibreboard	63
Chapter 3	Folding Cartons and Rigid Boxes	123
Chapter 4	Shipping Sacks	173
Chapter 5	Nailed Wooden Boxes and Crates	209
Chapter 6	Wirebound and Cleated Containers	251
Chapter 7	Cylindrical Shipping Containers	277
Chapter 8	Wrapping, Barrier, and Cushioning Materials	313
	PACKAGING SYSTEMS AND EQUIPMENT	
Chapter 9	Fastenings and Closures	359
Chapter 10	Reinforcing, Bundling and Unitizing Systems, Materials, and Equipment	407
Chapter 11	Easy-opening Devices	451
Chapter 12	Marking, Labeling, and Coding Systems, Materials, and Equipment	467
Chapter 13	Packaging Equipment	501
BIBLIOGRAPHY		527
INDEX		535

CHAPTER 1

Scope of Packaging for Distribution

From his earliest days man has recognized the need to provide means to carry, retain, collect, and store the products necessary for his existence. These products included water and other liquids, wild fruits and berries, wildlife he had hunted, and later, the agricultural surplus of a developed agrarian economy. The containers invented to satisfy these needs came from the land, as did man's complete livelihood, which was largely dependent upon the physical surroundings of the immediate environment. Examples of such primitive container materials included shells, leaves, hollowed wood sections, and animal skins. As man's skills developed, particularly handicrafts, woven, interlaced, and molded containers supplanted the earlier container forms. Many of these are still in use by primitive societies in undeveloped parts of the world.

Essentially, all of these earlier container forms existed as part of the household implements and were considered to be prized possessions. Because of the difficulties of fabrication, these containers never left the possession of the owner and were thus, strictly re-usable. In trading, the seller would transfer his merchandise from his own container to the buyer's container which had to be provided to accept the merchandise. As trading to far-flung areas developed, the feasibility of returning containers became less practical. Trading over land and sea routes introduced new problems in properly protecting the merchandise, since difficult environmental hazards were encountered, and provisions to transfer the merchandise from one mode of transportation to another had to be considered.

The establishment of trade is the result of specialization in the fabrication and cultivation of certain commodities and in the accumulation of surpluses which can be profitably disposed of in other geographic locations. Not only did trade permit a better distribution of produce and commodities, but it also served in the communication of new ideas and inventions, among which were early container types and modifications.

For instance, glassmaking was arrived at quite by accident some 6,000 years ago by a crew of Phoenician sailors on the sandy shores of what is now Israel. The basic ingredients producing this early glass has changed little during the course of time. Since the Phoenicians were energetic traders, the art of glassmaking was soon transferred to other areas of the civilized world.

As trade routes lengthened and surpluses increased, spoilage of perishable commodities became a serious problem. This spoilage in turn led to the adoption of preservation techniques for the commodity itself, such as the spicing, curing, pickling, and salting of food products. The distribution of liquids such as wine required suitable containers which eventually developed into a special craft of glass and earthenware producers.

In essence, trade is responsible for the development of packages, and the art of packaging is directly related to the development of the distribution of products. With the industrial revolution, and with the creation of manufacturing centers, a very extensive expansion of local and world-wide trade commenced. It was during this period that the requirements for a packaging industry were established and the fabrication of containers was evolved out of the handicraft era. The major emphasis of packaging at this time was protection, with little attention directed towards consumer appeal, although the marketing of luxury items had precipitated what is today known as consumer packaging.

At the time of the industrial revolution, packaging materials were primarily earthenware, wood, straw, and other natural fibres. Textile bags date back to the earliest days of civilization but were only widely introduced as a packaging media with the growth of the textile industry.

SCOPE OF PACKAGING FOR DISTRIBUTION

HISTORY OF PACKAGING IN THE UNITED STATES

In early American history, the balance of trade consisted of the barter of finished European goods for agricultural, mineral, and animal products. American export, which was largely bulk transfer due to the nature of the products traded, required little if any shipping protection. The imported material, however, representing high-value and fragile commodities, utilized European container types which were eventually adopted in this country. As this country became more self-sufficient economically and as domestic trade increased, the manufacture of packaging materials progressed. This self-sufficiency commenced with the gradual westward movement of population and the development of an improved transportation system. Much of the early trade in this country was water-borne, which was inadequate for complete geographic coverage. With the development of the railroads, the beginnings of a nation-wide transportation system was established, and the flow of goods was accelerated. The eastward flow of goods was predominantly in bulk and constituted natural resources and agricultural products. Packaging of these products was of relatively minor importance, and the railroad car, barge, or steamship hold was the shipping container. The westward flow consisted of manufactured materials produced in the newly established shops and mills along the Eastern Seaboard. These goods represented the work implements and luxury items required by the Western and Southern populations and shipping protection was needed.

The packaging materials utilized at this time included wood, glass and earthenware, and textiles. Lumber was plentiful and inexpensive and provided good protection during the many transfers a commodity experienced until it reached its final destination.

The first organized packaging industry in this country was the glass industry which had its origin in Salem, Massachusetts, where colonists started a plant to make glass bottles to transport New England rum and cider to the Carolinas and the West Indies. In 1739, one of the most successful of early American glass factories was founded in New Jersey.

Another early American packaging industry was the cooperage trade, which recognized as early as 1915 the need for detailed tight and slack barrel specifications.

At the time Columbus discovered America, paper manufacture had become established in Spain, France, Germany, and England. William Rittenhouse and three partners introduced paper into the American colonies in 1690 when he established a mill in Germantown, Pennsylvania. Paper at that time and until the early part of the nineteenth century was made by hand and therefore had limited applications for packaging. With the development of a machine for making paper in a continuous sheet, rapidly expanding production was possible. Originally, all papers were made from cotton and linen rags and straw as the raw materials. During the middle of the nineteenth century various processes for the manufacture of pulp from wood fibres were developed which reduced cost and influenced the sites for future paper mills. This development initiated the production and use of *coarse* papers for packaging.

With the expanded manufacture of paper and, subsequently, paperboard products, the birth of an important section of the packaging industry occurred.

Further historical background concerning principal packaging materials and container types is furnished in the introduction of the various chapters of the text.

THE PACKAGING INDUSTRY

Within less than a century, the packaging industry in the United States has grown into major proportions. The annual expenditures for packaging materials, containers, and equipment is estimated to approximate $27 billion. According to government and industry sources, the value of packaging materials alone was estimated to be in excess of $25.6 billion (Table 1-1) and for packaging machinery the comparable estimate was almost one half billion dollars (Table 1-2). The Bureau of Census estimates a major growth pattern as shown in Table 1-3. Five principal categories of containers and packaging materials representing nearly 65% of all materials used are estimated to have a value of shipments of $21.1 billion by 1980.

More detailed industry statistics relating to the five categories shown in Table 1-3 are reproduced in Tables 1-4, 1-5, 1-6, 1-7 and 1-8. This information on the packaging industry is compiled annually by the Bureau of the Census, Bureau of Labor Statistics, Bureau of Domestic Commerce.

The giant packaging industry is in no way homogeneous, nor are all products manufactured designed exclusively for packaging use. For instance, pulp and paper manufacture is one of the largest industries in the United States. Over half of the volume output of this industry is destined for packaging with an estimated value of $10.0 billion in 1980.

Other large industries produce packaging materials, but not exclusively. Notable among these are the steel, glass, textiles, and plastic industries.

SCOPE OF PACKAGING FOR DISTRIBUTION

The packaging industry can be segregated into the following major segments:

1. *Producers of raw materials.*
2. *Convertors of raw materials.*
3. *Packaging machinery manufacturers.*
4. *Packaging service organizations.*

The producers of raw materials include pulp and paper mills, steel mills, chemical plants, and lumber processors. The producers are usually located close to the basic raw materials required in the fabrication of the packaging and container types. Occasionally the producer of the raw material is also equipped in the same plant to perform some or all of the conversion.

In consideration of the freight costs of distributing containers and packaging materials, the convertors and their markets are in close vicinity. The convertors consist of corrugators, box shops, laminators, and extruders. When a producer owns raw material manufacturing and converting facilities, this is considered an *integrated* operation. The recent trend, for economic reasons, is for greater integration in the packaging industry.

There are literally hundreds of organizations supplying standard and specialized packaging tools and equipment, from simple cutting tools to complex filling, assembling, and packing devices. Some of these equipment manufacturers also convert packaging materials used in conjunction with their machinery, that is, strapping, stapling, labeling, bagging, etc.

Packaging service organizations include all other agencies directly concerned with the packaging function such as contract packagers, consultants, testing and research organizations, testing machine manufacturers, and trade associations.

Contract packagers are firms whose only function is to undertake the packaging operation for a producer or distributor of merchandise. The need to use this service is predicated on any one or several of the following conditions:

1. The need to package economically with specialized equipment when purchase is not economically justified. This condition may prevail when a new item is introduced and the sales potential is unknown or the normal output is limited.

2. To complement an existing packaging operation during peak activity, when space, personnel, or scheduling limitations would place a heavy burden on facilities.

TABLE 1-1

Value of packaging materials: 1960–1974

CONTAINER OR MATERIAL	SIC NUMBER	1960	1970	1971	1972	1973	1974
Paper and paperboard containers (total)		$ 4,705,937,000	$ 7,454,044,000	$ 7,632,891,000	$ 8,297,425,000	$ 9,050,100,000	$10,348,000,000
Grocery, variety and misc. bags	26431,26430/00	273,810,000	370,000,000	381,111,000	390,000,000	400,000,000	493,000,000
Specialty bags—paper and laminations		81,000,000	85,000,000	87,550,000	90,000,000	93,000,000	157,000,000
Glassine, waxed or parchment bags		35,564,000	30,000,000	31,080,000	32,000,000	33,000,000	30,000,000
Paper shipping sacks	26433	313,098,000	339,000,000	349,170,000	360,000,000	400,000,000	430,000,000
Folding paper boxes, cartons	2651	940,000,000	1,235,000,000	1,250,000,000	1,330,000,000	1,400,000,000	1,540,000,000
Rigid paper boxes (set-up)	2652	287,854,000	434,000,000	411,000,000	438,000,000	470,000,000	474,000,000
Molded pulp egg cartons		87,000,000	105,000,000	108,150,000	109,000,000	114,000,000	120,000,000
Die-cut fillers for egg cases	26451/98	25,000,000	20,000,000	21,000,000	22,000,000	23,000,000	24,000,000
Sanitary food containers	26541/11[4]	662,133,000	1,225,000,000	1,250,000,000	1,275,000,000	1,400,000,000	1,540,000,000
Fiber cans (composite)	26552/23-27	83,179,000	165,000,000	171,000,000	181,000,000	200,000,000	280,000,000
Fiber drums	26551/51	61,272,000	105,000,000	109,150,000	111,500,000	117,100,000	160,000,000
Solid fiber and corrugated shippers	2653	1,856,027,000	3,341,044,000	3,463,680,000	3,959,925,000	4,400,000,000	5,100,000,000
Flexible packaging materials (total)		$ 831,733,000	$ 1,575,500,000	$ 1,636,800,000	$ 1,763,500,000	$ 1,699,975,000	$ 2,100,000,000
Paper—Heavy wrapping	26216/01/42[5]		115,000,000	120,000,000	130,000,000	140,000,000	150,000,000
Wrapping tissue	26218/78		40,000,000	42,000,000	45,000,000	50,000,000	65,000,000
Laminated wraps		352,163,000	115,000,000	123,000,000	135,000,000	145,000,000	161,000,000
Waxed wraps (bread, candy, etc.)			110,500,000	75,500,000	70,000,000	65,000,000	70,000,000
Cigarette and gum wraps			106,000,000	109,000,000	112,000,000	115,000,000	124,000,000
Cellophane		276,570,000[7]	253,000,000[7]	254,000,000[7]	250,000,000[7]	245,000,000[7]	246,000,000[7]
Polyethylene		140,000,000[7]	600,000,000[8]	672,000,000[8]	770,000,000[8]	850,000,000[8]	983,000,000[8]
Others[1]		N.A.	160,000,000[8]	163,000,000[8]	170,000,000[8]	175,000[8]	200,000,000[8]
Metal foil, flexible		63,000,000	76,000,000	78,300,000	81,500,000	89,800,000	101,000,000
Metal containers and components (total)		$ 2,322,494,000	$ 4,265,320,000	$ 4,629,315,000	$ 4,997,712,000	$ 5,386,127,000	$ 6,001,000,000
Metal cans (steel and aluminum)	3411	1,753,901,000	3,312,000,000[9]	3,643,000,000[9]	3,930,000,000[9]	4,400,000,000	4,796,000,000[9]
Collapsible metal tubes	3496	40,521,000	71,320,000	65,815,000	59,830,000	62,527,000	65,000,000
Rigid and semirigid foil containers	34970/33	58,300,000	70,000,000	73,500,000	78,000,000	82,600,000	90,000,000
Steel shipping barrels, drums, pails	3491	248,790,000	353,000,000	383,000,000	336,118,000	370,000,000	480,000,000
Reconditioned barrels and drums		98,700,000	204,000,000	206,000,000	208,000,000	210,000,000	285,000,000
Steel strapping	34990/57	86,572,000	173,000,000	175,000,000	174,000,000	176,000,000	192,000,000
Gas cylinders	34434	35,710,000	82,000,000	83,000,000	84,000,000	85,000,000	93,000,000
Aerosols (total)[2]		$ 114,700,000	$ 430,000,000	$ 423,000,000	$ 449,000,000	$ 486,000,000	$ 525,000,000
Nonfood		107,200,000	406,000,000	400,000,000	425,000,000	460,000,000	497,000,000
Food		7,500,000	24,000,000	23,000,000	24,000,000	26,000,000	28,000,000
Glass containers (total)	3221	$ 937,428,000	$ 1,785,000,000	$ 1,930,000,000	$ 2,100,000,000	$ 2,290,000,000	$ 2,350,000,000

Category	Codes						
Closures (total)		$ 264,860,000	$ 505,100,000	$ 518,720,000	$ 525,600,000	$ 554,000,000	$ 604,000,000
Metal caps (includes milk bottle closures)	34616	143,012,000	175,000,000	164,500,000	161,000,000	155,000,000	160,000,000
Metal crowns	34617	74,277,000	90,000,000	94,020,000	91,500,000	90,000,000	90,000,000
Milk bottle caps, paper	26451/81	14,297,000	20,100,000	18,200,000	16,100,000	14,000,000	12,000,000
Plastic commercial closures	30794/71	33,274,000	220,000,000	242,000,000	257,000,000	295,000,000	342,000,000
Rigid & semirigid plastic containers (total)		$ 149,303,000	$ 660,200,000	$ 752,200,000	$ 810,680,000	$ 939,000,000	$ 1,108,000,000
Plastic bottles		64,000,000	345,000,000	405,000,000	428,000,000	514,000,000	607,000,000
Boxes and baskets		58,300,000	105,000,000	111,000,000	124,000,000	138,000,000	163,000,000
Tubes		N.A.	34,200,000	35,000,000	40,000,000	42,000,000	50,000,000
Jars and tubs		7,403,000	60,000,000	66,000,000	68,000,000	73,000,000	86,000,000
Plastic sheet[3]		15,100,000	80,000,000	92,000,000	101,000,000	112,000,000	132,000,000
Foamed plastics		4,500,000	36,000,000	43,200,000	49,680,000	60,000,000	70,000,000
Wooden containers (total)		$ 420,996,000	$ 615,000,000	$ 579,000,000	$ 572,000,000	$ 579,000,000	$ 637,000,000
Nailed and lock-corner boxes	2441	321,651,000	300,000,000	276,000,000	274,000,000	275,000,000	300,000,000
Wirebound boxes, crates	2442		210,000,000	205,000,000	200,000,000	205,000,000	227,000,000
Cooperage (tight and slack)	2445	69,885,000	70,000,000	66,000,000	64,000,000	63,000,000	70,000,000
Veneer and plywood	2443	29,460,000	35,000,000	32,000,000	34,000,000	36,000,000	40,000,000
Textile containers (total)	2393,22992/51	$ 185,450,000	$ 246,000,000	$ 239,000,000	$ 191,000,000	$ 185,000,000	$ 198,000,000
Cushioning materials (total)		$ 31,573,000	$ 32,900,000	$ 31,800,000	$ 32,500,000	$ 34,700,000	$ 37,000,000
Creped wadding	2692/65	13,000,000	13,600,000	13,100,000	13,000,000	13,500,000	14,000,000
Excelsior	24290/81/89	11,824,000	9,900,000	9,800,000	10,000,000	11,000,000	12,000,000
Flock	22940/41	4,590,000	7,500,000	7,000,000	7,500,000	8,000,000	8,500,000
Shredded paper for packing	26495/68 (1958)	2,159,000	1,900,000	1,900,000	2,000,000	2,200,000	2,500,000
Component materials (total)		$ 803,691,000	$ 1,061,000,000	$ 1,099,500,000	$ 1,123,500,000	$ 1,235,500,000	$ 1,387,000,000
Adhesives		67,000,000	104,000,000	108,000,000	110,000,000	115,000,000	165,000,000
Labels (paper or foil)	27512, 18, 19, 22	352,000,000	495,000,000	512,000,000	516,000,000	560,000,000	600,000,000
Tags, printed, unprinted	27516/75, 27526/78; 26495,82	46,000,000	60,000,000	62,000,000	64,000,000	70,000,000	86,000,000
Pressure-sensitive and gummed tape	26413/12, 26414	304,757,000	340,000,000	355,000,000	370,000,000	425,500,000	464,000,000
Cotton, soft- and hard-fiber twine	22981/32/39/00[6]	33,934,000	62,000,000	62,500,000	63,500,000	65,000,000	72,000,000
Cargo or bulk containers (total)		$ 95,000,000	$ 250,000,000	$ 267,000,000	$ 290,000,000	$ 320,000,000	$ 359,000,000
TOTAL		$10,863,165,000	$18,880,064,000	$19,739,226,000	$21,152,917,000	$22,746,402,000[10]	$25,654,000,000[10]

SOURCE: Based on information from Government and industry organizations. The 1973 entries are projections.
[1] Includes Pliofilm, polyester, saran, cellulose acetate, vinyl, etc. [2] Includes container, valve, cap and propellant. [3] For thermoformed or fabricated packages. Does not include forming. Does not include foams (listed separately below). [4] Includes 26542/31; 26543/41; 26543/51 & 61; 26543/31. [5] Less household. [6] Includes 22982/23/25/27/99, 22983/25. [7] These figures do not include the value added by conversion. [8] These figures include the value added to conversion. [9] Does not include aerosol containers. [10] It is generally accepted that addition of miscellaneous and unreported items would increase this figure to nearly $25 billion.
Reproduced by special permission of the publishers of modern packaging encyclopedia and planning guide, December 1974.

TABLE 1-2
Value and growth of packaging machinery—1972, 1973*

MACHINERY CATEGORIES	Dollar value of shipments 1972	Anticipated 1973
1. Filling:[1] Dry	$ 11,816,288	$ 12,842,000
Liquid, semiliquid, liquids/solids, milk, juices (except beverage)	28,261,606	32,319,000
Beverage (beer, soft drinks, liquor, wine)	11,904,243	13,411,000
Collapsible tube filling	1,011,000	1,249,000
Counting equipment	2,057,000	2,662,000
2. Capping and sealing[2]	9,278,785	10,140,000
3. Lidding equipment	1,697,500	2,153,500
4. Aerosols (fill, gas, close, crimp, etc.)	3,690,625	4,290,000
5. Form-fill-seal (except vac. or gas): Pouches, bags	35,516,109	40,196,385
Other, cups, trays, etc.	7,590,000	9,530,000
6. Vacuum and gas packaging: Pouch and bag	7,189,413	7,218,154
Rigid, semirigid (thermoformed, etc.)	5,014,000	4,990,000
7. Wrapping, banding (except shrink)	20,318,850	23,897,920
8. Bundling (except shrink)	200,000	200,000
9. Shrink wrapping: (consumer-type) (includes tunnel if integral)	5,266,000	7,813,000
Shrink case (except multipacks)	1,868,350	3,278,450
Shrink pallet equipment (includes tunnels if integral)	885,752	1,909,000
10. Shrink tunnels (free standing, except pallet tunnels)	508,000	739,500
Pallet tunnels	2,225,252	3,800,000
11. Bag handling, filling, closing: Loads under 25 lb.	3,575,800	4,190,000
Loads 25 lb. and over	4,277,000	4,945,000
12. Scales (used on packaging line) (free standing)	6,364,345	6,917,679
13. Checkweighers and integral control equipment	7,172,000	9,720,000
14. Cartoners (set-up and/or fill and/or close)	16,326,848	18,348,440
15. Multipackers, paperboard	3,139,173	3,343,000
Plastic	1,333,850	1,794,000
16. Labeling machines: Glue	16,753,113	17,312,088
Thermoplastic	1,096,000	828,000
Pressure-sensitive	2,230,000	2,790,000

17. Code marking and imprinting (on-line)	10,922,465	12,162,112
18. Code marking and imprinting (off-line)	837,435	840,000
19. Thermoformers: Blister, skin and related equipment for cardpacks	9,088,500	11,300,000
Sheet-formed trays, cartons, lids, tubs, bottle sections	5,049,000	7,008,000
20. Heat-sealers (separate units)	1,407,000	1,884,000
21. Fastening equipment:		
Staplers and stitchers (except case sealers)	4,785,000	5,962,000
Strapping, tying	1,650,000	1,750,000
Tape dispensing (except case sealers)	11,176,000	12,303,000
22. Case opening or forming (includes tray formers)	2,970,866	3,761,399
23. Case loading (includes opening, sealing, if integral)	14,476,502	20,136,385
24. Case sealing	5,254,084	7,630,000
25. Package handling: Case unloaders	3,007,841	3,636,000
Aligners, unscramblers, accumulators	4,089,032	4,757,700
Palletizers, depalletizers	4,001,000	5,709,000
Conveyors, transfer devices	12,160,909	16,597,800
Other conveyor types	3,671,935	4,460,641
26. Bottle cleaners: Liquid wash (caustic soak)	5,895,977	6,865,000
Rinsers	264,523	105,000
Air cleaners	279,100	300,500
27. Package-making (captive or packager use only):		
Box or carton making	4,017,000	2,054,000
Bag making	3,441,000	4,442,000
Can making—metal	1,535,000	2,160,000
Can making—paper, composite	350,000	410,000
Can sealing or seaming	2,675,000	2,675,000
Plastic bottle making (except thermoforming)	16,319,000	19,183,000
Miscellaneous, not classified	1,490,000	1,800,000
28. Miscellaneous (packaging line use):		
Inspection and quality control	4,038,846	5,455,000
Testing devices	5,175,000	5,325,000
Bottle/can warmers	701,708	808,000
Accessories	18,719,000	25,853,000
Other	4,087,400	4,596,000
TOTALS	**$382,103,025**	**$454,756,653**

*This study sponsored jointly by MODERN PACKAGING ENCYCLOPEDIA and the Packaging Machinery Manufacturers Institute. It is based on replies to a questionnaire received from manufacturers of packaging machinery. Replies were tabulated by McGraw-Hill Research.
[1] Except scale components and fillers for aerosols, bags, pouches. [2] Except aerosols.
Reproduced by Special Permission of the Publishers of Modern Packaging Encyclopedia and Planning Guide, December, 1973.

TABLE 1-3
CONTAINERS AND PACKAGING: PROJECTIONS 1972–80[1]
[Value of shipments in millions of dollars except as noted]

SIC code	Industry	1972	Percent increase 1971-72	1973	Percent increase 1972-73	1980	Percent increase 1972-80[2]
2651	Folding paperboard boxes	1,325	6	1,390	5	1,800	3.9
2653	Fibre boxes	4,200	12	4,600	9	8,500	9.0
3221	Glass containers	2,180	7	2,280	5	2,800	3.2
3411	Metal cans	4,600	9	4,900	7	7,200	5.8
3491	Metal barrels, drums, and pails	450	8	470	5	605	3.8
	Total Value	12,755		13,640		21,905	

[1] Estimated by Bureau of Domestic Commerce. [2] Compound annual rate of growth.

TABLE 1-4
Folding Paper Boxes: Trends and Projections 1967–73

[In millions of dollars, except as noted]

	1967	1969	1970	1971[1]	1972[1]	Percent increase 1971-72	1973[1]	Percent increase 1972-73
Product:[2]								
Value of shipments	1,109	1,229	1,225	1,250	1,325	6	1,390	5
Total employment (thousands)	49	54	48	44	44	0	---	---
Production workers (thousands)	39	43	38	35	35	0	---	---
Value added	532	590	575	610	NA	---	---	---
Value added per production worker man-hour (dollars)	6.48	6.59	7.26	8.37	NA	---	---	---
Quantity shipped (thousands of short tons)	2,510	2,627	2,490	2,445	2,500	2	2,575	3
Value of exports	4.3	4.4	4.9	6.5	6.0	-8	---	---
Wholesale price indexes (1967=100)[3]	100.0	104.4	108.2	111.6	115.3	3	---	---

[1] Estimated by Bureau of Domestic Commerce.
[2] Includes value of shipments of folding paper boxes made by all industries.
[3] Price index includes shipping containers.

NOTE.–NA = not available.
Source: Bureau of the Census, Bureau of Labor Statistics, Bureau of Domestic Commerce.

TABLE 1-5
FIBRE BOXES: TRENDS AND PROJECTIONS 1967-73
[In millions of dollars except as noted]

	1967	1969	1970	1971[1]	1972[1]	Percent increase 1971-72	1973[1]	Percent increase 1972-73
Industry[2]								
Value of shipments	2,960	3,500	3,508	3,650	4,200	12	4,600	9
Total employment (thousands)	97	106	104	104	106	2		
Production workers (thousands)	73	79	78	78	80	3		
Value added	1,130	1,419	1,459	1,505	NA			
Value added per production worker man-hour	$7.41	$8.50	$9.29	$9.76	NA			
Product[3]								
Value of shipments	2,893	3,401	3,435	3,575	4,115	15	4,505	9
Quantity shipped (billion square feet)	163	186	185	192	206	8	218	6
Value of exports	4	4	4	4.4	6	36		
Wholesale price indexes (1967=100)	100	105	108	112	116	3.5		

[1] Estimated by Bureau of Domestic Commerce.
[2] Includes value of all products and services sold by the fibre boxes industry (SIC 2653).
[3] Includes value of shipments of fibre boxes made by all industries.

NOTE.—NA=not available.
Source: Bureau of the Census, Bureau of Labor Statistics, Bureau of Domestic Commerce.

TABLE 1-6
GLASS CONTAINERS: TRENDS AND PROJECTIONS 1967-73

[In millions of dollars except as noted]

Industry:[2]	1967	1969	1970	1971 [p]	1972 [1]	Percent increase 1971-72	1973 [1]	Percent increase 1972-73
Value of shipments	1,352.4	1,664.7	1,863.9	2,046.2	2,180.0	7	2,280	5
Total employment (thousands)	66.7	71.5	74.6	76.6	77.5	1		
Production workers (thousands)	59.4	63.5	66.3	67.9	68.5	1		
Value added	842.2	1,112.4	1,222.6	1,358.6	NA	7		
Value added per production worker man-hour	$7.05	$8.94	$9.28	$10.21	NA	5		
Product:[3]								
Value of shipments	1,331.0	1,638.7	1,842.9	[1] 2,020.0	2,150.0	6	2,250	5
Quantity shipped (1,000 gross)	231,046	253,198	268,158	257,129	267,000	4	273,000	2
Value of imports	2.4	5.6	5.6	6.5	7.0	8	7	
Value of exports	21.1	20.5	23.1	22.4	22.0	2	22	
Wholesale price indexes (1967=100)								
Glass containers	100.0	114.8	120.4	131.6	135.1	3		
Wide mouth food	100.0	115.4	120.7	132.9	136.9	3		
Beer bottle, non-returnable	100.0	108.6	112.5	117.3	117.9	1		

[1] Estimated by Bureau of Domestic Commerce.
[2] Includes value of all products and services sold by the glass container industry (SIC 3221).
[3] Includes value of shipments of glass containers made by all industries.

[p] Preliminary.
NOTE.—NA=Not available.
Source: Bureau of the Census, Bureau of Labor Statistics, Bureau of Domestic Commerce.

TABLE 1-7
METAL CANS: TRENDS AND PROJECTIONS 1967–73
[In millions of dollars except as noted]

	1967	1969	1970	1971 P	1972 [1]	Percent increase 1971–72	1973 [1]	Percent increase 1972–73
Industry [2]								
Value of shipments	2,890.6	3,548.4	3,912.7	4,223.0	4,600.0	9	4,900	7
Total employment (thousands)	60.3	65.1	71.0	70.3	68.6	−2		
Production workers (thousands)	52.3	56.4	60.3	60.2	58.5	−3		
Value added	1,141.5	1,410.4	1,638.9	1,716.1	NA	NA		
Value added per production worker man-hour	$10.13	$11.87	$12.84	$13.77	NA	NA		
Product [3]								
Value of shipments	2,585.7	3,198.7	3,512.5	3,800.0 [1]	4,140.0	9	4,400	6
Quantity; total base boxes (thousands) [4]	133,980	152,617	159,969	161,890	168,400	4	175,000	4
Steel	126,141	140,248	142,825	140,475	144,000	3	148,000	3
Aluminum	7,839	12,369	17,144	21,415	24,400	14	27,000	11
Value of exports	12.3	11.1	12.3	12.4	13.0	5	13	
Wholesale price indexes (1967=100)								
Metal cans	100.0	106.8	112.3	122.0	129.3	6		
303×406 can	100.0	107.0	113.1	123.8	131.8	6		
12-ounce beer can	100.0	106.9	111.8	121.1	128.7	6		

[1] Estimated by Bureau of Domestic Commerce.
[2] Includes value of all products and services sold by the metal can industry (SIC 3411).
[3] Includes value of metal cans made by all industries.
[4] A base box is 31,360 square inches, equivalent to 112 sheets 14″ x 20″.

P Preliminary.
NOTE.—NA not available.
Source: Bureau of the Census, Bureau of Labor Statistics, Bureau of Domestic Commerce.

TABLE 1-8

METAL BARRELS, DRUMS, KEGS, AND PAILS: TRENDS AND PROJECTIONS 1967–73

[In millions of dollars except as noted]

	1967	1969	1970	1971 ᴾ	1972 [1]	Percent increase 1971-72	1973 [1]	Percent increase 1972-73
Industry: [2]								
Value of shipments	371	399	401	418	450	8	475	6
Total employment (thousands)	12	11	11	10	11	10		
Production workers (thousands)	9	9	9	8	9			
Value added	151	168	172	179	NA			
Value added per production worker man-hour	$7.76	$8.89	$9.33	$10.53	NA			
Product: [3]								
Value of shipments	378.2	414.9	408.6	[1] 430	465	8	490	5
Steel drums	212.6	236.7	237.5	250	268	7	280	4
Steel pails	111.6	119.5	113.3	120	132	10	140	6
Drums and pails	54.0	58.7	57.8	60	65	8	70	6
Quantity shipped (millions)	121	125	116	115	120	4	125	4
Steel drums	33	36	34	33	34	2	35	2
Steel pails	88	89	82	82	86	5	90	4
Value of exports	2.6	2.2	1.7	1.8	3	67		
Wholesale price indexes (1967=100):								
Steel drums	100	107.3	114.1	120.3	126.5	5.2		
Steel pails	100	108.9	115.2	121.0	127.9	5.7		

[1] Estimated by Bureau of Domestic Commerce.
[2] Includes value of all products and services sold by the metal barrels, drums, kegs and pails industry (SIC 3491).
[3] Includes value of shipments of metal shipper containers made by all industries.

ᴾ = Preliminary.
NOTE.—NA=not available.
Source: Bureau of the Census, Bureau of Labor Statistics, Bureau of Domestic Commerce.

3. When lack of skill and experience for a given packaging assignment makes contracting for the design, procurement, and packing economically attractive. This service is particularly useful with military and commercial export orders.

4. For organizations whose normal business procedure does not necessitate any packaging facilities whatsoever, but who may occasionally require packaging to be performed, that is, museums, importers, and experimental machine shops.

Consulting firms have developed due to the great complexity and diversity of the packaging industry to assist both the supplier of materials and equipment as well as the user in the formulation of packaging policies. Specifically, the following areas of activity characterize the packaging consultant's role:

1. To supplement the packaging personnel of a given firm for specific problems involving design, merchandising, productivity, protection, etc. In these instances, the consultants' temporary assignment obviates the need to enlarge an existing packaging staff.

2. To provide, either on a single job or retainer basis, technical and engineering skills for a firm not able to employ full-time packaging personnel.

3. To provide a fresh perspective in the resolution of a policy or problem and to benefit by the experience of others.

4. To assist in management decisions pertaining to major capital equipment expenditures, expansion, marketing problems, etc.

Testing and research organizations provide the theoretical and applied data for the development and evaluation of new and improved packaging materials, containers, methods, and applications. Many companies include research and testing facilities as part of their organizational structure. Such facilities are devoted exclusively to problems of the parent organization and are, as such, referred to as captive. Independent or noncaptive research and testing agencies exist both to supplement the work of captive agencies and to fulfill the needs of firms where such facilities are not justified. Research and testing are also carried out by academic institutions and by various branches of the government.

The following are briefly some of the major areas of research and testing organizations:

1. Basic research on materials and their properties.

2. Applied research pertaining to increased uses for existing or new materials, containers, equipment, and methods.

3. Problem solving for increased protection, cost reduction, improved appearance, and performance through scientific analysis.

SCOPE OF PACKAGING FOR DISTRIBUTION 17

4. To determine compliance with user, carrier, and governmental specifications or for referee purposes in disputes.

5. To measure and evaluate the quality of materials, containers, packages, etc.

Research and testing projects are initiated by individual firms, trade associations, government agencies, professional societies, academic institutions and foundations. Appropriations for the aforementioned purposes are constantly increasing, and the body of knowledge in the packaging field is therefore expanding at an accelerated rate.

The manufacturers of laboratory implements and controls make available to the packaging industry the tools to rate, control, test, and measure the properties and performance of materials, equipment, and methods. These tools are being used increasingly not only in research and testing but also in manufacturing processes for the attainment of precision and control.

There are more than fifty *trade associations* representing the various interests of the packaging industry. The primary function of the trade association is to act as a clearing house and as an industry representative on legal, public relations, economic, and technical matters. These groups serve the packaging field through education, research projects and through promotional means. Some of the smaller trade associations are organized for the collection and analysis of statistical data pertinent to the membership. Other trade groups work through many committees concerned with such matters as standards, specifications, new applications, and testing procedures.

PACKAGE ENGINEERING

The complexity of the packaging industry has grown to such a degree, encompassing such a wide scope, that specialization has become necessary. The specialized field dealing with packaging problems has become known as *package engineering*. By accepted definition, package engineering is the activity in which scientific and engineering principles are applied in solving the problem of functional design, formation, filling, closing, and/or preparation for shipment of containers (regardless of type or kind) or the product enclosed therein. Since package engineering includes the study of products, packages, materials and containers, structures, methods, machinery, and transportation, it deals with such disciplines as chemistry, physics, mechanics, machine design, industrial engineering, electronics, materials handling, and other specialized skills. The individual who is engaged in the packaging engineering function may have his major training in one of these specific areas, depending upon the main emphasis of his assignment. For example, a packaging engineer in the food processing industry might have a background essentially in chemistry, whereas, a packaging engineer in the electronics industry might have a background in physics and mechanics.

Because of the breadth of package engineering problems, the evolution of packaging engineering into a major profession has developed slowly. In the United States it had its inception about 1910 and expanded with greatest rapidity during World War II. The engineering approach to the solution of packaging problems is still not universally applied because of a general lack of recognition of the importance of this endeavor by some segments of industry. As a result, the personnel assigned to handle packaging problems are not always adequately trained to apply scientific and engineering principles.

However, modern industry is becoming increasingly aware of the large share of the total expense that packaging entails in the manufacture and distribution of a product. Consequently, industry is beginning to employ packaging specialists to apply package engineering principles in the solution of problems with resultant improvement in efficiency and reduction in costs.

Certain areas of packaging have been subjected to advanced research and development procedures which provide a sound theoretical basis for package engineering solutions. In other areas, the theoretical basis is inadequately established so that the package engineering approach must depend largely on empirical data.

Package engineering function. There are a number of areas of the package engineering function carried out in the supplier's as well as in the user's organization. The supplier's functions consist of research and development of new and improved packaging products and to assist customers in efficiently utilizing these products. The latter is referred to as customer technical service and may include designing, testing, and specialized consulting. This function complements the efforts of the sales department and has become an important factor in stimulating sales and in providing technical assistance to packaging product users. Technical customer service is provided either by highly trained sales representatives or by a separate staff of technical engineering specialists.

The functions covered by packaging engineering in the user's organization relate to many different phases of the company structure including purchasing, manufacturing, and sales. Specifically package engineering may be employed in the following areas:

SCOPE OF PACKAGING FOR DISTRIBUTION 19

1. In the *purchase* of packaging materials, containers, supplies, and equipment.

2. In the *development* and constant improvement of techniques and methods in the packaging operation, including filling, sealing, and labeling.

3. In the *control* of the quality of the packaging components and the finished packaging product.

4. In the *preparation* of merchandise for transportation and distribution including warehousing, materials handling, and carrier loading.

5. In the *rating* and control of packaging costs.

6. In the *merchandising,* advertising, and selling of the product.

In purchasing, packaging engineering provides the basis for decisions in the *selection* and the *rating* of purchases, as well as in the *specifying* of products for a clear understanding with the supplier of the detailed packaging requirements.

In the development of packaging for new products and the improvement of methods and techniques, packaging engineering principles are employed. It is in this area where the greatest skills and specialization are needed owing to the complexities of the product, processing, packing, and identifying.

After the exact details of the packaging system have been engineered and put into operation, the packaging engineering function extends to the control of quality and performance of the system. Feedback from control data may suggest improvements or revisions and forms the basis for standards.

Package engineering is employed in decisions involving the *distribution* of the product. These decisions concern basic protective requirements in warehousing and transit, as well as functional requirements of arrangement, size, weight, quantity, handling, disposal, re-use, etc. In addition, traffic and other carrier considerations frequently have a direct bearing on packaging engineering in the design or selection of an integrated packaging system.

The fulfillment of all objectives of a satisfactory packaging system at lowest possible cost is one of the most important functions of package engineering. This objective is accomplished through the technical and economic rating of alternates and the establishment of levels which usually provide a compromise solution. The establishment of levels relates to product type and value, damage rates, appearance, and convenience or may be based on customer requirements. For instance, the level of protection should be high for an expensive product, or for a product, regardless of cost, that causes delays or other customer inconveniences if damaged.

Appearance requirements vary widely by product and market. The cost of providing a high level of appearance must be justified. Consumer acceptance of pharmaceutical, personal, food, and cosmetic products is greatly influenced by appearance, and therefore the cost to provide a high level of appearance is comparatively great.

Apart from the economic aspects of merchandising, many technical considerations require packaging engineering solutions. These involve the selection of suitable printing surfaces, easy-opening or display features, package styles, and product identification.

Package engineering organization. It has been established that the package engineering function is an essential part of the operations of the main segments of an organization, that is, purchasing, manufacturing, sales, and so on. Frequently, therefore, packaging engineering as a separate entity loses its recognition by being combined into the operations of purchasing, engineering, industrial engineering, shipping, manufacturing, traffic, and sales. A specialist may be attached to any one of these departments or the function of packaging may be included as part of the responsibility of an individual within one or several of these departments.

When the performance of the package engineering function becomes too complicated and coordination between departments and individuals is not fully attained, a separate packaging department or a committee is usually organized.

PACKAGING COMMITTEES. Packaging committees operate in two basic manners:

1. *Policy formulating committee* consisting of representatives of different departments who collectively *develop* packaging policy and *whose members execute* the decisions which are reached.

2. *Packaging advisory committee* consisting of representatives of different departments advising an organized packaging department in the effective handling of packaging policy.

In Fig. 1-1, a typical packaging policy formulating committee organization chart illustrates the development of packaging policy through committee action. The representatives on the operating committees report to the policy committee and have a specific interest in its decisions and a responsibility for an appropriate phase of its work. For example, the packaging engineering department makes specific recommendations on types of protective packaging and available materials and prepares

SCOPE OF PACKAGING FOR DISTRIBUTION

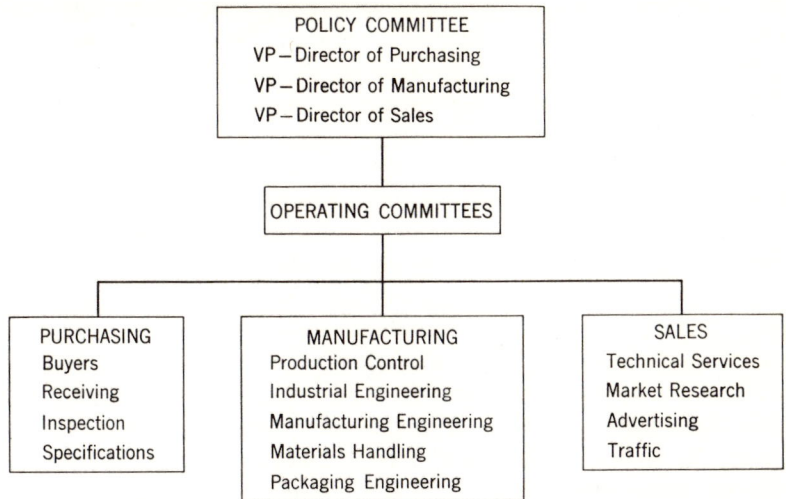

Fig. 1-1. Typical functional organization chart of packaging policy formulating committee, including committees implementing policy decision.

the necessary standards. The development of specifications for the materials to be purchased is the direct responsibility of the specification specialist attached to the purchasing department. The advertising department is responsible for the advertising material and art work that appears on the package. The production control department has the responsibility of providing information on quantities to be purchased and the delivery dates.

Similarly, other departments or individuals in these departments prepare basic information required for packaging changes. This information, in turn, is resolved in each of the operating committees before being submitted to the policy committee for a decision involving a major packaging change.

In a smaller company the packaging committee frequently consists of the plant or factory manager, sales manager, purchasing agent, and traffic manager. Each of these men carry out independent investigations in their specialized activity and contribute the data obtained towards the final resolution of the problem with top management who are thus provided with the basis for a decision.

In Fig. 1-2 the structure of a typical *packaging advisory committee* is shown. The head of the package engineering department is the permanent chairman of this group which discusses proposals, defines problems, and approves and disapproves recommendations of the package engineering department. All the detailed investigations are carried out by the package engineering department in cooperation with members of the committee. However, each major change is submitted, and thus arguments and disagreements are avoided since each department has been fully advised and has had the opportunity to offer comments.

In a smaller company in which no formal packaging department exists, the head of the department who has the packaging responsibility can similarly obtain the cooperation of other departments by the formation of a packaging advisory committee. This committee would function in the same manner, but on a more informal basis.

Committee action in packaging is to some degree related to the basic philosophy of the company. If policy on all levels is customarily formulated through committee action or if changes in general are collectively discussed, the likelihood of a packaging committee to make an important contribution is assured. On the other hand, in a more autocratic organization, a single individual, usually the head of a department, will take the initiative and therefore circumvent committee action. While this procedure is speedier, there is a greater likelihood for error and for friction among employees.

Justification for a packaging engineer or department. It is difficult to define the criteria justifying the establishment of a separate package engineering department. It is equally difficult to determine the exact place of a package engineering department in the hierarchy of an organization once the need has been established. The employment of a single

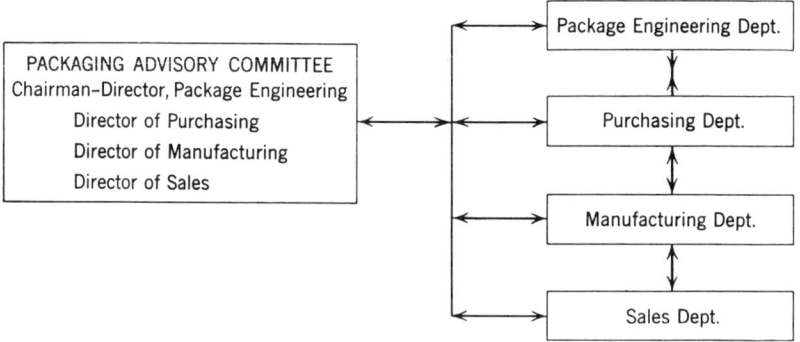

Fig. 1-2. Typical packaging advisory committee structure illustrating communications between packaging engineering department whose function it is to develop technical data in cooperation with other departments.

SCOPE OF PACKAGING FOR DISTRIBUTION   23

packaging engineer or the formation of a packaging section or department is usually the result of some or all of the following conditions prevailing in an organization:

Economic considerations in which (1) packaging per se represents a major material and labor cost item in relation to total expense, such as food processing, toiletries, and cosmetics, and in which (2) packaging, owing to the size of the company, is a major expense.

Product considerations in which (1) the product line is highly complex, and a wide variety of packages is used and (2) the nature of the business requires constant packaging changes for new models, deals, styles, etc.

Protective considerations in which (1) the packaging of the product requires constant attention in performing the distribution function at lowest cost, such as in furniture, glass products, and appliances and (2) the fragility of the product necessitates the application of package engineering principles, testing, and other evaluation techniques (instruments, missiles).

Sales considerations in which (1) consumer acceptance is largely dependent upon constant improvement in packaging style and product presentation and (2) compliance with the demands of the customer requires specialized attention, for example, military, major accounts, and mail order houses.

Some, if not all, of the above considerations apply to almost any manufacturing organization. However, the final justification for establishing package engineering as a separate function depends upon the following two factors:

1. The *total part-time effort expended* in the organization by a number of people, which must be rated against potential efficiencies attainable by specialization of one man or department in the handling of packaging problems. In addition, rating of the potential improvement in the efficiency of these individuals having part-time packaging assignments should be considered.

2. *The potential economies* through the employment of a specialist or a group of specialists through better skills and a more diversified background and the constant attention that packaging would be given should be rated against the costs involved.

In a recent management study of the packaging operations of a manufacturer of fragile sheet metal products, it was found that 20 per cent of the time was devoted to packaging by one of the product engineers, about 25 per cent by the chief draftsman, 10 per cent by the buyer of packaging materials, 100 per cent by a specification draftsman (supplemented by another draftsman 50 per cent of the time) and about 50 per cent of the time of one industrial engineer. It was discovered that all this effort was neither productive nor did it result in the effective solution of packaging problems through the application of modern package engineering techniques. It was therefore recommended that a packaging engineer be employed to relieve all personnel except one draftsman of packaging responsibility. The engineer's assignment consisted of the development of long range cost reduction projects, development of packaging for new products, and the maintenance of specifications and other records. It was found that substantial economies were obtained both in engineering effort, as well as in packaging materials and labor, with an overall improvement in packaging performance.

ORGANIZATION STRUCTURE

Once the need to establish centralized package engineering has been determined, the problem arises with respect to the logical and most effective placement of this department within the overall company structure. As a general rule, package engineering is a staff function under the supervision of any one of the following departmental heads:

1. *Industrial Engineering.* Packaging may be combined with materials handling, with primary emphasis on methods.
2. *Purchasing.* Emphasis is on rating, quality, and specifications.
3. *Physical Distribution.* Packaging may be combined with materials handling, warehousing, and/or shipping, with emphasis on integrated distribution logistics.
4. *Engineering.* Packaging needs to be closely correlated with product design and product changes.
5. *Research and Development.* The impact of packaging is closely related to consumer acceptance and much of the research is devoted to the development of new packages and applications.

Occasionally, the package engineering function is under the direct supervision of the sales or advertising department or directly under the plant or production manager in the manufacturing division of the company. Ideally, packaging engineering should report to a vice-president of the organization. Since packaging crosses many departmental lines, it requires sufficiently high-placed authority to formulate and execute policy. In multi-plant operations or in a very large company, a *packaging coordinator* is often assigned to integrate the packaging practices of

SCOPE OF PACKAGING FOR DISTRIBUTION

the various activities. The coordinator usually reports to a vice-president and the individual plant packaging engineers report to the plant manager or a department head.

THE PACKAGING ENGINEER

Any individual who is engaged in the package engineering function is in essence a packaging engineer. The title packaging engineer is not necessarily bestowed on the individual who is devoting all or a major share of his time to the package engineering function. The title depends on the recognition given this function by the company and to some extent, is based on the qualifications of the individual and whether or not he is truly applying package engineering principles. As a result, many men in industry who are dealing with package engineering functions have other titles such as Project Engineer, Industrial Engineer, Production or Distribution Analyst. In those companies in which the title of Packaging Engineer, Package Engineer or Packing Engineer is employed, the "engineer" in the title does not necessarily imply an engineering degree.

Today's practicing packaging engineers have backgrounds which vary widely in education and experience. The professional status of the packaging engineer has relatively recent origin, but now formal package engineering training is available to give the field the professional recognition it deserves. Some of today's practicing packaging engineers, who did not have the opportunity to obtain the desired basic training, had to develop the necessary skills and, consequently their professional status, through extensive work experience. In the absence of prerequisites to serve as a guide in the selection of a package engineer, industry emphasized aptitudes and related skills or training as closely as possible. Thus, the varied aspects of packaging permits participation of individuals in the field who have vastly different training and skills.

Despite the fact that the common title of packaging engineer is becoming widely used in industry, individual job assignments vary so widely that differences in background are necessary. Basically, packaging engineers in fulfilling the package engineering function in industry are concerned with the following major areas: (1) materials, (2) systems, and (3) equipment.

Materials packaging engineer. His field consists of the selection of the most suitable and economical materials and containers compatible with the product and satisfying all distribution interactions from purchasing to the final consumer. In the selection process, the materials packaging engineer modifies standard materials or alters their configuration through design to suit the individual product. Further, through evaluation processes, including laboratory, production and field testing, he establishes the suitability of the packaging and then develops specifications to assure the continued performance of these materials in use. Because of the enormous variety of products packaged and the differences in their individual requirements, the materials packaging engineer is dealing with different problems in different industries. In the chemical industry, the compatibility of the chemical with the packaging material is of primary importance. This decision requires a chemistry-oriented materials packaging engineer. In a company producing highly fragile electronic products, physical protection against shock and vibration are of primary importance. This situation warrants a materials packaging engineer oriented towards physics and mechanics. In the cosmetics industry, the materials packaging engineer is required to satisfy both functional and merchandising aspects of packaging. Experience in consumer acceptance techniques and advertising is important.

In some companies having a diversity of products that demand specialization with respect to orientation, the materials packaging engineer may act as a coordinator who is supplied by other specialists within the company with necessary technical data. If the company is large enough, materials packaging engineers specializing in different areas are employed.

The systems packaging engineer. This individual develops economical *procedures* to package the product and get it to the final consumer. As such, he is concerned with the methods of packing, unitizing, materials handling, warehousing, shipping, and identifying. Basically, the systems packaging engineer utilizes industrial engineering techniques in rating alternate systems as to economy, logistics compatibility, and protection. The importance of the systems packaging engineer's function depends on the diversity of products and the degree of mechanization attainable in the packaging, handling and distribution operations.

The equipment packaging engineer. The equipment packaging engineer develops new machines or modifies existing machines for the packaging operation. Such an assignment is necessary in mass-production industries where mechanization or automation of the packaging operation requires equipment which is not available commercially. The equipment packaging engineer's background is that of a mechanical engineer oriented towards machine design.

Since the development of a fully integrated packaging system requires the harmonious interplay of materials, methods, and equipment considerations, it is essential that the three packaging specialists work

SCOPE OF PACKAGING FOR DISTRIBUTION

closely together in solving any packaging problem. In a smaller company, where this degree of specialization is not possible, one individual is responsible to combine these functions drawing upon specialized skills from other individuals within and without his company since it is extremely difficult for one man to be fully experienced in materials, systems, and equipment. The company employing one man seeks the packaging engineer with the background most pertinent to its operations. Therefore a company having no feasible potential to mechanize its packaging, but has a very fragile, low volume product will employ a packaging engineer with a materials background.

PACKAGE ENGINEERING PROFESSIONAL ORGANIZATIONS

To enhance his professional status and to assist him in his job function the packaging engineer has organized and supports various professional societies. In summary professional societies aim towards the following:

1. To conduct and promote investigations of the principles and methods of packaging for products in distribution.
2. To encourage and initiate the development of packaging systems, methods, standards, test procedures, and specifications.
3. To provide the profession with current technical, statistical and economic data.
4. To promote training and education of members presently in the field, to encourage newcomers to the field.
5. To permit the free interchange of information through such media as seminars, conferences, clinics, and expositions.
6. To cooperate with governmental, carrier, and industry agencies in the formulation of packaging standards, codes, regulations, and practices.

The following paragraphs discuss the principal professional societies dealing in whole or in part with package engineering problems.

The American Society for Testing Materials (ASTM). It is the world's largest source of voluntary consensus standards for materials, products, systems, and services. Through the activities of some 118 main standards committees (about 2500 committees including all subcommittees and sections), ASTM has conducted investigations and researches leading to the development of more than 4700 widely used standards. Of particular interest to packaging engineers are the activities of Committees D-6 on Paper and

Paper Products, D-10 on Packaging, and F-2 on Flexible Barrier Materials. In addition to these three committees there are other ASTM committees investigating materials and testing procedures used for packaging.

The Technical Association of the Pulp and Paper Industry (TAPPI). It is a technical society organized for the development and dissemination of knowledge on the technology of pulp, paper, paperboard, converted paper products and allied industries. There are eleven major divisions of TAPPI; namely, Coating & Graphic Arts, Corrugated Containers, Engineering, Finishing, Management Science, Paper and Board Manufacture, Paper Synthetics, Pulp Manufacture, Research & Development, Environmental and Testing. TAPPI also functions through committee action with each division comprised of several committees which are frequently further subdivided into sub-committees.

Packaging Institute (PI). The national professional packaging society for all packaging people and those persons involved with packaging and its role in society representing also non-technical, social, legislative and managerial packaging aspects. The Packaging Institute, U.S.A. provides assistance to producers and users of packaged products in the solution of technical, engineering, economic and other package oriented problems. It is organized into industry committee interest groups (Pharmaceutical, Petroleum, Rigid Container, etc.), function groups (Production line, Equipment, etc.) and regional Chapters throughout the U.S.

The Society of Packaging and Handling Engineers (SPHE). This is a professional group with a membership consisting of both packaging and materials handling engineers and managers. This society is organized on a local chapter basis with primary emphasis on improving the technical and managerial proficiency and professional status of packaging and materials handling engineers and managers. The society has a certification program leading to the designations of "Certified Professional in Packaging" or "Certified Professional in Materials Handling" or both, based on examination. The society attempts to serve as the interface between packaging and materials handling.

There are other professional societies in whose membership the packaging engineer might benefit. These societies either have active packaging committees or are in a closely allied field to packaging such as the *International Material Management Society (IMMS)*.

The published matter of the professional societies in the form of proceedings, standards, specifications, and procedures forms much of the available basic technical data in the packaging field and is used authoritatively by members of the profession. National American standards of all types including matters pertaining to packaging are the function of the *American Standards Association* (ASA). This association is concerned with the development, promotion, coordination, and approval of American standards and represents American in-

SCOPE OF PACKAGING FOR DISTRIBUTION 29

terests in international standards works. In operation, the American Standards Association assists a group of companies and trade or professional organizations in the proper development of standards.

A large portion of the educational activity in packaging is initiated, co-sponsored, or supported by professional societies, and active participation in one or more of the professional societies depends upon the individual's field of specialization and his professional interests.

Other organizations. There are many other important agencies servicing the package engineering profession either wholly or in part. These groups either have a direct connection with the federal government or with carrier agencies or they serve a cross section of American industry on a functional basis. An excellent example of the latter is the *American Management Association* (AMA) which has recognized the importance of packaging to management by conducting an extensive educational program. Another such group, the American Standards Association, has previously been cited.

Governmental agencies dealing with packaging are under the jurisdiction of either the Department of Agriculture, the Department of Defense, Department of Transportation, the Department of Commerce, the Post Office Department, or under some of the regulatory agencies. Packaging work by these agencies is intended either to protect the public through safety measures, proper utilization of resources, assisting national defense efforts and so forth, or to propagate the field by acting as a clearing house as well as acting through research and development in certain areas. As one of the biggest purchasers in the world, the U.S Government must exercise safeguards in its procurement function. The government therefore develops and issues detailed specifications which include packaging requirements.

Important governmental groups concerned with package engineering research and development include the Forest Products Laboratory, U.S. Department of Agriculture; Engineering Research & Development Laboratories; U.S. Army Laboratories; Navy Materials Handling Laboratory; and Naval Logistics Engineering Group; Armament Development Test Center, Packaging Engineering Group; Wright-Air Development Center and others.

Among the carrier agencies there are several organizations assisting in the solution of package engineering problems. Included is The Freight Loss and Damage Prevention Section, Association of American Railroads and their laboratory and the Bureau of Explosives of the Department of Transportation.

The National Safe Transit Association constitutes a voluntary, cooperative effort to reduce the nation's physical distribution damage losses through a conscientiously applied packaged-product preshipment testing and identification program. Product manufacturers, packaging suppliers, carriers and receivers actively participate in this member companies managed and operated non-profit program.

PACKAGE ENGINEERING EDUCATION

The responsibility of providing for package engineering education in the past essentially has been under the auspices of the package engineering professional organizations complemented by extensive on-the-job training. With the increased need to provide fundamental training, academic institutions are beginning to include in their curricula, courses in various specialized areas of package engineering. Some of these academic courses are designed to supplement previously acquired engineering and/or science training, whereas, others are integrated into a special program with package engineering as the major emphasis. Other courses offered by academic institutions are designed to enhance the general knowledge of personnel practicing the package engineering or related functions. The latter courses are generally included in an institution's adult education program.

As in any other field of specialization, there are two general philosophical educational approaches in preparing the student for his professional role. One approach favors the sound indoctrination of basic fundamentals with subsequent specialization on a graduate level or through adult education courses. The other approach is to incorporate in the undergraduate curriculum sufficient free choice for the student to select a specialized field of activity, thus preparing the student on the undergraduate level for his work assignment upon graduation. The educator's general approach to training in package engineering at this time is to develop sound fundamentals rather than to encourage undergraduate specialization.

Largely through the efforts of the Packaging Education Foundation, Inc. eleven universities now offer formal packaging programs. Two of these award Masters degrees only, two award both undergraduate and graduate packaging degrees, four award regular four year degrees in packaging and the balance award minors in packaging.

At the vocational high school level, the Packaging Machinery Manufacturers Institute has developed training manuals to train packaging line mechanics. The Society of Packaging and Handling Engineers (SPHE) offers special preparation courses for candidates for its certification examination.

It should be recognized that the complexity of the packaging field itself and therefore the diversity of background training required, that is, chemistry, physics, mechanics, industrial engineering, etc., makes a

SCOPE OF PACKAGING FOR DISTRIBUTION

universally accepted academic curriculum extremely difficult to formulate. Perhaps, if the packaging field were more precisely subdivided into segments, training would be a more simplified task. One of the logical subdivisions of the package engineering field is by functional requirements.

DISTRIBUTION PACKAGING

As previously defined, package engineering is concerned with the materials, methods, systems and equipment required to prepare and deliver the product in the form required by the ultimate consumer. In this basic function the emphasis is either on point-of-sales in stimulating consumer acceptance or its primary emphasis is on performance in terms of the distribution process.

In packaging, emphasis is dictated by the product and by the merchandising and distribution strategy. If sales acceptance of a product is or should be influenced by packaging techniques, the activity towards the accomplishment of this objective is commonly referred to as *consumer packaging*. If the basic function of packaging is to provide for performance in the distribution process, so-called *distribution packaging* is being practiced.

In consumer packaging, as contrasted to distribution packaging, greater emphasis is placed on the following aspects:

1. *Surface design considerations* including color selection, patterns, background media, and product and brand identification.
2. *Configuration considerations* enhancing consumer acceptance including personal utility, and recognition factors.
3. *Utility considerations* including ease of opening, dispensing, closing, disposal and general use.
4. *Protective considerations* to retain the original attributes of the products, that is, odor, color, taste, and consistency.
5. *Marketing considerations* involving quantities, size, combinations, deals, displays, racks, etc.
6. *Production considerations* in filling, sealing, labeling, canning, wrapping, and so forth.
7. *Regulatory considerations* relating to safety, ecology, labeling, etc.
8. *Economic considerations* relating to all of the above in terms of the sales price of the product.

In distribution packaging, greatest emphasis is placed on the following:

1. *Protective considerations* to economically insure the safety of the

product in handling and shipping (selection of shipping containers, barriers, cushioning, etc.).

2. *Handling considerations* to permit the economical movement and storage of the product, such as size, weight, quantity, and cubage.

3. *Manufacturing considerations pertaining* to the economical preparation of the product for handling and shipment, that is, packing, sealing, reinforcing, bundling, unitizing, etc.

4. *Functional considerations* of opening, unpacking, reclosing, re-use, disposal, etc.

5. *Identification considerations* to identify the products and to provide instructions for use, protecting, handling, and routing of the merchandise.

6. *Shipping considerations* as related to carrier utilization, regulations, transit hazards, carrier rates, etc.

From the foregoing items it should be noted that consumer packaging is primarily sales-oriented because the packaging influences point-of-sales selection. Packages used for distribution do not customarily reach the ultimate consumer and as such are not designed to influence his selection. However, distribution packages are playing an increasingly important role in the selling of merchandise to the distributor, retailer, or industrial user. Distribution packaging includes considerations of transportation, warehousing and material handling factors with primary emphasis on performance and economy.

WHAT IS DISTRIBUTION PACKAGING?

Distribution packaging is an integrated approach which embraces the previously listed protective, handling, manufacturing, functional, identification, and shipping considerations. This basic definition does not preclude that the product reaches the consumer as in self-service mass merchandising. As a matter of fact, a substantial proportion of merchandise in distribution packages consists of prepackaged consumer items. However, the functions to be served effectively are removed from the consumer level because distribution packaging is concerned with considerations within the distribution network from the time the product is produced until it reaches its final destination.

At one time, the term *industrial packaging* became deeply entrenched in the jargon of industry, to describe this discipline. Other terms that could be applied are *transportation packaging, protective packaging, shipping packaging*, or most appropriately *distribution packaging*.

SCOPE OF PACKAGING FOR DISTRIBUTION

FUNDAMENTALS OF DISTRIBUTION PACKAGING

In the broadest sense, any product that is distributed has to be physically contained. The physical process of containing the product in distribution might be termed *distribution packaging*. However, *bulk* items distributed in pipe lines, tank cars, box cars, and dump trucks are customarily not included in distribution packaging considerations, but are more appropriately transportation and handling logistic problems. Portable devices for bulk items such as flour, chemicals, fertilizer, and cement, when they are suited not only for transportation but also for inplant handling and storage, are within the realm of distribution packaging. Examples of these portable devices include a cargo container of trailer size (Fig. 1-3), an inflatable rubber container (Fig. 1-4), or a steel, aluminum, wood or fibreboard "fork truck size" container (Fig. 1-5). In the selection of such bulk units, physical distribution management factors are of primary importance. The term *containerization* has been employed to define this practice, although all types of containers, whatever their size or construction materials, are properly within the scope of distribution packaging.

Another class of products requiring retention for economic handling and transportation includes self-supporting units such as ingots, brick, lumber, and pipe (Fig. 1-6). The bundle or unit load which is formed to combine individual items is also a distribution package, although it might be formed by use of a simple strap, or carrying base. Usually such items require minor protection.

Large unpackaged fabricated units such as tanks, machinery, and vehicles, frequently require skidding or reinforcing for handling and shipping (Fig. 1-7). The preparation of these items for shipment is in the scope of distribution packaging. Any additional blocking and bracing to retain the unit in transportation is also related to the distribution packaging function but is more properly considered as a *carrier loading practice*.

All the above distribution packaging requirements emphasize *retention* and *containment* to facilitate handling.

Of equal importance are the provisions made to protect merchandise in handling and shipment to avoid damage.

The fundamentals of *protection* involve (1) harm that can be done to the product and (2) harm that the product itself can do. The latter classification includes corrosives, explosives, poisons, radioactive materials, firearms, and other dangerous or hazardous products which require special packaging considerations. The former classification prevails

Fig. 1-3. Trailer-size cargo container.   Courtesy Pan-Atlantic Steamship Corp.

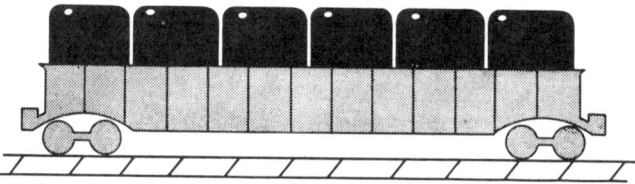

Fig. 1-4.   Inflatable rubber container for bulk shipments.   Courtesy U. S. Rubber Co.

SCOPE OF PACKAGING FOR DISTRIBUTION

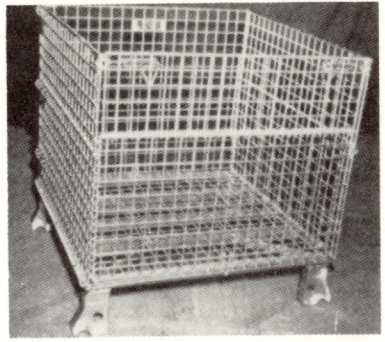

(*a*) Steel wire mesh construction. Courtesy Pittsburgh Steel Products Co.

(*b*) Re-usable and collapsible pallet box. Courtesy Bigelow-Garvey Lumber Co.

(*c*) Expendable fibreboard pallet box. Courtesy Signode Corporation.

Fig. 1-5. Fork truck size containers.

essentially for products which in themselves are not hazardous, but this category may also include hazardous products.

The harm that can be done to the products and the degree of protection which must be provided depends upon the product characteristics and the anticipated hazards in the distribution process.

Product characteristics vary widely and therefore the protection required also varies widely. For example, food products require protection for retention of flavor, odor, color, freshness, and consistency,

Fig. 1-6. Self supporting load of bricks. Courtesy Acme Steel Strapping Co.

whereas metal products require protection against corrosion, scratching, and denting.

The hazards encountered in distribution are (1) environmental hazards (2) physical hazards, and (3) miscellaneous hazards. Distribution packaging requires an understanding of the interrelated factors of the product and the anticipated hazards. The practical execution of this understanding is in the development of a distribution packaging system.

SCOPE OF PACKAGING FOR DISTRIBUTION

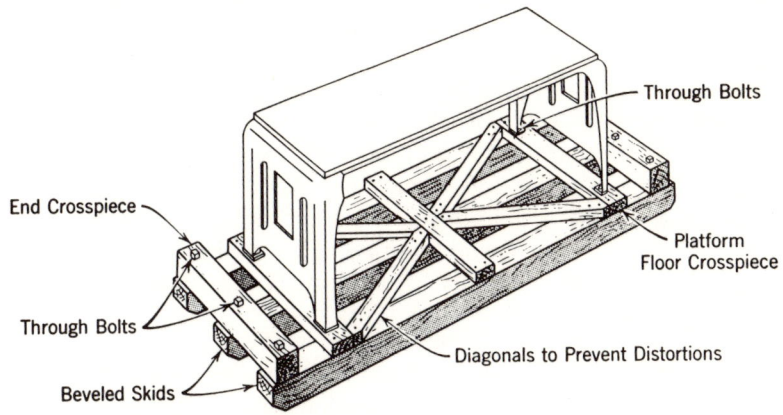

Fig. 1-7. Skidded unpackaged unit. Courtesy Association of American Railroads.

DISTRIBUTION HAZARDS

The distribution process embraces packing, handling, warehousing, and transportation activities. A product can be subjected to numerous cycles of these major activities prior to reaching its final destination. In this process the product is subjected to the following conditions requiring a measure of protection which must be provided entirely or in part by the packaging.

1. *Environmental Hazards* of temperature and humidity.
2. *Physical Hazards* of vibration, impact, compression, distortion, puncturing, etc.
3. *Miscellaneous Hazards* of infestation, pilferage, and contamination.

Environmental hazards. One of the most serious hazards of the distribution process involves the influence of environmental atmospheric conditions. Products may require protection against one or more of the following:

1. *Heat*—to prevent melting, spoilage, bleeding, blistering, peeling, fusing, discoloration.
2. *Cold*—to prevent cracking, freezing, brittleness.
3. *Water*—to prevent dissolving, dilution, separation, corrosion, illegibility, discoloration.

4. *Water Vapor*—to prevent corrosion, lumping, blocking, pitting.
5. *Pressure*—to prevent bursting, collapse, displacement.

The degree of protection required against environmental hazards is dependent upon the nature of the product and the distribution process. For instance, products for export generally require greater protection owing to the probability of adverse exposure. The locale to which merchandise is exported also influences the degree and type of protective packaging. For example, shipping to tropical climates presents far greater hazards especially if inadequate discharge and storage facilities are provided.

The United States Government is the largest exporter of merchandise of all types. In many cases the ultimate destination of the merchandise is not known at the time of packaging and stockpiling. Thus, the ultimate destination could be anywhere in the world, and the packaging must provide adequate protection, whether the climate is temperate, tropical, or frigid. This condition, in part, explains why government packaging requirements are rigidly controlled and why the product often appears to be overly protected. In an effort to economize, attempts are made whenever possible to predetermine the required level of protection by an analysis of the anticipated logistics of storage and destination.

The United States Government has established the following levels of packaging protection:

A—for overseas, long-term or unknown length of storage.

B—for intermediate length storage normally not greater than 180 days.

C—for immediate use.

In a study made by the National Safe Transit Association over a 5½ year period, it was disclosed that a seasonal trend of shipping damage was experienced in commercial shipments of major appliances and allied metal products (Table 1-7). Heaviest damages to merchandise occurred during the summer months. Although the reasons given pertain to some nonenvironmental hazards such as faster carrier speeds and seasonal labor factors, the importance of increased humidity and temperature as a contributing factor was stressed.

Even in the temperate zone, wide seasonal temperature and humidity variation is experienced. For example, in the United States, summer temperatures in excess of 100° F and winter temperatures of −40° F are not uncommon. Temperatures in excess of 130° F have been recorded in noninsulated or nonrefrigerated freight carriers exposed to direct sunlight. Occasionally these high temperatures may be ac-

TABLE 1-7
COURTESY NATIONAL SAFE TRANSIT ASSOCIATION

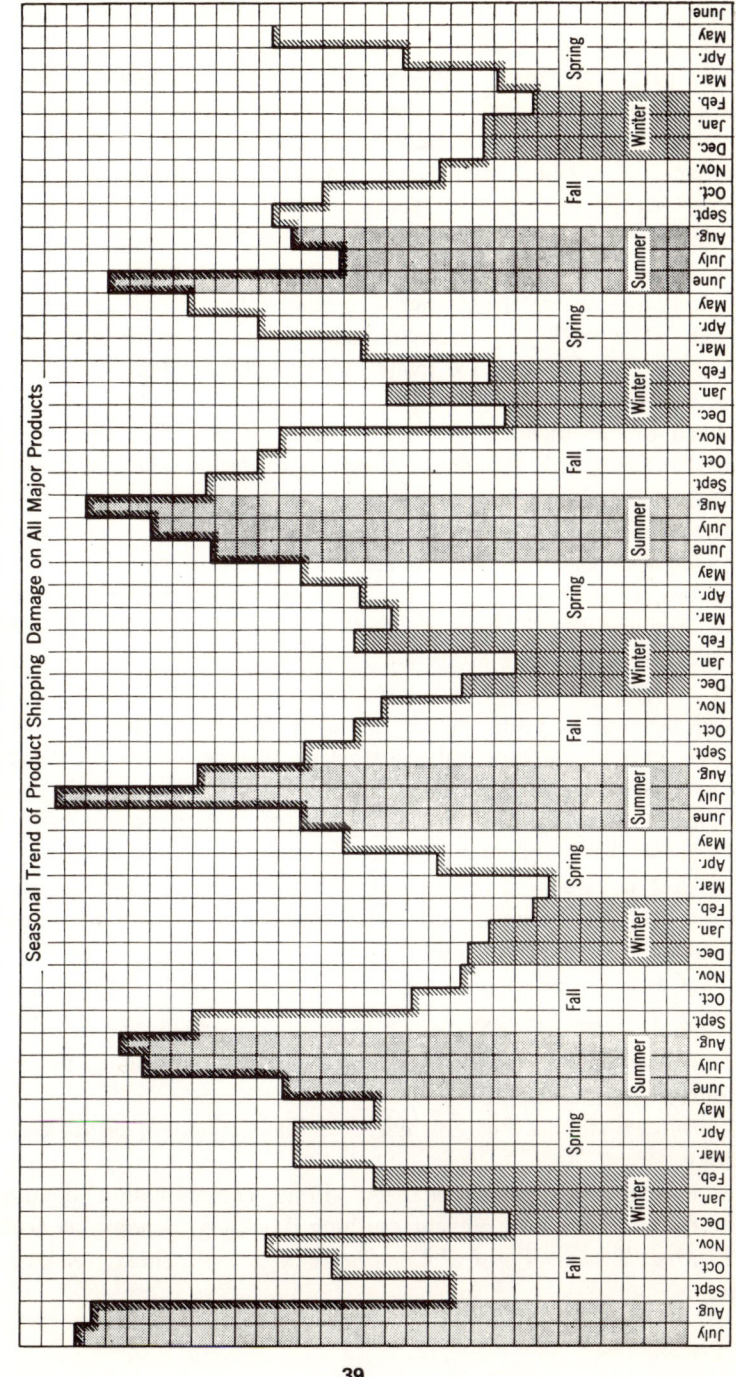

Seasonal Trend of Product Shipping Damage on All Major Products

companied by a major drop in temperature during the night which can cause a harmful temperature-cycling condition.

The previously cited National Safe Transit analysis revealed *regional differences* in product damage in the United States, which exhibited a similar seasonal pattern to the one shown in Table 1-7.

Whenever possible, environmental hazards are minimized by special precautions in handling, warehousing, and shipping. For example, perishable products subject to spoilage due to heat or cold, are shipped in insulated or refrigerated carriers. Such merchandise is also stored in warehouses where the temperature and/or humidity is carefully controlled. The packaging is so designed that air flow or heat transfer is facilitated and to provide temporary safeguards when optimum conditions do not prevail.

When carrier facilities are not available to protect against environmental hazards, it is frequently necessary to design a special packaging system which may be uneconomical.

The economies of protecting against environmental hazards through packaging compared to other means, such as special storage and carrier provisions, should be rated. For instance, outdoor storage can be far more economical with many products if these products lend themselves to inexpensive packaging protection against the elements. Similarly shipment in open cars or trucks offers handling advantages which frequently exceed the cost of protecting the products against the elements. Lumber, for example, wrapped in polyethylene film or waterproof paper on flat cars as illustrated in Chapter 8 (Fig. 8-6) results in over-all physical distribution economy even though it involves extra packaging labor and material cost that is not required in closed carriers.

Physical hazards. The physical hazards of the distribution process consist of both dynamic and static *stresses* caused by the physical transfer and storage of merchandise. Packaging must protect the product against one or more of the following hazards depending upon the fragility of the product:

1. *Vibration*—to prevent scuffing, marring, abrasion, loosening, fracturing, misalignment, etc.
2. *Impact*—to prevent crushing, breaking, cracking, distortion, and shifting.
3. *Puncture*—to prevent leakage, sifting, contamination, denting.
4. *Compression*—to prevent crushing, buckling, bending, deflection.
5. *Miscellaneous Physical Conditions* of tension, shear, torsion, and tear.

SCOPE OF PACKAGING FOR DISTRIBUTION 41

VIBRATION SHOCKS. Vibration shocks can be encountered both during inplant handling and in shipment. Internal handling on conveyors and industrial trucks can produce damage with such highly fragile products as electronic components. However, the vast majority of vibration damage occurs in transit.

There have been many studies conducted to determine the type and frequency of vibratory shocks experienced in different modes of shipment. With the improvement of accurate instrumentation it is now possible to better determine the effect of vibration shocks on merchandise. The studies have indicated that rail transportation produces shocks of a duration and intensity potentially most damaging to a wide class of products. Frequencies of 2½ to 5 cycles per sec with shock intensities up to 1.25 G are experienced depending upon the spring suspension system of the car, the speed of the car, the weight of the lading, and the condition of the road bed. Highway transportation produces vibration shocks of a lower magnitude subject to greater variation due to road and highway truck variables. Vibration shocks produced in marine transit vary widely in duration and intensity and are largely dependent upon weather conditions at sea.

IMPACT SHOCKS. Impact shocks are produced both during handling and in transit. Handling impacts primarily result from manual handling and potential abuses such as dropping, throwing, and tumbling. Transit

TABLE 1-8
SHOCKS EXPERIENCED IN AIR CARGO SHIPMENTS

Courtesy National Safe Transit Association

impact shocks are caused by sudden accelerations and decelerations of the carrier. Such impacts also occur in internal handling on conveyors and chutes, and with overhead and mobile floor-operated materials handling equipment. In rail transit major impact shocks are produced by the humping and switching operations of freight cars. These impacts are experienced when a moving car strikes one or more stationary cars at such speeds that the springs of the coupling fully compress and thus transmit shock to the car and the lading. At speeds up to 4 mph the spring action of the coupling absorbs sufficient energy to decelerate the moving car without creating major impact shocks. Above speeds of 4 mph and up to 11 mph direct shock transmission to the car occurs, which produces impact shocks potentially injurious to the car and lading. In normal car handling practice these potentially dangerous impact shocks are experienced thus necessitating provisions in car bracing and blocking techniques, as well as in protective packaging. Whenever possible, damage prone products should be transported in special "damage free" cars to minimize the cost of special bracing, blocking and protection.

As an example of the relative severity of handling and transit shocks in various modes of shipment, Tables 1-8 through 1-11 are reproduced. These data summarize averages of test shipments conducted by the

TABLE 1-9
SHOCKS EXPERIENCED IN TRUCK SHIPMENTS

Courtesy National Safe Transit Association

SCOPE OF PACKAGING FOR DISTRIBUTION

TABLE 1-10
Shocks Experienced in Railroad Freight Shipments

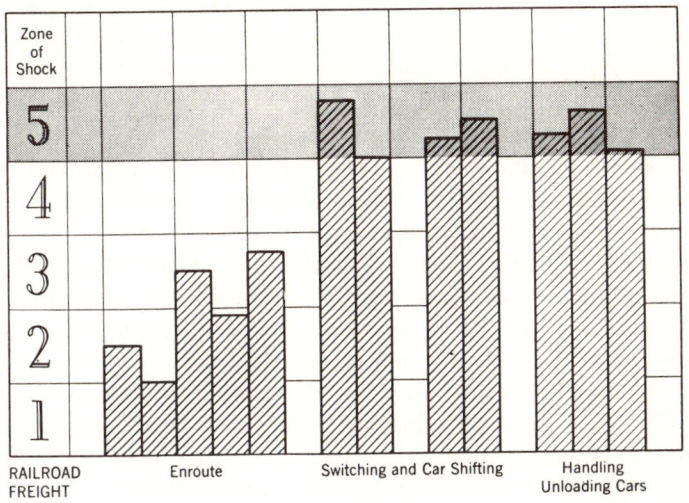

Courtesy National Safe Transit Association

TABLE 1-11
Shocks Experienced in Railway Express Shipments

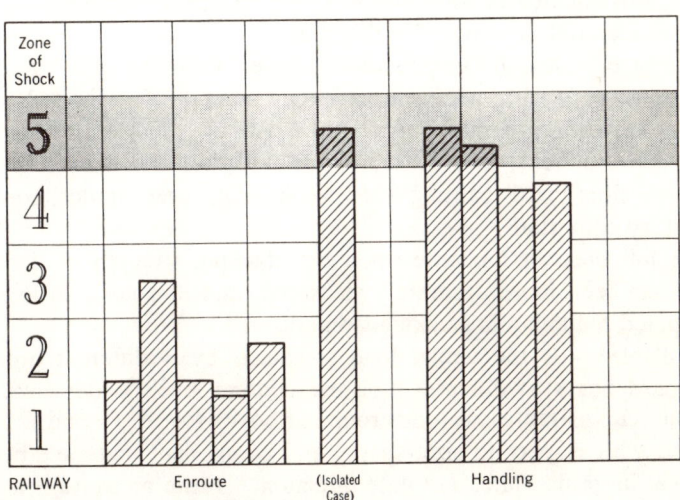

Courtesy National Safe Transit Association

National Safe Transit Association. Included are longitudinal shocks in the carrier as well as shocks incident to unloading, warehousing, and delivery (handling to ultimate destination). The unit of measurement is based on a shock-recording device which records by *zone* the intensity of the shock experienced. Fifth zone impacts (roughly 9 to 10 G) are considered unsatisfactory, fourth zone, borderline, and zones below considered less harmful.

The tables illustrate that with the exception of rail freight and isolated instances in truck shipments, shocks of greatest magnitude occur in the handling process and that shocks en route are less severe.

PUNCTURE HAZARDS. These hazards are caused by (1) improper handling procedure and (2) improper stowing and stacking procedures. The former involves careless operation of materials handling equipment such as fork lift trucks. The latter occurs when contact of merchandise with sharp objects is made in the carrier or during the process of handling. All carrier agencies as well as shippers are constantly educating their handling personnel to minimize abuses of this type.

COMPRESSION STRESSES. Compression stresses may be caused by (1) static conditions due to the superimposition of weight onto merchandise or by (2) dynamic stresses of impact shocks. Resistance against compressive stresses must be provided by the package in which the product itself is vulnerable to this type of hazard.

STATIC COMPRESSION. Static compression is primarily experienced in stacking. There are various factors influencing static stacking resistance including duration of stacking introducing factors of material fatigue, environmental influence affecting material strength, and other injury to materials as a result of handling.

Extensive research investigations have disclosed that fibreboard boxes designed with adequate stacking strength for limited storage periods fail owing to material fatigue resulting from continuous stress. It has been difficult to precisely assess the physical changes that take place in fibreboard causing this weakening, even under controlled laboratory conditions.

The influence of moisture upon the stacking strength of fibreboard boxes has been more accurately measured, and it is possible to predict anticipated reduction in stacking strength.

In Table 1-12 the actual load sustained by 4 different groups of fibreboard boxes at different moisture content levels is depicted. The separate curves represent four groups of boxes (A, B, C, and D) which differed with respect to size, shape, and component materials but were identical in grade, flute, and flute direction. From an average moisture content of 5.0 per cent produced in an atmosphere of 35 per cent rela-

SCOPE OF PACKAGING FOR DISTRIBUTION

TABLE 1-12
Effect of Moisture Content on Top-to-Bottom Fiberboard Box Compressive Rigidity

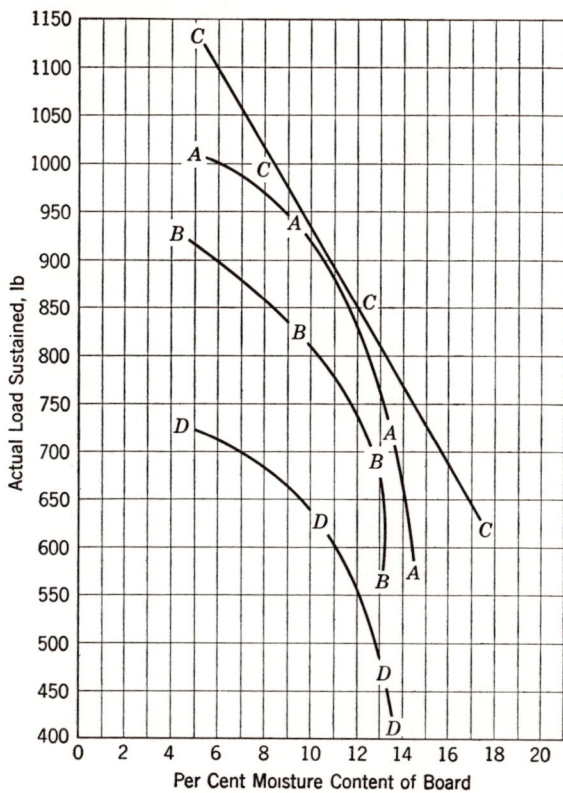

Courtesy Container Testing Laboratory, Inc.

tive humidity, 73° F, to an average moisture content of 14.5 per cent produced in an atmosphere of 90 per cent relative humidity, 100° F, the boxes sustained an average decrease of nearly 50 per cent in top-to-bottom compressive resistance.

Eccentric loading, concentrated stresses due to pallet runners or boards, and fractures or punctures caused by misuse of handling equipment are some of the additional factors which create static stacking hazards.

DYNAMIC COMPRESSIVE STRESSES. These stresses generally result from impacts, shock in handling, and transit. For example, shifting of load in a carrier imposes compressive stresses on merchandise other than in the normal stacking direction. Impacts in mechanical or

TABLE 1-13
Causes of Damage in Rail Transportation

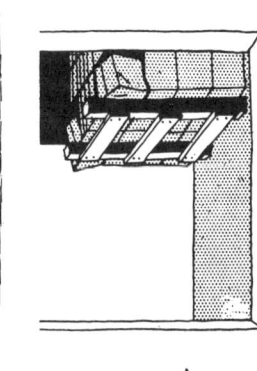

👉 **Inadequate or no bracing**
Generally applicable to partial layers, mixed loads, and to cars having vacant doorway areas.

Inadequate or no doorway protection
Failure to protect boxes from contact with doors and door posts, especially in cars where the doors are recessed.

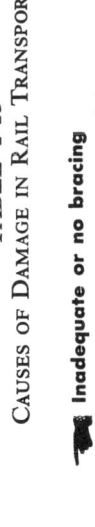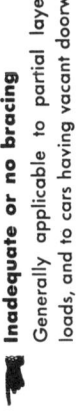

👉 **Inadequate or no doorway weather stripping**
Failure of shipper to insure that boxes in doorway area are protected against possible entry of rain or snow.

Boxes not protected from contact with blocking or bracing
Such protection should consist of suitable cushioning placed firmly between the boxes and bracing.

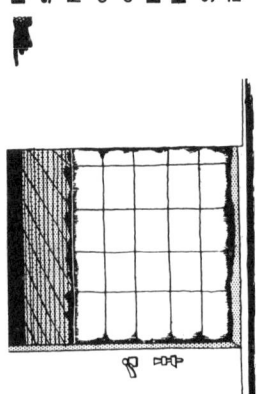

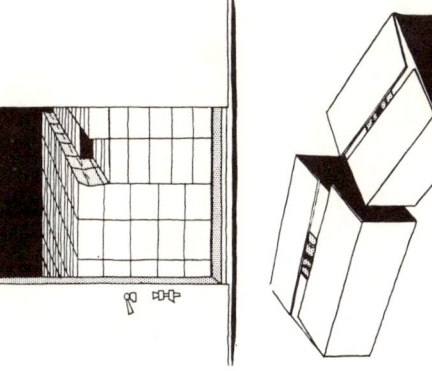

Failure to cover floor or floor racks properly
Proper protection such as the use of knocked-down corrugated boxes or corrugated sheets will usually prevent the snagging of boxes in the apertures between rack slats.

Poor arrangement of load
a) Non-segregation of different containers,
b) No separation of lightweight and heavy items,
c) Failure to level off the load,
d) Haphazard loading.

Overloading of car
Applicable to containers having a high density which are stacked to excessive heights within a car.

Improper closure of box
Failure to comply with the minimum requirements of Rule 41 with respect to closing and sealing the flaps of the box.

Shift in load due to loose loading
Excessive slack space due to a failure to provide a tight load.

Boxes and interiors which probably meet minimum requirements, but seem inadequate for contents
Concerns boxes and interior packing which should be redesigned to contain and protect contents more adequately.

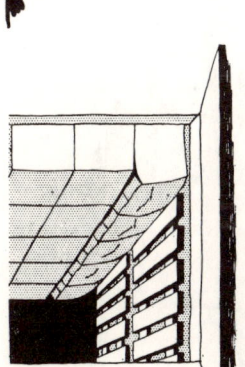

TABLE 1-13 (*Continued*)

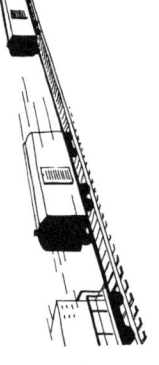

Dirty car
Presence of substances such as oil, lampblack, etc., within a car which may possibly contaminate or otherwise damage the contents.

Nails, wires, and boards not removed from previous use of car
Nails, wires, and boards not removed.

Boxes apparently handled roughly during or prior to loading; during or after unloading
Principally encountered where boxes have undergone a previous movement by highway or water. Also boxes damaged while being unloaded.

Shift in load due to improper handling of car in transit
Where a car in transit has been subjected to unusually severe shocks as in humping, flat switching, starting or stopping.

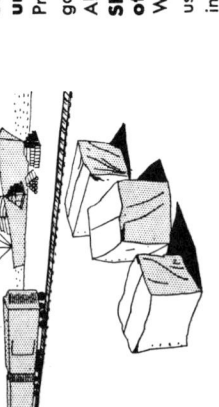

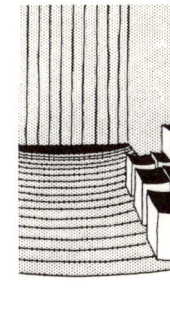
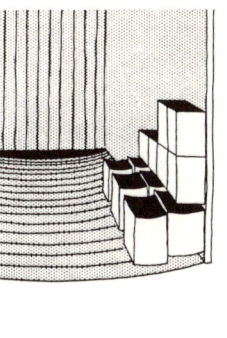

Leaky car
Due to defective condition of car permitting entry of rain or snow or seepage of ice water (remaining from previous loads) from bunkers.

Rough or broken car walls
Missing or splintered wall boards.

Defective floor or floor racks
Broken, punctured, or uneven boards or rack slats over which a covering of corrugated sheets, knocked-down boxes or similar material would be ineffective.

Other defective railroad equipment
Loose doors, loose or protruding bolts and brackets, warped walls, etc.

Defective product, including defective inner container
a) Where damage is due to an inherent defect in the product itself.
b) The existence of defective inner containers.

Shortage
To cover that portion of claims which arise from a short shipment or pilferage.

Cause unknown
Used when it was impossible for inspectors to assign the damage to any of the other 20 causes.

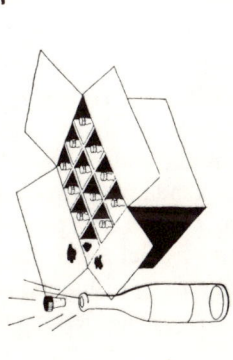

Courtesy Fibre Box Association

manual stacking or on conveyor lines create compressive stresses which must be counteracted by the packaging, unless the product itself has adequate structural strength to resist these forces without damage. The extent of compression caused by dynamic stresses is difficult to measure owing to the short duration and the variability of the input energy.

MISCELLANEOUS PHYSICAL CONDITIONS. Conditions of tension, shear, torsion, and tear result from the action of stresses on materials. For example, tension may be produced by suspending a container from its top, which causes the load to create tension on the packaging. Shear and tear are fundamentally caused by the interaction of divergent stresses. Torsion and distortion occur by the exertion of stresses not necessarily in an entirely horizontal or vertical plane.

Miscellaneous hazards. In addition to environmental and physical distribution hazards, packaging must provide for protection against several other external hazards and for preventing the product from being harmful to other merchandise or to persons and property. With some commodities protection against the following must be provided:

1. *Microorganisms:* to prevent attack by mold, bacteria, and other microorganisms.
2. *Insects and Rodents:* to prevent spoilage due to contamination from pests.
3. *Pilferage:* to prevent ease of removal of the product.

Contamination by microorganisms in storage and transportation is of relatively minor importance except for certain food products, soap, etc. In these instances, special treatments are available to treat the packaging material or the carrier.

Similarly, the attack of insects and rodents is restricted to certain products, although damage to packaging materials has been experienced in areas subject to infestation. Many *insecticides* and *insect repellents, rodenticides,* and *rodent repellents* are available when this is a problem.

Dangerous products include corrosives, explosives, flammable materials, poisons, and radioactive materials. Special provisions for the packing and handling of such materials are carefully prescribed to prevent injury to persons and property. In recognition of the problems involved, there are federal laws regulating the shipment of hazardous materials. Specifications for shipping containers for such items are published in a special tariff by the Department of Transportation.

SCOPE OF PACKAGING FOR DISTRIBUTION

TABLE 1-14
Amount of Claims Attributed to Damage Causes in Rail Transportation

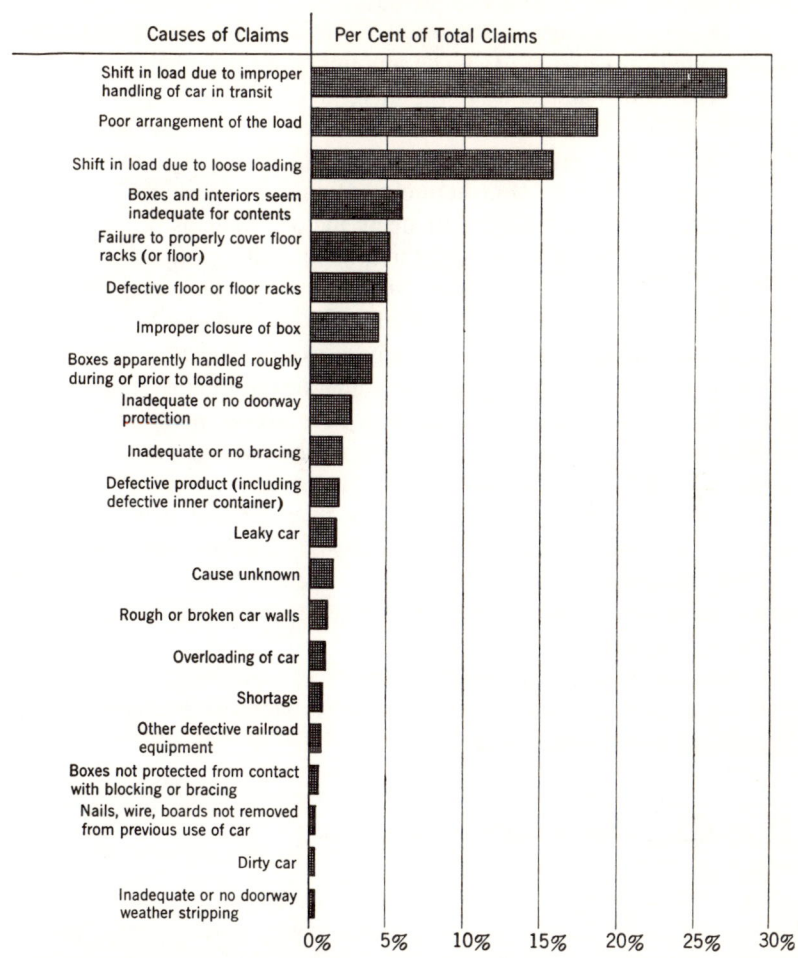

Courtesy Fibre Box Association

TRANSPORTATION AND PACKING SURVEY

In an attempt to analyze the various causes of damage experienced in rail transportation, an extensive survey was conducted a number of years ago, jointly sponsored by the railroads of the United States and the Fibre Box Association. This survey involved the inspection of merchandise in 3,340 freight cars over a three year period.

The causes of damage as the result of major rail distribution hazards are illustrated in Table 1-13. The amount of claims attributed to each of these causes of damage by the total value of the claim is summarized in Table 1-14. In addition to this data, other valuable information relating to transit, preparation of the load, packaging, and packing factors were developed. Furthermore, damage by selected shippers indicating groups of products experiencing major problems in distribution was disclosed. For instance, the three commodity groups found to be most susceptible to damage were wine, juices in tins, and fruit and vegetables in tins. The most interesting factor disclosed by the survey was that the principal causes of damage to goods packed in fibre boxes were: shift in load due to improper handling of the freight car in transit, shift in load due to loose loading, and poor arrangement of the load. Together this accounted for 62 per cent of the damage found.

Studies of this type demonstrate the attention required by shippers, carrier agencies, and material suppliers to determine the underlying causes of distribution damage and to develop the necessary economic remedial measures. Investigations of the effects of distribution hazards are a continuing practice and range from small trial field shipment to extensive, long-term shipping analysis.

In recognition of the anticipated hazards of distribution, the selection of alternate packaging to achieve the most economical packaging system is mandatory for continued progress.

ECONOMICS OF DISTRIBUTION PACKAGING

The total cost involved in the physical distribution of a product is influenced by many factors including freight costs, handling costs, packing and packaging labor, material costs, and storage costs. All of these cost factors are to a varying degree influenced by the distribution packaging system employed. For instance, the weight and/or size of the packaged product influences the *freight rate*. The weight of the package includes the weight of the product and the weight of the packaging called *tare weight*. The size of the package includes the product itself and its enclosing package. Depending upon the method of computing the freight rate, excessive weight and/or size of the package can substantially increase the total freight cost. The influence of packaging on freight economies through the selection of the most suitable containers and materials represents a major area of investigation covered in this section.

SCOPE OF PACKAGING FOR DISTRIBUTION

The cost of handling at all stages of physical distribution is related to the packaging system employed. The incorporation of features to facilitate mechanical pick-up or the design of packaging for ease of storage and handling is another important economic factor. Occasionally additional costs to include handling features are more than compensated for in lower overall physical distribution costs.

In the preparation of merchandise for shipment, a major cost factor is the labor required to package. Labor cost is influenced by the materials, methods and equipment of the specific packaging system. Sometimes it is difficult to assess the total labor cost of these operations since certain aspects of packaging may be considered to be a manufacturing expense, whereas other aspects are allocated to sales expense.

Storage cost is influenced by many factors, including some of the factors already discussed. Weight, cube, and ease of handling dictate some of the details of the storage systems. An additional major cost item is the labor and packaging material expense required for redistribution of merchandise if fractional quantities must be shipped. The cost of this operation is to a large degree influenced by the type and quantity contained in the incoming package.

The interaction of materials, methods, and equipment in providing lowest overall physical distribution cost must at all times be considered in the development of a distribution packaging system.

Materials. The material cost of the packaging system varies widely and depends upon the protective and merchandising requirements of the product. With some products, in which no merchandising considerations enter into the selection of the package, (as with sub-assemblies, semi-finished components of various types, or strictly industrial products) the packaging cost should be based upon the cost of the product. The packaging cost of a sophisticated business machine is a minor fraction of the total product cost. With a product of this type, the level of protection and therefore the material cost should be set so high that no damage to the product is to be anticipated. Conversely, with a product such as cement, in which the packaging material cost is high compared to the value of the product, the level of protection must be set to anticipate a certain percentage of product loss in shipment. The replacement of this loss is less than the expense of improved packaging to provide complete protection. Many attempts have been made to establish cost ratios of required protection with products of different value. Experience dictates the ratio that is both attainable and permissible, and, needless to say, this ratio varies widely. When merchandising considerations influence the packaging material details or when the product is of a critical nature, additional costs are incurred in distributing the merchandise over and above the basic protective requirements. In some cases, the additional cost may simply involve printing, whereas on other occasions, multiple purpose uses of the packaging may be required, such as a carrying case, a storage container, or a display rack. When multi-purpose requirements exist, the level of protection must be upgraded to preserve these features, regardless of the value of the product. Consequently, packaging costs can often far exceed the product cost, and the establishment of a cost ratio as a guide is more difficult to accomplish. Further, the distribution package is often designed to protect the features, including the appearance of another package (unit container), although the product itself is not fragile.

In the selection of packaging materials to accomplish protection and to fulfill any additional needed merchandising requirements, a variety of alternate materials is usually available which may differ considerably in price. The choice of the lowest price packaging material or container to satisfy the optimum objectives does not necessarily develop the lowest over-all physical distribution cost. Additional costs of packing, handling, storage, and freight may far exceed the savings attained in the purchase cost of these materials.

Some physical distribution systems lend themselves to the potential re-use of packaging. Typical areas where re-usable containers have found economical application are:

1. When merchandise is distributed in company owned vehicles which return empty to their originating points. In these instances no freight cost for the return of re-usable containers is incurred, that is, snack foods, dairy products, bakery products, etc.

2. When specialized containers facilitate extensive loading, handling, and unloading economies which exceed the added freight cost of return via common carrier, such as bins for bulk products, racks for assemblies and parts, and pallet boxes for a variety of products, or special freight rates apply.

3. When the cost of the container and/or its interior parts is high and repeated re-use saves in packaging material costs over and above return costs (hampers for telephone equipment, aircraft and missile containers).

In rating the economics of re-usable versus expendable packaging materials and containers, the following cost factors must be evaluated:

1. *Re-usable Containers.* Factors involved are: initial cost of container or packaging and anticipated number of return trips, content loading and unloading costs, the capital involved in furnishing an adequate supply, the assembly, disassembly and repair costs involved, loaded and empty handling and storage costs, freight costs of loaded and empty

SCOPE OF PACKAGING FOR DISTRIBUTION

units, protection, appearance and utility factors, accounting and inventory costs.

2. *Expendable Containers.* To be considered are: initial cost of container or packaging, content loading and unloading costs, freight costs, loaded and empty handling and storage costs, protection, appearance and utility factors, and disposal costs.

Studies have indicated that a re-usable container may benefit the manufacturer of a product in lowering his packaging costs, but the overall distribution costs may exceed these savings. Therefore, to serve the interests of the receiver or ultimate user, only a comparison of all cost factors determines the economic advantages of either system.

The cost of accumulating re-usable containers for return shipment in car or truck load quantities can impose a hardship on the recipient. In addition, the clerical and accounting aspects can be a further deterrent to the acceptance of re-usable shipping containers.

In general, a system employing re-usable containers functions best when a considerable flow of traffic in interplant shipments exists, especially if these plants belong to the same parent organization or where a back haul is feasible.

After the adequacy and suitability of a packaging material or container has been established on the basis of performance, the economies of this material can be rated against other suitable materials having similar performance. In the chapters that follow, the most widely used packaging materials and their characteristics are discussed. Wherever possible, comparative rating procedures have been included to assist in the selection of the most economical packaging system.

Systems and equipment. The ability to obtain lowest labor costs is a function of the container or packaging itself, and the methods employed in loading, sealing, reinforcing, identifying, and handling. The establishment of systems to accomplish these functions efficiently and economically is dependent on the procedures adaptable to either manual labor or mechanical aids. Systems that are employed range from simple manual operations to complex fully automated procedures. The success of a given system is influenced by the following major factors:

1. *Product Considerations.* Considerations are volume, configuration, fragility, special product characteristics such as consistency and density, number of sizes, styles, colors, and models.

2. *Manufacturing Considerations.* Rate, frequency, and location of output, inspection requirements, physical plant facilities, and materials handling equipment must be considered.

3. *Packaging Considerations.* Quantities, volume, weight and arrangement, type of package and interior components, sealing, reinforcing, and identifying requirements are factors involved in packaging.

4. *Distribution Considerations.* To be considered are compliance with customer, carrier or governmental specific packaging requirements, mode of shipment, that is, parcel post, export, etc., limitations of carrier vehicles, repacking, warehousing requirements, and order picking.

A deterrent to the accomplishment of an efficient system of packaging exists when a great variety of types of a product is produced—particularly when a large percentage of the total output is in relatively small lots. In order to reduce labor costs, systems should be investigated involving standardization, scheduling, and the use of packaging equipment which will permit flexibility. Essentially, extensive product variation is limited to predominantly manual methods owing to the unavailability of suitable packaging equipment or their excessive cost in view of the volume produced. Industrial engineering work simplification techniques as well as incentives and standards are tools to minimize labor costs of manual methods.

The influence of manufacturing facilities on improved systems is self-explanatory. Frequently economies are attainable through the revision of the flow of the product to be packaged and by consolidation of lines through accumulation of a central point where the use of mechanical equipment is feasible. Furthermore, the selection of equipment is dependent upon such plant factors as space, ceiling heights, floor-load capacity and other potentially limiting plant conditions.

Second only to the importance of the product itself insofar as systems are concerned is the type of packaging utilized. When product or manufacturing facilities limit the employment of mechanical aids, improvement in methods is frequently obtainable through changes in packaging material or container design. For example, the elimination of tissue wrapping of a fragile product through the use of an anti-abrasive coating or the elimination of interior components such as filler pads by changing box flap length are some of the many opportunities that exist to improve packaging procedures. The use of self-locking closures usually increases the container cost, but may reduce set-up and closure material and labor cost with an over-all net savings. Sometimes a simple change in the arrangement of the contents permits more rapid manual loading of containers.

Where conditions are conducive to the application of semi- or fully automatic equipment, container designs may have to be altered to make such equipment feasible. Examples of such container redesign include changing shape and opening to permit entry and withdrawal of case

SCOPE OF PACKAGING FOR DISTRIBUTION 57

packing equipment and replacing preformed bags by using roll stock applied with overwrapping equipment, etc. Frequently an entirely different packaging material is selected if mechanization of the container forming, packing, and/or closure is economically practicable.

The influence of distribution factors on systems is obvious. Special requirements for customers or mode of shipment may require separate packing lines or stations or additions on established lines or equipment. For instance, the separation of packing for parcel post and export shipments is necessitated by additional protective and delivery requirements. When sufficient volume exists for any special packaging requirements method improvement through specialization and the application of work simplification methods and equipment are attainable.

DYNAMICS OF DISTRIBUTION PACKAGING

The scope and magnitude of the packaging field has been established through an outline of historical evolution and areas of application in commerce and industry. It has been further established that a separate specialized profession has developed because of the magnitude and complexity of packaging. The package engineering function embracing the analytical and integrated approach in the solution of packaging problems is in itself divided into two major areas, consumer or retail packaging and distribution packaging. Although these two areas overlap in scope and are therefore not readily isolated as separate entities, this textbook is attempting to deal with distribution packaging as a distinct discipline.

The growth of distribution packaging is continuing at an accelerated rate, not only because of the general growth of the economy, but because of the increasing exploitation of new markets and new applications. New applications are the result of changing merchandising techniques which require some form of distribution packaging. These include such areas as the building industry, improved distribution of agricultural products, and the packaging of new products such as leisure craft and nuclear age hardware.

In addition to the normal growth and the field of new applications, current packaging is in a constant state of review and refinement for the attainment of greater efficiency. Existing materials constantly find new application in areas previously not considered suitable through changes in material processing, treatment, or design. Such changes include the transition of nailed wooden containers to fibreboard, wooden kegs for nails to fibreboard boxes for matresses to multi-wall paper or polyethelene bags, and for contemporary packaging changes such as fibreboard to unitized film packaging.

In almost every instance of distribution packaging change, the underlying motivation has been cost reduction or enhancing customer convenience. Some of these changes involved changes in the size and/or weight of the distribution package for greater economy in handling. On one hand, the size and/or weight of packages has been decreased whenever possible for easier manual handling; on the other hand, packages have been greatly increased to facilitate economical, mechanical handling. Changes of this type frequently dictate the selection of new packaging materials and/or container types.

The greatest change—the one which makes packaging such a dynamic field—is the constant addition of new materials, methods, and equipment.

Material and container development. Many new materials and container types are developed in the research laboratories of the packaging supply industry in an organized effort to create new markets through new products. Similar research is conducted for the improvement of the performance of existing products or their reduction in cost. Other developments include the utilization of waste and by-products for packaging potential, new methods for grading and recycling, the exploration of biodegradable materials and other techniques to conserve resources, and to protect the ecology.

Typical examples for distribution packaging application include the use of synthetic and inorganic fibres, largely replacing natural fibres, for such products as reinforced tape, barrier materials and for new uses in combination with plastics, paper, and composition materials. Coatings, treatments, laminations, impregnations and other techniques have greatly changed the characteristics of many existing packaging materials and have permitted more diverse application.

By-products extracted from agricultural waste have added to the available packaging material supply. For example, cane fibres have been converted into paper, fabricated paper products, and solid board for cushioning and blocking purposes.

Flexible and rigid plastic materials have long had wide consumer packaging acceptance. These are now being used extensively for distribution packaging applications. The flexible plastics in film form are used for barriers and to shrink or stretch-pack products and unit loads. Semi-rigid and rigid plastics are molded into such containers as carboys and drums as well as re-usable containers for the delivery of perishable food products. Plastics are linked to the supply and cost of petrochemicals and therefore factors of economy cannot be predicted.

Systems and equipment development. Allied with the development of new distribution packaging media is the constant improvement in productivity resulting from revised and new packing and packaging technology. Revisions in procedures which improve productivity may consist of simple changes in manual methods through improved scheduling, lay-out and flow, and the application of work simplification devices. Together with more advanced training and specialization, this has resulted in increased labor output without

SCOPE OF PACKAGING FOR DISTRIBUTION

the use of mechanized equipment. However, major labor economies in the distribution packaging operation have been and will to an increasing measure continue to be realized through the development and integration of semi- and fully-automatic machines. The advances attained in the mechanization of consumer packaging operations will certainly encourage the development of new equipment to accomplish distribution packaging. Although the problems involved in the latter are manifold because of generally larger size, heavier weight, and lower volume units, many examples of such mechanization already exist. Typical examples include wirebound box set-up and closing machines, fully automatic strapping machines, automatic industrial package forming, and loading and sealing machines. Further developments which will justify presently available equipment or the design of new equipment will be the refinement of integrated handling systems in conjunction with computer controls. Systems of this type, when interfaced with advanced material handling, warehousing, and transportation techniques will enable the fuller application of automation for many distribution packaging applications.

Packaging Materials and Containers

CHAPTER 2

Corrugated

and Solid Fibreboard

By virtue of remarkable utility, exceptional strength characteristics, light weight, and low cost, corrugated and solid fibreboard packaging materials are extensively used to accomplish industrial packaging functions. Fibreboard packaging materials are used to package a multitude of items differing with respect to shape, size, density, fragility, and weight. In addition to the excellent protective features of corrugated and solid fibreboard, other advantages of these materials include minimum storage space, low labor cost in preparation for use, and ease of disposal. The diversity of application and percentage of use is shown in Table 2-1.

The use of fibreboard packaging materials can hardly be hailed as a new development. Corrugated unlined paper was used many years ago as an interior packing and wrapping material, principally for fragile items such as bottles and incandescent lamps. Shortly before the beginning of the twentieth century, facings or liners were adhered to the corrugated paper to enhance rigidity. Ultimately, boxes were made from this combined material, and, in 1903, experimental freight ship-

TABLE 2-1
USE OF FIBRE BOXES BY INDUSTRY

Industry	Per Cent of Fibre Box Industry Production
Foods and beverages	28.7%
Paper products	12.1
Glass and ceramic products	11.0
Appliances and electrical equipment	5.5
Metal products	5.5
Clothing and textiles	4.2
Rubber and plastic products	3.9
Furniture and wood products	3.9
Chemical products	2.8
Machinery	2.6
Soaps, detergents and toilet articles	2.5
Transportation equipment	2.4
Toys and sporting goods	1.8
Printing and allied industries	1.4
All other	11.7
	100.0%

Source: The Fibre Box Association

ments were conducted. A period of litigation ensued as a result of conflict with other types of shipping containers accepted at that time. In 1906 the general use of fibreboard containers for freight shipments was authorized by the American railroads.

Since the beginning, the corrugated and solid fibre industry has had a phenomenal growth, even greater than the dynamic growth of the American economy. Literally billions of fibreboard containers are manufactured and employed annually. Fibreboard material is adaptable for use with other materials for the fabrication of cleated cases and has been used successfully as an insulating element, in weatherproof containers, for military packaging, interior cushioning, and in the construction of furniture, to name but a few of its wide range of uses.

The manufacture of corrugated and solid fibreboard is essentially a mass-production operation carried out with complicated machinery notable for its high speeds, size, and precise ingenuity of control. Fibreboard can be produced from a multiplicity of available component materials in numerous structural shapes. Fibreboard stock includes most of the heavier fibrous products made on Fourdrinier and cylinder paper machines. The paper may be formed as a *fluted board*, designated *corrugated fibreboard*, or formed *in a single flat sheet by the lamination of several thicknesses*, designated *solid fibreboard*. Of these

CORRUGATED AND SOLID FIBREBOARD

two materials, corrugated fibreboard is much more widely used, as is shown by the relationships established in Fig. 2-1.

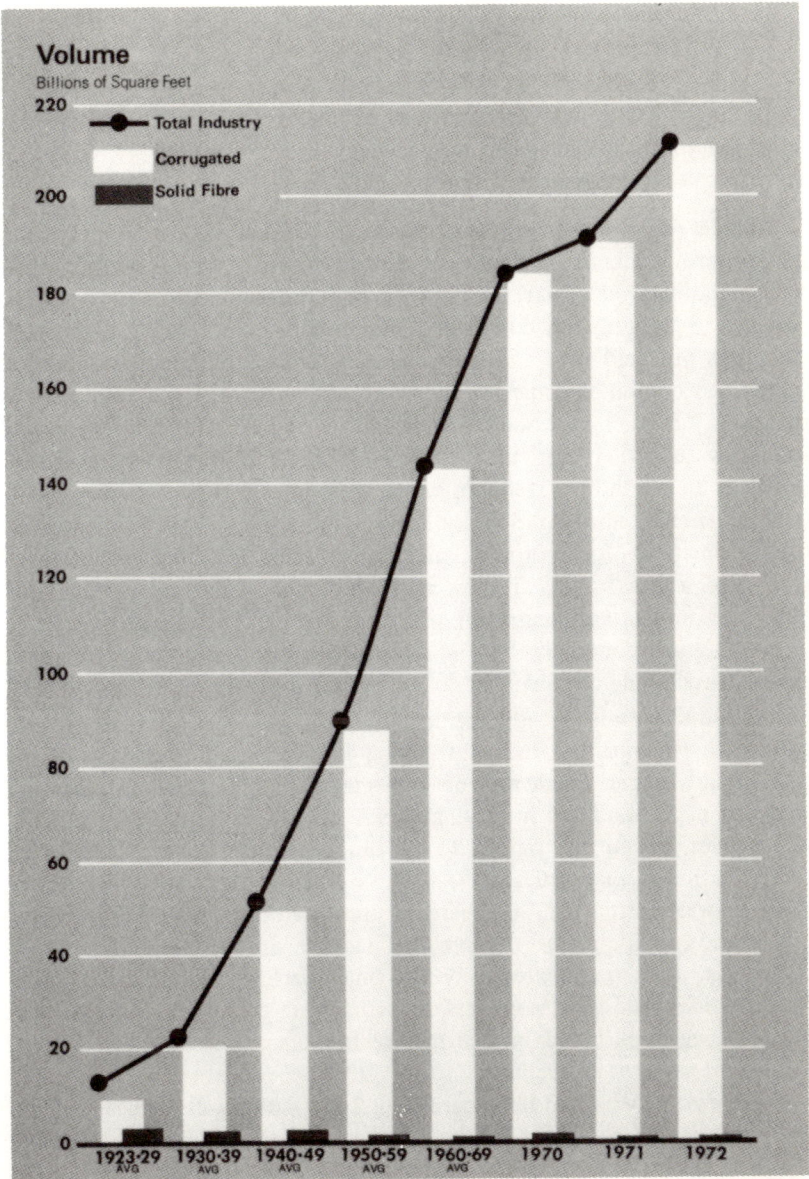

Fig. 2-1. Corrugated and solid fibre shipments. Courtesy Fibre Box Association.

CORRUGATED FIBREBOARD CONSTRUCTIONS

Corrugated fibreboard is produced in the following major constructions:

1. Unlined corrugated paperboard.
2. Single-faced corrugated fibreboard.
3. Single-wall (double-faced) corrugated fibreboard.
4. Double-wall corrugated fibreboard.
5. Triple-wall corrugated fibreboard.

Unlined corrugated paperboard (Fig. 2-2). Unlined corrugated paperboard is made in sheets or rolls in varying widths and lengths. It is manufactured of various kinds of paperboard and in varying thicknesses. By gluing the corrugated paperboard to a flat facing, the corrugations are held firm and are prevented from stretching or flattening unless the combined paperboard meets with a considerable amount of pressure. It may be obtained in rolls, in desired width, in sheets, or in circles and other particular shapes. When it is in rolls, the standard length is 250 ft and the standard width is 36 in.

Single-faced corrugated fibreboard (Fig. 2-3). This fibreboard is used principally for wrapping and for interior packing. Articles of glassware, other fragile articles, and articles in glass and earthenware containers are often encased in single-faced board before being packed in a shipping container. It is also used quite extensively by retail stores in wrapping merchandise for local delivery.

Single-wall (double-faced) corrugated fibreboard (Fig. 2-4). Single-wall corrugated fibreboard is made by gluing flat facings to each side of the corrugated member. Single-wall board is manufactured of various kinds and grades of paperboard and in varying thicknesses. The archtype construction, which gives strength and rigidity to the fibreboard, also provides cushioning and insulation characteristics.

Single-wall corrugated fibreboard is used for boxes, liners, pads, shells, tubes, and partitions for the interior packing of articles requiring such protection. It is used extensively for inner packing for glassware, other fragile articles, and articles in glass or earthenware containers. Miscellaneous uses for single-wall board include picture backing, wardrobe cabinets, toys, and freight car door protection.

Double-wall corrugated fibreboard (Fig. 2-5). Double-wall corrugated fibreboard consists of *three* flat facings and *two* corrugated members combined in the following sequence—a flat facing, a corrugated member, a center flat

CORRUGATED AND SOLID FIBREBOARD

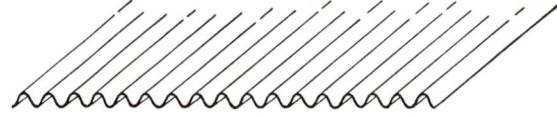

Fig. 2-2. Unlined corrugated paperboard.

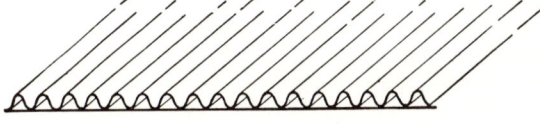

Fig. 2-3. Single-faced corrugated fibreboard.

Fig. 2-4. Single-wall corrugated fibreboard.

Fig. 2-5. Double-wall corrugated fibreboard.

Fig. 2-6. Triple-wall corrugated fibreboard.

facing, a corrugated member, and a flat facing. Double-wall board is manufactured of various kinds and grades of materials and in varying thicknesses. This type of board construction is used when greater strength or cushioning than that provided by single-wall corrugated fibreboard is required.

Triple-wall corrugated fibreboard (Fig. 2-6). Triple-wall corrugated fibreboard consists of *four* flat facings and *three* corrugated members. Triple-wall board is manufactured of various kinds and grades of materials and in varying thicknesses. This type of construction is most applicable for packages requiring exceptional rigidity and high puncture resistance.

CORRUGATIONS

Flutes. The flutes or arches of corrugated board are specified by size A, B, C or E. Figure 2-7 represents a cross section of single-wall corrugated fibreboard indicating the proportional relationship of the four flute constructions. Listed below are the dimensions for each flute:

	A Flute	B Flute	C Flute	E Flute
Number of flutes per lineal foot	36 ± 3	50 ± 3	42 ± 3	94 ± 4
Approximate flute height (not including thickness of facing) in in.	$3/16$	$3/32$	$9/64$	$3/64$

Three flutes (A, B, and C) are employed in different combinations to form *double-wall* and *triple-wall* corrugated board. Although all possible flute combinations are obtainable, the most commonly produced double-wall constructions are B with C. Double-wall board specified as B-C, for example, would be constructed with the B flute on the outside and the C flute on the inside. Similar flute combinations and designations apply to triple-wall corrugated fibreboard. In double or triple-wall construction the flutes are always *parallel* to each other.

In the fabrication of a container the flutes are aligned vertically or horizontally. Flute type and direction are extremely important as far as the durability and stacking characteristics of a corrugated fibreboard box are concerned.

Flute type is closely related to cushioning ability and the resistance of the board to crushing compressive forces. Owing to the arched structure and the engineering principles involved, B flute, having a greater number of flutes per foot, can support a greater weight than A or C flute when weight is applied, as is illustrated in Fig. 2-8a.

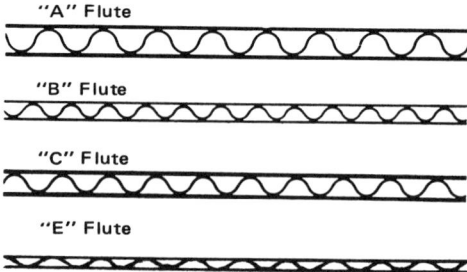

Fig. 2-7. Flute types.

CORRUGATED AND SOLID FIBREBOARD

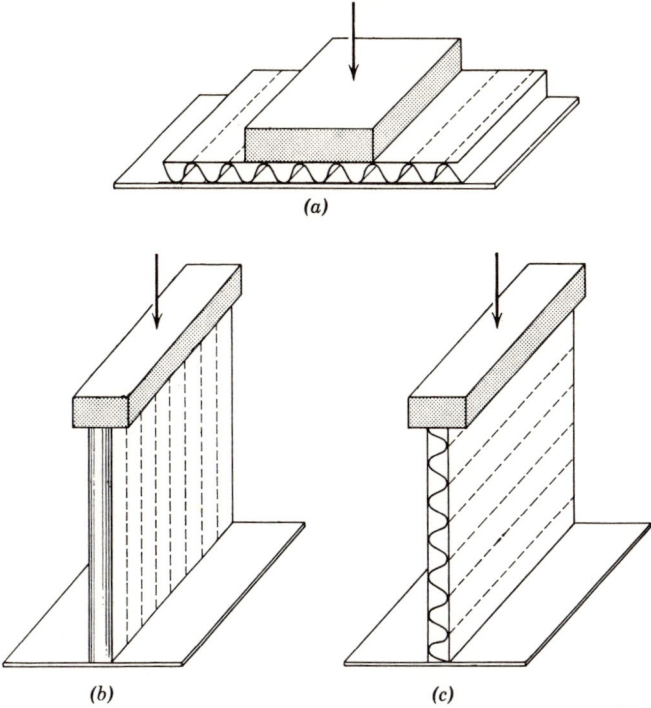

Fig. 2-8. Force application to corrugations.

On the other hand, if computations are based upon columnar engineering principles, A flute, having the largest columns, can support a greater weight than B or C flute when weight is applied in the other direction (Fig. 2-8b).

Finally, when pressure is applied in the direction illustrated in Fig. 2-8c, B flute is capable of supporting the greatest loads.

A flute has the best cushioning qualities because of its greater height or thickness. Conversely, A flute does not fold or crease as readily as B flute because of its greater thickness. Puncture and tearing resistance are influenced by flute height, with A flute having greater puncture resistance than B and B flute having somewhat better tearing resistance than A.

In all of these properties, C flute is about midway between the results obtained from A and B.

In view of these properties and varying flute characteristics, care must be exercised in the selection of the proper flute type for a specific

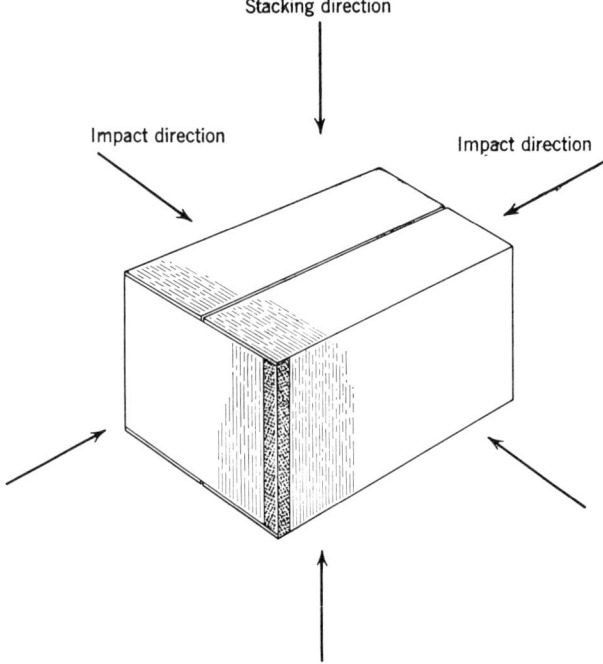

Fig. 2-9. Force application to shipping container.

packaging requirement. For example, canned goods are capable of sustaining great loads in the *top-to-bottom direction* and, therefore, do not require much assistance from the shipping container in this direction in conventional stacking. However, the side walls of the cans are quite vulnerable to denting and require greater lateral protection against *impacts* (Fig. 2-9). Further, because of the concentrated stress produced by the dense weight and sharp edges of the cans bearing against the scorelines of the box, good tearing resistance is required of the shipping container. Vertical B flute and C flute corrugations supply the required characteristics for the packaging of canned goods and these flute constructions are used to package this product.

For the packaging of glass containers the cushioning provided by A flute is highly desirable. Thus, A flute is generally specified in packaging for glass and similar fragile items. For contents incapable of supporting stacking loads without deformation, such as toilet tissue, containers having vertical A or C flute corrugations, or C flute alone are specified.

CORRUGATED AND SOLID FIBREBOARD
COMPONENT MATERIALS OF FIBREBOARD

Components of corrugated fibreboard. *Facings* are relatively heavy, coarse paperboard usually of kraft or jute* material. The term *kraft* is derived from a Swedish word meaning strong, and this term is applied to paperboard (usually made from spruce or pine chips digested by an "incomplete cook" process which preserves the long fibres of the original wood cellulose). This process results in a high yield and a product of exceptional tearing strength.

Kraft paperboard is designated as either *cylinder* or *Fourdrinier*, depending upon the type of machine used in its manufacture. Both of these types of paperboard have definite and unique properties which make one or the other well suited for certain specific applications. With respect to the relative usage of cylinder vs. Fourdrinier, recent figures disclosed a strong ratio in favor of the Fourdrinier product.

The term *jute* as applied to paperboard is a misnomer since this material no longer contains jute fibres but is made from varying amounts of virgin kraft pulp and from used boxes. All jute board is made on cylinder machines which form the pulp into stiff, smooth sheets. Such boards generally do not resist tearing as well as those made of virgin kraft, but for most shipping containers they are sufficiently tough and actually may be stiffer, and they have the advantage of a very smooth printing surface.

Important properties of facings which may be specified and checked by the user include the *basis weight* of the sheet, *bursting* (Mullen or Cady) *strength*, and *puncture resistance.*

Similarly, the properties of the *corrugating medium* contribute substantially to the properties of the finished board and the products manufactured from this board. The types of coarse paperboard most generally used as corrugating medium are *bogus, semi-chemical, Fourdrinier kraft, cylinder kraft,* and *chip*, each of which has characteristics of value for specific applications. Of these materials, semi-chemical is the most widely used corrugating material. The most important properties of the corrugating medium are specified indirectly by specifying properties of the finished board.

The careful formulation and application of the *adhesive* that welds the components of the board into a single unit is equally as important to the final board quality as any of the properties of the facings and corrugating medium. Laboratory tests indicate that defective application of the adhesive is among the most frequent causes of failure of shipping containers (Fig. 2-10).

*The term "jute" is being phased out in favor of "recycled".

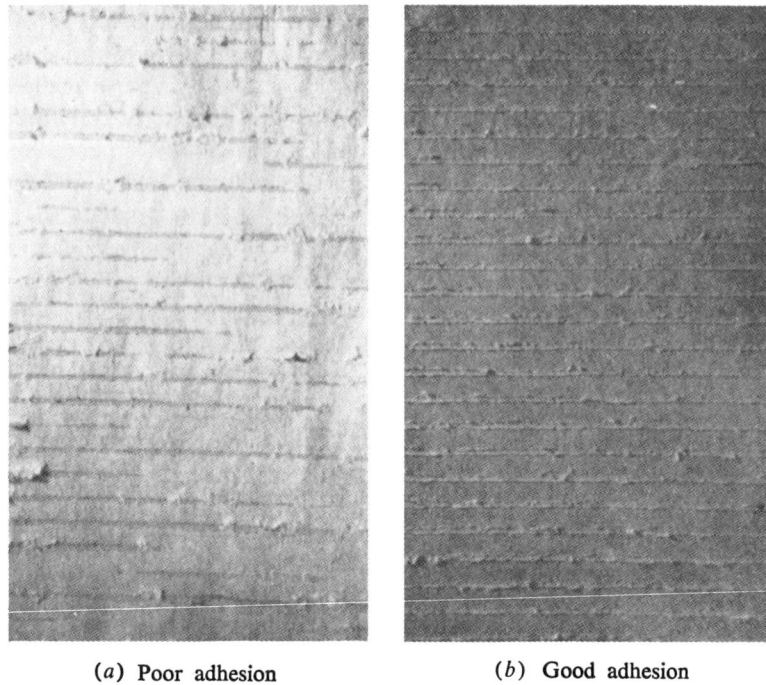

(a) Poor adhesion   (b) Good adhesion

Fig. 2-10. Adhesive patterns.

Both *sodium silicate* and *starch* adhesives are generally used for domestic corrugated fibreboard, and in some cases, both are employed. Impartial tests indicate that neither adhesive is outstandingly better than the other when used for the fabrication of corrugated fibreboard for shipping containers.

Because component materials in corrugated fibreboard boxes used for domestic shipments do not have a great resistance to the passage of water or water vapor, other forms have been devised known as *weatherproof* boards. These weatherproof boards, although highly effective for specific uses, largely in shipment to tropical countries and to the Armed Forces, are relatively high priced. Consequently their general use in domestic commerce is limited. Weatherproof boxes are made from dense, highly water-resistant components bonded together with waterproof adhesive.

CORRUGATED AND SOLID FIBREBOARD

Components of solid fibreboard. *Solid fibreboard* is essentially a pasted board consisting of two or more plies of paperboard bonded or laminated with a suitable adhesive. The paperboard may be *filler board, jute liners, cylinder kraft liners,* and *Fourdrinier kraft liners* in desired combination.

Domestic solid fibre boxes are only reasonably resistant to moisture penetration, and to enhance this characteristic various barrier components have been utilized. These components include special grades of *starch* combined with *urea formaldehyde resins*. Such boards are classed as weatherproof solid fibreboard and are obtainable in several grades (Fig. 2-11).

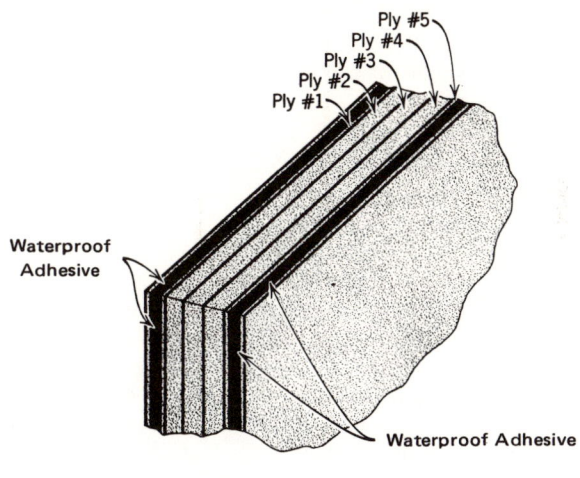

Fig. 2-11. A weatherproof solid fibreboard construction.

GRADES OF FIBREBOARD

The construction details of fibreboard have been described in preceding pages. This included a discussion of different types of component materials and flute and board constructions. Owing to the broad range in weight, size, and application of fibreboard containers (as demonstrated by the available styles and interior packing pieces), grading of the board with respect to its physical characteristics is necessary. In addition, rating of the finished container is also performed.

The most common and universally accepted method of grading fibreboard is by *bursting strength* expressed in pounds per square inch. The instruments used to determine bursting strength are known as Mullen or Cady Testers.

The grade is determined by the type and weight of components used in construction of the board. For corrugated fibreboard, the grade is controlled by the *combined weight of facing materials* used, with the weight of the corrugating material remaining generally constant. It is believed that the type of flute has little influence on the bursting strength. Table 2-2 lists the standard grades of corrugated and solid fibreboard plus the minimum combined weight of facings or plies required by the *Uniform and National Motor Freight Classification.*

It should be noted that the weights specified for corrugated fibreboard *do not* include the weight of the corrugating medium or adhesive. For

TABLE 2-2
STANDARD GRADES OF SOLID AND CORRUGATED FIBREBOARD
All Slotted Boxes, except Double-Wall Corrugated Boxes,
and All Other Fibre Boxes with or without Wooden Reinforcement

Minimum test per square inch of combined board, either solid or corrugated, lb	Minimum combined weight of component plies of solid fibreboard exclusive of adhesives, lb per 1000 sq ft	Minimum combined weight of facings of corrugated fibreboard, including center liner in double-wall boxes, lb per 1000 sq ft
125	114	52
175	149	75
200	190	84
275	237	138
350	283	180
500	330	
600	360	
Double-Wall Corrugated Boxes		
200		92
275		110
350		126
500		222
600		270
Puncture Test*	*Triple-Wall Corrugated Boxes*	
1100		264

*inch-oz per inch of tear

CORRUGATED AND SOLID FIBREBOARD

solid fibreboard the weight stipulated is exclusive of adhesives. The combined weights and not the weight of the individual facings or plies are specified.

For the packaging of hazardous materials or situations, which necessitate strength characteristics not supplied by a standard grade, interim grades are specified. Although rarely used, some of these interim grades are 150-lb test, 250-lb test, 300-lb test and 400-lb test. Although the combined weight of facings is not specified for the interim grades, weights have been established by common commercial practice. A few of these are listed below:

Bursting Strength, lb per sq in.	Construction	Common Minimum Combined Weight of Facings, lb per 1000 sq ft
150	Single-wall	66
250	Single-wall	111
300	Single-wall	159
400	Double-wall	153
450	Double-wall	174

A container made to interim grade specifications is subject to carrier regulations of the *nearest lower standard grade*.

It must be re-emphasized that the grading requirement stipulates the combined basis weight of the facings and that the supplier has a free choice in selecting the type (and to some extent the weight) of the individual facings. This procedure can be carried out provided that the bursting strength requirements are met.

Most manufacturers prefer to run what is known as a *balanced sheet*, in which the individual facings are of the same or nearly the same weight. Frequently, operational difficulties occur when *unbalanced facing* combinations are fabricated. As an example, of one convertor's preferences, Table 2-3 is included. It should be recognized that this table is based on kraft facings exclusively and involves common weights of kraft facings. Similar preferences are incorporated by convertors using jute facings exclusively. Others may use both jute and kraft facings depending upon their sources of supply.

If a container is made of unbalanced facing construction, and a good surface for printing is required, the *heavier facing should be on the outside*. On the other hand, some authorities claim that the use of the *heavier facing on the inside* increases the stacking strength of the box.

TABLE 2-3
REPRESENTATIVE WEIGHTS OF KRAFT FACINGS USED IN
CORRUGATED FIBREBOARD

Bursting Strength, lb per sq in.	Construction	Weight in Pounds of Individual Facings, per 1000 sq ft
125	Single-wall	26–26
125	Single-wall	33–26
150	Single-wall	33–33
175*	Single-wall	38–38
175	Single-wall	42–33
200*	Single-wall	42–42
250	Single-wall	69–42
275	Single-wall	69–69
300	Single-wall	90–69
350	Single-wall	90–90
200*	Double-wall	42–26–26
275*	Double-wall	42–26–42
350*	Double-wall	42–42–42
400	Double-wall	69–42–42
450	Double-wall	90–42–42
500*	Double-wall	90–42–90
600*	Double-wall	90–90–90

* Standard grades and common combinations using kraft facings.

Because of differences in fibre strength, heavier jute facings are required in order to obtain equivalent kraft facing bursting strength. For example, for 200-lb test single-wall board two kraft facings each weighing 42 lb per 1,000 sq ft are generally used, whereas for jute, facings weighing above 42 lb per 1,000 sq ft are usually used for the same purpose.

Formerly, for grading purposes, facing materials were specified by *thickness* in addition to weight. The thickness or caliper requirement has been eliminated except for corrugating materials which must be not less than *0.009 in. in thickness for all grades.* Furthermore, the carriers stipulate that the minimum weight for all corrugating materials shall be 26 lb per 1,000 sq ft. The basis weight of the corrugating material does not vary widely, regardless of the grade of board. The effect of increases in the basis weight of any given type of corrugating medium has less of an overall effect on the physical characteristics of the board and boxes than corresponding increases in the weight of the

CORRUGATED AND SOLID FIBREBOARD

facings. Generally, because of difficulties in the corrugating process the weight and thickness of corrugating material does not exceed 0.012 in. in thickness and 34 lb in basis weight. For weather resistant produce and ice-pack poultry boxes a 33 lb corrugating material is used without difficulty.

STYLES OF FIBRE BOXES

One of the major advantages of fibreboard is the ease by which it can be shaped into numerous styles of containers and interior packing pieces. The principal types of these are described below.

Regular slotted container (RSC). All flaps are the same length and the outer flaps meet (Figs. 2-12 and 2-13). This style of slotted container is in more general use than any other style because it is the most

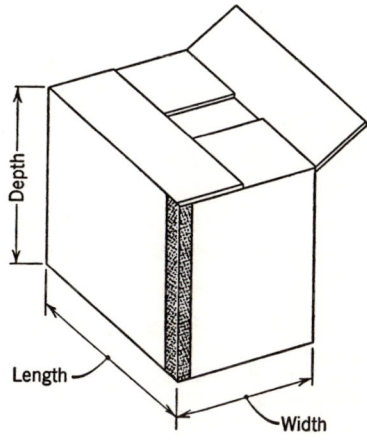

Fig. 2-12. Regular slotted container.

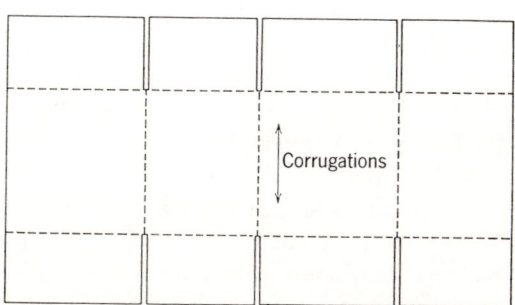

Fig. 2-13. Regular slotted box blank.

economical style of fibre box. All flaps are the same length and whereas the outer flaps meet at the center of the box, the inner flaps do not meet. The space between the inner flaps varies depending upon the relation of the length to the width of the box.

When the article to be shipped does not require the protection afforded by two thicknesses of fibreboard over the entire areas of top and bottom, this container is very safe, convenient, and satisfactory. When two thicknesses of fibreboard are required at top and bottom, fill-in pads between the inner flaps are often used.

Center special slotted container (CSSC). The length is greater than the width and all flaps meet (Fig. 2-14). This container gets its name from the fact that the inner flaps and the outer flaps meet at the center of the box. It is similar to the regular slotted container with the exception that the center special slotted container is stronger at the top and bottom because those areas are entirely covered by two thicknesses of fibreboard. It also requires more fibreboard square footage and an extra fabrication operation and is therefore more expensive than an RSC.

Overlap slotted container (OSC). All flaps are the same length and the outer flaps overlap a specified amount (Fig. 2-15). This style of slotted container is very similar to the regular and special slotted containers. The outer flaps overlap partially, and the inner flaps do not meet. The space between the inner flaps varies depending upon the relation of the length to the width of the box. This style is recommended when the box is to be reinforced with steel strapping or when self-supporting flaps for a stapled closure are advantageous.

Full overlap slotted container (FOC). All flaps are the same length, outer flaps overlap not less than inside width of box minus a maximum of one inch (Fig. 2-16). This style also is a variation of the regular slotted style of container. The dimensions of the box sometimes require this style to insure adequate strength, whereas at other times the box is used to provide additional strength at the top and bottom to afford greater protection to contents.

The outer flaps completely overlap, which makes an unusually strong container. Two thicknesses of fibreboard cover the entire top and bottom areas, and the areas of top and bottom covered by the inner flaps consist of three thicknesses of fibreboard. It is common practice to use this style for narrow width boxes.

Center special overlap slotted container (CSO). The inner flaps meet and the outer flaps overlap specified amount (Fig. 2-17). This style of box is exactly the same as the overlap slotted container, except that the inner flaps meet and provide three thicknesses of fibreboard over the entire areas of top and bottom.

CORRUGATED AND SOLID FIBREBOARD

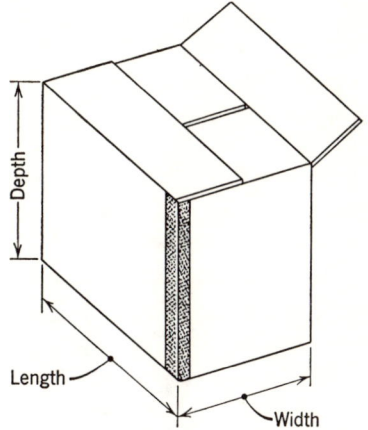

Fig. 2-14. Center special slotted container.

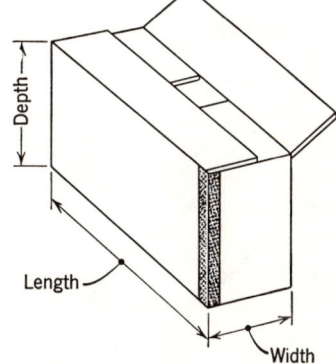

Fig. 2-15. Overlap slotted container.

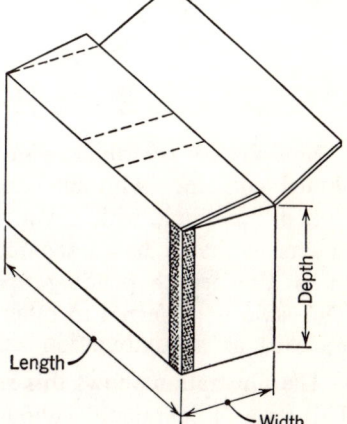

Fig. 2-16. Full overlap slotted container.

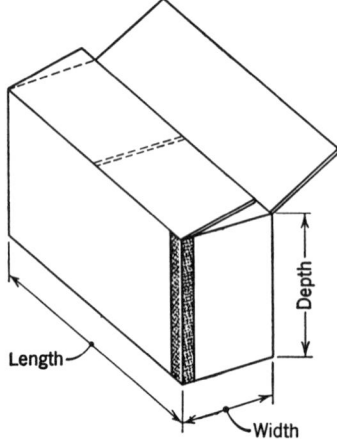

Fig. 2-17. Center special overlap slotted container.

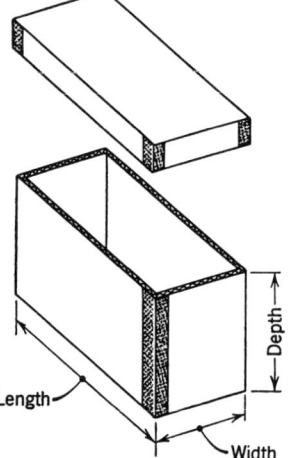

Fig. 2-18. Half-slotted container (with or without cover).

Half-slotted container with or without separate cover (HSC). A slotted container with one set of flaps only (Fig. 2-18). The half-slotted container, with cover, differs from other styles of slotted containers in that it has a separate flanged cover. The cover is separate or is attached, depending upon the particular use of the container. This style is preferred to other styles of slotted containers when it is to be used as a combination shipping and shelf package.

The illustration shows this style of container with cover not attached. This style of container, without cover, is used quite extensively as a slip

CORRUGATED AND SOLID FIBREBOARD

cover for the shipment of batteries. It is also used in the packaging of refrigerators, washing machines, furniture, and other articles. The bottoms of such packages consist of wooden frames with the half-slotted container securely fastened to them. The regulations often require that the open end of this container be constructed with flanges for fastening the container to the bottom of the wooden frame.

Double-cover box. The double-cover box is a three-piece box con-

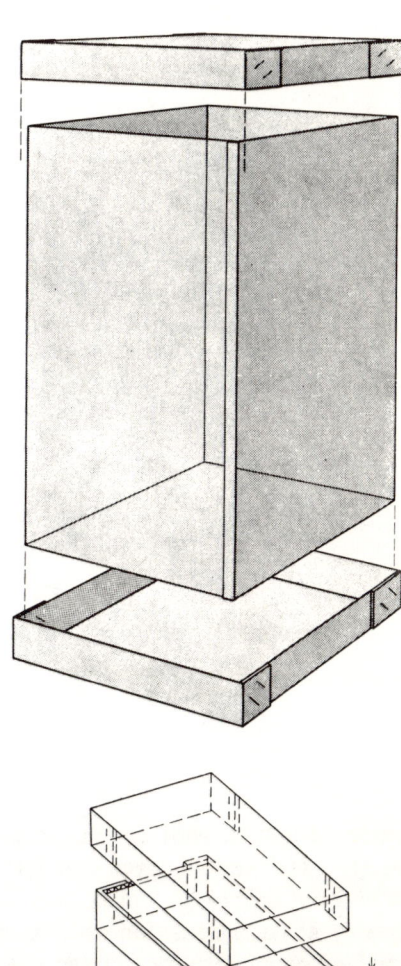

Fig. 2-19. Double-cover box.

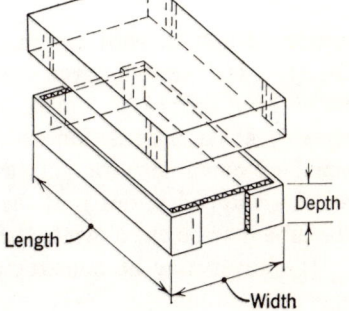

Fig. 2-20. Telescope box.

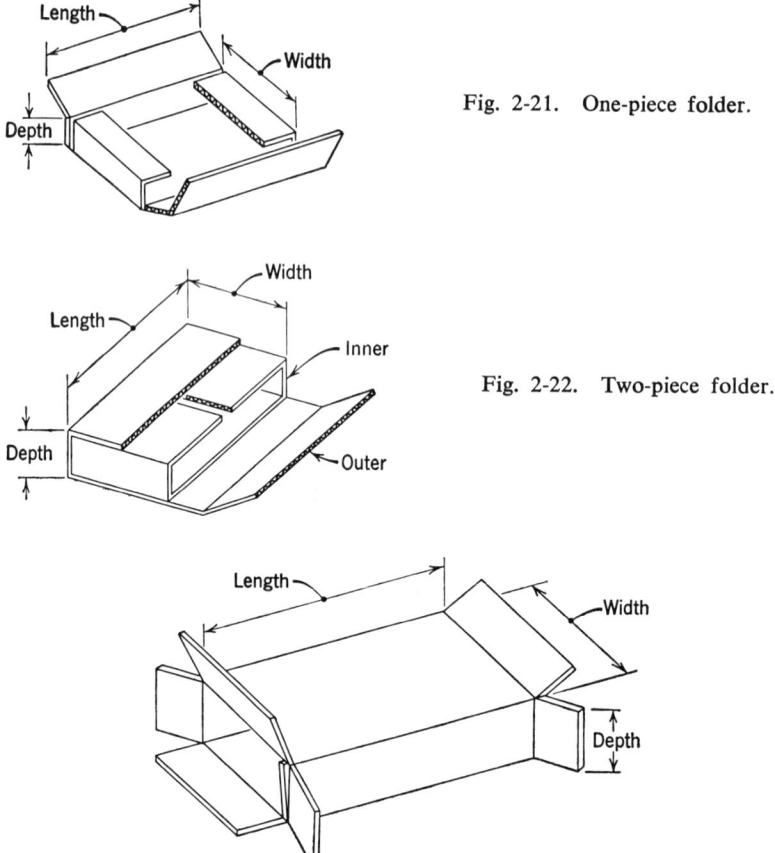

Fig. 2-21. One-piece folder.

Fig. 2-22. Two-piece folder.

Fig. 2-23. Five-panel folder.

sisting of a tube with two telescoping covers of specified depth (Fig. 2-19). The double-cover box has become increasingly popular as a pallet box and as a replacement for textile bales. In addition, this style is advantageous for articles which cannot be readily packed in standard containers for various reasons, often because the quantity of articles shipped varies. If the units being packed do not fill the box, the tube may be cut down to the level of the contents.

The cover may be constructed in such a way that the flanges form a

CORRUGATED AND SOLID FIBREBOARD

butt joint at each of the four corners or the flanged corners are overlapped for stapling or stitching.

Telescope box. A two-piece box with a full or partial depth cover (Fig. 2-20). The cover of this box may be made in any specified depth. This style of box is very convenient for use in shipping articles requiring a box of little depth or where reclosing is essential. It is an especially good box for commodities which need extra protection at the sides and ends as it provides a double thickness of fibreboard at these points which particularly increases stacking strength.

Shippers of paper, books, advertising material, pictures, cut-outs, and similar articles find that this style of box affords the needed protection to such contents.

The box is illustrated set up. It is usually shipped flat to the user and set up by him on his premises. Both sections of this box may be constructed in such a manner that a butt joint is formed at each of the four corners for taping, or the flanged corners are overlapped for stapling or stitching.

The telescope box with partial depth cover is used extensively for express shipment. It is especially adapted for the shipment of flowers, hats, chinaware, and glassware.

One-piece folder. This economical design is simply a scored and slotted sheet (Fig. 2-21). For the shipment of books, catalogues, or articles of that nature the one-piece folder is used extensively. It is shipped flat to the user, ease of packing is facilitated by the use of a jig or fixture, and closure is generally accomplished with tape.

The one-piece folder is used principally for parcel post and United Parcel Service shipments.

Two-piece folder. This style consists of two essentially identical scored sheets which are folded around the product (Fig. 2-22). The two-piece folder is stronger than the one-piece folder, as both top and bottom consist of two thicknesses of fibreboard.

This style is used principally for parcel post shipments. It may also be used for express shipments.

Three-piece folder. This style is quite similar to the two-piece folder consisting of three scored sheets with tucks of specified length. It finds wide use with items which are relatively long and flat. All folders are easily stored flat, set-up, packed and closed.

Five-panel folder. This style is basically a scored and slotted sheet of five panels, with one panel fully overlapping. The flaps of the fifth panel are cut out leaving four flaps for closure (Fig. 2-23). This style of box is shipped flat to the user and may be readily set up, packed, and

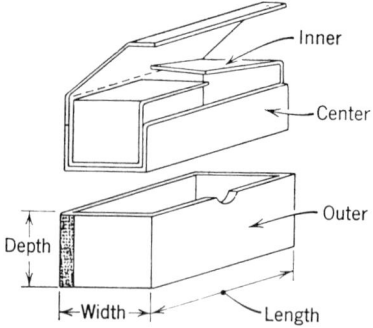

Fig. 2-24. Double-lined slide box.

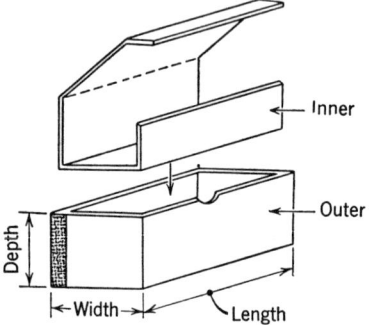

Fig. 2-25. Single-lined slide box.

closed for shipment. It is used principally for express and parcel post shipments.

It is an excellent box for the shipment of such articles as canes, rods, shade rollers, light fixtures, and umbrellas. The several thicknesses of fibreboard at each end provide considerable strength at the ends especially needed for long narrow and dense products, which might be readily damaged if forced through the ends of the container.

Double-lined slide box. (Also known as triple slide box.) This is a three-piece box consisting of inner and middle scored sheets and outer taped tube (Fig. 2-24). The three parts of the box, when assembled, provide a double thickness of corrugated fibreboard on all six faces of the box.

This box is made in any size or grade of material, although as a rule it is made in a comparatively small size and is generally used as an inner container or for parcel post shipment or for United Parcel Service shipments for fragile items.

CORRUGATED AND SOLID FIBREBOARD 85

Single-lined slide box. (Also known as double slide box.) This is a two-piece box consisting of inner scored sheet and an outer taped tube (Fig. 2-25). The difference between this box and the double-lined slide box (Fig. 2-24) is that there is no inner slide.

This style of construction provides double thickness of corrugated fibreboard on two sides and single thickness on top, bottom, and ends. It is used for the same purpose as the double-lined slide box when the extra strength provided by a third slide is not needed.

This box is made in comparatively small size and is generally used as an inner container or for express or parcel post shipment.

This box is made in comparatively small size and is generally used as an inner container or for United Parcel Service or parcel post shipment. It is easily opened or closed and provides the protection needed for shipment of such items as pharmaceutical products, books, etc.

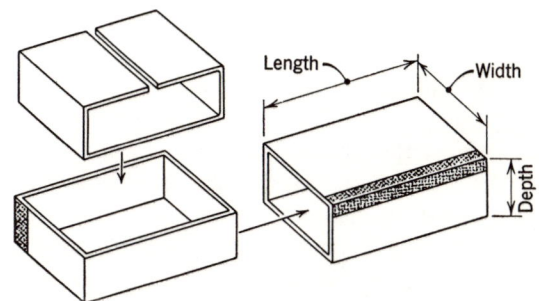

Fig. 2-26. Triple slide box.

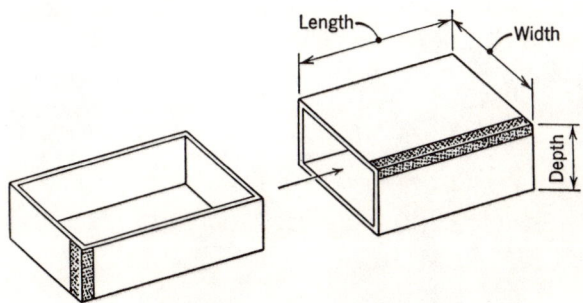

Fig. 2-27. Double slide box.

86   DISTRIBUTION PACKAGING

Triple slide box. (Also known as double lined slide box.) This is a three-piece box consisting of an inner scored sheet and middle and outer taped tubes (Fig. 2-26). The triple slide box, as is observed from the illustration, is similar to the double-lined slide box and, like the double-lined slide box, it has two thicknesses of fibreboard on all six faces. This box is used for freight, express, and parcel post shipment—principally for fragile articles.

Double slide box. (Also known as single lined slide box.) The style consists of an inner and an outer taped tube (Fig. 2-27). It is made exactly like the triple slide box, except that it does not have a third slide and, therefore, does not provide the same protection since it has two thicknesses of fibreboard on only two sides of the box.

This box is generally used as an inner container or for United Parcel or parcel post shipments.

Slide boxes of all styles are interchangeable, used either for the shipment of small items or as inner containers within an outer container for further distribution, or to provide extra protection.

SPECIAL STYLES OF SHIPPING CONTAINERS

In addition to the major styles described in Fig. 2-12 to Fig. 2-27 many special types of shipping containers are employed. These are either scored and slotted or die-cut and are made of one or more pieces of fibreboard. Frequently, a unique specialty is patented and can only

Fig. 2-28. Hardware container. Courtesy of Hinde and Dauch, Division of West Virginia Pulp and Paper Co.

CORRUGATED AND SOLID FIBREBOARD

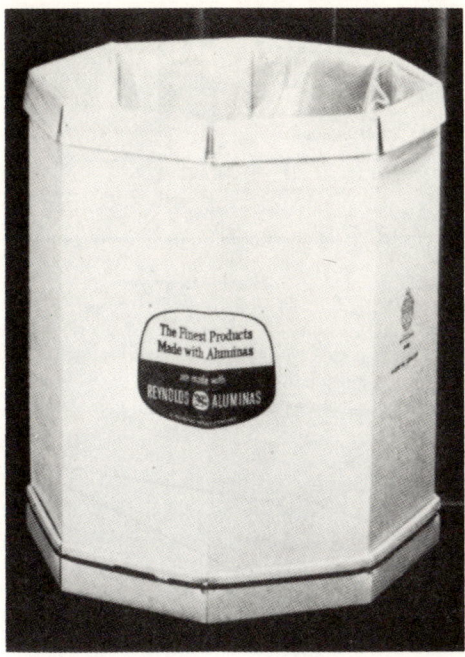

Fig. 2-29. Container with flanged tube and caps. Courtesy Gaylord Container Corp., Division of Crown-Zellerbach Corp.

be obtained from one supplier or its licensees. Owing to the diversification of fibreboard applications and entry into new fields such as the packaging of fruits and vegetables, nails and hardware items, and bulk chemicals, new styles and shapes are constantly being developed. Also the combined use of the shipper as a display or merchandiser has created the demand for originality, sales appeal, and utility. Furthermore, wide-scale expansion of prepackaging techniques has provided an opportunity to utilize the containers as a home storage receptacle for items such as irons, vacuum cleaners, and photographic equipment.

To illustrate some of these special styles and a few of their applications, Figs. 2-28 to 2-31 have been included.

Figure 2-28 illustrates a typical hardware container. This style is basically a five-piece container utilizing either a regular slotted interior container as shown, or a tube. It is used for very high density products such as nails, screws, and other bulk hardware, formerly packed in wooden kegs.

Figure 2-29 depicts a style of heavy duty container designed for bulk

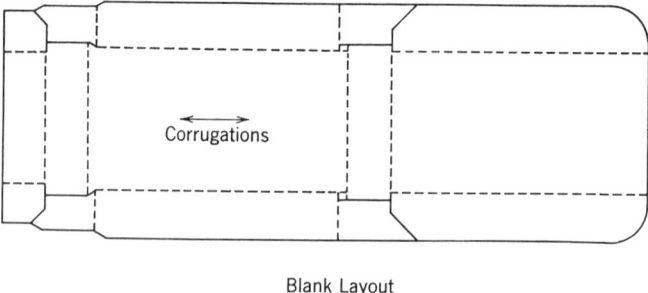

Blank Layout

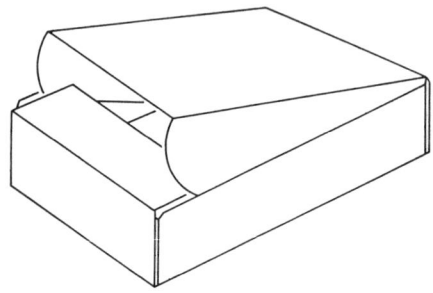

View of Partially Formed Container

Fig. 2-30. "FRAMERAP." Courtesy Continental Can Co., Inc., Fibre Drum and Corrugated Box Division.

or unit loads. The construction consists of three pieces, and the box when assembled is either square, rectangular, hexagonal, or octagonal in shape. The tops and bottoms are scored, slotted, or die-cut to form a double flange which interlocks with the flanges of the body section of the container. Steel straps applied over the interlocked flanges provide assembly for the container (Fig. 10-3, Chapter 10). Applications include appliances, yarn, and bulk chemicals.

Figure 2-30 depicts the patented "FRAMERAP" container of one-piece construction, die-cut and rectangular in shape, which when assembled provides an inexpensive and strong container for such items as books and samples. The maximum benefit from this design is obtained by the use of special forming equipment, which folds and seals the container in one operation.

Figure 2-31a and 2-31b illustrate two of many dual-purpose boxes. Frequently containers are designed to accomplish more than one func-

CORRUGATED AND SOLID FIBREBOARD

tion. The shipper may be used to display the product it has carried or may be used as a carry-home unit from the point of sale. The photograph illustrates a display-shipper. There are different applications of the dual-purpose principle combining die-cutting and printing to achieve maximum utility and merchandising appeal.

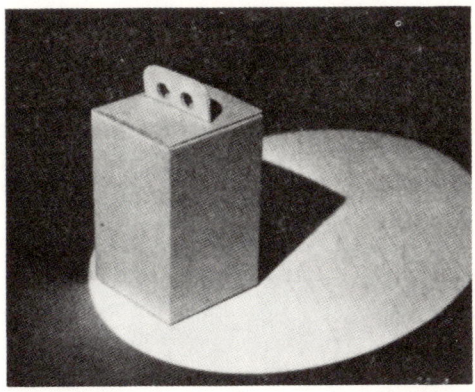

(*a*) Courtesy Hinde and Dauch Division of West Virginia Pulp & Paper Co.

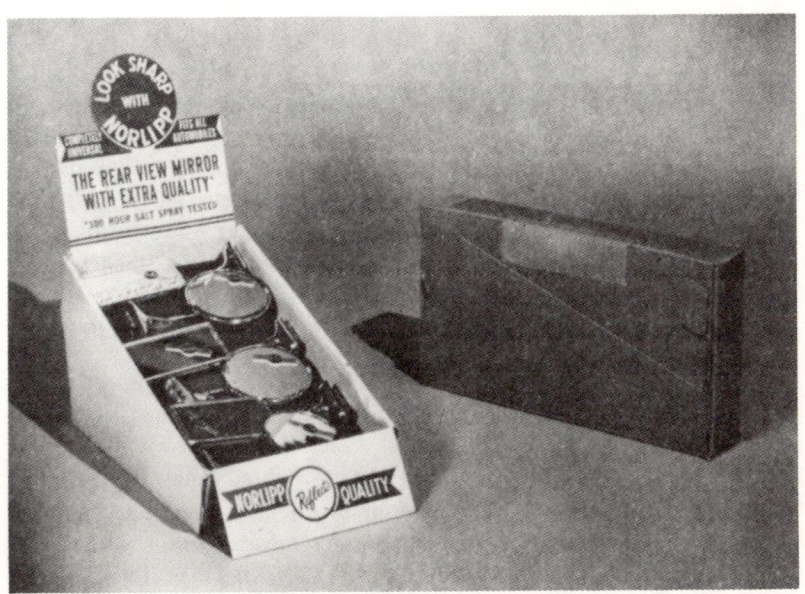

(*b*) Courtesy Fibre Box Association

Fig. 2-31. Dual-purpose boxes.

A precise reporting of all special styles of shipping containers would be voluminous. Container companies are excellent sources for details on such styles. Several other fibreboard specialty packs are referenced in other sections of this text.

INTERIOR PACKING PIECES

A further example of the versatility and remarkable utility of fibreboard is in its application as interior packing pieces. It is used in flat form, folded or die-cut and can be fabricated into a variety of forms, shapes, and patterns. Interior packing devices accomplish one or more of the following functions in the protection of a product.

1. Retention.
2. Separation.
3. Suspension.
4. Cushioning.
5. Protection against abrasion.
6. Insulation.
7. Blocking.
8. Clearance.
9. Positioning.

There are many different types of interior packing pieces as illustrated in Fig. 2-32. These pieces are fabricated from flat sheets of fibreboard by scoring, cut scoring, slotting, die-cutting and combinations thereof. Some interior pieces are self-supporting, others require stitching, taping, or gluing in order to obtain the desired shape and effect. Further, certain pieces do not achieve their full effectiveness until placed between the contents and the shipping container.

The following major groups of interior pieces are being used:

1. *Pads* (Item 1). Rectangular or square piece of fibreboard placed at the top and/or bottom of the container, or as a divider or separator between articles.

2. *Liners or collars* (Items 14, 41, 42). Scored sheets, either joined or unjoined, inserted to cover any four box panels resulting in double-thickness walls, added strength, and other reinforcement.

3. *Trays* (Items 16, 17, 35, 36, 38, 40). Scored and slotted, or die-cut sheets with two or more edges creased and turned at right angles to hold the product a specified distance away from the inside of a container.

4. *Slotted partitions* (Items 51, 54, 55, 56, 57, 58, 59, 60). Sets

CORRUGATED AND SOLID FIBREBOARD

of slotted sheets interlocked at right angles to form one or more air cells. The pieces can be of the same depth or of varying depth. The latter is exemplified by the shoulder height partition extensively used for glass bottle containers (Item 60). Extension partitions (Items 55, 56 and 59) have one additional member or half member in each direction extending beyond the edge compartments to provide additional air cells adjacent to the container walls.

5. *Radio, rat trap or rolled-up pads* (Items 2–9). Scored sheets designed to provide air-cell spaces at required points and surfaces while retaining their position with relation to the package interior. *Braced forms* are those in which certain panels are so folded as to provide triangular sections which brace themselves against deformation or collapse and thereby brace adjacent merchandise.

6. *Spring, folded or clearance pads* (Items 10, 11, 12, 13, 20, 27). Scored sheets folded to form one or more accordion-type pleats providing cushioning and clearance.

7. *Corner protectors* (Items 21-26, 28-31). Creased or slotted sheets folded to form an angle or a corner to provide cushioning along the edges or corners of the shipping containers. The protectors may be of *L* shape, triangular, or of the spring pad or radio pad type. They are placed between the shipping container and merchandise, such as furniture, appliances, etc.

8. *Built-up corrugated blocks or pads* (Fig. 2-33). Two or more layers of single- or double-wall corrugated board glued to each other. These pads are either attached to an interior component and are positioned to block or brace the merchandise, or they are positioned between the merchandise and/or shipping container. These pads may be prepared to conform to the required shape of the merchandise. They are generally used for heavy products.

There are several other types of interior packing pieces shown in Fig. 2-32. These examples demonstrate the many specialties that can be designed.

In the application of interior packing pieces simplicity of design is of utmost importance. This simplicity results in the minimum number of pieces required to perform the blocking, cushioning, bracing, or shielding function and will reduce the labor and material cost of packing. Frequently interior packing requirements can be fulfilled by modifications of the shipping container itself. As an example, the flaps of the container can be extended and slotted to form partitions for such items as glass jugs and paint cans (Fig. 2-34).

Sometimes a commodity may require several different types of in-

Fig. 2-32. Fibreboard interior packing devices. Courtesy Hinde & Dauch, Division of West Virginia Pulp and Paper Co.

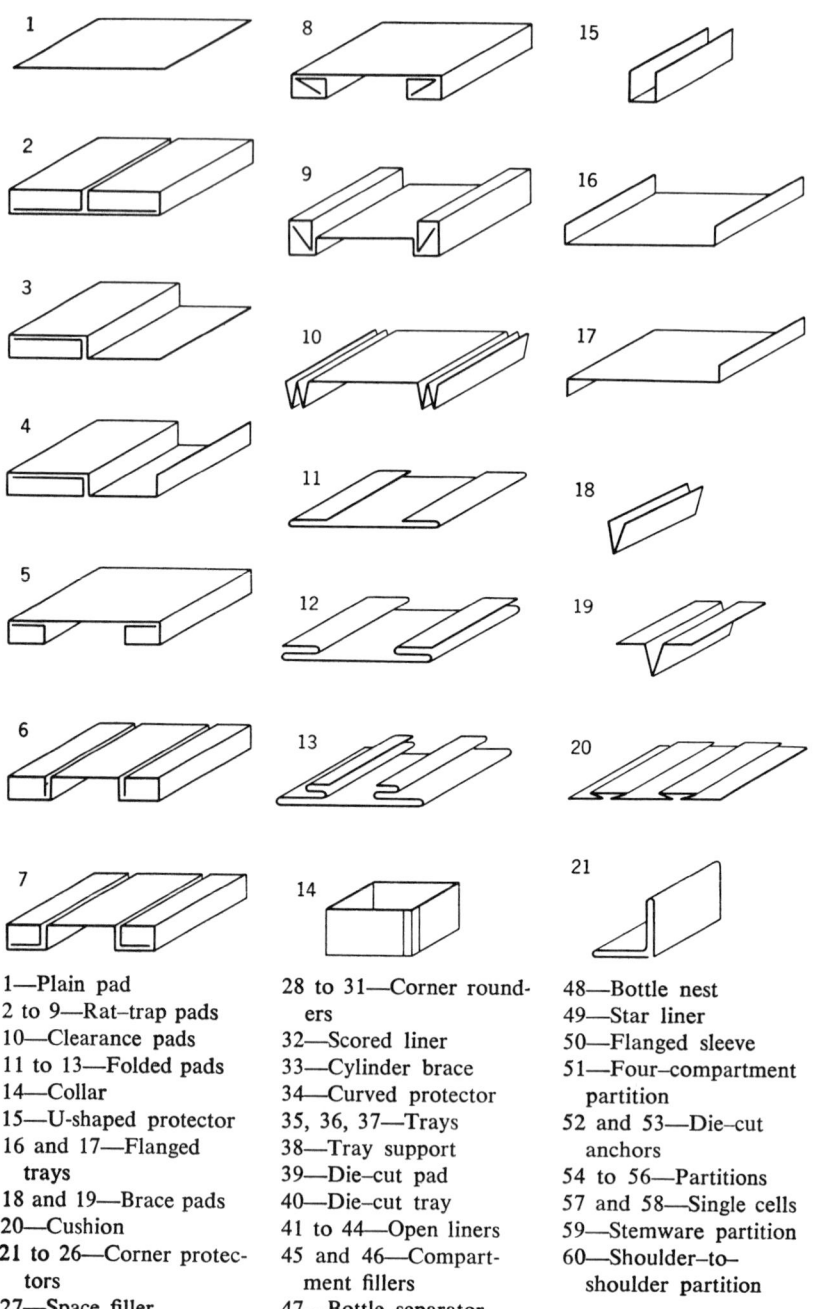

1—Plain pad
2 to 9—Rat-trap pads
10—Clearance pads
11 to 13—Folded pads
14—Collar
15—U-shaped protector
16 and 17—Flanged trays
18 and 19—Brace pads
20—Cushion
21 to 26—Corner protectors
27—Space filler
28 to 31—Corner rounders
32—Scored liner
33—Cylinder brace
34—Curved protector
35, 36, 37—Trays
38—Tray support
39—Die-cut pad
40—Die-cut tray
41 to 44—Open liners
45 and 46—Compartment fillers
47—Bottle separator
48—Bottle nest
49—Star liner
50—Flanged sleeve
51—Four-compartment partition
52 and 53—Die-cut anchors
54 to 56—Partitions
57 and 58—Single cells
59—Stemware partition
60—Shoulder-to-shoulder partition

CORRUGATED AND SOLID FIBREBOARD

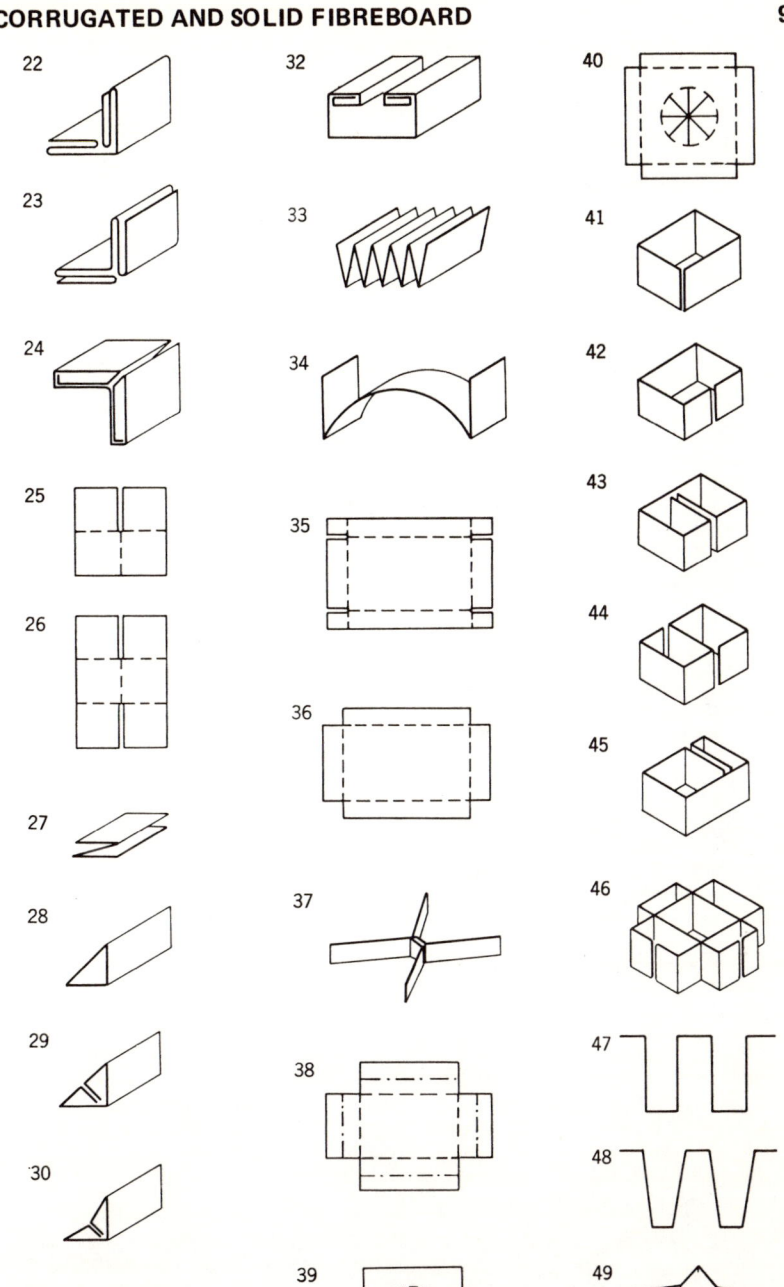

94   **DISTRIBUTION PACKAGING**

Fig. 2-32 (*Continued*)

50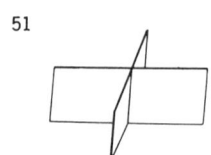

51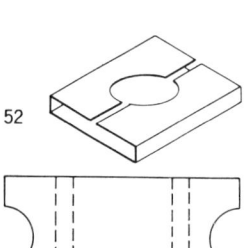

52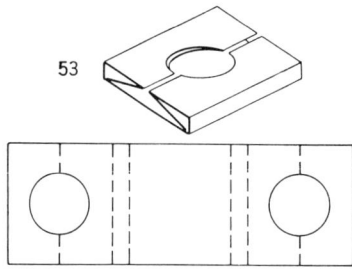

53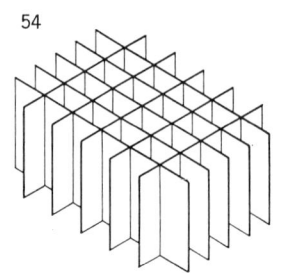

54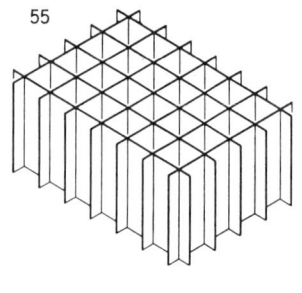

55

56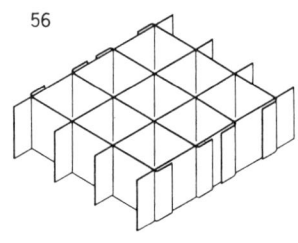

57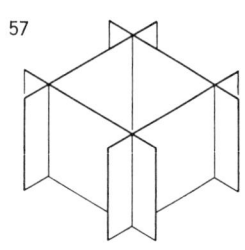

58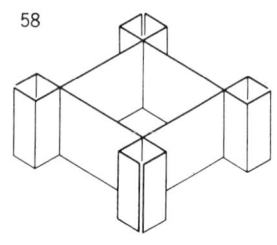

CORRUGATED AND SOLID FIBREBOARD

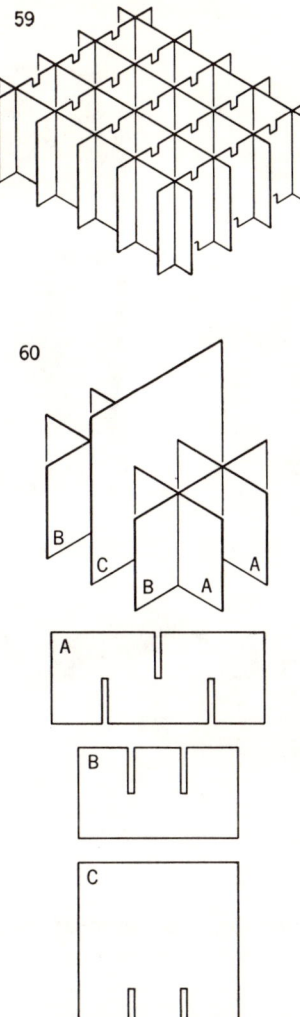

terior packing pieces for adequate protection. An example of a specification for a complex system is shown in Fig. 2-35. This illustrates four types of interior packing pieces in their flat form, and the exploded view shows their positioning in relation to the transmitter. It will be noted that the board dimensions, scoring, and flute direction are specified in detail for maximum strength and proper fit.

The application of a die-cut scored and folded sheet exhibiting desired simplicity is shown in Fig. 2-36. This single sheet fulfills the

96   **DISTRIBUTION PACKAGING**

Fig. 2-33. Built-up corrugated fibreboard blocks. Courtesy Delavans, Inc.

protective requirements for the safe shipment of this drill. This design reduced production-line labor and inventory control over previous methods used for this item.

For the shipment of a 25 hp. outboard motor, corrugated interior packing pieces replaced a wooden frame which was formerly required for support. The packing, shown in Fig. 2-37, features a die-cut assembled pad consisting of a scored sheet and a center and two end plugs. The plugs are glued in position on all contacting surfaces and form a rigid support as the sheet is rolled up.

The use of built-up corrugated blocks for the shipment of a cash register is illustrated in Fig. 2-38. The inner packing consists of a series of die-cut, built-up blocks which are laminated in a specific position on a base sheet. The die-cutting of all the internal parts is such that all pressure points on the cash register itself are eliminated and

CORRUGATED AND SOLID FIBREBOARD

complete flotation of this delicate machine in the container is accomplished. Previous to this method the register was supported by wooden blocks wrapped with excelsior padding, which resulted in the transfer of external blows to the contents. The register was then packed in a wooden container. The assembly of the new design is accomplished without nailing or strapping and 90 per cent of the labor involved with the original wooden boxes was eliminated.

BASIC DESIGN DATA

In the application of fibreboard for the packaging of any product, numerous basic design considerations must be evaluated. These pertain to the properties of the board itself, the properties of the fabricated container, the interior packing pieces, the type of manufacturer's joint, closure and reinforcement, and many other considerations. In addition, the selected design obviously influences the cost of the final package. The following paragraphs are a discussion of corrugated fibreboard and box properties vital in the economical development of a package design. Related design considerations are covered in subsequent chapters.

Physical characteristics of corrugated fibreboard. Various tests and determinations to measure the strength characteristics of fibreboard and to provide an indication of the quality of the finished board are used. In addition, they provide other valuable data not obtained by the re-

Fig. 2-34. Extension flaps forming partitions of fibre box. Courtesy Hinde & Dauch Division of West Virginia Pulp and Paper Co.

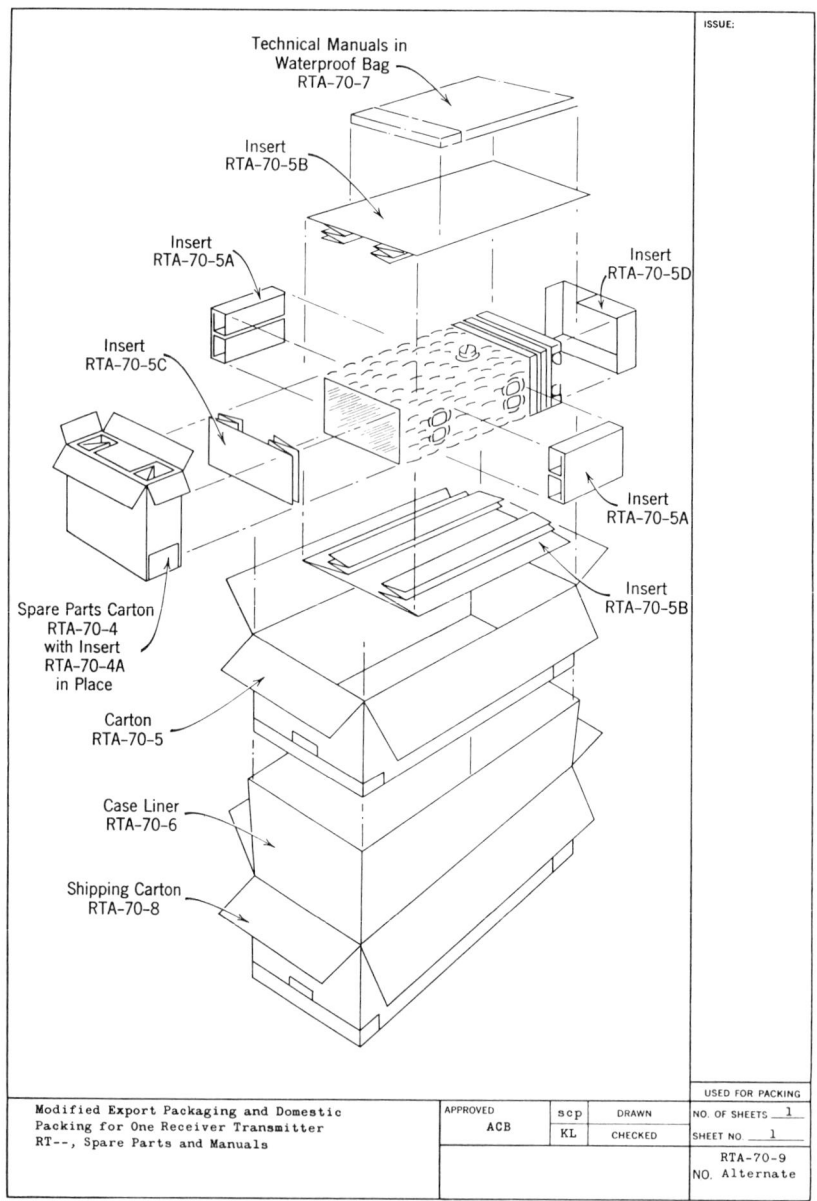

Fig. 2-35. Modified export packaging and domestic packing for one receiver transmitter. Courtesy Container Laboratories, Inc.

CORRUGATED AND SOLID FIBREBOARD

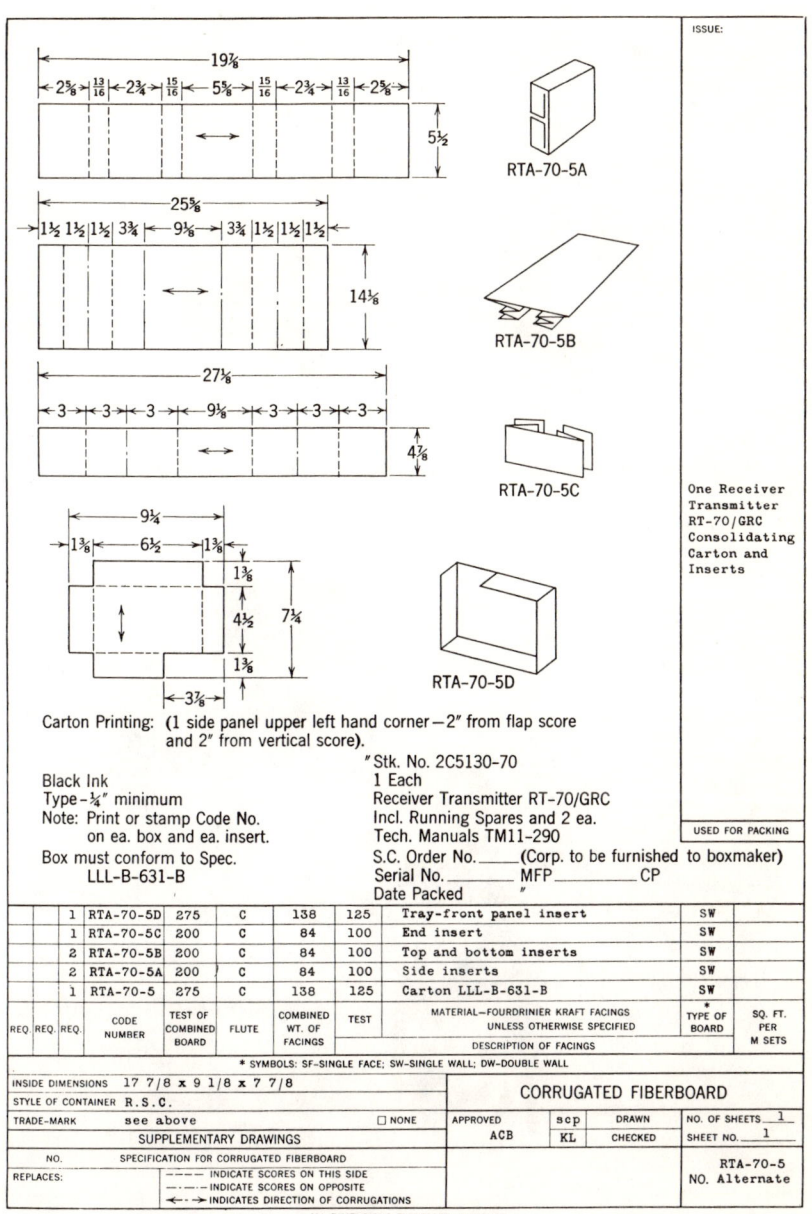

(Fig. 2-35 *continued*)

100 DISTRIBUTION PACKAGING

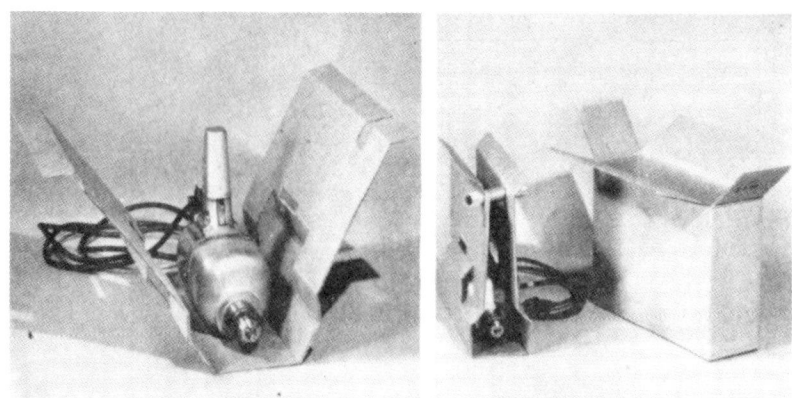

Fig. 2-36. Heavy-duty drill packaging. Courtesy Fibre Box Association.

Fig. 2-37. 25-h.p. outboard-motor packaging. Courtesy Fibre Box Association.

CORRUGATED AND SOLID FIBREBOARD

quired bursting test referred to earlier in this chapter. More accurate distinctions within a grade must be established to measure important strength characteristics of the board. These data permit the development of realistic specifications.

Tests commonly applied to combined corrugated fibreboard are now described.

BURSTING STRENGTH (MULLEN). As already discussed in the "Grades of Fibreboard" section of this chapter, the bursting-strength test furnishes a framework for the classification and identification of various grades of fibreboard. Although the Mullen test is the universally accepted basis for evaluating the quality of fibreboard, the test has

Fig. 2-38. Packaging for cash register. Courtesy Fibre Box Association.

many shortcomings. The bursting test does *not reflect the strength advantages* inherent in the structural form of a fluted or corrugated material nor is it *sensitive to deficiencies* in board fabrication such as improper bonding of the corrugating material to the facings. Nevertheless, as long as the bursting test is used as a basis for grading and is recognized by carrier regulations, it will continue to be a major factor in fibreboard evaluation.

The bursting-strength tester exerts hydraulic pressure upon a rubber diaphragm causing gradual distention of the diaphragm. This distention is transmitted to the sample being tested and causes it to burst or rupture. The results in pounds per square inch are indicated on a calibrated gauge. This type of duress to which the board is subjected is not experienced in the actual use cycle of the container. However, in addition to its value for grading and identification purposes, the bursting test provides a fair indication of the expected performance of the container. Generally speaking, there is *fair* correlation between good bursting strength and other important board properties such as puncture and flat crush resistance. Thus, if the bursting strength is high for a particular grade the *probability* is that the container is of good quality. Conversely, if a board is deficient in bursting strength other board properties have invariably suffered. Unfortunately, these indications can be badly misleading. For this reason additional tests are performed on corrugated fibreboard to supplement the bursting strength data.

PUNCTURE RESISTANCE. The puncture test measures the resistance of board to puncturing, which is a function of the tearing resistance of the board, its stiffness, its thickness, type of components, and efficiency of fabrication.

Although similar in some respects to the bursting test, the puncture test differs in that it reflects the structural nature of the corrugated board and also reflects to a certain extent the quality of adhesion of the facing material to the corrugated medium. Unlike bursting strength, puncture resistance is influenced by the type of flute used in the fabrication of the corrugated board. Thus, by inference, puncture resistance is sensitive to the overall thickness of the fluted structure. Because of these characteristics the puncture test is a more accurate measure of quality, efficiency of fabrication, and expected performance. Therefore, puncture resistance values are frequently incorporated in purchase specifications and are being introduced in the carrier regulations. For example puncture requirements have been established for triple-wall corrugated fibreboard.

The puncture test is a measure of the energy required to force a puncture head of designated size and shape completely through a sample of the material to be tested. The puncture head has the shape of a right-angle triangular pyramid, is affixed on a pendulum incorporating an arm in the form of an arc, and is released by trigger action from a horizontal position. The resultant puncture is similar to the damage which might be caused by the corner of a box striking the side wall of a container.

There are no standards for puncture resistance. However, certain minimum expectancies have been developed through extensive testing. Minimum puncture values generally commercially acceptable, are listed in Table 2-4 below for each of the standard and interim grades of corrugated fibreboard. Puncture values are reported either as puncture units or more properly as inch ounces per inch of tear, with the former more widely used.

TABLE 2-4
REPRESENTATIVE PUNCTURE TEST VALUES

Grade and Test of Board		Puncture Resistance, puncture units
125	test double-faced board	120
150	test double-faced board	143
175	test double-faced board	175
200	test double-faced board	200
250	test double-faced board	235
275	test double-faced board	290
350	test double-faced board	335
275	test double-wall board	315
350	test double-wall board	330
500	test double-wall board	600
600	test double-wall board	700

FLAT CRUSH RESISTANCE. Flat crush resistance is the ability of the corrugations to withstand a *force exerted perpendicular* to the facings. It represents a more precise evaluation of the property which is approximated by squeezing combined board between the thumb and fingers. The basic procedure for determining flat crush resistance consists of cutting circular or square sections of a known area from the combined board either 5, 9 or 10 sq in. in area, and then applying pressure over the entire area with a small compression machine.

Flat crush resistance is influenced primarily by the flute of the corrugations, the type of corrugating material used, and the efficiency with which the board has been combined. Laboratory tests have shown that in this determination the grade of the facings, that is basis weight, does not significantly affect the values in single-wall construction.

Flat crush resistance is expressed in lb per sq in. The values are classified by flute since different corrugating materials produce widely different results. Like puncture resistance, there are no official standards for flat crush resistance. However, certain minimum expectancies for the applicable flutes have general acceptance. In Table 2-5 these values are reproduced.

Flutes in double-wall construction generally collapse consecutively or simultaneously. In the former, the flute having the lowest resistance collapses first. In the latter, the flutes may have approximately the same resistance. In reporting the results of the flat crush resistance of

TABLE 2-5
FLAT CRUSH VALUES

Flute in Single-Wall Construction	Flat Crush Resistance, lb per sq in.
A	32
C	40
B	50
E	80

double-wall board the value for each flute is reported separately and is not totalled. Flat crush tests are customarily not performed on double-wall board.

The principal value of this property is in quality evaluation. Furthermore, flat crush resistance provides a measure of the rigidity of the board and affords valuable cushioning data.

Another measure of the strength characteristics of fibreboard is the edge-crush test which is gaining in popularity.

PROPERTIES OF CORRUGATED FIBREBOARD CONTAINERS

One of the most important functions of a shipping container is to withstand superimposed loads encountered in warehousing and transportation. Recent trends in increased warehouse heights (20, 24 and 30 ft) permissible through the use of palletized handling, have greatly increased the importance of stacking strength. In addition, modern automatic conveying systems subject shipping containers to impacts during handling. Impacts of a more severe nature are also experienced by shipping containers in transportation during the humping and switching operations of freight cars. Because of these hazards, the compressive resistance of the container *on all of its faces* is of primary importance.

Compression strength importance. Compressive resistance is measured by subjecting the containers to compressive forces which attempt to simulate field conditions. The container is tested individually in a compression machine consisting of a platform scale over which a power-driven crossbeam is mounted. The container is tested between two flat surfaces (the scale platform and a plate under the crossbeam), and the applied pressure is read on a dial as the opposite surfaces of the container are forced together.

CORRUGATED AND SOLID FIBREBOARD

The compression test is performed on boxes set up and sealed, with or without contents. The latter test is widely used as a quality evaluation tool, to predict the expected performance of the container in actual stacking. The great virtue of compression testing in quality evaluation procedures stems from the fact that any one of a host of rigidity reducing fabrication deficiencies such as improper adhesion, deformation of the flutes, misapplication of the manufacturers' joint, and incorrect depth of slotting can be detected. Thus the compression test discloses weaknesses which may escape visual examination.

The compressive strength of a fibreboard box is dependent upon physical characteristics such as:

1. Size and load bearing perimeter.
2. Shape—relationship of dimensions.
3. Flute and flute direction.
4. Grade of components.
5. Design details such as the distance between the inner flaps of the container.
6. The amount and type of printing.

In addition to the above, the compressive resistance of a container is influenced by these environmental conditions:

1. Atmospheric conditions affecting the moisture content of the fibreboard (Table 1-12).
2. Duration of load—*container fatigue*.
3. Method of stacking—such as palletizing.
4. Damages caused by handling and shipment.

Compression values, as determined relatively instantaneously in the laboratory under controlled conditions, discount the influence of environmental conditions. For this reason, the values obtained on the testing machine are far greater than the actual stacking resistance of the box. For practical estimating purposes, the *actual stacking strength is considered to be one-third to one-quarter of the test value.*

Over the years, the data obtained in thousands of compression tests have been organized to provide standards of expected performance. These standards are based upon the physical characteristics of the particular container and provide a means to determine the expected compressive resistance of a box of good commercial quality. These standards provide the packaging engineer with important design information and permit selection and theoretical evaluation, which can be confirmed later by the compression test.

A typical form used to record and analyze the compressive resistance

106   **DISTRIBUTION PACKAGING**

Fig. 2-39. Typical form used for reporting of compression tests and other fibreboard characteristics.   Courtesy Container Testing Laboratories, Inc.

CORRUGATED AND SOLID FIBREBOARD

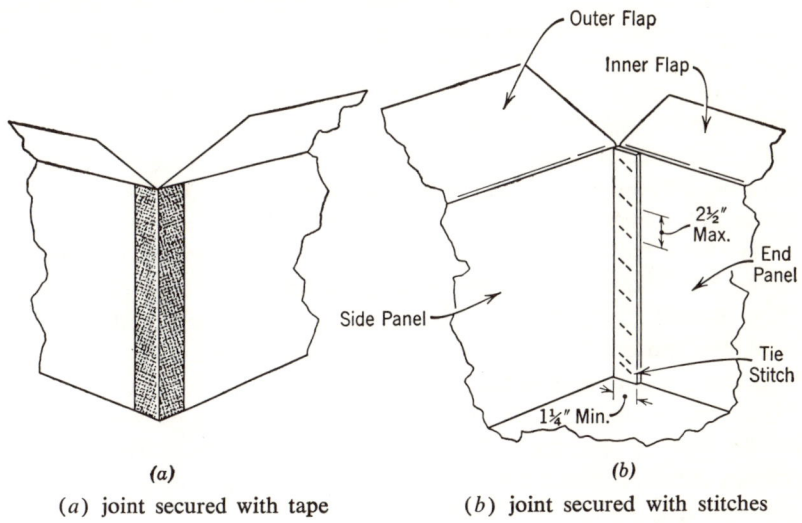

(a) joint secured with tape   (b) joint secured with stitches

Fig. 2-40. Manufacturer's joint constructions.

of empty corrugated boxes and other fibreboard characteristics is illustrated in Fig. 2-39.

Manufacturer's joints. Most corrugated shipping containers such as regular or full overlap slotted construction are joined together where the two edges of the box blank meet. The box manufacturer joins the edges by *stitching, taping, or gluing*. Certain other container styles, such as one-piece folders, are supplied to the user in box blank form and do not require a joint for formation. These containers obtain their final shape by their closure with either tape, stitching or stapling, strapping, gluing, or combinations of these methods which are carried out by the user in the packing operation.

The selection of the type of joint is either left to the discretion of the manufacturer or is specified for economy or special requirements by the user. Three types of joints will be discussed in the following paragraphs.

TAPED JOINTS   (Fig. 2-40a). Taped joints are made with either *gummed paper, reinforced paper*, or *cloth*, in 2- and 3-in. widths. *Paper tapes* are made of *gumming kraft* and have a basis weight ungummed of not less than 60 lb per 500 sheets 24 in. × 36 in.

REINFORCED TAPES. These tapes are generally made of two layers of paper sandwiched together with reinforcing fibres running in the crosswise direction and imbedded in a nonasphaltic laminant. Several longitudinal fibres are frequently incorporated in the tape to prevent breakage on high-speed taping equipment. *Cloth tapes* are made of a textile fabric (cambric and Osnaburg) adhered to a paper backing. There are other special types of joint tapes available. For example, one tape is marketed in which only a single thickness of paper is employed, with the reinforcing fibres and the gumming adhered to the underside of the single sheet. Other specialties are marketed in which layers of paper are used as the reinforcing medium.

STITCHED JOINTS. Stitched joints are fabricated by joining one edge of the box blank to a tab or extension of the other edge of the blank (Fig. 2-40*b*). The tab can be placed on the inside or outside of the container and can be an extension of the side or end panel. Stitching is performed on high-speed equipment which forms and applies the stitches from a continuous coil of wire. The wire is galvanized and is mostly flat or arcuate. If appearance is of prime consideration, colored stitching wire may be used. When an extra strong joint is required it may be accomplished by means of an extra staple or stitch at the top and bottom of the joint. This practice is known as a tie-stitch joint. Figure 2-40*b* illustrates a tie-stitch at the bottom of the joint.

GLUED JOINTS. These joints are made by gluing the extension of the blank or tab to the other edge of the blank. The adhesive, applied over the entire area of contact, is waterproof. A special glued joint is constructed by removing the corrugated medium from one end of the blank and splicing the inner and outer facings over the other end of the blank.

Carrier regulations. Certain minimum requirements with respect to manufacturer's joint constructions are specified as follows:

TAPING. Paper tape of not less than 60 lb per 500 sheets 24 in. × 36 in. in 2 in. width is permissible up to gross weights of 65 lb. Generally, nonreinforced paper tape is used for interior containers or for boxes for very lightweight products. Most tapes used for boxes under 65 lb are of reinforced or cloth construction. The former utilizes glass fibres for added strength. They are permissible in 3 in. width above 65 lb gross weight, although boxes made of double-wall board are generally stitched.

STITCHING. The sides of the box forming the stitched joint must lap not less than $1\frac{1}{4}$ in. and must be fastened with metal rivets, staples, or stitches spaced not more than $2\frac{1}{2}$ in. apart. This construction requirement is applicable for boxes both under and exceeding 65 lb gross weight. Tie-stitching, defined previously, is not specified by the classification committees. For solid fibre boxes the spacing may not be more than 3 in. apart, but when the joint exceeds 18 in. the stitches must be not more than $2\frac{1}{2}$ in. apart.

GLUING. The sides of the box forming the glued joint must lap not less than 1¼ inches. The joint must be firmly glued throughout the entire area of contact with a glue or adhesive which cannot be dissolved in water after the film application dries. Gluing is permissible for boxes both under and exceeding 65 lb gross weight.

Joint performance and evaluation. Although all three types of joint constructions are permissible, certain preferences for one type or another are found in actual use. Most corrugated boxes are glued with lesser amounts being stitched or taped. The glued joint construction, which is the more recent development, has rapidly increased in popularity. The introduction of new high-speed machines that combine several finishing operations with the formation of the glued joint has been responsible for this growth.

Examination of actual practices indicates that *taping* and *gluing* are generally preferred for *light gross weights and small containers*. Conversely, *stitching* is ordinarily used for *large containers and boxes having dense loads*. Because of this, high-test boxes, including most double-wall and triple-wall containers, are almost always made with stitched joints. Further, *solid fibre boxes* and *all weather-proof boxes* have *stitched or glued* joints.

Generally, the user permits the box manufacturer to exercise his own judgment in the selection of the joint construction for the container. The boxmaker is guided by the classification requirements, his own judgment and past experience, existing prices, and his equipment limitations. By the same token, when stitched or glued joints are used, the box manufacturer must also take into consideration the positioning of the lap extension. The manufacturer is guided in this matter by such considerations as: (1) whether the merchandise is machine or hand packed, so that the lap will not interfere with loading, (2) whether a square edge is mandatory owing to the nature of the product, (3) considerations of appearance for printing, and (4) whether snagging may occur during the user's handling operations if the lap is on the outside of the box.

Although the selection of the particular joint construction to be used is left to the boxmaker's discretion in many instances, there are special circumstances in which the preparation of a detailed joint specification is required. Such conditions may be as follows:

1. *Shipping and testing experiences.* Indicating better performance of one type over another. An example is most canned goods boxes, which are stitched for this reason.

2. *Outdoor storage or excessive humidity hazards.* Necessitating stitched or glued joints or the use of waterproof tapes.

3. *High temperature.* Necessitating the use of non-asphaltic laminants for reinforced tape construction.

4. *Mold growth.* Experienced with such items as soap, which requires a mold-resistant tape.

5. *Storage experience.* Experience with knocked-down containers indicating that a more stable and smaller stack is obtainable with taped joints. Glued and stitched joint boxes are occasionally compressed at the joint to minimize this problem. However, this operation must be carefully controlled to avoid damage to the container itself by crushing.

When boxes are very large and must be fabricated from two separate sheets, two separate joints are required per container. This situation arises when the blank width exceeds the width of the corrugating machine. Since this entails an additional operation, a surcharge is made.

The workmanship, type, and size of joint construction is evaluated visually. If doubt concerning compliance with ordering specifications arises, the following tests for tapes may be applied:*

Kind of Tape	Test
Nonreinforced paper	Basis weight, bursting test
Reinforced paper	Basis weight of paper components, fiber examination (type and weight), spacing
Cloth	Tearing resistance of the crosswide (filler) threads

Stitching and Gluing. The strength of these constructions can be checked by tensile tests on sections of the joint with tension applied perpendicular to the joint. Rupture of the specimen should preferably occur in the fibreboard instead of failure of the stitching or gluing.

All joint constructions can be evaluated by making tests on loaded containers. Special tests utilizing the revolving drum and or the drop tester have been developed to evaluate joint performance for comparative purposes.

JOINT COSTS. Basically there are two major cost factors involved in manufacturer's joint construction: (1) material cost and (2) fabrication cost.

As a general rule tape material cost per inch of depth of the container is higher than the stitching wire or glue cost, the cost differential is generally counter-balanced by the extra square footage required for the lap on glued or stitched boxes. Fabrication costs outweigh material cost factors and are influenced by the type of manufacturing equipment used, i.e., semi-automatic

---

*The specifications pertinent to these tapes are found in Section 11 of Rule 41 of the *Uniform Freight Classification.*

CORRUGATED AND SOLID FIBREBOARD

versus fully automatic, the size, grade and shape of the box blank, the size of the order, and prevailing labor rates. Therefore, the computation of a specific joint cost involves materials, conversion, set up and special up charges.

DESIGN AND COST FACTORS

One of the principal factors of fibreboard container design is optimum shape for protection, ease of packing, and handling including warehousing and palletizing at the lowest possible cost. It must be recognized, however, that the prime function of the container is first and always to provide adequate protection to the contents.

Shape of fibreboard boxes. The shape of the box and the relation of its three dimensions can profoundly influence its utility, its cost, and the cost of packing. In many instances the shape of the box is fixed by the shape of the contents. However, there are circumstances in which a certain amount of flexibility is permissible in a box shape. This is particularly true with bulk items or contents consisting of a number of relatively small units.

There are three major aspects to consider in the selection of container shape: (1) the optimum shape for content protection, (2) the optimum shape with relation to cost, and (3) the optimum shape for ease of handling. Unfortunately, optimum shapes satisfying all of these aspects do not necessarily coincide, and compromises are sometimes called for. A *long narrow* box is not desirable because of the heavy impacts generated when the container tips over. On the other hand, *a cubical* box sustains the least amount of damage in handling, since the box has a tendency to roll with a fall. However, it has the undesirable characteristic of being unsuitable for interlocked stacking in palletizing and warehousing.

Some packaging authorities contend that the best-shaped box for handling and storage is one in which the length is approximately 1½ times the width, and the depth is a little less than the width. Such a box stacks well, has satisfactory strength characteristics, and is ideally suited for palletizing.

An equally important factor in determining the shape of a fibre box is the optimum utilization of board square footage. It has been mathematically proven that the *most economical shape in relation to cubic capacity is the half-cube*. This is a box the length of which is twice the width and equal to the depth; or $L = 2W = D$.

Since the length and depth are double the width, it is only necessary to determine the width in order to develop the other two dimensions. If the volume of a box is known, the following procedure is followed in order to compute the most economical regular slotted container from the standpoint of board consumption:

Since $V = L \times W \times D$
and $L = 2W = D$
then $V = 2W \times W \times 2W = 4W^3$
and $W^3 = V/4$
and $W = \sqrt[3]{V/4}$
where V = Volume
W = Width
L = Length
D = Depth

For other container styles, different formulae are required which can be similarly developed. For instance, for a full-overlap slotted container the proportions are $2L = 4W = D$, so:

Since $V = L \times W \times D$
and $2L = 4W = D$
then $V = 2W \times W \times 4W = 8W^3$
and $W = \sqrt[3]{V/8}$

To determine the square footage of board used for a container, the box is measured in its blank form. This measurement is made by opening the box at the manufacturer's joint. Such a box blank is shown in Fig. 2-13.

The blank area in square footage is computed by *multiplying the sheet length by the sheet width* in inches and dividing by 144. For estimating purposes, one inch is generally added to the theoretical length and width of the blank obtained from the inside dimensions of the box to compensate for scoring allowances.

In determining the most economical container from square footage considerations the following example excluding scoring allowances is cited:

Present box size = 16 in. $L \times$ 12 in. $W \times$ 8 in. D (Regular Slotted Container)
Volume = 1536 cu in.

Square footage (see comment on scoring allowances) =

$$\frac{(2L + 2W)(W + D)}{144} = \frac{(32 + 24)(12 + 8)}{144}$$

$$= \frac{(56)(20)}{144} = \frac{1120}{144} = 7.78 \text{ sq ft}$$

CORRUGATED AND SOLID FIBREBOARD

To determine dimensions utilizing the least square footage with the same volume (1536 cu in.), the following computations are made:

$$V = 4W^3 = 1536 \text{ cu in.}$$

Therefore
$$W = \sqrt[3]{\frac{1536}{4}} = 7\frac{1}{4} \text{ in.}$$

Since
$$L = 2W = D$$
$$L = 14\frac{1}{2} \text{ in.}$$
$$W = 7\frac{1}{4} \text{ in.}$$
$$D = 14\frac{1}{2} \text{ in.}$$

Thus proposed box size = $14\frac{1}{2}$ in. (L) $\times$ $7\frac{1}{4}$ in. (W) $\times$ $14\frac{1}{2}$ in. (D)
Square footage (see comment on scoring allowances) =

$$\frac{(29 + 14\frac{1}{2})(7\frac{1}{4} + 14\frac{1}{2})}{144} = 6.57 \text{ sq ft}$$

This represents a 16 per cent saving in board footage required to package the same volume of merchandise. Since the board square footage used in a container represents, generally, the largest single cost factor, this saving would reflect a major reduction in the over-all cost of the container.

TABLE 2-6
Formulas for Determining Board Footage *

Style of Container	Blank Length	Blank Width	Blank Area
Regular slotted	$2L + 2W$	$W + D$	$\dfrac{(2L + 2W)(W + D)}{144}$
Center special slotted container	$2L + 2W$	$L + D$	$\dfrac{(2L + 2W)(L + D)}{144}$
Full-flap slotted container	$2L + 2W$	$2W + D$	$\dfrac{(2L + 2W)(2W + D)}{144}$
Special full-flap slotted container	$2L + 2W$	$L + W$	$\dfrac{(2L + 2W)(L + W)}{144}$
Half-slotted container	$2L + 2W$	$W/2 + D$	$\dfrac{(2L + 2W)(W/2 + D)}{144}$

The shape of the box and the location of the flap opening influences the method of packing the container. Packing and sealing equipment

*See comment on scoring allowances.

has been developed which is described in Chapter 9 (Fig. 9-6) which handles an "end-opening" box which approaches the optimum in minimum board consumption. The so-called end-opening box is basically a box with the optimum shape, which has the flap opening on the smallest box panel. Since this shape box is stacked on its side, it is called an "end-opener."

Another economical design is the so-called "wrap around" blank. This style is a variation of the five-panel folder with the distinction that the "fifth panel" represents the manufacturers joint tab and is, therefore, minimally 1¼ in. wide. It is formed with special equipment in the box users plant, which, generally, collates the product in the form of cans, folding cartons or bottles, then automatically wraps the blank around the assembled product and effects the joint and closure of the flaps by adhesive.

Cost estimating. In addition to the square footage of board consumed, other factors influence the final purchase price of a fibreboard container. These factors include set-up charges, running charges, manufacturer's joint costs, and special manufacturing charges such as die-cutting, flap-cutting, and stripping.

There are other important considerations influencing the ultimate price. These include the size of the order, the square footage of the order, waste or trim considerations, delivery distance, and other operations which consume more time and/or labor.

All pricing systems for fibreboard containers are expressed in cost per thousand boxes, regardless of the quantity involved. For crude estimating and for comparative economic evaluations the following procedure may be used: the price of an existing container is obtained and is divided by the amount of its square footage. This develops an approximate square footage price which includes all fabrication and other charges. A new square footage figure developed for another size or shape box is then multiplied by this hypothetical base figure. In this example, it is assumed that the grade and type of board *remain the same* and that the style of the container would require similar fabrication techniques. When differences in grades are being considered, similar calculations can be performed by utilizing actual box prices of these grades and computing the hypothetical base price per square foot.

In developing the final price for the container, the container manufacturer must be consulted.

SPECIAL TREATMENTS AND CONSTRUCTIONS

To obtain special performance characteristics, treatments of the board, additives, and special board constructions have been developed.

CORRUGATED AND SOLID FIBREBOARD

These characteristics can be produced by *additives to the pulp itself* during the *finishing of the paper,* as accomplished through *impregnations, sprays, and coatings,* and during *board combining* operations as typified by laminating. The purpose of these treatments is to provide resistance to abrasion and scuffing, water or water vapor, rodents and insects, mold, spoilage, etc., or to impart added strength, attractiveness, better cushioning, and insulation to the finished board.

Although this textbook is not primarily concerned with the esthetics of packaging, it is appropriate to mention some of the techniques which have been developed to enhance the appearance of board. This includes the use of colored, coated, and bleached facings and special sizing and finishing techniques applied to the outside facing. These techniques provide a suitable background or better quality for printing and provide distinct contrasts important in merchandising.

For increased resistance to abrasion or scuffing or to enhance release characteristics, fibreboard is treated with paraffin or a combination of paraffin and polyethylene by lining, laminating, coating, or impregnating the inside of the box. If separate packing components such as tissue, wax paper, or cellulose wadding are required for the protection of finished surfaces (as in the case of appliances, cabinets, etc.), treatment of the board of the interior packing pieces and/or the container generally results in reduced packaging costs. This reduction is attained by the elimination of separate materials, reduced inventory, and less labor for packing. In applications of this type, each individual packaging problem must be carefully analyzed to determine the compatability of the treatment with the finish of the product. As a release agent, treatments of this type have been successful in the packaging of tacky or sticky substances such as synthetic rubber.

Conversely, one development consists of single-face corrugated fibreboard which is coated with pressure-sensitive adhesive on the tips of the corrugations to secure a product or prevent movement in shipment. Its main field of application is for lightweight, fragile items which adhere to the board. Partitions or other interior packing components are thus eliminated.

Boxes can be treated to impart skidding or sliding resistance in order to minimize potential damage in handling and shipping. These treatments are administered either by the box manufacturer or by the user at the end of his packing line. Some non-skid processes include mechanical dimpling, pimpling, or embossing, spray or roller coating of colloidal solution of hydrated silica, compounded copolymer resin emulsions, synthetic resin dispersions, or latex emulsions; resistance is also obtainable with non-skid printing inks. Another development indicates the feasibility of providing non-skid paperboard properties in the papermaking process.

Reference has already been made to weatherproof boards which utilize special components and adhesives. In addition to these components, there are other treatments designed to enhance the water or weather resistance of fibreboard. These include paraffin and paraffin-polyethylene coatings, impregnations and laminations of such materials as plastics and resins. As an example, a manufacturer of premium gummed stamps reduced stamp blocking under high humidity conditions through the use of a paraffin-coated box. An alternate method to counteract blocking would have been to use a moisture-proof bag or liner, which would necessitate greater material and labor costs.

To increase the strength of the corrugating medium, one technique used is a simple process by which one layer of medium is superimposed upon a second and corrugated as a single sheet. This technique essentially produces a laminated corrugating medium without using a laminat-

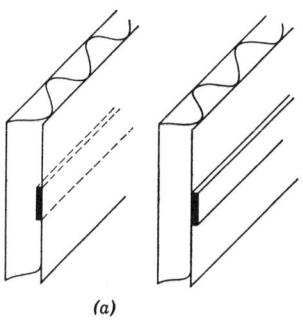

(a)

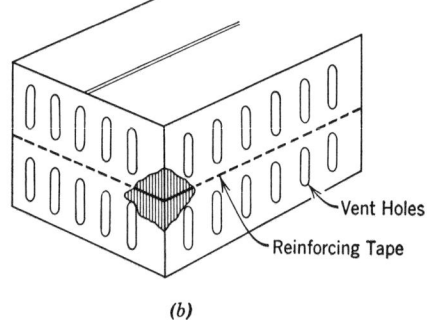

(b)

(a) Reinforcing strips applied either between liner and medium or on inside surface of combined board

(b) Reinforcing of vegetable box having ventilation slots

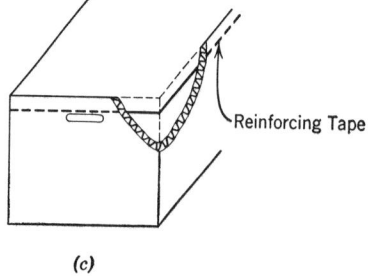

(c)

(c) Reinforcing of box with hand hole cut-outs

Fig. 2-41. Structural box reinforcements.

ing agent. This dual-arch construction is primarily used to increase the compressive resistance of the container.

The application of extensible paper in the manufacture of corrugated fibreboard boxes showed promise at one time. It was believed that this paper would assist in eliminating scoring and folding fractures of the facings during the converting process. However, it did not prove successful in corrugated box manufacture and its application remains for multi-wall shipping sacks. (Refer to Chapter 4 for further details.)

Experimentation with structural reinforcement of fibreboard boxes is applied constantly during the conversion process. This development work is directed toward strengthening weak elements of a container such as scorelines without an over-all major container cost increase. Examples of such techniques which have been used are paper reinforcement along horizontal box scores for canned goods boxes and tape applied on vertical score lines or as reinforcement of ventilation and hand holes (Fig. 2-41). Reinforcement applied during the packaging process is covered in Chapter 10.

Various specialty boards have been constructed and used for industrial packaging applications throughout the years. A few examples are discussed here.

1. Multi-corrugating materials either composed of multiple thicknesses of thin corrugated paper (which forms a soft and flexible protective blanket for the wrapping and protection of fragile, difficult-to-handle products) or made in more rigid form are available to serve in lieu of built-up blocks (Fig. 2-42).

2. A special fibreboard construction in the form of honeycomb has been used for packaging. Simply, this material is a kraft paper expanded to produce a large-volume, low-density cellular structure. It is available in unexpanded form (Fig. 2-43*a*) or as a structural sandwich (Fig. 2-43*b*). The latter structure results from bonding expanded kraft material between thin facing sheets of such products as wood, metals, plastics, paper, asbestos, gypsum wall boards, hardboard, and fiberglass.

3. Another material consists of foam plastic sandwiched between layers of kraft paper (Fig. 2-44). The expanded plastic board, which provides excellent insulation qualities and high moisture vapor protection, is ideal for withstanding high-humidity conditions. Its light weight enhances its versatility.

4. Molded pulp has been formed in the shape of corrugations and has been used plain, or in combination with one or more facing materials for cushioning, spacing, and even as a shipping container. A variety of molded pulp constructions are available. Unlined molded pulp is used extensively for wrapping fragile items and for other cushioning purposes (Fig. 2-45).

Fig. 2-42. Multi-corrugating material. Courtesy Hinde and Dauch, Division of West Virginia Pulp and Paper Co.

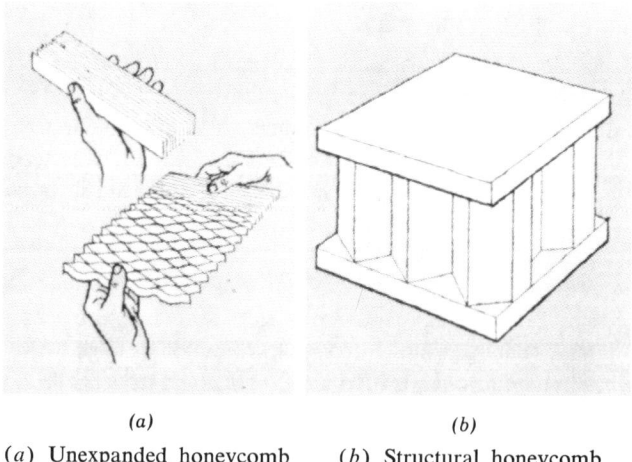

(a)   (b)

(a) Unexpanded honeycomb   (b) Structural honeycomb

Fig. 2-43. Honeycomb paper structures. Courtesy Union Camp Corp.

CORRUGATED AND SOLID FIBREBOARD

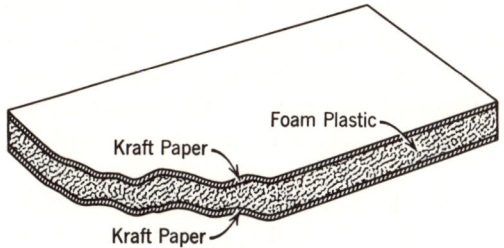

Fig. 2-44. Kraft paper laminated foam plastic. Courtesy St. Regis Paper Co.

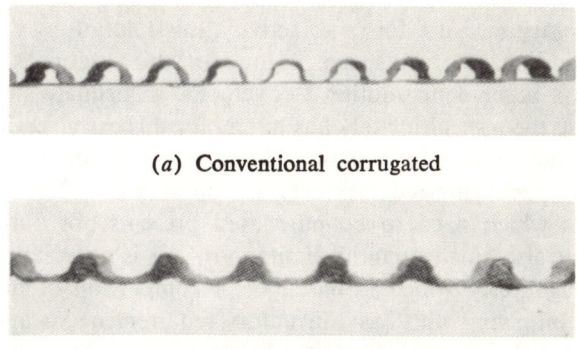

Fig. 2-45. Molded pulp corrugated construction ("Pillo-Pak"). Courtesy Packaging Materials Corp.

USER BOX FABRICATION

It has already been mentioned that the size of an order influences the per unit cost in the purchase of fibreboard boxes and parts. The unit cost consists of fixed as well as variable costs, with the fixed costs essentially constant regardless of the size of the order. As the size of the order increases, the fixed costs are prorated over a larger number of units. Therefore, the per-unit cost decreases.

In order to take advantage of this economic factor, it is desirable for the user to purchase in large lots that are consistent with his inventory requirements and storage facilities. Consolidation of orders of similar items may also result in some economies by the converter and the reduction of shipping cost. The major limiting factors in the economic ordering of corrugated boxes are (1) the available storage space, (2) the variety of different sizes, styles and grades, and (3) the required quantity.

Storage space considerations include provisions for working stock in close proximity to the user's packaging stations and provisions for reserve stock representing the packaging material inventory. The latter may be stored in unit loads at any other suitable plant site. When inadequate storage facilities are available for reserve stock, the convertor may, at times, perform part of the warehouse function for the user by permitting economical production runs and consolidated order delivery. As a temporary measure for peak activity, this solution may obviate the need for additional storage space. However, this practice is not recommended as a permanent solution to overcome inadequate storage facilities, because the user ultimately has to pay for this extra service.

In companies that manufacture a large group of products which are dissimilar in packaging requirements, the number of different packaging components which need to be purchased presents not only a storage problem but also small quantity limitations. It is necessary to develop a purchasing policy which is based on a comparison of unit costs in quantities complying with basic inventory requirements as against larger quantities benefiting from quantity discounts. For example, if inventory limits are a period of ninety days, purchases of a specific box size may have to be made in 500 quantities. On the other hand, if the same box is purchased in 2,000 lots, a lower unit cost is attained. However, the additional cost to warehouse this extra supply—adequate for a one-year period—plus the cost of the invested capital must be rated, as well as possible material deterioration and waste costs.

To overcome the inherent problems of small order purchasing of many items, some users have justified the purchase of *specialized equipment* to fabricate all or a portion of their *own* required fibre boxes and components. The factors to be considered by the user, prior to this decision, are discussed under "Selection Criteria."

Selection criteria. The decision to fabricate containers or packaging materials of any type by the user is one that is difficult to make, since it is dependent upon highly individualized considerations. For this reason, even within the same industry some companies have resorted to packaging material fabrication, whereas others have not found this practice to be advantageous.

CORRUGATED AND SOLID FIBREBOARD

In developing the feasibility and practicability of packaging material fabrication, many factors must be analyzed and rated.

In essence, comparable cost factors must be estimated for either purchased or user-fabricated packaging materials. In addition, numerous intangible factors will favor one or the other practice in the ultimate decision.

The following factors pertain to the rating of costs of *purchased* containers:

1. The unit or annual purchase price.
2. The cost of storage space utilized for packaging materials.
3. Material handling costs to transport and store purchased materials.
4. Clerical costs including purchase orders, production scheduling, and stock and inventory control.
5. Waste and obsolescence costs.
6. Capital expense of inventory.

The above cost factors will permit the computation of a total per unit or annual packaging material expense and will provide a basis for comparison with *self-fabricated* packaging materials whose cost elements consist of the following:

1. Purchase cost of sheet stock, tape, stitches, dies, inks, etc.
2. Space costs consisting of the space required for the fabrication equipment, storage of sheet stock, and storage of finished containers.
3. Material handling costs of the sheet stock and in-process handling.
4. Fabrication costs including direct and indirect labor costs, operating and maintenance costs, and capital costs for the required fabrication equipment.
5. Clerical costs including purchase orders of sheet stock, work order records, production scheduling, and stock and inventory control.
6. Fabrication waste.

In the purchase of sheet stock for user-fabricated containers, standardization must be employed to minimize the number and grades of material to be stocked. This standardization may present a substantial waste problem and in some instances may result in either overprotection or in a compromise in quality. The necessary utilization of a given size and grade of sheet for as many boxes and interior components as possible may result in the use of the improper type and direction of flute or the improper certificate grade of board and excess trim. Similarly, because of the limitations of the equipment, the selection of box styles and construction is restricted.

Additional factors to be considered in the comparison of convertor-supplied and user-fabricated boxes are:

1. Ability to provide an equivalent level of quality in box performance and appearance. This factor is influenced by the combination of equipment limitations and skill of personnel in accomplishing printing, scoring, slotting, joint formation, die cutting, and other fabrication operations.
2. Personnel requirements including the training or hiring of operating, supervisory, and inspection personnel.
3. Material availability in times of shortages.

There are no basic guide lines which can be applied, and a complete analysis of all of the above factors must be carried out. In many companies, a compromise solution has been found to be most economical in the continuation of the outside purchasing of large orders and the self-fabrication of small orders.

User box-fabrication equipment. Fundamentally, user fabrication techniques are similar to the convertor's processes, except that the raw material is in combined sheet form, and the equipment does not possess the mass production features needed by the convertor.

CHAPTER 3

Folding Cartons and Rigid Boxes

Folding cartons and rigid boxes might be included more appropriately in the realm of consumer packaging than industrial packaging. Nevertheless, both of these container types are used fairly extensively in industrial packaging operations as intermediate as well as shipping containers for small items in mailing.

Folding cartons and rigid boxes are produced in many sizes and shapes, with many different combinations of materials. Modern merchandising methods are dependent to a large degree on the use of these containers. End-use classifications for rigid boxes and folding cartons are listed in Table 3-1.

HISTORICAL DEVELOPMENT AND APPLICATIONS

Early in the nineteenth century tack manufacturers began using paper packages for ¼ lb, ½ lb, and 1 lb quantities. These containers were shaped by the hardware clerk on a wooden form and were held together with

TABLE 3-1
END-USE CLASSIFICATIONS

Rigid Boxes (Percent of Total Rigid Paper Box Sales)		Folding Cartons Breakdown Showing End Use of Folding Paper Boxes (Based on Tonnage)	
Electronics	3.1%	Medicinal	3.7%
Shoes and leather goods	7.2	Cosmetics	2.2
Confections	8.6	Soap	9.0
Cosmetics, including soap	6.7	Food	34.1
Department stores and other retail	14.7	Candy	4.7
		Bakery	10.1
Drugs, chemicals, pharmaceuticals	7.7	Tobacco	2.9
		Hardware	5.4
Food and beverages	1.0	Sporting goods and toys	1.6
Hardware, household, auto supplies	4.2	Textiles	3.4
		Retail and laundry	4.0
Jewelry and silverware	9.2	Beverages	8.3
Sporting goods	2.0	Paper goods	7.0
Stationery and office supplies	5.9	Misc.	3.6
Textiles, including hosiery	8.7		
Toys and games	3.9		
Educational	3.0		
Other	14.1		
	100.0%		100.0%

Courtesy of the National Paper Box Association, Inc.

Courtesy of Paperboard Packaging Council

tacks or tied with string. This crude package of tacks was the ancestor of the folding carton.

In 1860 the folding carton as we know it today made its debut. Carried only by better specialty shops, the original cartons were hand-made and were set up after each purchase by the retail clerk.

In 1894, a national organization of cracker bakers developed a new cracker which required protection from air and moisture to preserve its flavor and texture. The crackers were packed in a folding carton with a waxed paper inner liner and overpacked with a printed outer wrapper. With the success of the campaign and the package, other manufacturers rapidly adopted the folding paper box.

The folding box has developed into one of the principal prepacking media. An analysis of folding carton volume indicates that about 49 per cent of the output is used to package food products such as cereal, crackers, candy, flour,

FOLDING CARTONS AND RIGID BOXES

fresh meats, butter, fruits, and vegetables. Bottled and canned beverages account for about 8 per cent, hardware about 5 per cent, tobacco 3 per cent, soap and cosmetics about 11 per cent, paper and paper products 7 per cent, and the remaining 17 per cent is used to package medicinal products, sporting goods, textiles, appliances, etc. (See Table 3-1.)

RIGID BOX DEVELOPMENT

The Chinese are generally credited with the manufacture of the first heavy paper, or cardboard, during the sixteenth century. This discovery made paper box making possible, and again it was the Chinese who pioneered in this art. The first rigid boxes were used in packing tea and were fashioned entirely by hand.

The industrial development of rigid boxes in the United States, which took place in Boston, Massachusetts in 1839, was brought about by dissatisfaction with long delays in the delivery of German-made rigid boxes used in the jewelry trade.

In 1850 the mechanization of the industry was initiated with the invention of a paper cutter and a hand shear. From this point the industry made rapid progress, with the output of paper boxes passing the $1,000,000 mark by 1869. Currently the value of rigid boxes produced approximates $500,000,000.

Although they share certain basic similarities and common markets, the folding carton and rigid box also differ considerably—so considerably that they require separate discussion for clarity. It will be shown that each type of carton has particular advantages and disadvantages which necessitate careful analysis in selecting a container for a specific packaging application. However, since folding cartons and rigid boxes both are made of paperboard, a discussion of the raw material will precede the separate analysis of each type.

PAPERBOARD

The principal material used in the manufacture of folding cartons and rigid boxes is called paperboard. Paperboard is manufactured principally on cylinder board machines. The pulp is produced by various chemical and mechanical processes depending on the required end properties of the board. After refining, screening, and agitating operations the pulp flows by gravity to the cylinder mold vats where formation of the paperboard web takes place. A wide, flat belt of wool travels over the surface of the cylinders and picks up the fibres which have formed on the top surface of each cylinder mold. The individual layers are then combined by a process of adhesion and pressure, and are dried on steam-filled drying rolls. The board is then calendered to the desired finish. Fourdrinier machines are also used in the manufacture of paperboard.

Board may be coated by applying a fine white clay (blended with other white materials with starch or casein as a binder) after the web leaves the drying rolls and before calendering. A number of other operations may also be performed on the completed board prior to fabrication into boxes. These operations include laminations with special papers or films such as kraft, bond, glassine, or polyethylene, combinations with another sheet of board to produce extra heavy stock, or the introduction of a special moisture-resistant filler between two sheets.

Folding cartons are made of *bending board*, and rigid boxes are generally made of *non-bending board*. A full *bending board*, when properly scored, will sustain a 180-degree fold without showing pronounced failure in the top liner surface fibres. A semi-bending or creasing board does not have the bending qualities of a full-bending board; but when properly scored, it will sustain a 90-degree fold without showing pronounced failure in top liner surface fibres. *Non-bending boards* are generally thicker so they will crack or split when folded and require special cutting to facilitate straight-line bending. Within these major boxboard classifications there are many different types which vary widely in their properties.

Non-bending boards. Whereas the non-bending grades of boxboard are used primarily in the manufacture of rigid boxes, they are also used for other end use products in which the important characteristics of stiffness and rigidity are desired, for example display cards, interior trays, etc. Basically, non-bending board is produced in four major grades, which include: (1) *plain chipboard*, (2) *news vat lined chip*, (3) *filled news*, and (4) *white vat lined chip*.

According to the definitions adopted by the American Paper Institute* the following describes these major grades:

Plain chipboard: A non-bending paperboard which has variations in appearance as to color, shade, specks, etc. and is composed of mixed reclaimed fibres.

News vat lined chip: A non-bending paperboard which has the general characteristics outlined above, except that one side of the sheet has a vat liner, made from a furnish of reclaimed news, or other clean fibres, giving that side a fairly uniform and clean appearance.

Filled news: A non-bending paperboard with both sides having a vat liner made from a furnish of reclaimed news and other clean fibres.

White vat lined chip: A non-bending paperboard in which one side of

---

*Combination Paperboard Standards, adopted by members of Combination Paper board Division, Paperboard Group, American Paper Institute (effective January 1969).

FOLDING CARTONS AND RIGID BOXES

the sheet has a white vat liner made from a furnish resulting in a brightness index customarily in the range of 56-60.

The type of raw material used in the manufacture of these grades is the influencing factor as to strength characteristics, appearance, and cost. For example, chip, which uses as raw material mixed, low-grade waste paper, has the lowest quality. News, which is made from waste newspaper pulp has greater uniformity due to the use of a selected raw material. It is therefore slightly more expensive than chip. White vat lined chip has one liner of top-grade waste paper pulp, which usually consists of envelope clippings. This grade is still a further improvement in quality but correspondingly increases the cost.

Non-bending boards produced on the board machine are available in various thicknesses 0.014 in., weighing about 60 lb per 1,000 sq ft, to 0.065 in., weighing about 206 lb per 1,000 sq ft. If greater board thickness is desired, two or more sheets can be pasted together.

Bending boards. Whereas the bending grades of paperboard are used primarily for the fabrication of folding cartons, these boards have other packaging and non-packaging uses. For folding cartons full-bending qualities and a suitable printing surface are highly desirable. The American Paper Institute classifies bending board into the following principal divisions:

1. Bending chip.
2. Colored lined.
3. Mist colored.
4. White lined 60.
5. White lined 65.

According to the association classification, there are three principal grades of the white lined boxboards, (1) *bleached manila lined chip,* (2) *white patent coated news,* and (3) *white clay coated boxboard.*

The commercially popular names and the descriptions of the major bending boxboard grades are listed in the next paragraphs.

BENDING CHIPBOARD. Bending chipboard is made of reclaimed mixed papers with sulphate or sulphite wood pulp added to give it the necessary bending qualities. Because of its characteristic dark grey color, it is not suitable for multi-color printing.

BENDING NEWSBOARD. This board is made of reclaimed newspaper with sulphate or sulphite wood pulp added to give it the necessary bending qualities. Like chipboard, newsboard has a dark grey color and is not suitable for multi-color printing.

SINGLE MANILA LINED BOARD. Single manila lined board is a commercial grade of boxboard with a top liner of unbleached sulphite and groundwood pulp. The rest of the board is composed of news or mixed papers. Its top liner has a light buff color.

BLEACHED MANILA LINED BOARD. This slightly higher grade is produced from much the same basic ingredients as single manila lined board. However, since it is whiter in color, it is more extensively used for folding boxes. This board has a customary brightness range on one side of 56–60.

WHITE PATENT COATED BOARD. White patent coated, oddly named since it is neither patent nor coated, is made with a top liner of bleached sulphite wood pulp mixed with white shavings, aspen, or soda pulp, to produce a good, white surface for multi-color printing. Its customary brightness range on the white side is 66–70 and the board has good permanence against fading.

CLAY COATED BOARD. This board has an exceptional surface for high quality printing. When desired, its top liner may be coated with a mixture of white mineral powder, an organic binder, and water. This type of board is divided into two classifications: board coated on the paper machine, and board coated after the usual machine operations are completed. Machine-coated board is produced by removing a section of the drier rolls and applying the coat at that point. The drying and calendering then continues and produces a smooth surface. In the traditional clay-coating process, the application is accomplished after calendering. White clay-coated board has a brightness range of 74–78 and good resistance to fading.

SOLID BLEACHED SULFATE PAPERBOARD. Solid bleached board is used for packaging certain moist, liquid and oily foods, especially those susceptible to contamination. It is produced entirely from virgin material from the sulfate process. It is made primarily on Fourdrinier machines.

BOXBOARD TRADE DEFINITIONS

For the purposes of standardization and to develop means of uniform measurements, the American Paper Institute has defined various terms applied to boxboard. These terms reflect commonly accepted trade practices and form the basis for the purchase and rating of all types of boxboards. The following are the terms and boxboard formulas and computations employed.*

1. *Standard Sheet:*

A sheet having an area of 1,000 square inches, such as 25 × 40 inches, is

---

*Combination Paperboard Standards, adopted by members of Combination Paperboard Division, Paperboard Group, American Paper Institute (effective January 1969).

FOLDING CARTONS AND RIGID BOXES 129

used as a means of computing regular number, count and weight of paperboard, and hereafter is referred to as standard area sheet.

2. *Bundle:*

A bundle is a unit containing 50 pounds. The quantity of sheets referred to as count varies with the size, weight and the caliper, but the weight of 50 pounds of board in a bundle is fixed.

3. *Regular Number:*

The regular number is the quantity of sheets of 1,000 square inches each of paperboard required to make a bundle of 50 pounds.

4. *Count:*

The count is the actual quantity of sheets of a given size, weight and caliper required to make a bundle of 50 pounds.

The area in square inches of the board in a bundle of a regular number is the number multiplied by 1,000 (25 × 40). The square inch area of an odd-sized sheet is the width multiplied by the length, and by dividing this figure into the area of the corresponding regular number, the count for the odd-sized sheet is found.

To determine the count of odd-sized sheets in a bundle, proceed as follows:

(1) Given the kind, finish and caliper of board desired, ascertain from the proper gauge list the regular number of such board;

(2) Divide the square inches in a bundle of that particular regular number (regular number × 1,000) by the square inches in the odd-sized sheet, and the result is the count.

Example:

Given: Plain chipboard 30" × 39", .041 caliper Medium Finish
55 = regular number (see gauge list No. 1)
55 × (25 × 40) = 55,000 sq. in. in bundles of regular number 55
30 × 39 = 1,170 sq. in. in odd-sized sheet
55,000 divided by 1,170 = 47 count of odd-sized sheet, or 47 sheets in bundle.

5. *To Determine Regular Number:*

Given the count of odd-sized sheets in a bundle, ascertain the corresponding regular number as follows:

(1) Compute the square inches in the odd-sized sheet;
(2) Multiply by the count of the odd-sized sheets in a bundle;
(3) Divide by 1,000—the square inches in a standard area sheet, and the result is the regular number.

Example:

Given: Plain chipboard 30" × 39" count 47 Medium Finish
30 × 39 = 1,170 sq. in. in odd-sized sheet
1,170 × 47 = 55,000 sq. in. in bundle
55,000 divided by 1,000 = 55 regular number

DISTRIBUTION PACKAGING

6. *Paperboard Formulas:*

C = Count
RN = Regular Number
A = Area of Sheet in square inches
W = Basis weight in Lbs. M sq. ft.
$LBS.$ = Lbs. per M Sheets
D = Density
F = Finish
$Caliper$ = Thickness in thousandths of an inch

$$\frac{RN}{A} = \frac{C}{1,000}$$

$$C \times W = \frac{72 \times 100,000}{A}$$

$$RN \times W = 7,200$$

$$RN = \frac{7,200}{W}$$

$$RN = \frac{C \times A}{1,000}$$

$$C = \frac{RN \times 1,000}{A}$$

$$C = \frac{72 \times 100,000}{A \times W}$$

$$W = \frac{7,200}{RN}$$

$$W = \frac{72 \times 100,000}{A \times C}$$

$$D = \frac{W}{Caliper}$$

$$D = Caliper \times RN$$

$$F = Caliper \times RN$$

$$LBS. = W \times 6.944$$

$$LBS. = \frac{50,000}{C}$$

$$LBS. = \frac{A}{.02 \times RN}$$

$$PRICE/MSF = \frac{\text{Price per ton} \times (\text{Wt./MSF})}{2,000}$$

$$PRICE/TON = \frac{\text{Price per MSF} \times 2,000}{\text{Wt./MSF}}$$

FOLDING CARTONS AND RIGID BOXES

7. *Caliper:*

Caliper is the thickness of board expressed in $1/1000$ of an inch and written as a decimal of an inch. Board with a caliper of .025, for example, is generally referred to as twenty-five "point"—points being $1/1000$ of an inch.

8. *Density (Finish):*

The term "finish" actually refers to the density of paperboard and can be expressed as: Pounds per point of caliper by dividing the weight per thousand square feet by the caliper.

Example:

Pounds per Point of Caliper for .020 White Lined 65, Gauge List No. 2.

$$\text{Density as Pounds per Point of Caliper} = \frac{80 \text{ LBS./MSF}}{.020 \text{ Caliper}} = 4 \text{ LBS. per Point}$$

It can also be expressed as Points of Finish by multiplying caliper times regular number.

Example:

Points of Finish for .030 Plain Chipboard, Gauge List No. 1.
Density as Points of Finish = .030 Caliper × 80 Reg. No. = 2400 Points of Finish.

A machine producing combination paperboard can produce two varying densities, generally speaking, the greater the density, the better the surface smoothness quality.

9. *Dimensions:*

In stating the dimensions of a sheet, the width in inches is given first and the length next. Width is the measurement across the machine, i.e., space between slitter knives. Length is always the measurement in the direction of the grain, i.e., space between knife cut-off.

10. *Bending:*

A full bending board, when properly scored, will sustain a 180 degree fold without showing pronounced failure in the top liner surface fibres. A semi-bending or creasing board does not have the bending qualities of a full bending board; but, when properly scored, will sustain a 90 degree fold without showing pronounced failure in the top liner surface fibres.

11. *Gauge Lists:* (Table 3-2)

COMBINATION PAPERBOARD—NON-BENDING

The non-bending grades of paperboard are used in the manufacture of set-up boxes and for many other end use products where the qualities of importance are stiffness, rigidity, etc. Finish (i.e., Density) may be varied to obtain more or less surface smoothness.

Gauge List No. 1—Combination Paperboard—Non-Bending, Rough, Medium and Smooth Finish

    Plain Chipboard: A non-bending paperboard which has variations in appearance as to color, shade, specks, etc., and is composed of mixed reclaimed fibres.

    News Vat Lined Chip: A non-bending paperboard which has the general characteristics outlined above, except that one side of the sheet has a vat liner, made from a furnish of reclaimed news, or other clean fibres giving that side a fairly uniform and clean appearance.

Filled News: A non-bending paperboard with both sides having vat liners made from a furnish of reclaimed news and other clean fibres.

White Vat Lined Chip: A non-bending paperboard in which one side of the sheet has a white vat liner made from a furnish resulting in a brightness index customarily in the range of 56-60.

Gauge List No. 4–Combination Paperboard–Plain Chipboard
Two or more plies of plain chipboard, pasted or laminated together to form a multi-ply sheet, greater in thickness than can be produced on the paperboard machine.

COMBINATION PAPERBOARD–BENDING

The bending grades of paperboard are used primarily for the manufacture of folding cartons. A wide range of physical qualities can be obtained in these boards because of the great variety of raw materials which can be "combined" on the paperboard machine in their manufacture.

TABLE 3-2
GAUGE LISTS*
No. 1 Gauge List–Combination Paperboard–Non-Bending**

Regular Number 50 Lb. Bundle	Finish (Density)			Weight per M Sq. Ft.	Weight per M Sheets 25 × 40
	Rough 1	Medium 2	Smooth 3		
35	.068	.064	.060	206	1430
40	.060	.056	.052	180	1250
45	.053	.050	.046	160	1111
50	.048	.045	.042	144	1000
55	.044	.041	.038	131	910
60	.040	.037	.035	120	833
65	.037	.034	.032	111	771
70	.034	.031	.029	103	715
75	.032	.029	.027	96	667
80		.027	.025	90	625
85		.025	.024	85	590
90		.023	.022	80	556
95			.021	76	528
100			.020	72	500
110			.018	65	451
120			.016	60	417

NOTE: For extra smooth finish for above grades of Combination Paperboard, refer to No. 2 Gauge List.
**Plain chipboard, news vat lined chip, filled news, and white vat lined chip.
*Combination Paperboard Standards, Paperboard Group, American Paper Institute.

FOLDING CARTONS AND RIGID BOXES

Exceptional strength, variety in color and brightness, fine functional coatings and laminations, printing surfaces, precise machinability, economy are among the many qualities which may be "custom tailored" into these grades by taking advantage of the "combination of materials" principle.

Gauge List No. 2–Combination Paperboard–Bending
    Bending Chip   White Lined 60
    Colored Lined   White Lined 65
    Mist Colored

Gauge List No. 6–Combination Paperboard–White Lined and Clay Coated
    White Lined 70   Clay Coated 70
    White Lined 75   Clay Coated 75
                                Clay Coated 80

All White Lined 60 and White Lined 65 Bending Paperboard grades made with extra strength jute, kraft or pulp back are to be included on this list.

Gauge List No. 7–Combination Paperboard–Double White Lined, White Lined Extra Strength Back and Clay Coated Extra Strength Back Paperboard
    Double White Lined 70
    Double White Lined 75
    White Lined 70, Extra Strength Back
    White Lined 75, Extra Strength Back
    Clay Coated 70, Extra Strength Back
    Clay Coated 75, Extra Strength Back
    Clay Coated 80, Extra Strength Back

No. 2 Gauge List–Combination Paperboard–Bending*

Caliper	Weight per M Sq. Ft.	Weight per M Sheets 25 x 40	Regular Number 50 Lb. Bundle
.040	144	1000	50
.036	131	910	55
.033	120	833	60
.030	111	771	65
.028	103	715	70
.026	96	667	75
.024	90	625	80
.022	85	590	85
.020	80	556	90
.018	72	500	100
.016	65	451	110
.015	60	417	120

*Bending chip, colored lined, mist colored, white lined 60, white lined 65. Any of the above listed grades made with extra strength jute, kraft or pulp backs require No. 6 Gauge List.

No. 4 Gauge List—Combination Paperboard—Pasted Chipboard

Caliper	Weight per M Sq. Ft.	Weight per M Sheets 25 × 40	Regular Number 50 Lb. Bundle
.060	206	1430	35
.070	240	1667	30
.080	277	1923	26
.085	288	2000	25
.090	313	2173	23
.095	327	2271	22
.100	343	2382	21
.110	379	2632	19
.120	423	2937	17
.150	514	3569	14

No. 6 Gauge List—Combination Paperboard—White Lined and Clay Coated*

Caliper	Weight per M Sq. Ft.	Weight per M Sheets 25 × 40	Regular Number 50 Lb. Bundle
.040	160	1111	45
.038	152	1055	47
.036	144	1000	50
.034	136	944	53
.032	128	889	56
.030	120	833	60
.028	112	778	64
.026	104	722	69
.024	96	667	75
.022	88	611	82
.020	82	569	88
.018	77	535	93
.016	69	479	104
.015	65	451	111
.014	63	437	114
.013	60	417	120
.012	56	389	128
.011	51	354	141

*White Lined 70, White Lined 75, Clay Coated 70, Clay Coated 75, Clay Coated 80 All White Lined 60 and White Lined 65 Bending Paperboard Grades made with extra strength jute, kraft or pulp back are to be included on this list.

FOLDING CARTONS AND RIGID BOXES

12. *Brightness:*

As commonly used in the paper industry, the reflectivity of a sheet of pulp, paper or paperboard for specified light measured in index numbers, under standardized conditions on a particular instrument designed and calibrated for the purpose. It is sometimes called G. E. Brightness after the manufacturer who made the original instruments.

Brightness tests shall be made in accordance with Technical Association of the Pulp and Paper Industry (TAPPI) Standard T 452 m—"Brightness of Paper and Paperboard."

No. 7 Gauge List—Combination Paperboard—Double White Lined, White Lined Extra Strength Back and Clay Coated Extra Strength Back*

Caliper	Weight per M Sq. Ft.	Weight per M Sheets 25 × 40	Regular Number 50 Lb. Bundle
.040	164	1139	44
.038	156	1083	46
.036	148	1028	49
.034	140	972	51
.032	132	917	54
.030	124	861	58
.028	116	806	62
.026	108	750	67
.024	100	694	72
.022	92	639	78
.020	85	590	85
.018	80	556	90
.016	72	500	100
.015	69	479	104
.014	65	451	111
.013	63	437	114
.012	58	403	124
.011	53	368	136

* Double White Lined 70
  Double White Lined 75
  White Lined 70, Extra Strength Back
  White Lined 75, Extra Strength Back
  Clay Coated 70, Extra Strength Back
  Clay Coated 75, Extra Strength Back
  Clay Coated 80, Extra Strength Back

*Combination Paperboard Standards, adopted by members of Combination Paperboard Group, American Paper Institute (effective January 1969).

PHYSICAL PROPERTIES OF BOXBOARD

It should be recognized that the boxboard standards reproduced in Table 3-2 only present purchase guide lines for the most popular grades of available boxboards. Any one of these "standard grades" can vary widely in performance even while complying with all of the requirements listed. This variation is mainly due to differences in the raw material used. In addition, there are literally hundreds of different types of board formulations which make the development of standards that much more difficult. These different types of boards have been developed to provide specific performance such as is required for beer carrier stock, special food boards, and other specialized items.

The convertor is essentially concerned with boxboard characteristics as they relate to his fabricating and finishing operations including the efficiency of adhesion, printing, etc. The user, on the other hand, is concerned with boxboard characteristics which influence his packaging operations, such as setting up, bending, and sealing and with those characteristics which influence quality, appearance, and utility.

For purchase specifications, the generally accepted means of identification are (1) grade or type, (2) basis weight, (3) caliper, (4) finish, (5) bursting strength, and (6) brightness (for coated and bleached boards). In addition, certain physical properties may be of importance to assure consistent performance during conversion, packaging and sales. These properties include (1) water and ink penetration (absorption), (2) stiffness and rigidity, (3) adhesion, (4) scoring efficiency, (5) tearing resistance, (6) tensile strength, (7) curling, (8) rub resistance, (9) gloss, (10) softness, smoothness, and porosity, (11) puncture resistance, (12) bending, and many others.

Of the six items generally included in purchase specifications, the weight, caliper, bursting, and brightness are determined by testing. Grade or type may be determined either visually or by testing and the finish is established by the relationship of weight to caliper. A brief description of these basic test procedures follows:

1. Basis weight determinations are carried out by weighing a known area and making the necessary calculations (TAPPI Standard T410).

2. Caliper (thickness) determinations are made with a standard micrometer (TAPPI Standard T411).

3. Bursting strength as described in Chapter 1 is performed with either a Mullen or Cady testing machine (TAPPI Standard T403).

4. Brightness is measured by comparing the light reflection of a test specimen against standard reference panels (TAPPI Standard T452).

FOLDING CARTONS AND RIGID BOXES

This property is of importance in controlling printing quality and appearance.

Procedures used to establish some of the more important supplementary properties listed previously are as follows:

1. *Tearing resistance* provides a measure of the internal tearing resistance of the boxboard. It is determined on a standard Elmendorf Tearing Tester (TAPPI Standard T414).

2. *Tensile strength* expresses the tenacity of the fibres comprising the boxboard. This property is determined on any suitable tensile testing device (TAPPI Standard T404).

3. *Bending characteristics* reflect the ease by which bending boxboard may be folded, formed, or otherwise assembled. These may be determined by manual folding or by the use of special equipment.

4. *Absorption, smoothness, and other surface characteristics* provide an indication of the receptivity of board to printing inks, adhesives, special coatings, etc. There are many techniques and instruments available which measure these characteristics, some of which are covered by TAPPI Standards.

5. *Stiffness and rigidity data* furnish an indication of the rigidity to be provided by the finished box or carton. There are many different techniques used to evaluate this property.

6. *Rub resistance* expresses the resistance of paperboard to scuffing or surface peeling as the result of abrasive action.

At the present time, universally accepted standards or expectancies are not available for reference purposes. It is frequently necessary for the user or supplier to develop his own information as related to acceptable performance.

Until more data has been accummulated on board properties other than those currently accepted (grade or type, basis weight, caliper, finish, bursting, and brightness) other properties should not be arbitrarily included in purchase specifications.

RIGID PAPER BOXES

By accepted definition, a rigid box is a package usually made of non-bending boxboard having an approximate thickness between 0.020 in. to 0.120 in. The rigid box is delivered to the user in a rigid, ready-to-use form or the blank is assembled by the user prior to packing.

In its simplest form, the rigid box is square-cornered, of two-piece construction with reinforcing at the vertical corner edges. In manufacturing the rigid box, the flat sheets of boxboard are first cut and scored to the size required for the particular box. The large sheet of board is then separated into individual blanks and the corners are cut out. The forming is accomplished by bending the sides to right angles and staying the corners.

The stays of the rigid box are usually of 90 lb (500 sheets 24 in. X 36 in.) weight paper, although muslin, fibre glass reinforced tape, metal clips or other materials may be used for this purpose. Some larger boxes are *self-stayed* or *ended* by using a flap from the box sides for the stay. Envelope style boxes are often made in this way.

For added strength and decorative reasons, the assembled box may be overwrapped with paper, cloth, or leather.

STYLES

The basic construction form of the rigid box permits the design of a variety of shapes and styles. Rigid boxes can be rectangular, conical, oval, round, star-shaped, or oblique. The rigid box is usually designed with a base or bottom and a cover or lid. Other basic modifications include boxes with no cover, a flat lid, died-out openings, or other variations. Platforms or partitions, straight or slanted, can be included, and shoulders, lips, extension edges, drop trays, and special covers add further possible design variations.

The most common style of rigid box is the full-telescope type (Fig. 3-1*a*), in which the lid fits down over the base. In the partial-telescope style (Fig. 3-1*b*), the lid is not as deep as the base. A variation of this is the neck or shoulder style (Fig. 3-1*c*), in which the outside of the base fits flush with the lid.

Another common style of rigid box is the slide style box illustrated in Fig. 3-1*d*, consisting of an outer sleeve into which the base slides. Other styles of rigid boxes, including some specialty boxes, are also illustrated in Fig. 3-1.

Several types of rigid boxes that vary considerably from the basic styles are sufficiently common to be classified as special set-up boxes. Round boxes—from wig boxes to new cosmetic creations—are examples of these specialties. The use of acetate and other plastic materials has added considerably to the variety of available styles. Combinations of acetate lids and paperboard bases, and the use of formed plastic inserts and bases as well as box *windows* have increased considerably. Many rigid box plants operate transparent box departments, as the manufacturing methods for both types are similar.

FOLDING CARTONS AND RIGID BOXES 139

Fig. 3-1 (*a-q*). Various styles of rigid boxes.

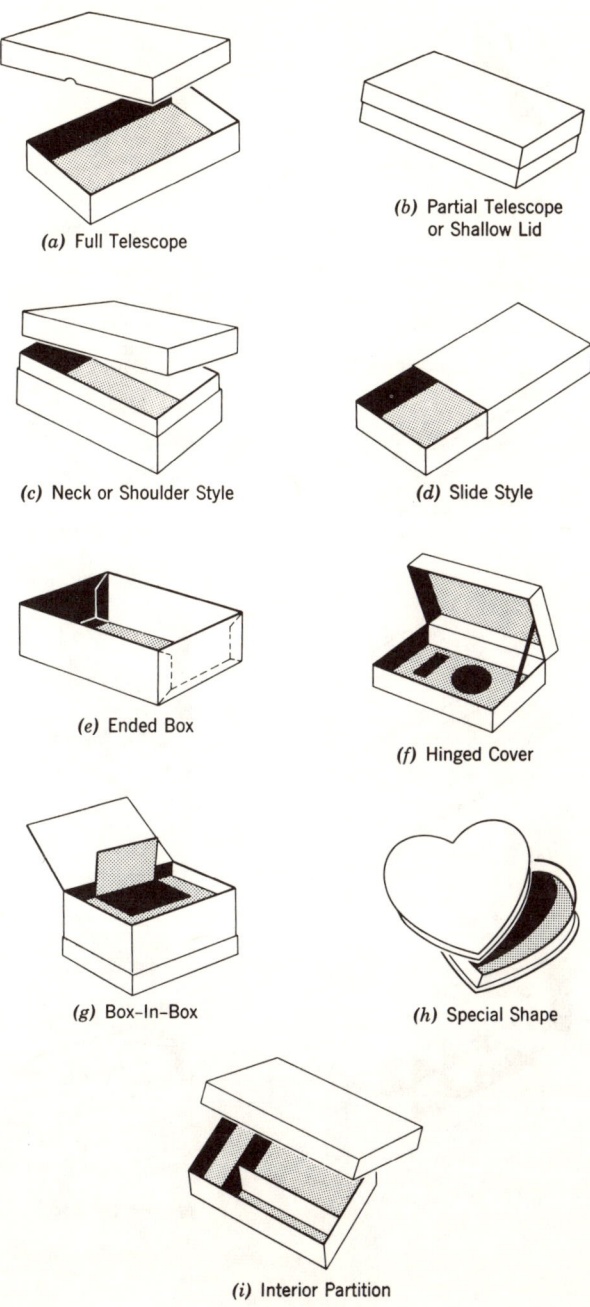

(*a*) Full Telescope

(*b*) Partial Telescope or Shallow Lid

(*c*) Neck or Shoulder Style

(*d*) Slide Style

(*e*) Ended Box

(*f*) Hinged Cover

(*g*) Box-In-Box

(*h*) Special Shape

(*i*) Interior Partition

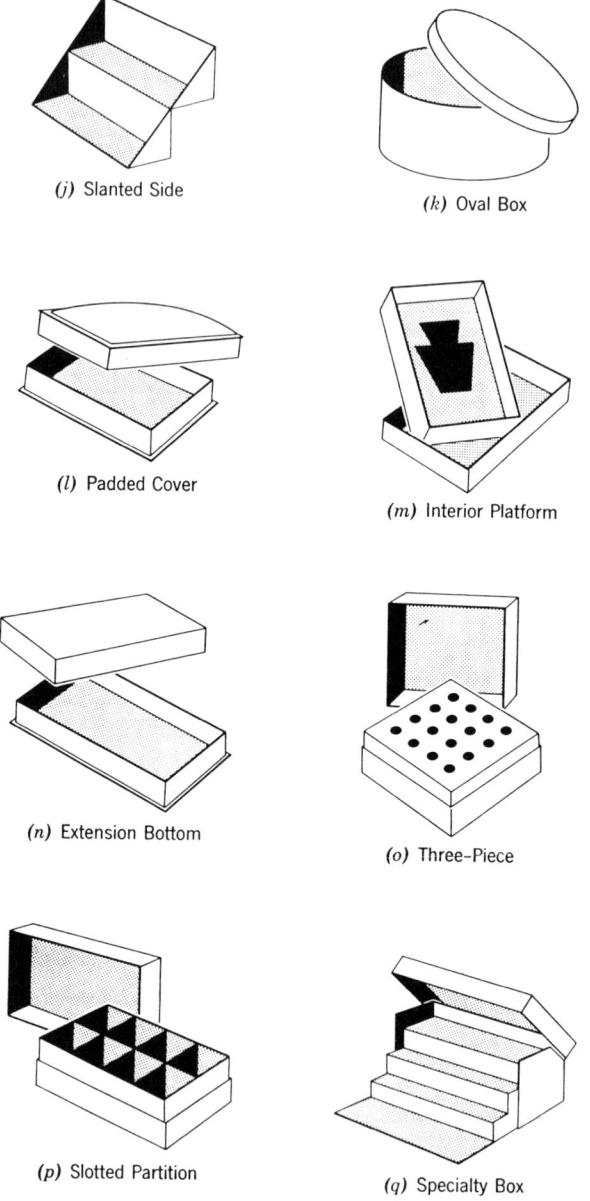

FOLDING CARTONS AND RIGID BOXES

COVERING MATERIALS AND METHODS

The covering material (or wrap) is the paper or other covering that is used for the base, lid, or other component parts of the box. There are four basic applications of wraps:

1. *One-piece tight wrapping* is the predominantly used process of entirely covering the base, lid, or other component parts with paper or other covering material on wrapping machines. The wrap is in one piece and has its entire surface covered with adhesive (Fig. 3-2d).

2. *One-piece loose wrapping* is a variation of tight wrapping and results in a pillowed effect of the covering paper resembling a hand wrapped package. The paper is applied by a wrapping machine and adhesive is applied to the sides only.

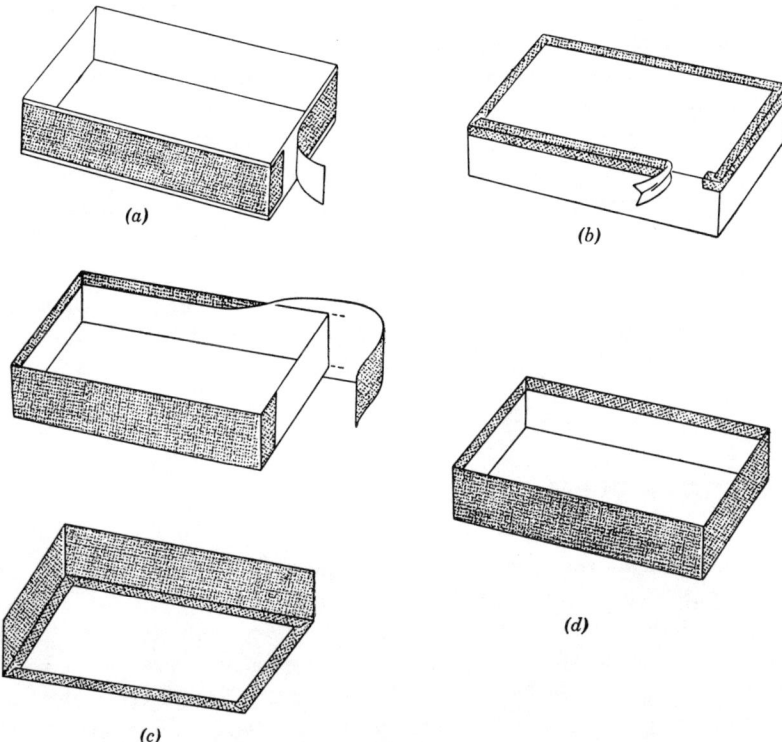

Fig. 3-2. Four basic applications of rigid box wrapping.
- (a) Banding or strip staying
- (b) Trimming
- (c) Strip covering
- (d) One-piece tight wrapping

Other infrequently used methods include:

3. *Strip covering* consists of covering the sides and ends of the base or lid with paper or other material and turning in over the top edge to a distance of approximately ½ in. (Fig. 3-2c). When strip covering is used, bottom or top covers, or both, may be applied to complete covering of base, lid, or both.

4. *Banding or strip staying* is the process of covering the sides and ends of the box or parts with paper or other material without turning in (Fig. 3-2a).

5. *Trimming* is the process of applying a narrow strip of paper or cloth centered over the length and width scores of the base and/or other component parts (Fig. 3-2b).

Most box wraps or labels are printed, lithographed, embossed and debossed or otherwise decorated with the design applied on 60 to 70 lb number 1 white litho paper, basis size 25 in. X 38 in. per ream of 500 sheets.

Other box covering papers range from uncoated colored books, krafts, foils, cheviots, machine calendered plates, glazes, micas, and flints, to the more recent types of resin and lacquer-spray coated sheets. Most of these can be obtained in various colors and embossings, plus over-prints in one or more complementary colors. Cloth, leatherette, vinyls and other more exotic coverings are also used.

Uncoated papers usually weigh 20 to 22 lb per ream, basis size 20/26. Good glazed- and flint-coated papers are made from special formulated 28 lb book stock with approximately 6 lb of coating added, making a finished sheet of 34 lb per ream of 500 sheets.

DESIGN CONSIDERATIONS

In the selection of a rigid box, three basic design considerations are of paramount importance. These are (1) appearance and utility, (2) structural attributes, and (3) economic factors. Recognition of the above should be given in determining the suitability of a rigid box for the packaging of a given product in relation to other types of containers. If the suitability of a rigid box is established, specific details of the box should also be rated through an analysis of these major considerations.

Appearance and utility. One of the most important assets of the rigid box from an appearance standpoint is its ability to be overwrapped with decorative materials which greatly enhance consumer appeal, particularly with luxury items. Of equal importance is the freedom of form and shape permissible in rigid box manufacture. Because of this versatility of form, it is possible to construct packages which are individually designed to enhance the over-all appearance of the product. For example, heart-shaped candy or perfume boxes convey the sentiment attached to the product.

After the merchandising of the product, the rigid box can effectively enhance utility by serving as a storage unit, dispenser, or carrying case. For example, such items as jewelry frequently are kept in the original sales unit by the purchaser. Such items as face or bath powder are gradually dispensed from the original rigid box with the lid serving as an effective reclosing device. If repeated use is indicated, the opening and closing feature should be designed with sufficient strength to remain effective.

Rigid boxes can be used effectively for the display of the product without detracting from its reclosing feature. In window or counter displays the lid of the box can be removed or hinged back to display the product. In its subsequent sale, the rigid box can be reclosed and used. With folding cartons and corrugated boxes, similar display ease usually requires more elaborate design features, and reclosing can present difficulties.

Structural attributes. The rigid box is particularly suited for packaging purposes when substantial rigidity is required. Rigid box rigidity is a function of two major characteristics: (1) the basic rigidity of the non-bending boxboard used in its fabrication and (2) the non-collapsible feature of rigid box construction. The full-telescope style set-up box, by providing two thicknesses of boxboard on the side and end box panels, with light reinforced corner posts is particularly suitable for the packaging of dense products subject to settling in handling and storage. This style, because of its rigidity, is extensively used as a mailing box.

Experience has indicated that the two potentially weakest structural elements of a rigid box are the corner stays and the scorelines. Distortion of the box may cause rupturing of the stay tape and fracturing along the creases which are partially cut in manufacture to permit rightangle folding of non-bending board. To strengthen these potential weak areas, the depth of scoring must be controlled and reinforced stay tapes may be used.

The overwrapping of the boxboard structure, using any of the techniques described in Fig. 3-2 automatically add strength. Ended boxes made of bending boxboard eliminate the need for corner stays and minimize potential corner and edge rupture.

In order to provide a guide in the selection of the proper grade of boxboard to be used for a given application, Table 3-3 has been reproduced. This table is excerpted from Federal Specification PPP-B-676C—"Boxes, Set-up"—and relates volume and weight of contents to thickness and Regular Number. The exact boxboard requirements for a given style, size, and type of contents must be determined through field experience or laboratory testing since optimum structural design criteria for rigid boxes have not been developed. Because of this lack of design data, cost reduction is frequently possible through analysis of the essential materials used. For example, a manufacturer of carbon papers used an overwrapped rigid box to pack and dispense his product. Extensive laboratory and shipping tests disclosed that a boxboard reduction of the plain chipboard in a number 1 finish was possible from a caliper of 0.035 in. to 0.030 in. with a corresponding reduction in the basis weight from 111 lb per 1,000 sq ft to 96 lb per 1,000 sq ft. This reduction in grade resulted in a significant cost savings through a 13.5 per cent reduction in the boxboard weight. The structural attributes and the dispensing feature of this rigid box was not adversely affected.

Economic factors. In the consideration of the economics of rigid box use, several important factors must be rated, including (1) sales, (2) unit box cost, and (3) storage and handling. Marketing aspects can dictate the selection of a rigid box regardless of the economics involved, provided that the advantages of this style of container, in terms of sales, have been established. This is particularly true with high-value products for which the packaging cost represents a small percentage of the total product cost or where gift appeal or special glamour is required. For many other items, notably cosmetics, the value of the product is greatly enhanced by the selection of ornate, luxury-type packaging. Even in these circumstances, when marketing dictates packaging selections, economics in rigid box procurement or fabrication cannot be overlooked.

The *unit box cost* essentially consists of fabrication and material costs. The latter includes the type and grade of boxboard, the staying and overwrapping material, and material for auxiliary components such as platforms and partitions. Fabrication techniques of rigid box manufacture are not influenced by the factor of quantity to the same degree as folding cartons. Folding cartons require more complex machinery with corresponding greater set-up costs. For this reason, when limited quantities are required, the rigid box is more economical per unit purchased than a folding carton.

If large quantities of rigid boxes are used, the handling and storage of the fully assembled container is a major expense. The non-collapsible feature, although it eliminates the need for a separate assembly operation on the packing line, requires excessive handling labor and a large storage capacity. Similarly, since the supplier is limited in the quantity that can be loaded into a carrier, transportation costs increase and precipitate an increase in the unit cost of the container. For these reasons, some rigid box users have found it economically attractive to fabricate rigid boxes in their own plant. The extent of fabrication engaged in by the user is influenced by conditions peculiar to the user's operations. For instance, a certain cosmetic manufacturer has developed cost data which establish the economic feasibility of cutting and scoring his own box blanks from sheet stock of boxboard. These blanks are then stored until they are needed for a particular packing

TABLE 3-3

Relationship of volume and weight content to thickness and regular number of paperboard for the fabrication of rigid paperboard boxes.

Volume of box	Weight of contents	Regular number	Thickness			Basis weight per 1000 square feet
			No. 2 finish	No. 3 finish	No. 4 finish	
Cubic inches	*Pounds*	*Sheets per 50 pound bundle 25 x 40*	*Inch*	*Inch*	*Inch*	*Pounds*
10 or less	1/4 or less	80	0.026	—	—	90
over 10—20	over 1/4—1/2	75	0.028	0.026	—	96
over 20—40	over 1/2—3/4	70	0.030	0.028	0.026	103
over 40—60	over 3/4—1-1/2	65	0.032	0.030	0.028	111
over 60—80	over 1-1/2—3	60	0.035	0.033	0.030	120
over 80—110	over 3—5	55	0.038	0.036	0.033	131
over 110—150	over 5—7-1/2	50	0.043	0.040	0.037	144
over 150	over 7-1/2—10	45	0.048	0.045	0.041	160
		40	—		0.046	180

Federal Specification PPP-B-676C

Fig. 3-3. Metal edge staying machine. Courtesy National Metal Edge Box Company.

run. Through proper scheduling, the staying, overwrapping, and platform-insertion operations are geared to the rate of packing, so that a direct flow between the box shop and the packaging department is possible. Other companies purchase precut and scored blanks and simply assemble the box by forming the corners. Either a direct flow to the packaging department may be established, or a small inventory of the different box sizes and types is carried, from which the box requirements are filled.

User fabrication equipment of many different types is available. If large quantities of boxes are fabricated, the investment in high-speed equipment of the type used by the rigid box supplier is justified. When limited quantities and great flexibility is required because of many different varying sizes and styles, semi-automatic equipment is employed. In Fig. 3-3 a metal edge staying machine is illustrated. This machine, fabricating a box suitable for many style variations requires a simple length adjustment of the metal edge stay and no further adjustment for variations in the box blank. In Fig. 3-4, several basic box styles with metal stays are illustrated. Two models of fully automatic rigid boxmaking machinery are shown in Fig. 3-5. These models represent available fabrication equipment designed for both short and long production runs.

FOLDING CARTONS AND RIGID BOXES

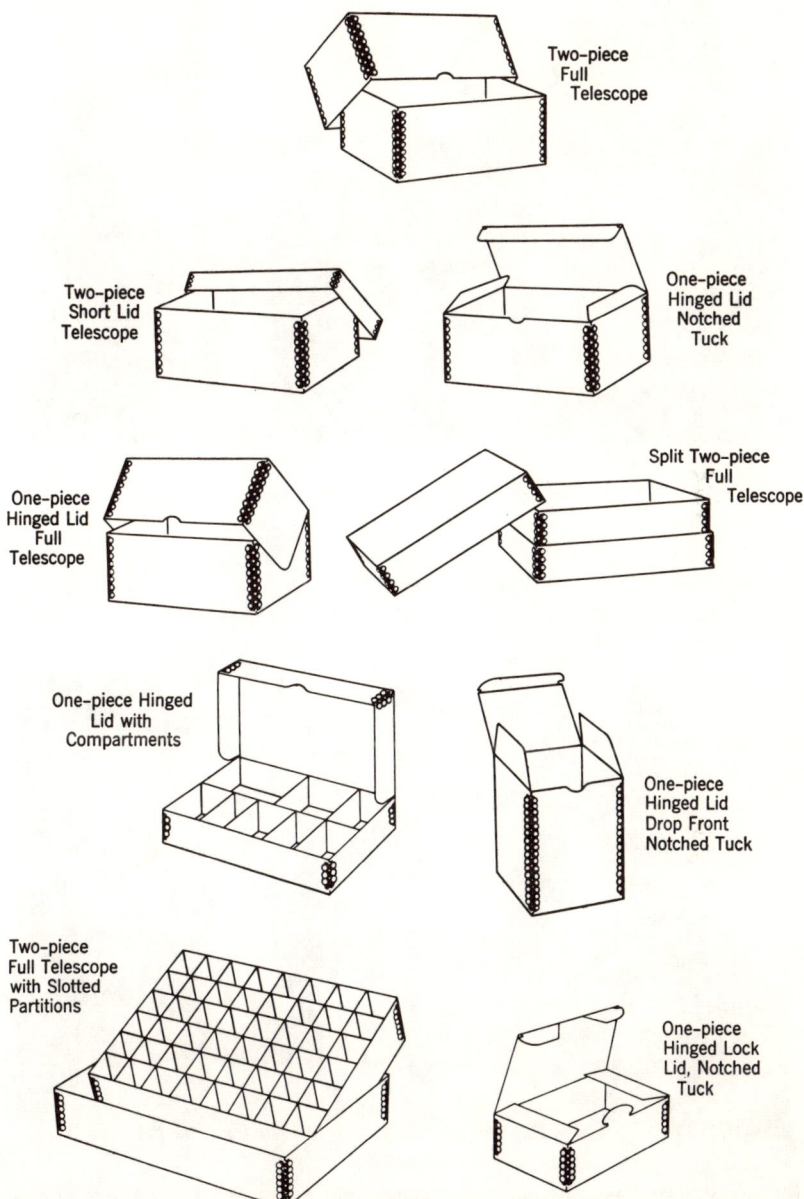

Fig. 3-4. Various styles of metal-stayed boxes. Courtesy National Metal Edge Box Co.

(a) Highspeed fully automatic machine. Courtesy FMC Corp.

(b) Basic semi-automatic machine converted into fully automatic type. Courtesy NJM, Inc.

Fig. 3-5 Fabrication equipment for rigid boxes.

FOLDING CARTONS AND RIGID BOXES

FOLDING CARTONS

A folding carton is defined as a closed container made of bending grades of paperboard, plain or printed, cut and creased, in a variety of sizes and shapes, folded and delivered in flat, or glued and collapsed form by the maker, and to be set up, filled, and closed by the user.

The success of the folding carton is based upon several important considerations. The folding carton, like the rigid box, is capable of structural versatility. The paperboard, from which folding cartons are made is available in a number of grades which can be readily treated, coated or combined to accommodate a wide range of product requirements. The folding carton is an economical package, capable of meeting a wide range of technical and merchandising requirements at a low-unit cost. It offers many economies through its capacity to be formed and loaded at high speeds on machines requiring a low labor cost per unit package. The fact that folding cartons are shipped from boxmaker to user in a flat or knocked-down form offers savings in storage space and transportation costs.

The production of folding paper cartons involves two complex manufacturing processes: (1) the production of the paperboard that is the basic raw material from which all cartons are manufactured and (2) the production of the cartons. As such, the folding carton industry is divided into companies that are strictly convertors, producing cartons only, and integrated firms making both cartons and paperboard. Whereas there are available an almost infinite variety of paperboards with unique characteristics, the user should know at least the basic types and general characteristics of these raw materials as described earlier in this chapter.

The characteristics of any given paperboard are themselves subject to considerable modification, both during the manufacturing process and in later operations, through the application of special additives, surface treatments, or combining with other materials. In view of the many paperboard formulations available, it is usually more helpful for the user to explain all conditions of use to the supplier and to permit the supplier to determine the proper paperboard characteristics for the given product.

For the production of folding cartons, a number of separate and specialized manufacturing processes are involved. These include printing, if required, cutting and creasing, stripping the waste material from the sheet, and gluing when required. In addition, various auxiliary steps are often introduced at different stages in the basic process, depending on the final package requirements. Special processes include film laminating, waxing, the application of transparent windows, and embossing. Paperboard may be processed through the printing, cutting, creasing, and stripping stages in either roll or sheet form.

DISTRIBUTION PACKAGING

Here again it is not necessary for the purchaser of cartons to be intimately familiar with the technical details by which the various manufacturing steps are achieved. The user should be generally familiar with what can be done in terms of his own packaging problem by the folding carton supplier.

STYLES

Despite the many varieties of folding carton constructions available, all cartons can be classified in one of three main groups.

Tray-type construction. Tray-type construction cartons have an unbroken paperboard bottom panel, with side- and end-wall panel members hingedly connected to this bottom panel. Each side- and end-wall panel member is connected to the adjacent wall member by means of a glued flap, a hook which engages a slit in the adjacent wall or some other form of connection such as a paper strip or a metal corner edge.

In addition to this basic tray construction, carton forms in this group may have a wide variety of cover and flap members extending from the wall panels. Two trays, one slightly smaller than the other, may be used as the base and cover, respectively, of a two-piece construction. Typical trays and tray-type cartons with different construction variations are shown in Figs. 3-6 and 3-7.

Fig. 3-6. Basic tray style constructions. Windows can be incorporated in these styles either in the cover or cover and side panel.

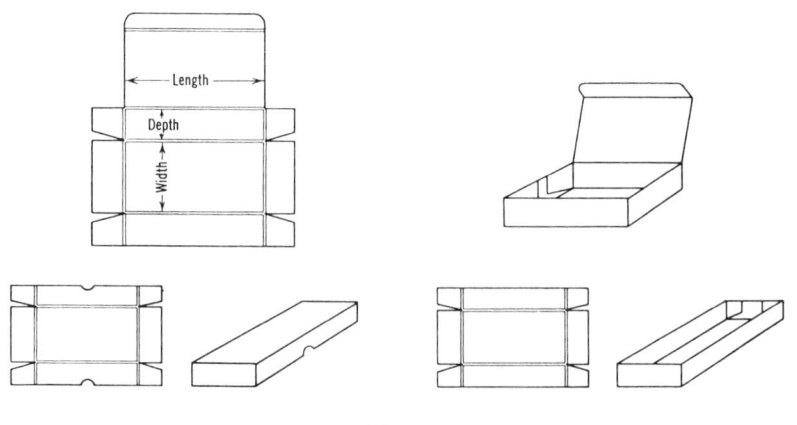

(a)

(a) One- and two-piece "Brightwood" tray constructions

FOLDING CARTONS AND RIGID BOXES

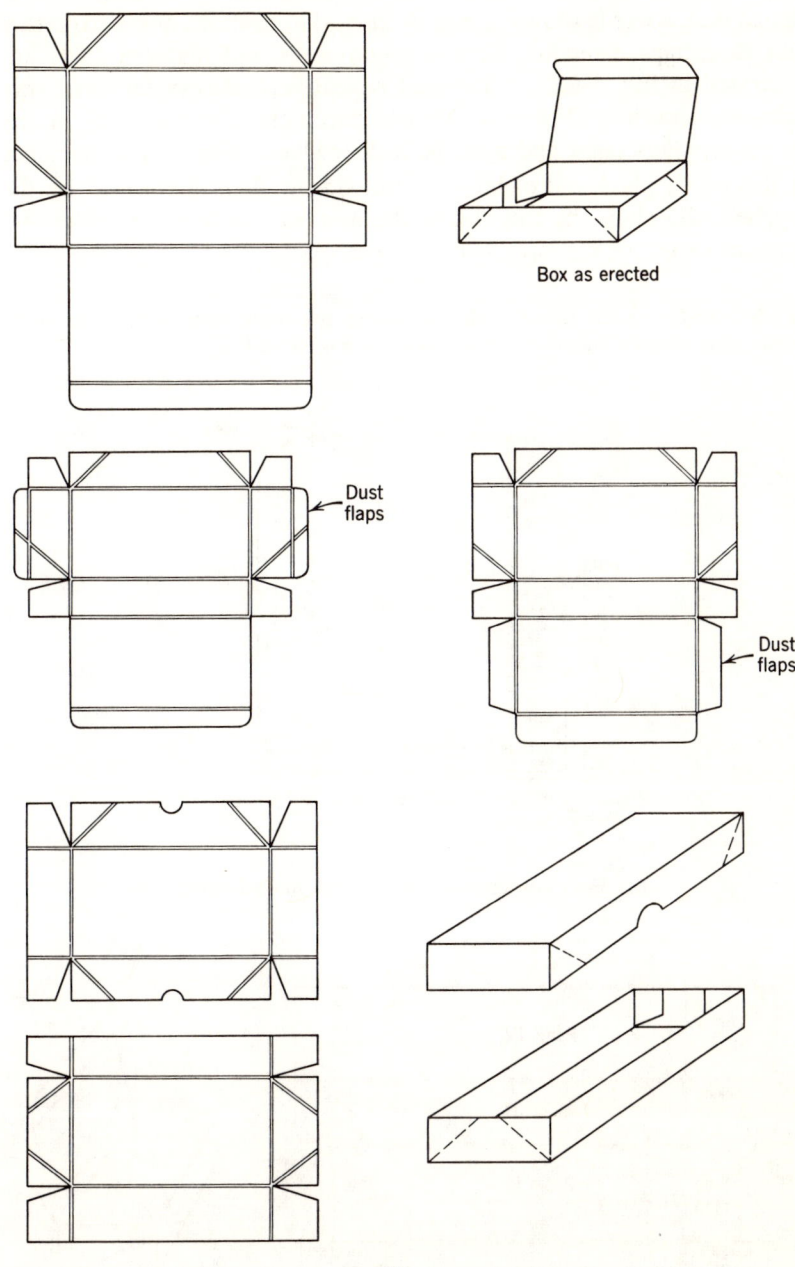

(b)

(b) One- and two-piece diagonal fold style trays

152   **DISTRIBUTION PACKAGING**

The basic tray style of glued corner construction with and without hinged tuck cover is shown in Fig. 3-6a. Modifications of the basic tray with dust flaps extending from the end walls and diagonal folds are illustrated in Fig. 3-6b. Additional typical variations of the basic tray style are shown in Fig. 3-7. Other variations are pre-glued in the carton suppliers plant and must be further glued on cartoning machines in the user's plant. Pre-gluing of the side walls is performed by the supplier who ships the tray flat to the user who in turn assembles and seals the carton on his forming equipment.

Fig. 3-7 (a-f). Tray carton styles including two-piece constructions in blank form. Courtesy Folding Paper Box Association of America.

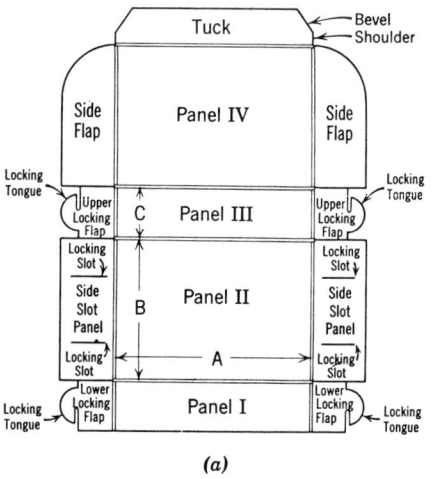

(a)

(a) Four corner lock bottom with hinged cover

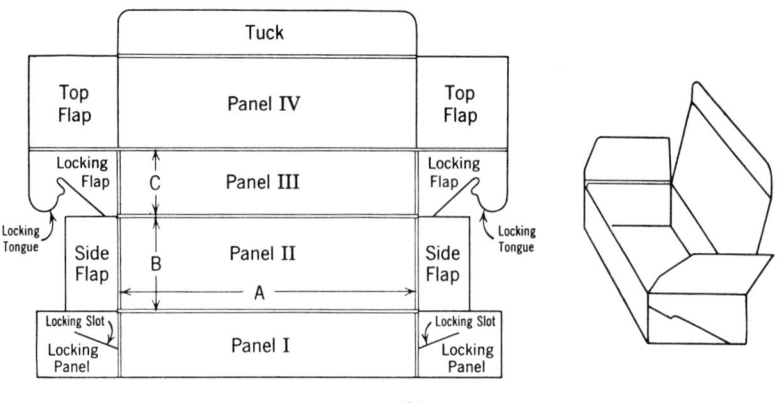

(b)

(b) Cracker shell style

FOLDING CARTONS AND RIGID BOXES

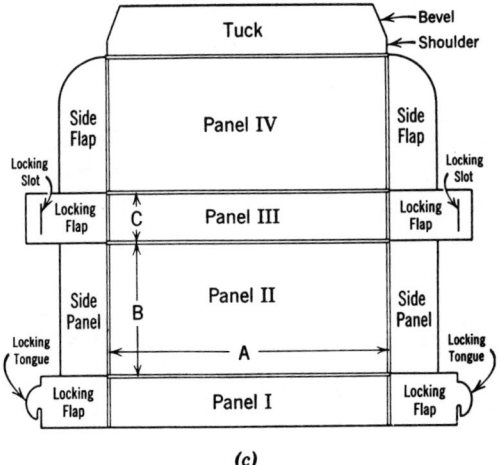

(c) Single lock cake box style

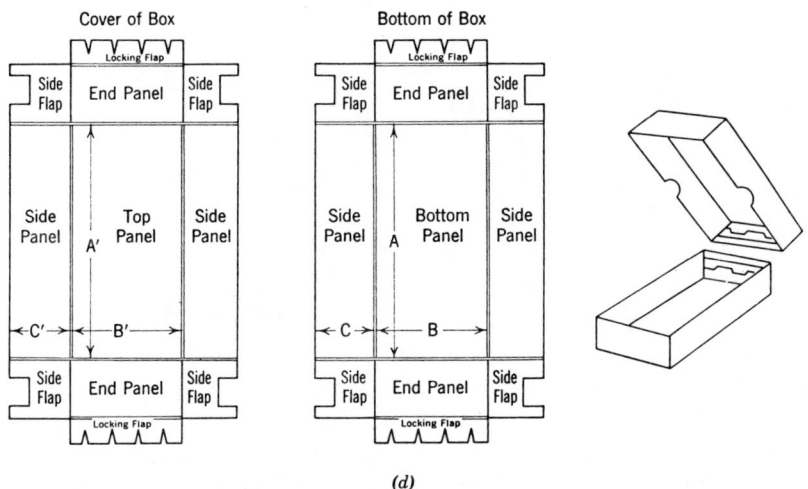

(d) Telescope hardware lock end

Lock corner tray styles are shown in Fig. 3-7a–e. Once again the structural variations attainable with the glued corner and corner stayed styles are applicable to the lock corner construction.

Lined tray cartons other than those made of a barrier material laminated to paperboard are those with an inner barrier material such as waxed paper, glassine, vegetable parchment, or a corrugated lining material formed inside the tray. This lining material may be inserted manually or automatically in the preformed tray, or tray and lining material may be formed together.

Separate lining of the tray does not limit the selection of the tray style or construction. However, when tray and lining material is simultaneously formed, the most commonly used style is of lock-end tray construction.

Tube-type construction. Cartons in this category with a few exceptions have a seam or lap which is glued in the manufacturer's plant. Tubular cartons have a number of wall panels hingedly connected to each other. At one end of the flat blank, a shorter panel is provided

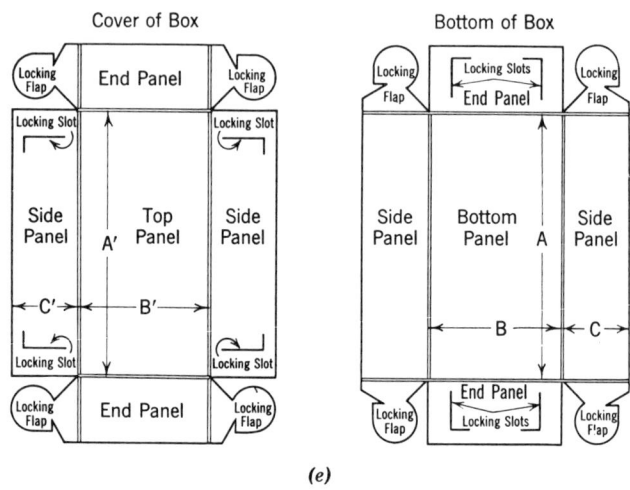

(e)

(e) Telescope lock corner

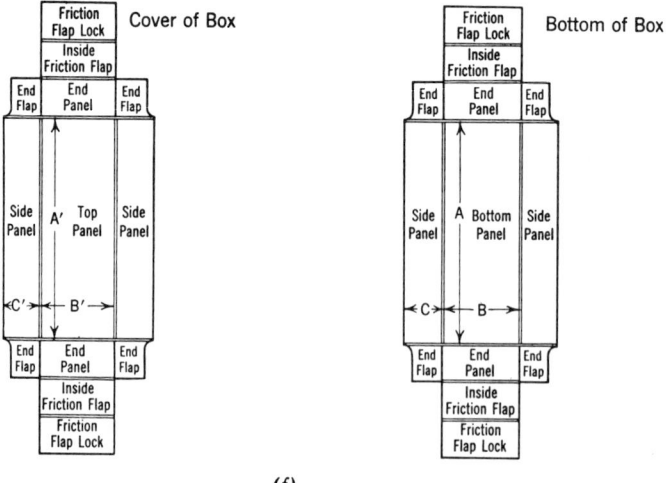

(f)

(f) Two-piece telescope friction end

FOLDING CARTONS AND RIGID BOXES

which is attached to the free edge of the wall at the opposite end of the blank. Pressure along the crease lines at opposite ends of the folded and glued carton forms or erects the walls.

The open ends of the tube may then be provided with a wide variety of end-closure flaps connected to the walls of the tube. Four basic types of end closures are used in tubular cartons: (1) tuck-end closures,

Fig. 3-8 (*a-j*). Tubular carton styles. Courtesy Folding Paper Box Association of America.

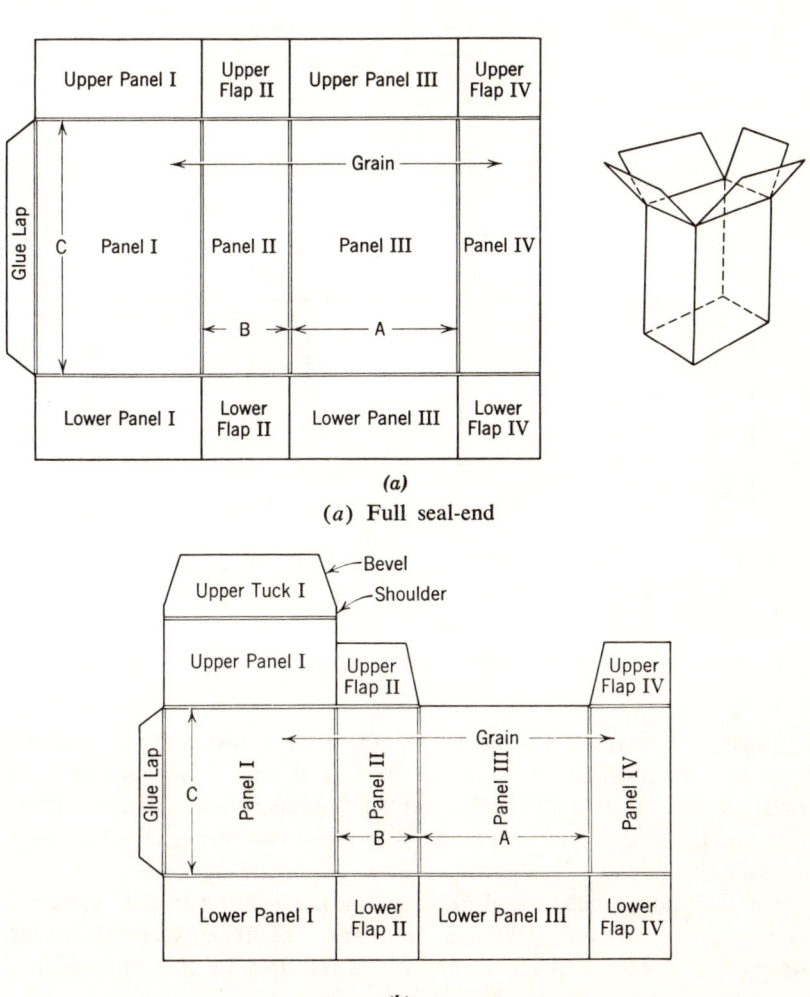

(*a*)
(*a*) Full seal-end

(*b*)
(*b*) Tuck-end, seal-end

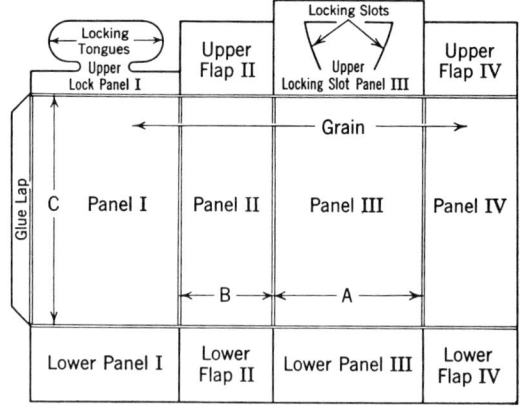

(c) Lock-top, seal-bottom

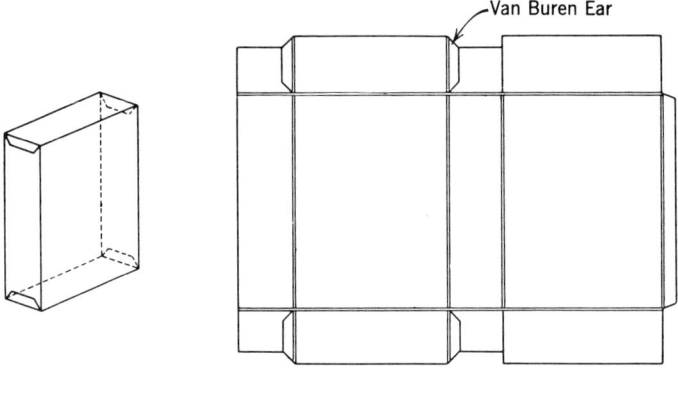

(d) Siftproof variation of seal-end closure tubular carton (Van Buren Ears)

(2) lock-end closures, (3) combinations of these two, and (4) seal-end closures. In the first three types, the end closure is accomplished by friction between or an interlocking of the flap members. In the remaining type, the end closure is glued. Combinations of these end closures are also common.

The common procedure of loading a product into a tubular carton is by insertion. In some instances the carton blank is wrapped around the product. Use of the latter technique requires a fifth panel which is interlocked between the product and the carton walls.

In Fig. 3-8 typical tubular carton constructions of the seal, tuck, and

FOLDING CARTONS AND RIGID BOXES

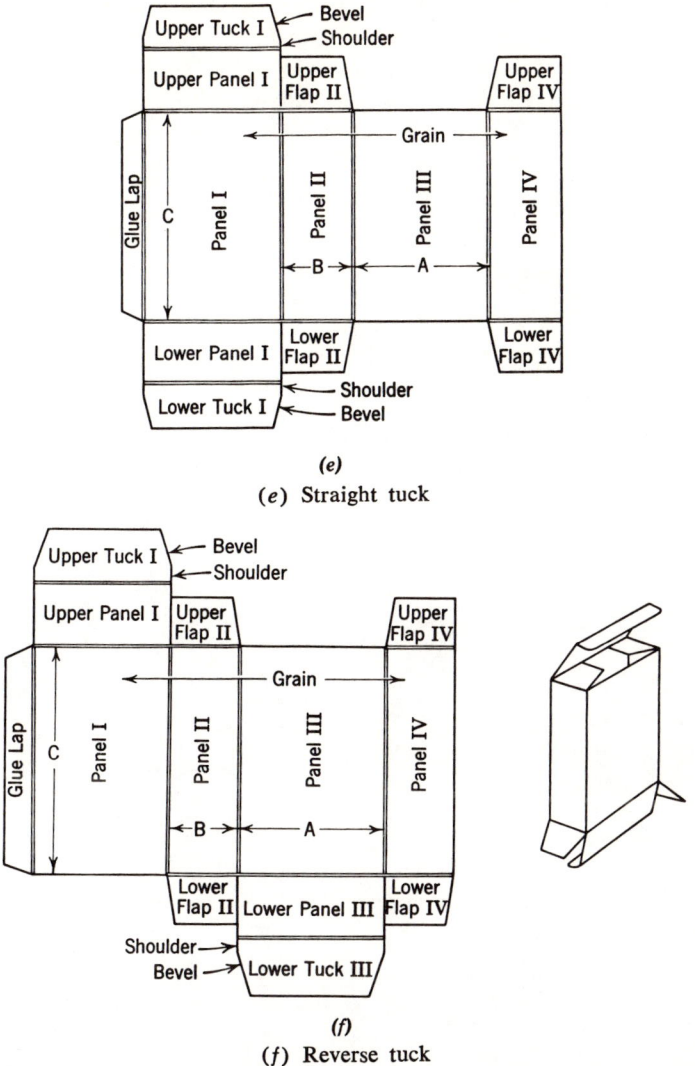

(e) Straight tuck

(f) Reverse tuck

lock-end type are reproduced. Most of these are suitable to be formed on cartoning equipment.

In Figs. 3-8a, b, and c, various seal end constructions are reproduced. The basic style of seal end construction, (Fig. 3-8a) represents one of the simplest forms of tubular folding cartons used. Its sift-proof variation (Fig. 3-8d) represents an example of simple modifications of this basic style.

Tuck styles are shown in Fig. 3-8*e*, *f*, *g*, and *h* with the two basic types being illustrated in blank form in Fig. 3-8*e* and the reverse tuck style in blank and partially assembled form in Fig. 3-8*f*. Variations of possible tuck and tuck-lock ends are sketched in Fig. 3-9*a*.

Lock end style tubular cartons are illustrated in Fig. 3-8*i* and *j*.

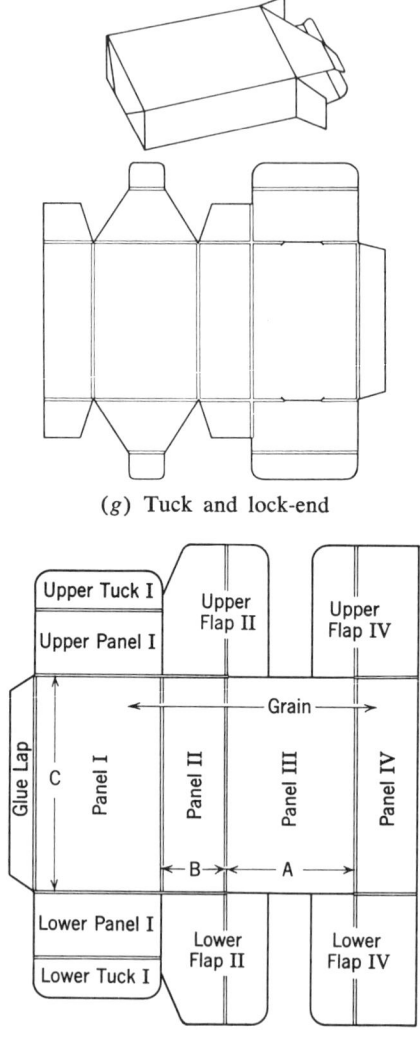

(*g*) Tuck and lock-end

(*h*)
(*h*) Straight tuck end with full overlapping flap and side tucks

FOLDING CARTONS AND RIGID BOXES

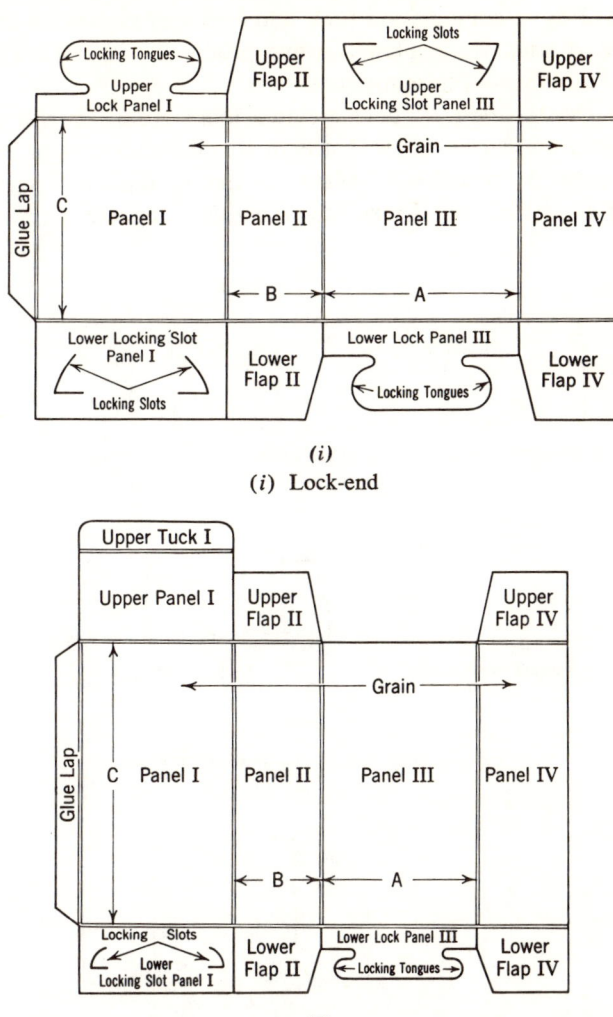

(i) Lock-end

(j) Tuck-top, lock-bottom

Additional details of locks are shown in Fig. 3-9b. Lock-end closures offer a particular advantage by permitting rapid manual setting up of tubular folding cartons. However, there are carton-forming machines available for these and other lock-end closures performing carton-forming either separately or in combination with product insertion mechanisms. There are important differences in lock-end closures with respect to retentive ability as well as boxboard requirements.

In Fig. 3-8g a combination tuck and lock-end closure is illustrated. This style is used for mailing various products such as books, small parts, and merchandising samples.

Special constructions. There are two main groups in this category. *Hybrid types* are carton styles that embody characteristics of both tube- and tray-type constructions. Many of the multiple-unit carriers for bottles of beer and soft drinks fall into this category. *Structures* are

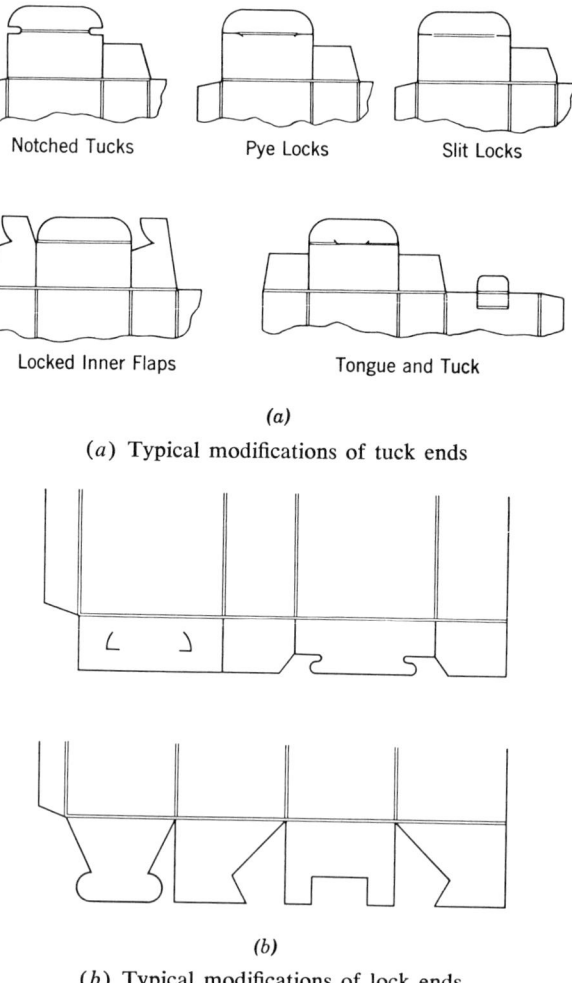

(a) Typical modifications of tuck ends

(b) Typical modifications of lock ends

Fig. 3-9. Tubular carton end modifications.

FOLDING CARTONS AND RIGID BOXES

paperboard constructions that serve the packaging function but have no clearly defined tube or tray characteristics. In this group, some multi-unit carriers, various types of displays, certain bakery packages, and specialty packs are included.

DESIGN CONSIDERATIONS

Because the folding carton is suited for high-volume packaging operations, the details of a particular design are not only related to appearance and utility but also to production. The selection of a particular folding carton style and the required boxboard are highly interrelated with box fabrication as well as user forming and loading methods and equipment. The over-all unit packaging cost of a given folding carton design must include the following factors:

1. The purchase cost of the carton which is a function of the type of board and footage required as well as the fabrication processes needed to produce a given style including finishing operations such as printing, pre-gluing, and window inserting.

2. The packaging cost of the carton which includes the forming, loading, sealing, and other operations such as overwrapping and unitizing.

Since an integrated approach to folding carton design is essential, a separate discussion of carton and packaging costs is presented below for clarity.

Carton cost. Basically, the cost of the carton consists of the board costs and the fabrication costs. The board cost depends on the type and grade of board and the required board footage.

The selection of a given type of board, such as bending chip or manila depends upon performance and appearance requirements. The performance of a type of board is influenced by its physical properties and relates to carton fabrication, carton forming, and subsequent carton use. Factors of carton use may include product protection, that is, resistance to moisture, contamination, and sifting and utility features such as opening, reclosing, display, and carrying. The appearance requirements relate to printability and desired merchandising appeal. Because of these many factors, the various types of boards previously discussed, as well as many other specialty boards and combinations, have been developed.

The *grade of board* (basis weight, caliper, finish, bursting strength) of a board type is dependent upon the structural requirements of the final package and upon considerations of efficient performance on the

TABLE 3-4

Physical requirements of nontest and test grades of paperboard and corrugated board, relationship of box volume and container weight to thickness of nontest paperboard and to thickness and bursting strength of test paperboard and corrugated board.

Volume of box	Weight of contents	Nontest paperboard or corrugated board supporting loads					Test paperboard or corrugated board			
		Thickness[2]	Basis Weight pounds per 1000 square feet[2]				Semisupporting loads		Nonsupporting loads	
			Group I	Group II	Group III		Thickness[2]	Minimum bursting strength[3]	Thickness[2]	Minimum bursting strength[3]
Cubic inch	Pound	Inch[1]	Pound[1]	Pound[1]	Pound		Inch[1]	Point	Inch[1]	Point
20 or less	1/4 or less	0.012	---	56	(Facings minimum)		---	---	---	---
20 or less	1/4 or less	.014	---	63	---		---	---	---	---
20 or less	1/4 or less	.016	65	69	---		0.016	48	0.018	54
20+ to 40	1/4+ to 1/2	.018	72	77	---		.018	54	.020	60
40+ to 60	1/2+ to 3/4	.020	80	82	---		.020	60	.022	66
60+ to 80	3/4+ to 1	.022	85	88	---		.022	66	.024	72
80+ to 110	1+ to 1-1/4	.024	90	96	---		.024	72	.026	91
110+ to 150	1-1/4+ to 1-1/2	.026	96	104	---		.026	78	.028	98
150+ to 200	1-1/2+ to 2	.028	103	112	---		.028	84	.030	105
200+ to 250	2+ to 2-1/2	.030	111	120	---		.030	90	.032	112
250+ to 300	2-1/2+ to 3-3/4	.032	117	128	---		.032	96	.036	144
300+ to 375	3-3/4+ to 5	.036	131	144	---		.036	108	.040	160
375+ to 500	5+ to 7-1/2	.040	144	160	---		.040	120	---	---
500+ to 750	7-1/2+ to 10	.045	160	---	---		.045	144	---	---
750+ to 2000	10+ to 20	---	---	---	52		---	125	---	125

[1] Tolerance— Basis weight ± 5 percent: thickness ± 0.001 inch thru 0.026 inch and ± 0.0015 greater than 0.026 inch.
[2] Add 11 pounds to basis weight values and 0.002 inch to thickness values when paperboard is glassine or greaseproof lined. Use a higher capacity instrument to obtain results in thickness greater than 0.026 inches.
[3] Also applicable to glassine or greaseproof lined paperboard.
Federal Specification PPP-B-566d.

supplier's and user's equipment. The style of the carton and the carton closure can also influence the grade of board to be selected. The structural requirements of the folding carton are influenced by the weight of the product, the size of the carton, the contour or shape of the product, the support (if any) provided by the contents, the type of contents, and the rigidity (if any) provided by the closure. In addition, the type of shipping protection, including the shipping container and any interior components used, affect the structural requirements of the folding carton board. Finally, end use factors such as long shelf life, re-usability, and anticipated environmental hazards can also determine the grade and type of board to be used.

There are relatively few guides available to assist in specifying the needed caliper, basis weight, finish, and bursting strength of board to satisfy needed structural requirements for a given product. One such guide is contained in Federal Specification PPP-B-566d — "Boxes, Folding Paperboard" — in which a relationship of cube and weight to required thickness (caliper) is listed (Table 3-4). This table also includes bursting strength, based upon the characteristics of the product (that is supporting, non-supporting, or semi-supporting loads).

Even when the structural requirements have been met, problems with equipment may be experienced. In the converters plant production can be severely hindered by such problems as cracking, excessive spring, inadequate adhesion, and peeling. Part or all of these problems can be directly related to the type and grade of board selected. In the user's forming, loading, and sealing operations similar board problems can reduce production rates, and any defects produced can severely influence carton appearance and utility. Examples of such defects include impairment of the efficiency of lock and tuck type closures and misalignment. In general, carton styles utilizing a lock-type closure are more sensitive as to proper board grade selection than glued-type closures, since good tear resistance is mandatory.

Since the cost of different board types varies widely and since board grade (caliper, basis weight, finish, and bursting strength) is obtainable in fairly wide ranges, optimum costs are obtained through specifications which fulfill the necessary requirements and are not arbitrarily chosen. Major economies are thus occasionally attainable through a review of board specifications in consideration of requirements.

The *amount of board footage* consumed for a carton of a given cube content is a function of the carton style. Different styles require varying amounts of board footage as a result of various types and positioning of closures. For economy in board footage, selection of a style should be made not only in consideration of the blank size of a single carton but also as a combination of a number of the same or other cartons for fullest utilization of the board sheet width. There are significant differences in the board requirements of a particular style, which is apparent in the blank layouts illustrated in this chapter. If changes in style are permissible for appearance or utility reasons, major carton costs savings are sometimes possible. For example, one company packaging food containers used a tubular carton having a reverse tuck closure with locking inner flaps on both top and bottom. By a change of the bottom closure to a seal end, a significant board saving was obtained with a desirable improvement through a more effective bottom closure.

Since folding cartons are die-cut, any complexities of the carton blank generally do not increase the per-unit carton cost, except by the added initial cost of the die. Variations in printing layout or printing coverage similarly do not influence unit carton costs except for the cost of the die. However, there are cost differences in producing different levels of printing quality and number of colors. Any special carton features requiring extra fabrication operations such as window gluing or reinforcing are up charges which are directly reflected in the purchase cost of the carton.

Packaging cost. The packaging cost of folding cartons relates to the particular method which is employed to form the carton, load the product, and seal the filled package. For maximum economy the carton design should be selected to conform to the specific method employed in performing these elements of packaging. Basically, three methods can be utilized: (1) manual, (2) semi-automatic, and (3) fully-automatic. The method to be selected is dependent upon the following factors:

1. Volume and rate of the specific size, style, or type of carton to be packed.
2. Volume and rate of other sizes, types, and styles of cartons in the same facility.
3. The type of product, that is, size, weight, physical configurations, etc.
4. The quantity of product and its arrangement in the carton, plus the inclusion of any other inserts such as literature, platforms, and cushioning.
5. Flow of the product to the packing station, that is, rate, continuous or intermittent, product orientation, etc.

FOLDING CARTONS AND RIGID BOXES

6. Special carton and product characteristics including supplementary packaging operations such as overwrapping, combination packs or deals, and easy-opening features.

After a comprehensive engineering analysis has been conducted, the feasibility, practicability, and desirability of a given method of operation becomes apparent. For instance, if the product runs are too short and wide variation in carton size is required, an essentially manual method may be most economical. In another instance the volume of a given product may be sufficiently great for fully automatic operation provided that several product lines can be combined to feed into centrally located packaging machinery (Fig. 3-10). Inserts required to position the product in the carton may prohibit fully automatic operation but permit automatic set-up of the carton with subsequent manual loading followed by automatic sealing. If sufficient volume justifies the expenditure, specialized equipment or attachments are obtainable for automatic insertion (Fig. 3-11).

Following the basic selection of either a manual, semi-automatic, or fully automatic packaging system, considerations of the details of carton

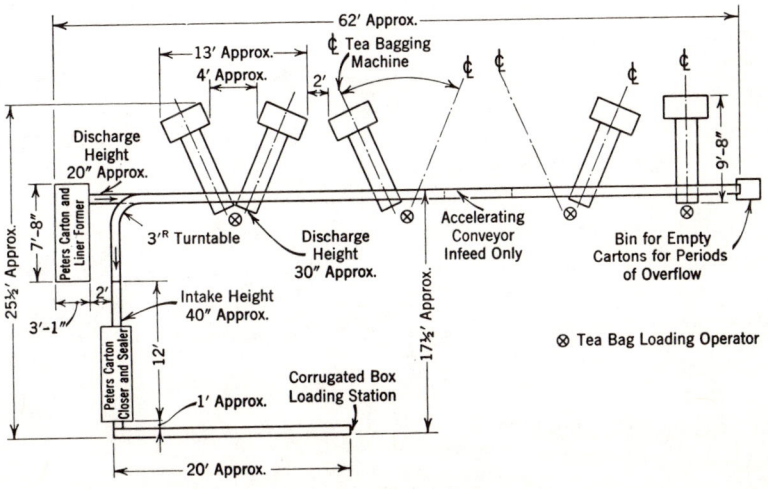

Fig. 3-10. Centralized carton lining and forming operation serving seven tea bag making machines. Loaded cartons are conveyed in the opposite direction, and are closed and sealed on centrally located packaging machine.

Fig. 3-11. Mechanism for automatic insertion of folded circulars, booklets, or other enclosures around or over the ends of bottles or other products. Courtesy F. B. Redington Co.

design must be fully integrated with the method and the particular equipment to be employed. With entirely manual packaging (Fig. 3-12) the speed of the operator is directly influenced by the ease of carton forming, ease of product insertion, and simplicity of closure. The grade of boxboard, influencing such factors as forming and bending, the type and size of carton opening, influencing speed of product insertion, and the type of closure, affecting the rate of closing, all have a direct bearing on labor productivity in a manual operation. When packaging equipment is considered suitable, the features of the equipment may dictate a particular carton design. Changes in carton style to permit mechanized packaging may influence the per-unit carton cost. The *over-all* packaging cost must be rated in terms of the carton cost and the packaging cost to develop the *actual* cost of the operation.

PACKAGING EQUIPMENT

A cartoning operation consists of three basic functions: (1) forming, (2) filling, and (3) closing. Equipment can be selected to perform any one of these functions mechanically, either separately or as a combined operation. Therefore, a packing line may utilize one machine for the complete packaging operation, or three separate machines may be arranged in tandem to accomplish the same purpose. In other in-

FOLDING CARTONS AND RIGID BOXES 167

stances, the filling or loading function may be performed manually (Fig. 3-13) with equipment providing automatic or semi-automatic forming and closing.

Cartoning equipment is classified by the style of carton used: (1) tray, (2) tube, and (3) special. Equipment for trays consists of various types of forming machines, which deliver the formed tray to a manual or automatic packing station. For tubular cartons equipment

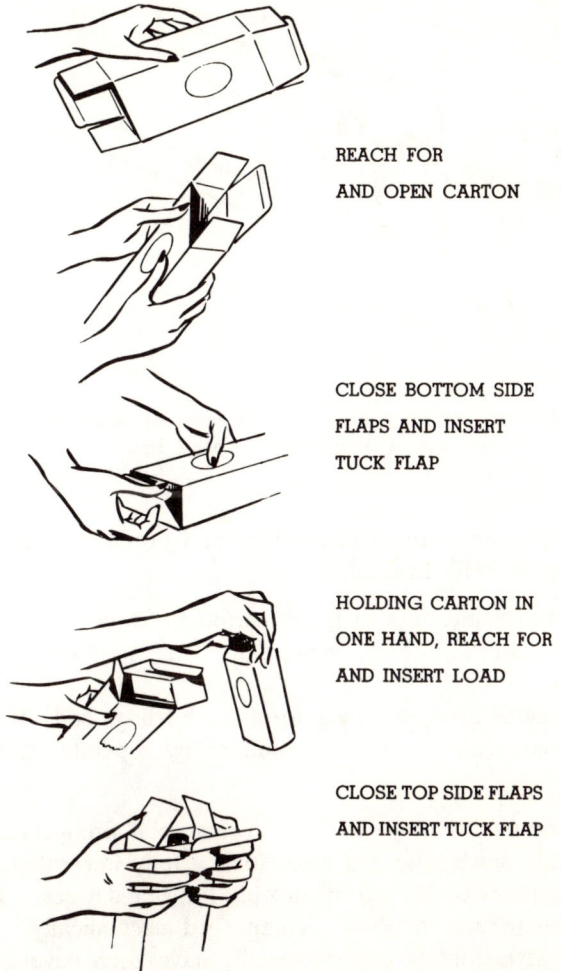

REACH FOR
AND OPEN CARTON

CLOSE BOTTOM SIDE
FLAPS AND INSERT
TUCK FLAP

HOLDING CARTON IN
ONE HAND, REACH FOR
AND INSERT LOAD

CLOSE TOP SIDE FLAPS
AND INSERT TUCK FLAP

Fig. 3-12. Basic elements of manual folding carton packaging. Courtesy R. A. Jones and Company, Inc.

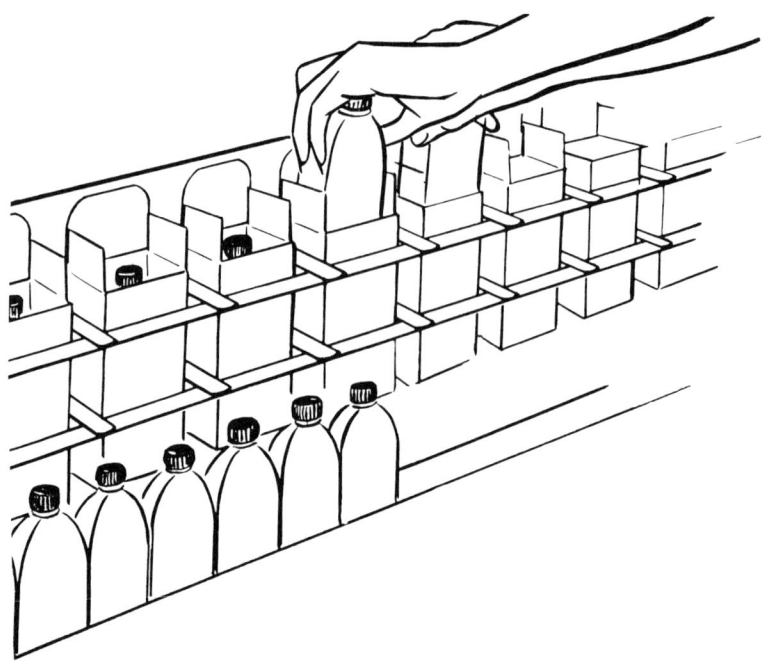

Fig. 3-13. Manual bottle loading into an automatically formed and bottom sealed carton. Courtesy R. A. Jones and Company, Inc.

is available for unit- and free-flowing solid products. Equipment for unit products is divided as follows:

1. Semi-automatic, in which the carton is formed and closed automatically but insertion of one or more products into the carton is manual (Fig. 3-14).
2. Fully automatic, in which forming, loading, and closing is performed automatically by a single unit or by separate machines (Fig. 3-15).

Most equipment that packs free-flowing solid products into tubular cartons is fully automatic and uses either a fully integrated machine or separate machines connected by suitable transfer devices. In addition, lining of tubular cartons can be accomplished mechanically.

Specialty style folding cartons usually have been developed in connection with specific equipment including beer-carrier forming and loading machines and spout insertion machines. Other machines are available for proprietary carton designs.

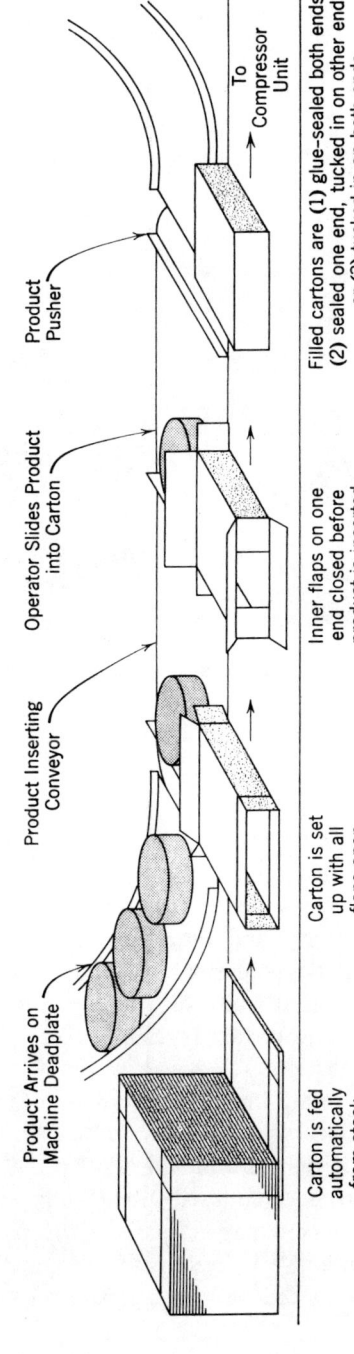

Fig. 3-14. Schematic diagram showing the functions of semi-automatic cartoning (manual insertion of product). Courtesy Container Equipment Corp.

Fig. 3-15. One of the principles employed in automatic carton loading. Courtesy R. A. Jones and Company, Inc.

FOLDING CARTON APPLICATIONS

Free-flowing solids such as soap, detergents, sugar, cake mixes, baking soda, gelatin products, and breakfast cereals are almost universally packaged in tubular-type cartons with adhesive sealed end closures to prevent sifting. Dried fruits such as apricots, prunes, raisins, many types of seasonings, and certain types of bakery products are also packaged in seal-end cartons.

When it is necessary to retain the flavor or aroma of the product or to protect it from exterior hazards such as water vapor, or for products with an extremely small particle size in which sift-proof features are desired, the seal-end carton is often used along with an inner liner or barrier material such as waxed paper, glassine, or parchment. This lining material can be laminated directly to the paperboard. Occasionally, the inner liner takes the form of a separate bag which is made

FOLDING CARTONS AND RIGID BOXES

from a roll of the barrier material directly on the carton forming and loading equipment. The bag is then inserted into the carton or the carton blank is formed around it.

The tuck-end carton, also of tubular construction, is widely used for the packaging of such high-volume items as collapsible tubes of toothpaste, shaving cream, medicinal products, bottled products such as drugs, patent medicines, liquid shoe polish, and cosmetic preparations. Tubular cartons with a seal-end bottom closure and a tuck-top closure are preferred for use in packaging when the product weight requires the security of a glued bottom closure.

Tubular carton constructions are suitable for packaging a wide range of products either manually or on high-speed, automatic equipment. Tuck-end styles lend themselves to manual forming and loading as well as for semi-automatic cartoning systems. Seal-end cartons are used almost exclusively in conjunction with mechanical systems, either fully or semi-automatic. Lock-end tubular styles are particularly suited for manual forming and loading operations, although in specialized cases, cartons of this type are mechanically formed.

Tray-type styles are suitable for both mechanical and manual cartoning operations. The glued-corner tray is universally used for ten-pack cigarette cartons. Trays of this type are generally formed on automatic or semi-automatic equipment with one basic exception. Trays with diagonal creases in the side-wall panels can be glued in a collapsed form in the carton manufacturer's plant. Constructions of this type provide the advantages of a glued carton with respect to corner strength and permit manual forming by the user.

Lock-corner tray constructions, as previously defined, are those in which the adjacent wall members are connected by a hook attached to one panel which engages a slit in the other. This tray style is widely used in bakery packaging in which both manual and mechanical methods can be employed. Most trays of this type are printed and overwrapped with a transparent film which permits excellent product visibility.

The applications for reinforced tray constructions with double-thickness wall panels are primarily for products requiring a more rigid unit container than is provided by other folding carton styles. Typical products packaged in reinforced tray constructions are hardware, automotive parts, candy packs, and soft goods such as towels, blankets, and sheets. This style also lends itself for use as display pack for such items as cough-drop cartons, toys, drugs, and cosmetics. In combination with die-cut inner platform members, reinforced trays provide positioning in addition to rigidity for display purposes. Both manual and mechanized methods of forming and/or loading are possible with this style of tray.

Specialty carton constructions are used whenever tray or tube styles do not provide a package that meets all the structural or marketing requirements of a given product. Examples include six-bottle carriers for beer and soft drinks, luxury perfume packages, and cartons for special promotional items.

Other applications, embodying special construction features, include flip-top and slide-top boxes for cigarette packaging and built-in metering devices for the dispensing of soaps and detergents.

CHAPTER 4

Shipping Sacks

The sack as a shipping container has the inherent advantages of low tare weight ratio (the ratio of the weight of the container to the weight of the contents), flexibility, ease of filling and handling, small required storage space, and low cost of materials.

DEFINITIONS AND FIELD OF APPLICATIONS

The terms *bag* and *sack* are often used synonymously. The term *sack,* however, generally refers to the heavier duty bags or shipping bags which are covered in this chapter. An acceptable definition for a sack is a *preformed container made of flexible material, generally enclosed on all sides except the one which forms an opening that may or may not be sealed after filling.* A sack or bag may be made of any flexible material, multiple plies of flexible materials, or a combination of two or more materials, such as paper, metal foil, and textiles, any of which may be coated, laminated, or treated in other ways to provide the properties required for the packaging, storage, and distribution of a product.

The bag or sack is chronologically probably one of the oldest forms of packaging used by man. However, significant usage of bags and sacks has paralleled the development of filling, closing, sewing, and other production machinery. The recent trend back to bulk packaging has served to focus greater attention upon the principles of sack packaging and to foster further improvements and refinements in this field. Through the proper selection and combinations of materials and special additives the range of sack packaging applications has been extended. Sacks can be reinforced with fibres to reduce tearing hazards and can be treated to resist vermin, oils, odors, and moisture.

Sacks are used for the bulk shipment of such items as chemicals, fertilizers, cement, grains, feed, flour, sand, dog food and insecticides. In general, the major products packed in sacks can be divided into five broad classes:

1. Agricultural and food.
2. Building materials.
3. Chemicals.
4. Minerals.
5. Miscellaneous.

The selection of a shipping sack for the packaging of products within these major classifications depends upon the following criteria:

1. End-use requirements, related to further processing and marketing of the product, which include considerations of the quantity required, the type of container suitable for unloading, handling, and storage facilities.
2. The suitability of the product for sack packaging including ability to fill, provisions for adequate transit and storage protection, etc.
3. Economic factors including the value of the product in relation to the cost of packaging and its replacement value in case of damage loss.

The products packed in sacks, although essentially in bulk, vary greatly in characteristics. They range from finely powdered materials to large-size chunks, and they may be either free flowing or non-free flowing. For adaptability to sack packaging, the product must not only lend itself to economical filling but also be readily dispensed or dumped for use. Usually sacks are best adapted for end use conditions in which the contents of a complete sack are consumed in a mixing, batching, grinding, melting, or other processing operations. If large quantities of a bulk product are consumed, other forms of containers in con-

SHIPPING SACKS

junction with gravity or pneumatic handling equipment may be more economical. Larger container types for products in bulk include pallet boxes or bins, inflatable rubber containers, and covered hopper cars.

A sack, being a flexible container, does not provide support for the product against the superimposition of loads or resistance to impact shocks. Therefore, the product itself must not be susceptible to damage caused by compression or shock. Whereas most bulk products satisfy these requirements, some bulk products need protection to prevent breakdown or build up of particle size. Of even greater importance are provisions against environmental hazards which may change the physical or chemical state of the product. Changes of this type may create difficulties in unloading due to hardening, lumping, or to processing equipment. Furthermore, the desired characteristics of the original merchandise can be altered.

Since a flexible container does not provide the same degree of protection against punctures, tears, and snags as does a rigid container, a greater percentage of product loss may be encountered in sack packaging. Owing to the greater risk involved, insofar as protection against physical hazards is concerned, products shipped in sacks are generally of low value. Another factor which is to be considered is the potential contamination of other products in the shipment because of sifting or spilling of the material. Despite some of these limitations, sacks are widely used, since they are inexpensive bulk packages which can be readily integrated into mechanized packaging operations.

This chapter discusses the various available sack types and important design and selection criteria.

For industrial packaging purposes there are three basic sack types, namely, (1) textile, (2) textile laminated, and (3) multi-wall paper. The greatest percentage of shipping sacks currently used are made of multi-wall paper construction. This type has largely replaced textile sacks.

TEXTILE AND TEXTILE LAMINATED SACKS

Since textile sacks can be constructed of greater strength than paper sacks, a preference for textile still prevails with some shippers. The potential re-usability of these sack types offers economies which however must be rated against the cost to clean, repair, and re-distribute the sacks. Textile laminated bags combine the strength characteristics of cloth with the advantages of other materials including paper, polyethylene, and various laminants and therefore extend the area of sack application.

TYPES OF MATERIALS

Burlap and cotton are the two principal textiles fabricated into sacks for distribution packaging purposes. Burlap cloth is woven from imported jute fibre. The weight of the cloth, the twist of the yarn, and the method of weaving influence the strength characteristics of the end product. Burlap cloth also serves many other packaging applications such as are involved in the packaging of felts and carpets.

Cotton is more attractive than burlap and better resists deterioration resulting from exposure to the elements. The popular types of cotton fabric used in the manufacture of sacks are known as sheetings, drills, print cloths, and Osnaburgs. Cotton sheeting is a material which compares favorably to burlap in strength and has the advantage of lower weight.

Osnaburg cloth is the coarsest plain woven cotton material commercially available. Osnaburg is fabricated in medium and heavy weights, with the latter suitable for heavy-duty sack applications. Cotton drills are of a twill weave, and the close weave thus provided makes this material particularly suitable for the construction of sacks for products requiring protection against sifting. Cotton print cloths are used for small and lightweight packages.

Textile laminated materials include combinations which impart properties to the sack which a single material alone may not provide. The construction may consist of alternate plies of textile, paper, or films joined together with asphalt, latex, or other laminants. Examples of laminations used in sack construction are paper-asphalt-textile, paper-asphalt-textile-asphalt-paper, polyethylene-adhesive-paper-adhesive-textile, and many other combinations. When paper is laminated to textile either a plain kraft or wet strength kraft can be used.

The paper must be able to expand or contract with the textile covering as the bag or sack is handled. Creping or corrugating of the paper allows for this movement and minimizes tearing. The paper may be creped in one direction of the paper, creped in two directions, or creped in one direction of the paper and corrugated or pleated in the other. Plain kraft paper is treated for moisture resistance by the use of rosin sizing. Plain paper in this instance refers to paper that has not been treated for special qualities such as resistance to water, scuffing, molten

SHIPPING SACKS 177

resins, hot asphalt, and grease. Wet-strength paper used in the manufacture of textile and paper laminated shipping sacks is generally marked for identification purposes by striping.

A laminated material for shipping-sack application derives its main strength from the layer or layers of textile used, whether cotton sheeting, cotton Osnaburg cloth or burlap.

Waterproofness, resistance against product contamination and insect infestation, loss of aroma or absorption of undesirable odors, as well as other needed properties, are obtained through the selection of the proper type and number of plies of other suitable materials and laminants.

SEWING TECHNIQUES

Textile and textile laminated sacks are fabricated by sewing the sides and bottom seams (Fig. 4-1). To eliminate needle holes for water-

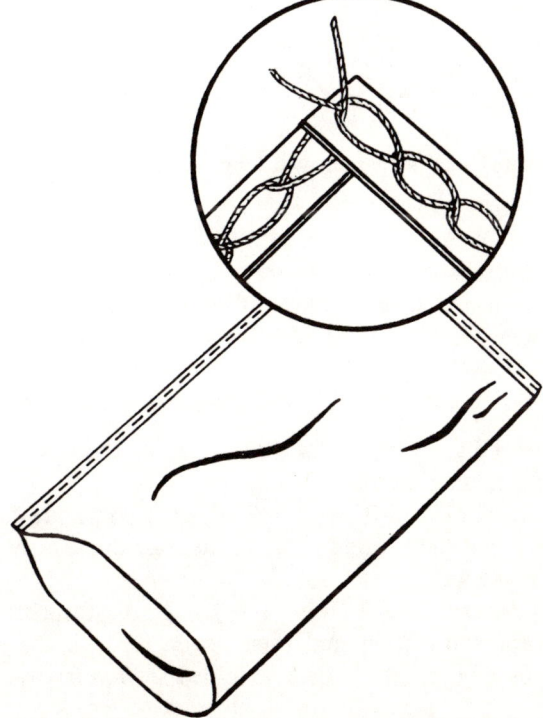

Fig. 4-1. Laminated sack with sewn side and bottom seam.

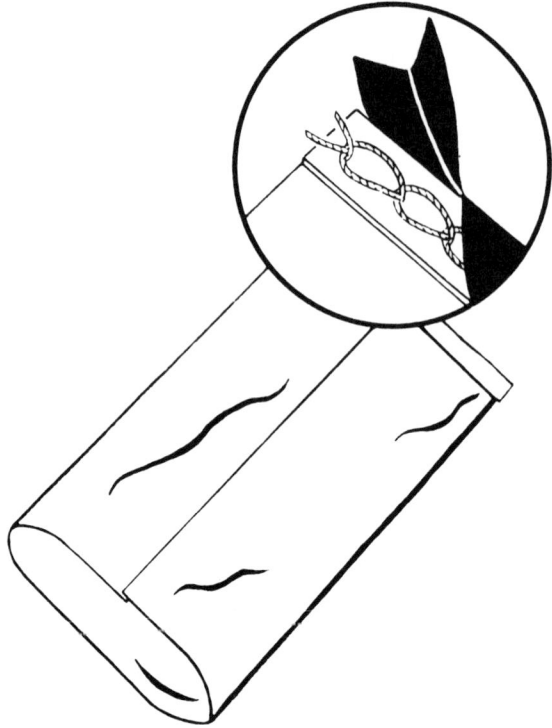

Fig. 4-2. Laminated sack with cemented longitudinal seam and sewn and taped bottom.

proof sacks, textile laminated sacks are made with cemented longitudinal seams and a sewn and taped bottom (Fig. 4-2).

There are five basic methods of sewing. In the first four methods illustrated in Fig. 4-3, the sack is turned inside out after sewing so that the seams are inside the bag. The following seams are formed:

1. *Flat Seam* (Fig. 4-3a). Two flat edges of the cloth are stitched through without folding them.

2. *Export Seam* (Fig. 4-3b). Two adjacent edges are folded away from each other and then sewed through the four thicknesses to prevent unraveling of raw edges.

3. *Single-Turnover Seam* (Fig. 4-3c). Two adjacent edges are folded in the same direction and then sewed through the four thicknesses. This method is also known as folded seam, turnover seam, and lap seam.

4. *California Seam* (Fig. 4-3d). The California seam is a combina-

SHIPPING SACKS

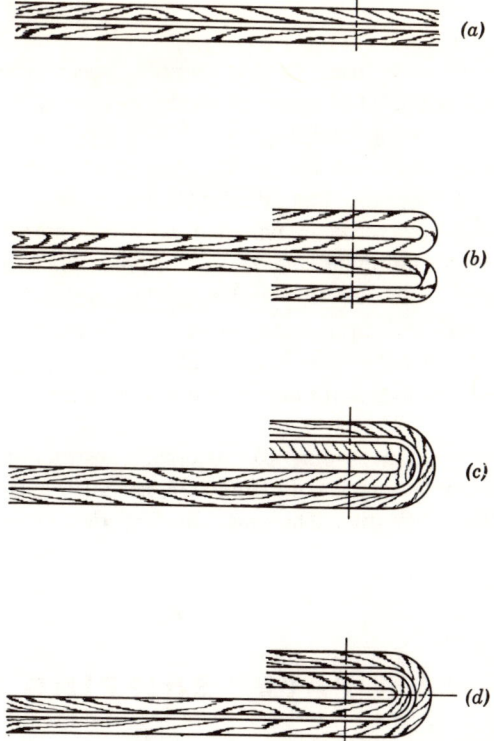

Fig. 4-3. Seam constructions.

tion of flat seam and single turnover seam. The edges are first stitched through, and the flat seam thus formed is then folded on the thread line and stitched again (as in a single-turnover seam) through the four thicknesses of cloth.

5. *Overstitched Seam.* Two adjacent edges are brought together flat and are stitched in a manner that rolls the combined edges while stitching in a zig-zag pattern. Such bags are not turned.

Cemented longitudinal seams are constructed with an overlap and are cemented with a bonding agent which provides a joint equivalent in strength to the material itself. The sewn and taped bottom seam required for waterproof-sack construction consists first of a flat seam formed by stitching, with the tape adhered over the stitches. The tape extends along the seam and protrudes beyond both edges of the sack as illustrated in Fig. 4-2.

CLOSURES

The top closure to be used for the shipping sack is dependent upon the style of sack used. Open-mouth sacks (Fig. 4-4h) are closed by sewing, tying, or sewing and taping. Valve-type sacks (Fig. 4-4b) require no top closure since the sack is furnished with the top closed, except for the valve area. The latter may consist of a tuck in or inner sleeve. In special instances in which a positive waterproof closure is required for textile laminated sacks, all sewn seams are dipped in a sealing compound fully covering and extending beyond the tape.

Closure methods and equipment are treated in greater detail later in this chapter in the discussion of multi-wall paper shipping sacks. Some of the techniques described are applicable to textile and textile laminated sacks.

An excellent guide pertaining to material, construction, and end use requirements of textile and textile laminated sacks is Federal Specification PPP-B-35b — "Bags, Textile, Shipping, Burlap, Cotton and Waterproof Laminated."

MULTI-WALL PAPER SHIPPING SACKS

The origin of the multi-wall paper sack dates back to the late nineteenth century when the need to increase the filling and packing speeds for salt demanded a better flexible container. In 1898, the basic valve-type sack and the machinery to fill it was patented. The early paper sack construction consisted of a single wall made of heavy-duty rope stock paper. This type of sack continued to be used until World War I, when a shortage of supplies of hemp and salvage rope forced the sack manufacturers to experiment with kraft mixed with rope fibres, sisal, and jute. This research eventually led to sacks made entirely of heavy, inflexible sheets of kraft paper. Further experimentation developed the advantages in strength and flexibility of two walls or plies of lighter kraft paper when combined with corn, potato paste, or tapioca.

Although the advantages of more than two plies for sack construction were realized, it was not until 1925 that this construction became feasible through the development of multi-ply sack-forming equipment. This equipment permitted the sewing of ends to supplement the pasted satchel-bottom sacks then used, a technique limiting the number of plies combined.

From this point on the use of multi-wall paper shipping sacks was

SHIPPING SACKS

rapidly expanded and an approximate yearly consumption of three billion units has been reached today.

CONSTRUCTIONS

There are two distinct types of paper shipping sack: the light-duty shipping sack constructed from one or two plies of paper and the heavy-duty shipping sack known as multi-wall, made of three or more plies of paper and used largely for bulk shipments.

Each sack type is made in a wide variety of sizes and constructions to satisfy particular requirements of the contents and the shipping and storage conditions. Paper sacks are made with each ply arranged and fabricated in tube form. These tubes are assembled one within the other so that each contributes to the required task. It has been found that greater flexibility and strength are obtained by using several walls of relatively light weight, rather than a fewer number of heavier paper walls. For this reason, the average heavy-duty multi-wall sack is constructed of a number of sheets ranging from 40 to 70 lb in basis weight. (ream weight–500 sheets 24 in. $\times$ 36 in.). The basis weights of kraft papers commonly used in multi-wall sack construction are 40, 50, and 60 pounds. Heavier papers are generally used in single- and double-wall pasted type paper shipping sacks for domestic shipment of some commodities.

Different weights of paper are frequently combined in one multi-wall sack, and special kraft papers such as waxed, asphalt-laminated, polyethylene-coated, wet strength, and extensible, are included as a separate ply

TABLE 4-1

SHIPPING BAG KRAFT PAPER PLAIN OTHER THAN WET-STRENGTH

Basis Weight 24x36-500	Minimum Average Dry Tearing Strength		Minimum Average Dry Tensile Strength per Inch Width	
	M.D.	Total M.D. Plus C. D.	C.D.	Total C. D. plus M.D.
Pounds	Grams	Grams	Pounds	Pounds
40	88	188	14	41
50	110	235	19	53
60	132	282	23	64
70	154	329	27	74

Uniform Freight Classification

TABLE 4-2

Shipping Bag Kraft Paper Plain Wet-Strength

Basis Weight 24x36-500	Minimum Average Dry Tearing Strength		Minimum Average Dry Tensile Strength per Inch Width		Minimum Average Wet Tensile Strength per Inch Width	
	M.D.	Total M.D. plus C.D.	C.D.	Total C.D. plus M.D.	C.D.	Total C.D. plus M.D.
Pounds	Grams	Grams	Pounds	Pounds	Pounds	Pounds
40	75	160	16	45	3.8	11.1
50	94	200	21	58	5.1	14.6
60	113	240	25	70	6.2	17.8
70	132	280	29	80	7.3	20.5

Uniform Freight Classification

or plies in constructions in which certain protective features are needed. Other specialty papers, films, and foils may be used as an inner ply or liner, or protective sheets of this type may be laminated to a sheet of kraft paper. (See Chapter 7.)

The type of paper used for the manufacture of shipping sacks contains long fibers properly interlaced by the Fourdrinier process to obtain maximum balanced strength and durability. It is distinguished from other types of kraft paper by its name, *shipping sack kraft*. Typical truck and rail shipment requirements for shipping sack kraft appear in Tables 4-1 and 4-2. For additional information consult Federal Specification UU-S-48 and Rule 40 of the Uniform Freight Classification or Rule 200 of the National Motor Freight Classification.

TYPES OF PAPER SHIPPING SACKS

Paper shipping sacks are furnished in six basic types. These types differ mainly in the manner of closure as follows:

1. Pasted bottom, open mouth.
2. Sewed bottom, open mouth.
3. Pasted valve.
4. Sewed valve.
5. Sewed open corners.
6. Pinch type.

SHIPPING SACKS

The *open mouth sack* may be filled manually or by machinery, and the closure is accomplished after the sack has been filled. The valve-type sack is constructed with the top closed except for a small opening left for machine filling. The open corner sack is machine loaded and is sewed and taped after it has been filled.

Two additional terms are applied to paper shipping sacks, namely *baler* and *bulk* sacks. A *baler* sack is a paper shipping sack constructed to enclose smaller units packed in multiples. A *bulk* sack is a sack into which the contents are placed without employing interior packages.

Pasted bottom, open mouth baler and bulk sacks are made without a thumb cut. The factory-closed end is adequately and securely pasted with a waterproof adhesive. This adhesive is used in fastening the several plies of paper together to prevent sifting and to provide a strong bottom closure, as well as to secure the bottom folds of the sack. Spot pasting between the various plies of paper may be provided at or near the top or open end of the sack to facilitate opening of the sack for filling.

Sewed bottom, open mouth baler and bulk sacks are made without a thumb cut. The bottom end is sewed through all plies of paper at a minimum distance of $\frac{1}{2}$ in. and not more than $\frac{3}{4}$ in. from the end. A strip of tape is frequently incorporated into or over the sewed end of the sack in such a way that it folds over the end to control sifting.

Pasted valve sacks are factory closed on both ends by means of a waterproof adhesive, which is used on all seams in such a way that the folds of the closures on each end are adequately secured. The successive plies of paper forming the sack are securely nested one to the other, except that the folded closure at one corner of the sack is not pasted shut.

Sewed valve sacks are factory sewed at both ends, except that a filling valve is provided in one corner of the sack. The sewing on both ends, as furnished by the manufacturer, is through all plies of paper at a distance of $\frac{1}{2}$ in. to $\frac{3}{4}$ in. from the ends. A strip of tape is also incorporated into or over the ends of the sack in such a way that it folds over the ends to control sifting.

Sewed open corner sacks are factory closed by the manufacturer at both ends, except that the sewing of the top closure does not close the mouth of the sack for a distance suitable to permit filling. The sewing on both ends of the sack, as furnished by the manufacturer, is performed similarly to the sewed valve sack. If the sewing does not close the mouth of the sack (to provide the open corner), the tape is continued across the end of the sack to serve as a reinforcement for the closure to be applied when the sack is filled.

In Fig. 4-4, ten common styles of multi-wall paper shipping sacks are

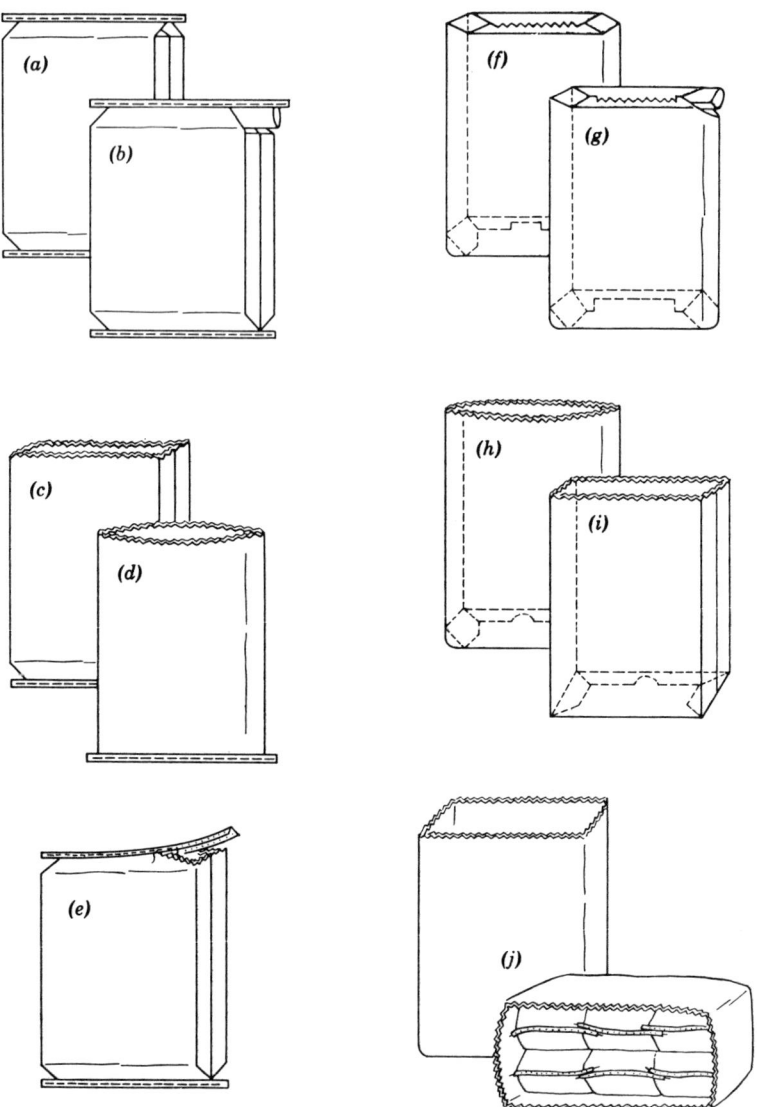

Fig. 4-4. Common styles of multi-wall shipping sacks.

illustrated. All of these are made with weight capacities up to 100 lb and differ in the method in which they are manufactured with respect to filling and sealing requirements and basic variations in style. A brief description of each style is presented below:

SEWN, GUSSETED, VALVE-TYPE SACK (FIG. 4-4a). This sack is completely fabricated in the bag factory and is made with a small opening or valve in one corner through which the sack is filled. The valve is

SHIPPING SACKS

made with an inner flap which operates as a self-closing check valve as soon as the sack has been filled.

SEWN, GUSSETED, VALVE-TYPE SACK WITH OUTER SLEEVE (TUCK-IN-SLEEVE) (FIG. 4-4b). This sack is similar to the previous one, except that after the sack is filled a sleeve is folded manually down and back into the pocket under the sleeve making a securely held positive closure against sifting.

SEWN BOTTOM, GUSSETED, OPEN MOUTH TYPE SACK (FIG. 4-4c). Only the bottom closure is fabricated by the sack manufacturer. After filling, the top closure is usually made by sewing, although wire-tying, stapling, gluing, or gummed tape may also be used.

SEWN BOTTOM, FLAT TUBE, OPEN MOUTH TYPE SACK (FIG. 4-4d). This sack is similar in construction to the foregoing type, except that it is a flat tube. The top may be closed by any bag-sealing technique.

SEWN BOTTOM, SEWN TOP, OPEN CORNER SACK (FIG. 4-4e). These sacks are used for commodities that are liquid when poured hot and that solidify into a block which takes the shape and size of the sack when cooled. Filling is usually accomplished through a pipe inserted in the open corner of the sack. The plies at the open corner are then brought together and stapled, with tape applied over the closure for further reinforcement.

PASTED VALVE-TYPE SACK (FIG. 4-4f). Similar to the sewn, gusseted, valve-type bag, this sack is completely fabricated in the bag factory. A corner valve facilitates filling with automatic valve-filling equipment.

PASTED VALVE-TYPE SACK WITH OUTER SLEEVE (TUCK-IN-SLEEVE) (FIG. 4-4g). This sack is similar to the sewn bottom, gusseted, open valve-type sack except that it is of pasted construction.

PASTED SATCHEL BOTTOM, OPEN MOUTH SACK (FIG. 4-4h). This sack is made from a flat tube, non-gusseted, with a factory-made satchel-type pasted bottom. After the sack is filled, closure is made with twine or with wire which can be twisted manually or by machine. Other types of closure such as gluing, stapling, taping, or sewing can also be made.

AUTOMATIC PASTED BOTTOM, OPEN MOUTH SACK (FIG. 4-4i). This sack is gusseted with a factory-made, square-type, pasted bottom. After filling, a top closure is made similar to other top-open bags.

PINCH TYPE SACK. This is a relatively new sack construction still in its early stages of development. The underside of the stepped folds of the manufacturer's closure of the sack are continuously sealed with an adhesive to the opposite face of the sack. The outer wall of the sack is stepped at the top and bottom fold over flap beyond all inner plies to provide a positive seal over the ends of the inner walls. The fold line is 1-5/8 inches ± 1/4 inch below the top edge of the long side of the sack and not less than 1/4 inch below the top edge of the innermost ply of the short side of the sack. The outer edge of the long side of the sack extends not less than 3/4 inch beyond the topmost edge of the short side of the sack. The closure may not have peeled corner or corners, open channels, or unsealed edges greater than 3/16 inch.

BALER (FIG. 4-4j). As described previously, this is a pasted bottom, open mouth sack designed for shipment of groups of small paper shipping sacks or packets of such things as sugar, flour, salt, rice, and beans usually of 2-lb to 10-lb capacity. The individual packets are closed by sewing or pasting. The capacity of the baler is generally not more than 60 lb. The advantage of the baler is that it provides a tight solid compact unit load with little opportunity of movement of the packets within the baler during handling and shipment. Baler closures are made by sewing or pasting and must hold the packets firmly in place.

FILLING AND WEIGHING OF PAPER SACKS

The ease by which multi-wall paper shipping sacks can be filled is partly responsible for their extensive use. There are three general types of bag-filling equipment: (1) gravity feed, (2) auger feed, and (3) valve packer.

Gravity feed. In this system the commodity falls by its own weight into the open mouth bag. This method of filling is generally only suitable for free-flowing commodities and when high speed is not required. Any type of hopper device may be used for gravity feeding.

An integrated bagging station operating on the gravity-feed principle and utilizing open mouth bags is illustrated in Fig. 4-5. The material flows from a supply hopper through an agitator to the feeder conveyor, which carries the material to the weigh hopper. This hopper is suspended from scale levels. When approximately 90 to 95 per cent of the desired package weight is contained in the weigh hopper, the material flow is reduced by approximately 75 per cent. After the exact weight (with certain allowances) is reached, the complete material flow into the weigh hopper is stopped. The hopper gates are then opened (simultaneous with this action the feeder conveyor is stopped) and the material drops through the bag spout into the bag. The bag is held on this spout lightly with one hand while the operator reaches for another bag with the other hand. After the bag is filled, the bag drops onto a moving conveyor which transports the bag to the closing station. Closure on this unit is accomplished by sewing. Rates up to 20 bags per minute are attainable with this integrated system.

Auger packing. Here the commodity is pushed by means of a power-driven auger into an open mouth bag. This type is used for non-free-flowing materials such as flour (Fig. 4-6).

Valve packers. These are used only with valve bags and are generally

SHIPPING SACKS

made of four basic types. These include *gravity feed type, belt-type, auger-type,* and the *impeller-type.*

The *gravity feed* valve-packing machine is used for free-flowing commodities such as silica sand, grits, beans, and peas. The belt-type is suitable for such items as sugar, fertilizer, and miscellaneous granular material. It operates by pouring the material into a channel between a moving valve and a rotating pulley. The movement of the valve and pulley throws the entrapped material into the bag.

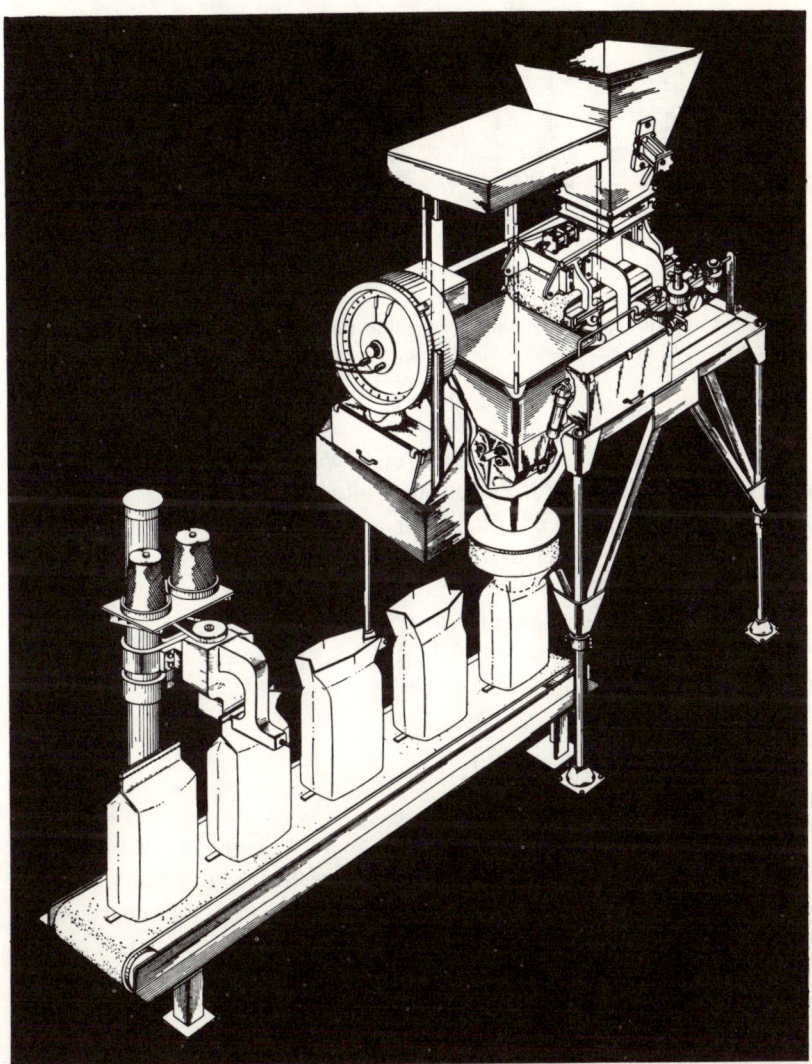

Fig. 4-5. Typical integrated filling, weighing, and closing station. Courtesy Union Camp Corp.

Fig. 4-6. Integrated, semi-automatic mechanism for filling, weighing, and closing open mouth multi-wall shipping sacks using the auger principle. Courtesy Bemis Co.

The *auger-type* is used for non-free-flowing commodities and uses a horizontal auger which forces the item into the bag.

The *impeller type* is obtainable in a horizontal or vertical model in which an impeller wheel drives the product into the bag through the valve-filling spout. A cradle supports the bag and is generally balanced on a scale which is so designed that when the proper weight is packed into the bag the flow is cut off. This type is generally used for cement, limestone, lime, and other similar materials.

WEIGHING

Weighing is frequently incorporated as part of the filling operation, as has been described with the impeller-type filling equipment.

There are basically two types of weighing devices: (1) *net weighers,* which provide a means of weighing the commodity before it is loaded into bags and (2) *gross weighers,* which weigh the material and the bag as it is being filled.

Both of these types may be fed by either gravity or power.

Net weighers are made to accommodate many sizes, ranging from a

SHIPPING SACKS

fraction of an ounce to 200 lb or more and may be designed for manual or automatic discharge. Gross weighers are normally used for medium and heavy weight packages.

In Fig. 4-7 a typical net weigher is shown. This unit can be used singly or as a multiple batch unit proportioning system. In multi-unit systems it can be interlocked with other scales in the installation to discharge only when all weighings of various ingredients have been completed. The scale may also be arranged mechanically for continuous weighing, which increases productivity owing to a definite cycling time requiring the operator to keep up with the machine pace.

Fig. 4-7. Typical automatic net-weighing scale which can be incorporated with various types of packers, take-away conveyors, and sewing machines for a completely coordinated bagging system. Courtesy Consolidated Packaging Machinery Corp.

190 DISTRIBUTION PACKAGING

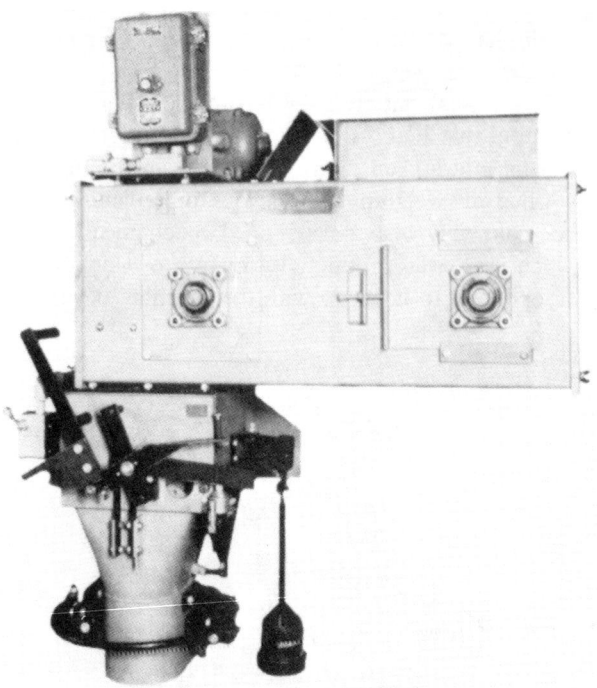

Fig. 4-8. Typical gross weigher with feeding mechanism. Courtesy Richardson Scale Co.

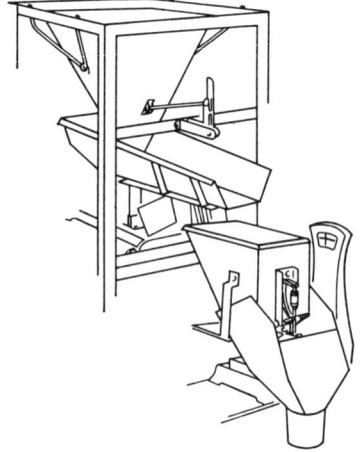

Fig. 4-9. Simple gross weigher with vibratory feeder. Courtesy The Exact Weight Scale Co.

SHIPPING SACKS

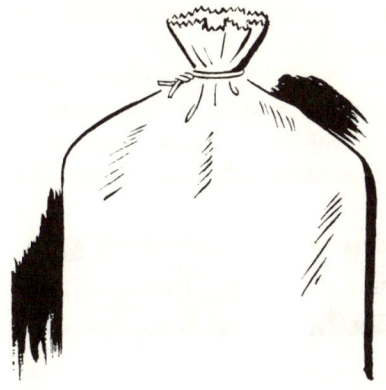

Fig. 4-10. Tied sack closure.

In Fig. 4-8 a typical gross weighing unit is demonstrated. The sack is manually slipped over the bag spout and is gripped in position by a bag holder. The operator then pulls the gate handle which also starts the belt feeder. When the beam of the scale comes to balance, the belt feeder stops and the cut-off gate is closed. Such a machine is completely automatic except for manual application, removal of the sack, and initiation of the cycle by opening of the inlet gate.

A simple gross weigher in conjunction with a vibratory feeder is illustrated in Fig. 4-9. This scale is similar in operation to the previously described type.

MULTI-WALL SACK CLOSURES

The principal types of sack closures have already been discussed briefly in the description of the various styles of shipping sacks. A further description follows.

The three types of closures accepted by industry are *tying, sewing,* and

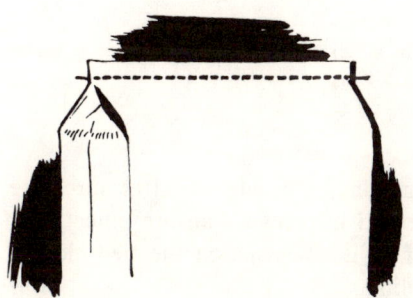

Fig. 4-11. Sewn sack closure.

192 **DISTRIBUTION PACKAGING**

pasting. Others such as stapling or using gummed or pressure-sensitive tape may occasionally be used in special instances.

Tied closure. This closure (Fig. 4-10) is diminishing in popularity and is generally preferred for small-volume operations in which the investment in bag-sewing equipment is not warranted or in which valve-packing methods are not suitable. Either twine, which is always manually tied, or wire, which may

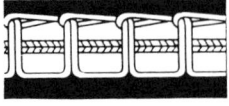

Single thread chain stitch for closing textile and lightweight paper bags.

Two-thread double lock chain stitch for more secure closure of textile and multiwall paper bags.

Two-thread double lock chain stitch equipped for sewing through rip cord placed on the looper side of the bag closure. When ripped, all broken thread strands remain in rip cord and not in bag top.

Single thread chain stitch using twisted kraft paper reinforcing cord laced into stitch on needle side to strengthen closure. For closing light and heavy duty paper bags.

Single thread chain stitch applied through cotton filler cord on needle side of closure restricting minor sifting while strengthening seam. For closing light weight paper bags.

Single thread chain stitch incorporating flat paper filler cord under needle side of seam. For closing light weight paper bags.

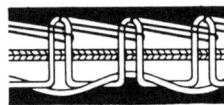

Two-thread double lock chain stitch using twisted kraft paper reinforcing cord laced into stitch on needle side. For closing all types of paper bags.

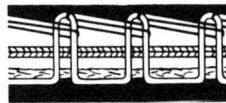

Two-thread chain stitch with needle thread applied through cotton filler cord. For closing all types of paper bags.

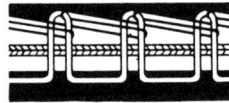

Two-thread double lock chain stitch having flat paper filler cord under needle thread restricting sifting and strengthening seam. For closing all types of paper bags.

Fig. 4-12. Typical examples of available stitch types for paper shipping sacks. Courtesy Consolidated Packaging Machinery Corp.

be either twisted manually or by machine, is used. Tied closures are more subject to damage in handling and in transit than are other types of closures. Frequently on bags of light construction the tied closure may be torn off from the bag in handling.

SHIPPING SACKS

Sewn closure. This closure provides for a tighter seal especially for products in which sift-proofness is important. A single tape bound over the top edge of the bag adds sift-resistance as well as strength. A variety of tapes are available for this purpose. Most sewn closures are made with cotton thread. (Fig. 4-11.) A number of different stitch types are available depending upon the type of sewing machine used and the purpose of the stitching operation. (Fig. 4-12.)

The above methods are adapted to open mouth, pasted or sewn bags. The valve bags, of course, do not require any closing method.

Essentially there are two types of bag sewing machines: (1) stationary machines and (2) portable machines. The stationary machines may be incorporated as part of an integrated filling and weighing system with a direct flow provided to the closing station. Such machines are de-

Fig. 4-13. Bag closer (sewing) for open-mouth paper bags. Courtesy Consolidated Packaging Machinery Corp.

Fig. 4-14. Portable bag sewing machine. Courtesy Dave Fischbein Co.

signed for high-speed operation with minimum labor effort. (Fig. 4-13.) A typical portable bag sewing machine is illustrated in Fig. 4-14. Tape may be applied prior to sewing as a separate operation, if required. For stationary use the machine can be suspended with a suitable counterweight.

Sewn closure comparisons. In an attempt to compare the effectiveness of various types of closures for multi-wall bags, experimental field work together with original research was summarized in a paper presented by Mr. T. E. Dowling, American Cyanamid Company.*

The characteristics which are of utmost importance in sack closure are moisture vapor resistance and strength. The moisture vapor resist-

* Presented for the Bag Committee of the Packaging Institute at the 20th Annual Packaging Institute Forum.

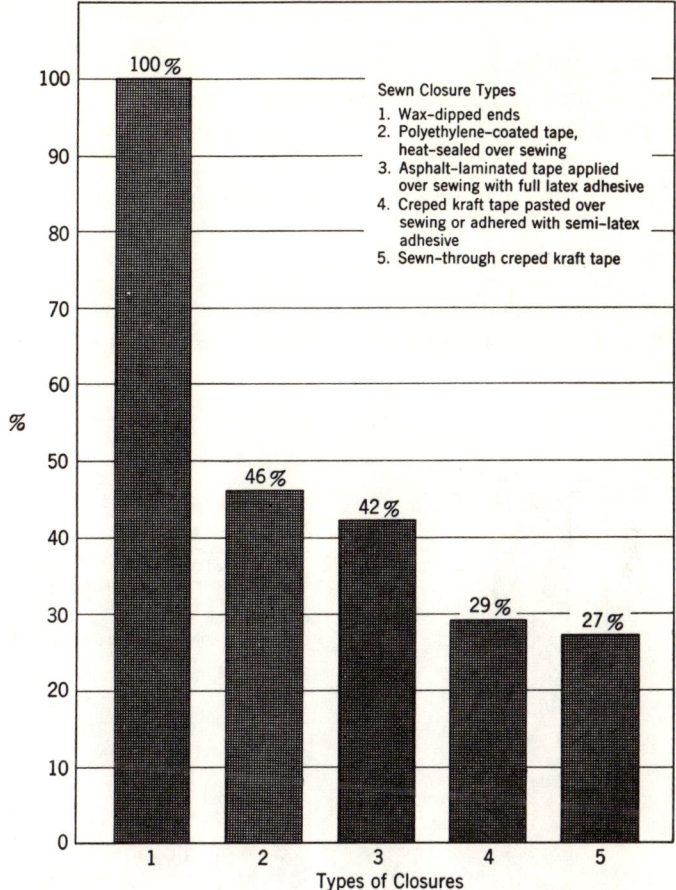

Fig. 4-15. Comparative moisture vapor resistance of five types of sewn bag closures. Courtesy Mr. T. E. Dowling, American Cyanamid Co.

ance is determined by developing moisture vapor transmission rates. In this instance test pouches filled with calcium chloride were subjected to controlled temperature and humidity to develop this information along with cut-out sections of the bag closure itself. For the latter, the procedure employed was a modification of the method customarily used to determine moisture vapor transmission rates of flat sheet stock such as paper, films, and foils.

The strength of the closure was established by means of tensile tests made on sections of the closure. Other methods, of course, are available to rate and compare the effectiveness of sack closures. Tensile tests, however, were judged to be most reliable for this purpose.

The following types of sack closures were investigated:

1. Wax-dipped ends.
2. Polyethylene-coated tape, heat sealed over sewing.
3. Asphalt-laminated tape, applied over sewing with full latex adhesive.
4. Creped kraft tape, pasted over sewing or adhered with semi-latex adhesives.
5. Sewn-through, creped kraft tape.

The moisture vapor tests disclosed that the resistance to moisture penetration was highest with the wax-dipped closure and lowest with the

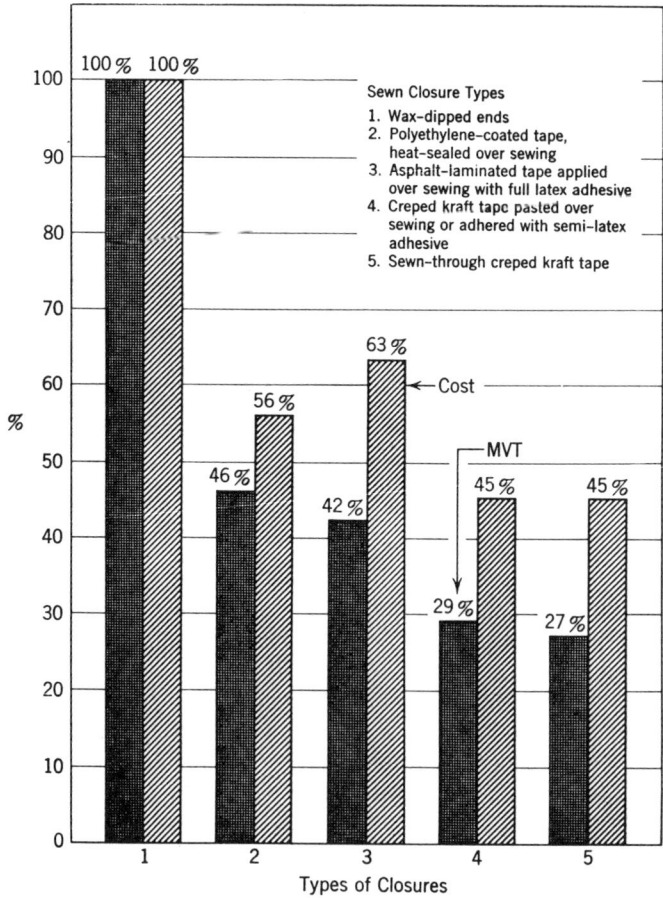

Fig. 4-16. Comparison of moisture vapor transmission rates of five sewn bag closures with cost. Courtesy Mr. T. E. Dowling, American Cyanamid Co.

SHIPPING SACKS

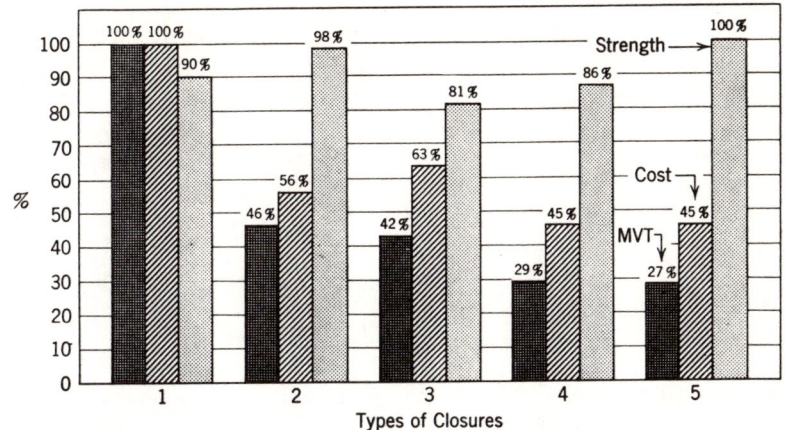

Sewn Closure Types
1. Wax-dipped ends
2. Polyethylene-coated tape, heat-sealed over sewing
3. Asphalt-laminated tape applied over sewing with full latex adhesive
4. Creped kraft tape pasted over sewing or adhered with semi-latex adhesive
5. Sewn-through creped kraft tape

Fig. 4-17. Comparison of moisture vapor transmission rates of five sewn bag closures with cost and tensile strength. Courtesy Mr. T. E. Dowling, American Cyanamid Co.

sewn-through, creped kraft tape closure. In Fig. 4-15 the comparison expressed as a percentage of the most effective closure (wax-dipped considered as 100 per cent), is graphically plotted.

To compare the effectiveness of the closure to cost, additional data, based on a typical sewn valve bag 19 in. wide, were assembled. In Fig. 4-16 cost information, expressed as a percentage of the wax-dipped closure cost, is plotted against the moisture vapor rate performance.

The comparison was further expanded by including the tensile strength data in the graph shown in Fig. 4-17. It will be noted that the sewn-through, creped kraft tape closure had the highest strength rating. Therefore, the strength values are expressed as a percentage of the closure type.

This work exemplifies a sound approach to select a particular packaging technique through a comparison of both performance and cost data. It must be recognized, however, that experimental data of this type are frequently difficult to relate to practical use. Therefore, additional laboratory tests including drop tests and puncture tests should be conducted to confirm such findings and finally, field trials should also be carried out.

TABLE 4-3
Typical Products Packed in Multi-Wall Paper Sacks of Common Construction

Commodity	Weight Packed lb	Bag Construction	Commodity	Weight Packed lb	Bag Construction
Aluminum sulphate	100	$2/40-3/60$	Insecticides (Non poisonous)	50	$1/60$ A.L.*$-1/40-1/60$
Ammonium phosphate	100	$4/40-1/60$	Limestone (Ground)	50	$1/50-1/60$
Asbestos cement	50	$1/50-1/60$	Limestone (Ground)	80	$1/40-2/60$
Barium carbonate	100	$3/40-1/60$	Limestone (Ground)	100	$4/40$
Barytes (Barium sulphate)	100	$2/40-3/60$	Lithopone	50	$2/60-2/60$
Bone meal	100	$3/40-1/60$	Manganese ores	100	$4/40-1/60$
Calcium chloride	100	$2/60-3/75$ A.L.*$-1/40$	Meat scrap	100	$3/40-1/60$
Calcium magnesium chloride	100	$1/60$ A.L.*$-3/40-1/60$	Pebble lime	80	$1/70$ A.L.$-2/40-1/60$
Calcium phosphate	100	$3/60$	Phosphate rock	100	$3/40-1/60$
Casein	100	$4/40-1/60$	Plaster	80	$3/40-1/60$
Cement	94	$3/40-2/60$	Plaster	100	$1/40-3/60$
Cement	94	$1/40-3/60$	Quick lime	50	$1/75$ A.L.*$-1/40-1/60$
Cocoa	100	$1/40-3/60$	Quick lime	80	$1/40-1/75$ A.L.$-2/60$
Corn starch	100	$4/40-1/60$	Rice	100	$4/40-1/60$
Copper sulphate	100	$1/60$ A.L.*$-3/60-1/60$	Rosin sizing	70	$4/40-1/60$
Disodium phosphate	100	$4/40-1/60$	Salt	100	$2/40-1/60$ A.L.*$-1/40$
Dough improver	100	$3/40-1/30-30$ Sisal $-1/60$	Salt	100	$3/40-1/60$
Feldspar	100	$3/40-1/60$	Silica	100	$3/40-1/60$
Fertilizer	100	$2/60$ A.L.*$-3/60-1/60$	Silica	50	$2/40-1/60$
Fertilizer	100	$3/40-1/60$	Soda ash	100	$4/40-1/60$
Fertilizer (Superphosphates)	100	$1/60$ A.L.*$-2/40-1/60$	Sodium bicarbonate	100	$3/40-1/60$
Filter clay	50	$3/40-1/60$	Sugar (Cane, beet, corn, and grape)	100	$4/40-1/60$
Fire clay	100	$2/40-2/60$	Sugar (Granulated & powdered)	100	$1/40-1/40$ Wax$-3/40$ $1/60$
Flour	98	$3/40-1/60$	Talc	50	$1/50-1/60$
Flour	98	$4/40-1/60$	Tankage	100	$3/40-1/60$
Glue	100	$3/40-1/60-1/60$	Whiting	50	$2/40-1/60$
Graphite	80	$2/40-1/60$	Zinc oxide	50	$2/60-1/60$
Ground sulphur	100	$2/40-1/60$			
Ground sulphur	50	$2/40-1/60$			
Hydrated lime	50	$2/50$			
Hydrated lime	40	$1/40-1/60$			

* A.L. indicates asphalt laminated sheet, and the basis weight shown for the sheet includes the asphalt.

MULTI-WALL SACK SPECIFICATIONS

A typical multi-wall sack specification is reproduced in Fig. 4-18. As will be noted, this specification makes reference to compliance with both the *Uniform Freight Classification* and federal requirements and supplements the specific information for the particular sack.

Attention is directed to the instructions concerning the proper method of specifying sack dimensions. Additional information on standard sack measuring for all types of multi-wall shipping sacks is illustrated in Figs. 4-19 through Fig. 4-24, inclusive.

SHIPPING SACKS

The multi-wall sack construction to be specified depends upon the nature of the product, the weight, and the method of shipment. Many construction variations are possible, some of which are developed in Table 4-3 for a number of typical products. In specifying sack construction the plies are oriented from the inside of the sack to the outside. For instance two 40-lb plies, one 50-lb ply, one 60-lb ply (2/40-

Instructions For Obtaining Multi-Wall Bag Dimensions

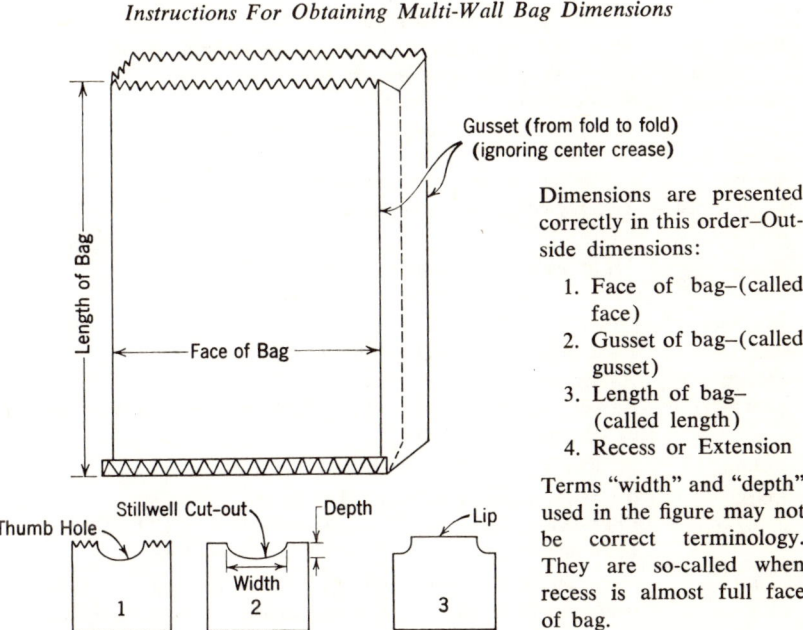

Dimensions are presented correctly in this order—Outside dimensions:

1. Face of bag–(called face)
2. Gusset of bag–(called gusset)
3. Length of bag– (called length)
4. Recess or Extension

Terms "width" and "depth" used in the figure may not be correct terminology. They are so-called when recess is almost full face of bag.

Ignore recess or extension in determining basic bag dimensions.

When bags contain either conditions above (1), (2), or (3)—give the dimensions of the recess or extension as the last figure with appropriate explanation. This will include max "width" and "depth" of recess or extension.

Valve Bag: Give dimensions ordinary way—full length; with an "x" in. upper right-hand valve as the case may be. "x" = distance from line of sewing to bottom crease of valve.

Pasted Open Mouth: (loosely termed "satchel–bottom"). Give "face" by "length" by "bottom." (Tube size = 4 in.). Tube size is length from upper serrated edge to lower serrated edge which is approximately in center of "satchel bottom." It is measured from upper serrated edge to bottom fold of bag on side without "bottom" and other from this fold to the serrated edge within the "bottom."

Fig. 4-18. Typical multi-wall sack specification. Courtesy Stauffer Chemical Co.

STAUFFER CHEMICAL COMPANY	PRODUCT–TOXAPHENE 20 DUST
PACKAGING SECTION	SPEC. NO.–1–1–071R1
TRAFFIC DEPARTMENT	50 POUNDS NET
CONTAINER SPECIFICATION	MFRD. AT TAMPA
	SHEET 1 OF 1

Multiwall Sack

Size: 15½ in. × 3 in. × 30 in.

Type: Sewn Valve (gusseted, tuck-in sleeve)

Construction: Shall comply with U.F.C. Rule 40, Section 10 (c)
  Plies – $2/40$, $1/50$, $1/60$, all natural kraft
  Paper – Shall conform to the requirements of Federal Specifications UU–S–48 of the latest issue in effect
  Glue – Standard glue throughout
  Ends – Sewn, standard stitch over tape
  Tape – 70# flat natural kraft
  Thread – $5/12$ cotton needle, $4/12$ cotton looper
  Filler Cord – Two ply cotton on needle side

Valve:
  Size – 3¾ in. width, ⅞ in. notch, 2¾ in. extension
  Sleeve – Tuck-in type, 60# creped natural kraft
  Location – Upper right hand

Printed: Face, gussets, and back printed in three colors
  Design – As per approved present Format "E"
  Text – As per approved present copy of file with the Labeling Section at Chauncey, New York
  Colors – Stauffer YELLOW, BLACK and WHITE
  Marks – To be printed near the top of left gusset –
    (1) U.F.C. stamp   (3) SCC 1–1–071R1
    (2) 4/190 ANK   (4) Purchase Order No.

Packing: For C/L or T/L shipments: Palletize 3,500 bags per pallet, overwrap each pallet and mark clearly all four sides with the following information:
  (1) Tox. 20 Dust   (4) Quantity of bags on pallet
  (2) SCC 1–1071R1   (5) Date of manufacture
  (3) Purchase Order No.

Weight: 506# per 1,000 bags, approximately

Quantity: Approximately 65,000 bags per minimum carload

Requirement:
1. Proof of text and design to be approved by Labeling Section, Chauncey, N. Y., before production is started.
2. Supplier to send to Stauffer Chemical Company, Chauncey, N. Y.
  Att: Packaging and Container Section–2 outturn samples
  Att: Labeling Section–100 tear sheets
3. Adherence to all provisions of this Specification

Revised: To change ply, tape, thread, valve, and printing specification.

Fig. 4-18. (*Continued*)

SHIPPING SACKS

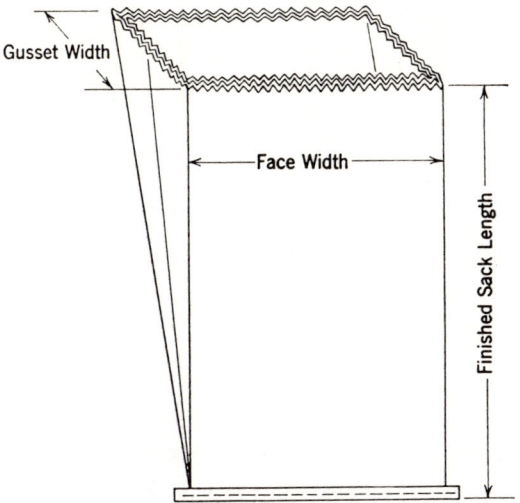

Outside measurement of sack should be specified as follows:

Face Width___, Gusset Width___, Finished Sack Length___,

Fig. 4-19. Specifying outside measurements of sewn open-mouth *gusseted* shipping sack. Courtesy Paper Shipping Sack Manufacturers Association, Inc.

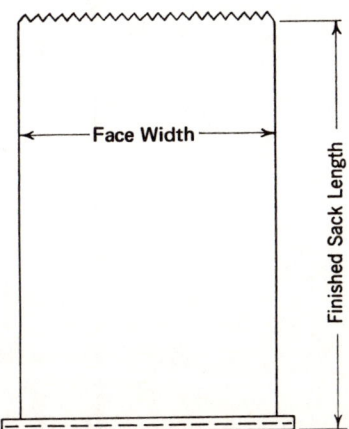

Outside measurement of sack should be specified as follows:

Face Width___, Finished Sack Length___,

Fig. 4-20. Specifying outside measurements of sewn open-mouth *flat tube* shipping sack. Courtesy Paper Shipping Sack Manufacturers Association, Inc.

202 **DISTRIBUTION PACKAGING**

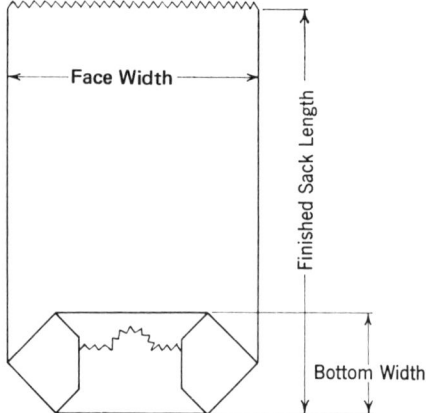

Outside measurement of sack should be specified as follows:

Face Width___, Finished Sack Length___, Bottom Width___,

Fig. 4-21. Specifying outside measurements of pasted open-mouth *satchel-bottom* shipping sack. Courtesy Paper Shipping Sack Manufacturers Association, Inc.

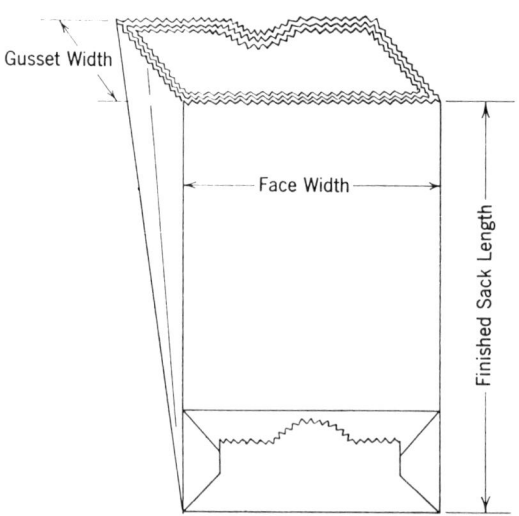

Outside measurement of sack should be specified as follows:

Face Width___, Gusset Width___, Finished Sack Length___,

Fig. 4-22. Specifying outside measurements of pasted open-mouth *automatic* (S.O.S.) shipping sack. Courtesy Paper Shipping Sack Manufacturers Association, Inc.

SHIPPING SACKS

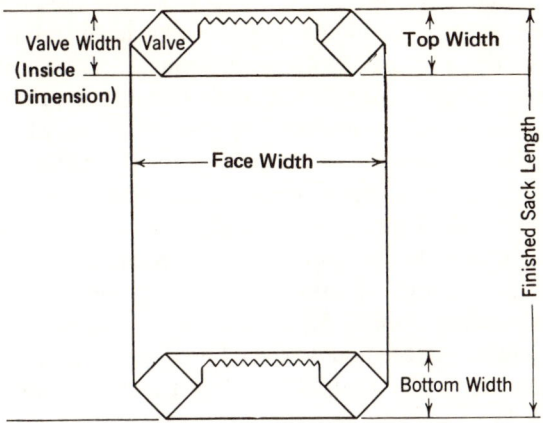

Outside measurement of sack should be specified as follows:
Face Width___, Valve Width (Inside Dimension___,) Top Width___,
Finished Sack Length___, Bottom Width___,

Fig. 4-23. Specifying outside measurements of pasted *valve* shipping sack (without tuck-in sleeve). Courtesy Paper Shipping Sack Manufacturers Association, Inc.

1/50-1/60) construction consists of two inner plies of 40 lb (24 in. × 36 in., 500 sheets) kraft, one intermediate 50 lb ply and an outer ply of 60 lb. As a general rule the heavier weight ply is outermost. Treated or laminated plies are usually oriented either innermost or in an intermediate location.

Some specifications incorporate performance requirements of the sack in addition to the test requirements for the paper from which the sack is made. Performance tests for loaded sacks are almost exclusively drop tests. This test rates all elements of the completed package as prepared for shipment.

STORAGE OF EMPTY PAPER SACKS

Paper shipping sacks provide optimum service in packing and closing operations when the moisture content of the paper is 6 to 7 per cent. To insure that empty sacks retain their moisture they should be stored in a fairly humid atmosphere. The recommended average good storage conditions are 70° F at 50 to 60 per cent relative humidity. Somewhat higher humidities are not necessarily deleterious. At 70° F, 50–60 per cent relative humidity, paper sacks have about 6.5 per cent moisture content and are flexible and strong. As shipping sack paper loses moisture below 6 per cent it tends to become brittle. Brittleness can be detected by the rattling dry paper sound made when the sacks are shaken briskly by hand. When brittleness is encountered, the sacks should be placed in a mechanically humidified room or placed on a covered loading platform in rainy weather for one or two days to permit the sacks to regain the lost moisture. As mentioned in previous chapters, paper is sensitive to atmospheric conditions and *loses or gains* moisture depending upon the moisture present in the surrounding air. Therefore, empty sacks should never be stored in hot, dry rooms, near heating units, nor under roofs where dry attic heat absorbs the moisture from the sacks.

In addition, empty sacks should never be stored at temperatures below freezing. In extremely cold weather, the moisture content of the air itself is generally very low. Paper sacks stored under these conditions surrender moisture.

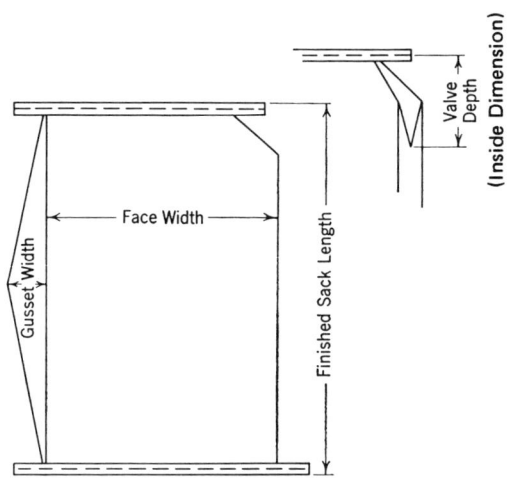

Outside measurement of sack should be specified as follows:
Face Width___, Gusset Width___, Finished Sack Length___,
Valve Depth (Inside Dimension)___,

Fig. 4-24. Specifying outside measurements of sewn *valve gusseted* shipping sack (without tuck-in sleeve). Courtesy Paper Shipping Sack Manufacturers Association, Inc.

SHIPPING SACKS

If weather conditions or building limitations make it impossible to have ideally moist, warm storage, it is economical for users of paper shipping sacks to install humidifiers in their sack storage rooms.

GUIDES TO THE SELECTION OF PAPER SACKS

The major factors that must be considered in the final selection of the sack type to be used are as follows:

1. Product characteristics.
2. Economy.
3. Filling, weighing, and closing.

Product characteristics. Product characteristics include such considerations as the flowability of the product, which affects bag type, the hydroscopicity of the product, which affects material selection, and the siftability of the contents, which affects construction and closing.

In determining the suitability of a multi-wall paper sack for the shipment of a given product, the value of the product is often used as the main criteria.

Fig. 4-25. Natural frequency vibrating bag flattener. Courtesy Carrier Conveyor Corp.

One chemical company has selected an arbitrary value of their bulk chemicals of $3.50 per lb as the limit for shipping products in multi-wall shipping sacks. Higher priced chemicals are packed in rigid containers such as fibre drums. This company has recognized the potential danger of product loss due to puncturing and sifting which can occur with sack packaging. Similarly, an extra paper ply or heavier plies may be used to safeguard more expensive merchandise. For overseas shipments, most sacks are constructed with at least one ply of moisture-proof asphalt-laminated kraft paper.

Economical factors. Economical factors include such considerations as carload purchasing, proper warehousing, inventory, and proper sizing, and the selection of the lightest grade sack to satisfactorily carry the product.

Filling, weighing, and closing. In these operations the greatest savings in cost can be realized. Several machine types have been illustrated previously. Additional equipment to prepare sacks for shipment is available. This equipment includes bag flatteners of the vibratory (Fig. 4-25) and compression type and other aids for palletizing. Glue applicators to secure palletized or bulk-loaded shipments assist in reducing shifting in transit and minimize damage resulting therefrom.

SACK SPECIALTIES

Several developments and improvements in the paper sack field have further enhanced the performance characteristics and potential of paper shipping sacks.

One sack manufacturer has developed a construction technique which applies additional paper reinforcement to the vulnerable gusset and edge areas of conventional multi-wall shipping sacks. The application of this type of reinforcement is claimed to permit the use of one less ply than that required for a conventional sack designed for the same purpose. It is further claimed that there is no sacrifice in performance and that a considerable cost savings results.

Another specialty sack jointly developed by three industrial firms has exhibited widespread application. This paper sack is reinforced by a mesh of glass fibre yarn which is incorporated into the paper itself. Special resins are also incorporated into the paper to provide wet strength characteristics. The glass yarn, introduced when the paper is in the pulp state, thus becomes an integral part of the paper. These reinforced sacks have been considered replacements for conventional

SHIPPING SACKS

canvas mail sacks and acceptable for the export packaging of chemicals and cement. For further information on reinforced paper shipping sacks refer to *Federal Specification* PPP-S-50a.

The development of extensible paper and its application to sack fabrication has greatly improved the performance of sacks at lower cost. The stretchability of this paper assists in the reduction of puncture failures and permits either lightweight plies or fewer plies to be specified. Similarly the drop resistance of sacks made from extensible paper is higher since the paper yields under impact shocks reducing rupture hazards.

Another specialty-type bag is the stepped-end bag. This bag type is constructed by cutting back each succeeding ply so that when the ends of the tube are folded and adhesive is applied to form the valve end and the bottom, each ply on the face of the tube is pasted to its corresponding ply on the back of the tube. It is claimed that this construction provides greater flexibility of the valve end and valve-closing devices, better sift-proofness, and greater opportunity to employ special barrier sheets not adaptable to standard pasted valve paper shipping sacks.

CHAPTER 5

Nailed Wooden Boxes

and Crates

    The wooden box and crate was the first type of modern shipping container used for the distribution of all types of manufactured products and many raw materials. In the early days of modern shipping, lumber was plentiful and inexpensive and box manufacturers, like nearly all users of wood, demanded high grades and ignored inferior materials. Little attention was given to designing boxes so as to obtain the maximum strength with the minimum amount of materials. The constant depleting of forest areas and the ever-increasing demand for lumber raised the prices of higher lumber grades and forced the boxmaker to use the lower and cheaper grades. In addition, the competition from other types of materials, primarily fibreboards and composition materials such as plywood and paper-overlaid veneer, has further been responsible for the design of wooden containers at the least possible cost. Furthermore, ever-increasing freight costs are demanding the least tare weight and cubic capacity of shipping containers.

    In spite of the many competitive types of containers and materials which have gradually reduced the over-all consumption of nailed wooden boxes and crates in the United States, many millions of such containers are still being used. Products carried in such containers vary from fruits and vegetables to large machinery and other materials requiring superior protection in shipment and distribution. Wooden boxes and crates are manufactured throughout the United States, with large mills primarily located in the southern and western lumber regions close to the source of the grades utilized for box construction. Modern box manufacturers are equipped with wood-fabrication machinery which permits small tolerances in the fabrication of these containers. These are shipped in knocked-down form, generally termed *shook,* to the user, who assembles the container prior to packing the commodity into it.

Wooden boxes and crates are also quite frequently manufactured by the shipper himself. Usually this practice exists only when small quantities of containers are required, such as for export or for tailor-made containers. Plant box shop facilities range from well-organized box departments operating at high efficiency with modern equipment to small operations in which a carpenter, often an all-around maintenance man, uses hand saws and hammers to fabricate the required container. The local availability of lumber places the making of such containers within the reach of any manufacturer. When box shops are undesirable or uneconomical for small quantities of containers, local box container manufacturers are available to supply small quantities to the container user. Such boxes are generally delivered set up, inasmuch as the freight constitutes a minor expense, in relation to the over-all cost of the container.

Through the efforts of the Forest Service the first recorded laboratory tests for the improvement of wooden shipping containers were made in 1905 in cooperation with Purdue University. After the establishment of the Forest Products Laboratory at Madison, Wisconsin, in 1910, research on container design was greatly expanded. During World War I and again during World War II a great deal of research by the military was conducted on proper wooden container designs. For this reason a great deal of basic information concerning the most economical construction of such containers and the required design characteristics for the protection of a given commodity has been made available.

WOOD CHARACTERISTICS

Wood constitutes the basic raw material from which wooden containers are fabricated. The cost of a container and its strength characteristics are greatly related to the *type of wood, its quality, its thickness,* and *the workmanship* that is used in fabricating the raw material into

NAILED WOODEN BOXES AND CRATES

the finished container. For this reason a basic understanding of lumber characteristics as used in shipping container construction is mandatory.

WOOD GROUPINGS

Tests and shipping experience have shown that the strength of a wooden container is very largely dependent on the type of wood used in its construction. The durability of wood and its ability to withstand shock and impact stresses are important properties in the selection of wood for boxes and crates.

All woods fall into two general categories: (1) *soft woods,* which come from coniferous or needle-bearing trees, and (2) *hardwoods,* which come from deciduous broad-leaf trees. Further classifications of species has been made dividing the species into four large groups. They have been arranged so that all the species of wood in any given group have approximately the same properties, including *nail holding power, shock resistance,* and *resistance to splitting.* When one group is specified, any species in that group may be used. These groups are listed in Table 5-1.

TABLE 5-1
GROUPING OF COMMERCIAL BOX WOODS

	Group I	
Alpine Fir	Cottonwood	Redwood
Aspen	Cucumber	Spruce
Balsam Fir	Cypress	Sugar Pine
Basswood	Jack Pine	Western Yellow Pine
Buckeye	Lodgepole Pine	White Fir
Butternut	Magnolia	White Pine
Cedar	Noble Fir	Willow
Chestnut	Norway Pine	Yellow Poplar
	Group II	
Douglas Fir	Larch	Southern Yellow Pine
Hemlock	North Carolina Pine	Tamarack
	Group III	
Black Ash	Pumpkin Ash	Sycamore
Black Gum	Red Gum	Tupelo
Maple (soft or silver)	Sap Gum	White Elm
	Group IV	
Beech	Hickory	Rock Elm
Birch	Maple (hard)	White Ash
Hackberry	Oak	

Group I woods. *Group I woods have little tendency to split as a result of nailing;* they have *moderate nail-holding power, moderate strength as a beam,* and *moderate shock resisting capacity.* They are soft, light in weight, easy to work, hold their shape well after manufacture, and, as a rule, are easy to dry.

This group, like Group II, requires thicker material than Groups III and IV for comparable strength characteristics. Its principal difference, as compared with Group II, is that the woods have less tendency to split than the Group II woods. Hence, the Group II woods require smaller nails, which, in compensation, must be spaced at shorter intervals.

Group II woods. These woods consist entirely of the *heavier coniferous* types, which have a pronounced difference in hardness between the spring wood (the lighter colored portion of each ring of annual growth) and summer wood (the darker portion of the ring). They have greater nail-holding power than Group I woods but greater tendency to split in nailing, since both nail-holding and splitting increase with hardness. The hard bands of summer wood sometimes have a tendency to deflect nails and cause them to run out at the side of the cleats.

Group III woods. These are *hard woods of medium density.* They have approximately the *same nail-holding power and strength as a beam* as the *Group II woods,* but they are *less* inclined to *split* or shatter on impacts. These are the most useful woods for box ends and cleats. They also furnish most of the rotary-cut veneer for wirebound boxes.

Group IV woods. Group IV is made up of the *dense hardwoods species.* These species have great *shock-resisting capacity* and *nail-holding power,* but because of their extreme hardness they present difficulties in the driving of nails. They also have the greatest tendency to split at the nails. Group IV woods are generally used in manufactured wooden containers for which high-speed wood working equipment is used. The problems encountered with driving of nails and splitting of lumber is less severe when proper fabrication equipment is used.

FACTORS INFLUENCING THE STRENGTH OF WOOD

The performance of wood sections, when fabricated into a box, in resisting stresses in tension, in compression parallel or perpendicular to the grain, in shear, in bending or in shock-resisting ability, is influenced

NAILED WOODEN BOXES AND CRATES

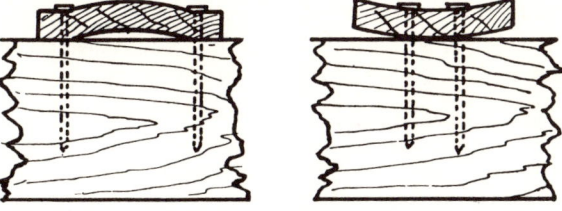

Fig. 5-1. Checking and cupping of green lumber.

by a number of factors. These factors are important, regardless of the wood grouping.

Moisture content. The moisture content of wood is one of the principal factors affecting its strength. As wood dries, most of its strength properties are increased. This increase in strength, however, does not occur until the drying has reached the fibre-saturation point, the condition in which the cell cavities are empty, but the cell walls are fully saturated. This condition usually prevails when the lumber has a moisture content of 25 to 30 per cent. Clear material of the thicknesses ordinarily used for containers dried to 12 per cent moisture content may be twice as strong in bending as green material, which may have a moisture content as high as 250 per cent. If the lumber is kiln-dried to 5 per cent, its bending strength may be tripled.

When green lumber is used for wooden boxes and crates, it may *check* or *cup,* as shown in Fig. 5-1. Checking or cupping causes loosening of the nails and consequent weakening of the container. Green lumber also increases freight charges. For example, a wooden box having a tare weight of 12 lb when made with green lumber with a moisture content of 100 per cent would be reduced in weight to 6.9 lb, if seasoned lumber with a moisture content of 15 per cent were used. This equates to a saving of 5.1 lb in tare weight with a corresponding savings in freight charges.

Moisture content can be readily determined by the following technique:

1. Establish the weight of the section of lumber (preferably not less than $\frac{1}{2}$ in. thick).
2. Remove all moisture from the wood by suitable drying processes. This removal may consist of oven drying or maintaining the sample at a temperature of 212° F for the required time.
3. Weight determination should be made periodically until all moisture is driven from the wood. When several repeated identical weights

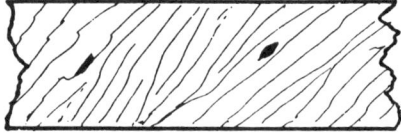

Fig. 5-2. Extensive cross grain of a lumber section.

are recorded, the sample is considered *bone dry,* that is, without moisture and at equilibrium with a zero moisture content at 212° F.

4. Weigh the bone dry specimen and subtract this from the original weight of the lumber sample. The difference in weight can then be either divided by the bone dry weight (*dry weight basis*) or the original weight of lumber (*wet weight basis*) to establish the moisture content in per cent of the original sample.

Defects in lumber. Boxes and crates are ordinarily made of low-grade lumber containing defects, some of which seriously affect the strength properties important in container construction. These defects should be kept within reasonable tolerances, as discussed here.

CROSS GRAIN. In Fig. 5-2, a bad cross grain section of lumber is illustrated. Cross grain is defined as wood cells or fibres which do not run parallel with the axis or sides of a piece of lumber. It is one of the most serious defects reducing strength in bending and increasing the susceptibility of wood to splitting in nailing. Divergence of the grain in any board more than 1 in. in 10 in. of length is not desirable.

KNOTS. The weakening of the bending strength of lumber is nearly proportional to the effective knot diameter as measured across the width of the board. A knot is a portion of a branch or limb that has become incorporated in another branch or in the body of the tree. The wood fibres running out into the limb and those passing around the limb and continuing in the main body produce cross grain. The weakening effect of knots results mainly from the cross grain around them. Knots weaken board most if they are in the *middle third* of the length of the board. At no time should the knot diameter exceed *one third the width* of the board. In Figure 5-3, excessively large knots are illustrated. Knots of this size should be eliminated from all construction pieces.

CHECKING. Checking is caused by stresses introduced by non-uniform shrinkage. End checking is caused by wood drying more rapidly at the ends than away from the ends. This condition, can often be avoided by painting or coating the ends to retard their drying or by reducing the circulation of air around the ends. Checks reduce the holding power of nails and may result in splits running the full length of the piece.

NAILED WOODEN BOXES AND CRATES

MISCELLANEOUS WOOD DEFECTS. Defects of this type include *bark* present on the lumber, *cupping, bowing* and *twisting, case hardening, collapse, discoloration, decay,* and *insect attack.*

Bark on lumber is not entirely objectionable except when used for frame members or diagonal crate braces in which the full cross section of the board is essential in resisting bending stresses. *Cupping* is the curvature of lumber across the grain or width of a piece, which gives it a trough-like appearance. Cupping results when one side of the board dries more rapidly than the other. This condition is usually temporary. When plain, sawed lumber is dried with insufficient weight on it or is improperly *stickered,* permanent cupping takes place.

Improper drying techniques are also responsible for conditions of board *bowing* and *twisting,* which may occur in lumber having spiral or interlocked grain. *Case hardening* or *surface hardening* in lumber is caused by too rapid surface drying. It results in the setting up of stresses in the piece, which may cause warping when the lumber is resawed.

Collapse is an abnormal type of shrinkage that takes place in drying certain types of lumber. The surfaces of collapsed lumber have a caved-in or corrugated appearance when the lumber is dried. *Discoloration* is caused either by the sapwood, which gives the wood a bluish stain, or during the air-seasoning or kiln-drying process, which produces a brown stain called *yard* or *kiln brown stain.*

Decay is a disintegration of the wood substance resulting from the action of wood-destroying fungi. Wood dried below 20 per cent moisture content and kept from re-absorbing moisture, rarely decays.

Certain woods are subject to insect attack in the green lumber, some in dry lumber, and some in insufficiently seasoned lumber. When small worm holes are found in lumber, they have a very slight deleterious effect on the strength, and if the material is otherwise satisfactory it is still satisfactory for wooden container use.

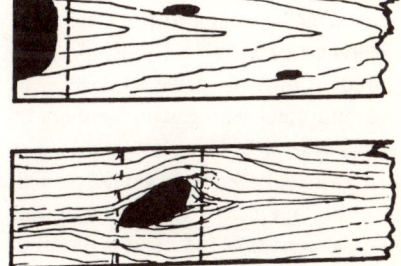

Fig. 5-3. Excessively large knots in lumber.

WOOD FASTENERS

The type and quality of fastening devices used in the fabrication of lumber into wooden shipping containers is extremely important and is one of the prime factors contributing to their ultimate strength. The primary fasteners used are *nails, staples, corrugated fasteners, screws, and bolts*. For expendable-type wooden containers, nails and staples are the most popular fasteners.

Nails. Nails are classified by primary function, special shapes and/or coatings, gauges, and sizes and types of heads. These variations have been developed through research and usage to satisfy a wide range of uses, such as box nails and pallet nails. An excellent source of details of nail dimensions, count per pound, and so forth is the Catalog of the American Steel and Wire Division of United States Steel Corp.

TYPES OF NAILS. Basically, several types of nails are used for nailed wooden containers: *Common or bright* nails, *barbed nails, cement-coated* nails, *plastic-polymer coated* nails, and *threaded* nails.

The withdrawal resistance of cement-coated nails is approximately 40 per cent greater than that of common nails immediately after driving. This increase, as attributed to cement coating, however, drops off in the heavier woods to practically no advantage over non-coated nails. The initial increase in withdrawal resistance may partly disappear in time, particularly if the nails are driven into green or partially air-seasoned woods.

Barbed nails have a withdrawal resistance 40 per cent less than common or bright nails. The shallow indentations along the shank provide a surface roughness which actually decreases the tightness of the contact with wood fibers. Barbing is most effective when only limited moisture changes occur in the wood.

As a result of the bond achieved between the nail shank and the surrounding wood, partial coating with plastic polymers of gun-driven nails increased the immediate holding power of collated plain-shank nails 1¼ times and their delayed holding power 1¾ times.

Threaded nails provide up to 2½ times the holding power of same-size plain-shank nails immediately after driving and up to ten times the holding power after seasoning of the assembled lumber.

The comparative relationship of non-threaded nails was established in tests conducted on nails driven 1 1/8 in. into white pine. In Table 5-2, the dimensions and steel wire-gauge numbers of four types of cement-coated nails are given. These include *sinkers, coolers, standard box nails,* and *corkers*. A sinker nail has a slightly lower gauge than the common nail, with the underside of the head tapered to the body of the nail. The cooler nail is the same as a sinker, except that the head is flat underneath and of slightly greater diameter than a sinker of the same penny size. A corker has a flat-countersunk head

TABLE 5-2
DIMENSIONS OF CEMENT-COATED NAILS

Size of nail	Sinkers			Coolers			Standard box nails			Corkers		
	Length, in.	Diameter, in.	Steel wire gage no.	Length, in.	Diameter, in.	Steel wire gage no.	Length, in.	Diameter, in.	Steel wire gage no.	Length, in.	Diameter, in.	Steel wire gage no.
Two-penny				1	0.0625	16						
Three-penny	1⅛	0.0673	15½	1⅛	0.0673	15½	1⅛	0.0625	16			
Four-penny	1⅜	0.0800	14	1⅜	0.0800	14	1⅜	0.0673	15½			
Five-penny	1⅝	0.0858	13½	1⅝	0.0858	13½	1⅝	0.0720	15			
Six-penny	1⅞	0.0915	13	1⅞	0.0915	13	1½	0.0858	13½	1⅞	0.0985	12½
Seven-penny	2⅛	0.0985	12½	2⅛	0.0985	12½	2⅛	0.0858	13½			
Eight-penny	2⅜	0.1130	11½	2⅜	0.1130	11½	2⅜	0.0985	12½	2⅜	0.1205	11
Nine-penny	2⅝	0.1130	11½				2⅝	0.0985	12½	2⅝	0.1205	11
Ten-penny	2⅞	0.1200	11	2⅞	0.1200	11	2½	0.1130	11½	2⅞	0.1350	10
Twelve-penny	3⅛	0.1350	10									
Sixteen-penny	3¼	0.1483	9									
Twenty-penny	3¾	0.1770	7							3⅞	0.1770	7
Thirty-penny	4¼	0.1920	6							4⅜	0.1920	6
Forty-penny	4¾	0.2070	5							4⅞	0.2070	5

with a heavier diameter shank than corresponding sinkers and common nails. Box nails are generally used with woods of Group I, whereas corkers and sinkers are used with Groups II and III. To increase the holding power of nails, chemical etching is also used. The holding power produced by chemical etching is similar to that of cement coating.

A number of years ago new nails were developed to further improve holding power. The two principal types of nails in this category are *helically and annularly threaded* nails. The thread shoulders may be single or may consist of both primary and secondary shoulders of different heights with varying upsets and thread flanks. The thread roots may be rounded or flat, and the thread spacing and thread angle may vary.

Helically threaded nails with long pitch turn as they are driven into the wood, and the threads displace the wood fibres to form a thread in the wood. In this manner, the frictional resistance between the wood and the shank surface is increased as the surrounding fibres are compressed. Annularly threaded nails force the fibres over the ring shoulders into their annular grooves like wedges that are driven into wood. In Table 5-3, a comparison of

TABLE 5-3
COMPARISON OF EFFECTIVENESS OF NAILS*

Wood Species	Driven When	Tested When	Property	Plain-Shank Nail	Helically Threaded Screw-tite † Nail	Annularly Threaded Strong-hold † Nail
White Oak	Green	Green	Withdrawal resistance	100%	161%	183%
			Lateral load-carrying capacity	100%	114%	85%
Beech	Green	Green	Withdrawal resistance	100%	136%	152%
			Lateral load-carrying capacity	100%	136%	180%
Southern Pine	Part. Air-dry	Part. Air-dry	Withdrawal resistance	100%	125%	244%
			Lateral load-carrying capacity	100%	150%	124%
White Oak	Green	Air-dry	Delayed withdrawal resistance	100%	228%	245%
			Delayed lateral load-carrying capacity	100%	134%	114%

Courtesy of the Wood Research and Wood Construction Laboratory, Virginia Polytechnic Institute and State University.

* Comparison of the effectiveness of low-carbon-steel, diamond-point, 2½-in. long × 0.135-in. diameter, plain-shank, helically threaded *Screwtite*, and annularly threaded *Stronghold* nails on the basis of comparison with respective plain-shank nail.

† Tradename of products of the Independent Nail Corporation.

NAILED WOODEN BOXES AND CRATES

the effectiveness of plain-shank nails, helically threaded "Screwtite", and annularly threaded "Stronghold" nails is illustrated. From this comparison, it is evident that, in the four wood species tested, the holding power is increased in all cases except one, and in some instances, by a very substantial amount.

Owing to the added cost of these special nails, they are primarily recommended for severe handling and shipping conditions, thus, for ammunition boxes and reusable containers as well as for reusable pallets and pallet containers. Pallet nails should have satisfactory head pull-through resistance and shank withdrawal resistance, both of which need to be ascertained by means of performance data or on the basis of descriptive specifications. Pallet nails are also classified on the basis of average MIBANT (Morgan Impact Bend-Angle Nail Tester) test angles for 25 random nail samples from a single lot of nails. Three basic nail classifications and three sub-classifications were developed as shown below:

1. *"Hardened-steel nails"* having bend angles ranging from $8°$ to $28°$:
    1a. *"Very tough hardened-steel nails"* having bend angles ranging from $8°$ to $12°$.
    1b. *"Tough hardened-steel nails"* having bend angles ranging from $13°$ to $18°$.
    1c. *"Hardened-steel nails"* having bend angles ranging from $19°$ to $28°$.
2. *"Stiff-stock nails"* having bend angles ranging from $19°$ to $28°$.
3. *"Soft nails"* having bend angles higher than $46°$.

Pallet nails are acceptable for the assembly of quality warehouse pallets if they meet the following MIBANT test criteria; provided the nails also meet such other criteria as are specified in addition to the MIBANT test criteria:

1. None of the 25 random samples of a single lot of nails show partial or complete head failure.

2. Not more than 8% of the nails tested, that is, not more than two of the 25 random samples of a single lot of nails show partial or complete shank failure.

3. The 25 random samples of a single lot of nails do not yield an average MIBANT bend angle higher than $46°$.

SIZE OF NAILS. The size of nails is designated by the term *penny*, which identifies a certain length of nail and diameter of wire, according to the type of nail. The penny system of designating nails originated in England, with the abbreviation "d", signifying both pound and penny. One thousand nails weighing 10 lb. were designated as "10d nails" or "10-penny nails".

The proper size of nails to use for any wooden box is determined by the thickness and species of the wood member in which the point of the nail is held or from which the clinched nail point protrudes. The size of nails recommended for different thicknesses and for the specific wood groupings are listed in Table 5-4.

TABLE 5-4
RECOMMENDED NAIL SIZES BY WOOD THICKNESS AND WOOD GROUP

Species of Wood Holding the Points of the Nails	Size of Nails in Penny — Thickness of Ends or Cleats to which Sides, Top, and Bottom are Nailed Inches											
	$3/8$	$7/16$	$1/2$	$9/16$	$5/8$	$11/16$, $3/4$	$13/16$	$7/8$	1	$1\,1/4$	$1\,1/2$	$1\,3/4$
Group I	4	5	5	6	7	8	8	9	10	12	16	20
Group II	4	4	5	5	6	7	7	8	9	10	12	16
Group III	3	4	4	5	5	6	7	7	8	9	12	12
Group IV	3	3	4	4	4	5	6	7	7	9	10	12

Courtesy of the Association of American Railroads.

Mechanical fasteners are also used to reinforce lumber members, to provide increased resistance to shear and splitting, by using sufficiently long nails or staples which penetrate the member all the way or overlap where maximum shear and tension-perpendicular-to-grain stresses may be developed.

A certain minimum spacing is recommended to be followed, to prevent splitting of the sides, tops or ends. Since nails driven into the *end grain* have less holding power than those nails driven into the *side grain,* a larger number of nails with a correspondingly closer spacing is required for end-grain nailing. Table 5-5 describes the recommended nail spacings for side- and end-grain nailing.

TABLE 5-5
RECOMMENDED SPACING FOR SIDE- AND END-GRAIN NAILING

	Spacing, In Inches, When Driven Into	
Size of Nail Penny	Side Grain of End Inches	End Grain of End Inches
6 or less	2	$1\,3/4$
7	$2\,1/4$	2
8	$2\,1/2$	$2\,1/4$
9	$2\,3/4$	$2\,1/2$
10	3	$2\,3/4$
12	$3\,1/2$	3
16	4	$3\,1/2$

Courtesy of the Association of American Railroads.

NAILED WOODEN BOXES AND CRATES

No piece or part of a board should have less than *two nails at each end*. Except in tongued and grooved tapered boards, one nail at each end should never be permitted since such nailing does not produce the desired rigidity of the box. Nail heads should always be driven flush with the board. Overdriving of the nails materially weakens the container by destroying the grain of the wood. Under-driving also weakens the container by reducing effectiveness of the nails. The protruding nail heads are potential injury hazards to handling personnel as well as to adjacent packages. In Table 5-6, typical nailing errors are illustrated.

Staples. There is a new breed of slender, plastic-polymer coated, collated staples which are gun-driven into the wood at high speed. Because of this speed of driving, their legs can be much more slender than those of hammer-driven staples, since the danger of fastener buckling during driving is reduced considerably, if not eliminated altogether, during high-speed driving. As a result of the frictional heat developed during high-speed driving, the coating is melted. After cooling, it bonds the staple leg to the surrounding wood.

The two slender staple legs act like two slender nail shanks. Because of their slenderness, they reduce lumber splitting to such an extent that the staples can be driven much closer to the lumber end than the nails commonly used in the assembly of boxes and crates, pallets and pallet containers. Also because of their slenderness, these slender staples can be spaced much closer together than the nails. It is well known that a large number of slender fasteners can be considerably more effective than a small number of stout fasteners. Because of this, closely spaced slender staples are much more effective in the assembly of containers than the out-dated stout staples and the relatively heavy nails. In the light of today's high-speed staple driving, the fastening with a larger number of slender staples than nails required does not present any problem. As a matter of fact, the ease of high-speed gun-driving often results in the driving of more staples than are actually specified.

A problem encountered at times is the appearance of staple-leg shiners, since the slender staple legs deviate more readily with the springwood than the stout staple legs and nail shanks. To eliminate shiners, the points of the staple legs have to be fully symmetrical and the staples have to be driven truly perpendicular to the lumber surface or angular into the direction away from the lumber edge from which they might protrude as shiners.

Staples should be driven with their crowns across the grain of the lumber or face-grain of the plywood. Whenever two lumber members are joined with their grain at $90°$ to each other, the staples should be driven in such a way that the staple crowns are at a $45°$ angle to the face-grain direction. Then, both staple legs are off-set in the nailed member as well as in the nailing member.

With a few exceptions, collated gun-driven staples are up to 2½" long.

TABLE 5-6
TYPICAL NAILING ERRORS

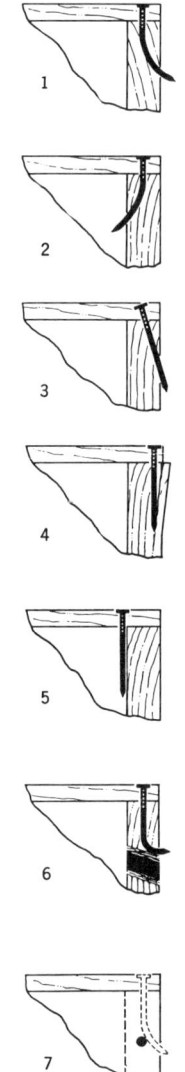

No. 1. Properly centered and started nail turned outward by cross-grained board. Protrusion of nail point called a shiner. If undetected, possible results could be injury to hands of handlers, damage to other boxes and difficult stacking of containers.

No. 2. Properly centered and started nail turned inward by cross-grained board. If defect not eliminated there might result damage to box contents, puncture of waterproof lining, and injury to person packing the box.

No. 3. Nail started at an angle causing it to break out on the outside of the box. Adverse consequences are the same as with the first type of error.

No. 4. Nail set too near the edge of the box end or cleat resulting in the nail splitting out of the board. With defect uncorrected, the nail is devoid of holding power, and after the end edge breaks off with handling, hazards similar to those encountered with the first type of error result.

No. 5. Nail has entirely missed penetrating the box end or cleat and is therefore devoid of holding power. If uncorrected, hazards described for the second type of error are present.

No. 6. Nail centered correctly and driven straight has been diverted by a knot too hard to penetrate causing a shiner. If not replaced by a second nail offset to miss the knot, adverse consequences described in the first or second type of error, depending upon whether the nail is deflected inward or outward, would result.

No. 7. Nails have collided, causing deflection of the last nail driven. Final result often is a shiner with consequences similar to those set forth in the first or second types of errors. Collision of nails can be avoided in either hand or machine nailing by offsetting. Both nails likely to collide should be offset an equal amount in order to keep both as near center as possible.

Courtesy National Wooden Box Association.

NAILED WOODEN BOXES AND CRATES

Such 2½" long pallet staples are made of 15-gauge zinc-coated steel wire. They have a rounded 0.067" by 0.073" leg cross-section, a 0.448"-wide crown, and short symmetrical points. They are coated with an effective plastic polymer inside and outside of the staple legs for a length of at least 1 1/8" from the tips of the points. During gun-driving, these staples are countersunk at least 1/16" and driven at an angle of the crown of 45° to the grain of the stapled pallet components. Two such pallet staples are to be used for the fastening of nominal 4" or narrower intermediate deckboards, four for nominal 6" intermediate deckboards, five for nominal 6" end deckboards, six for nominal 8" intermediate deckboards, and seven for nominal 8" end deckboards to each stringer.

Screws. Common wood screws are available in many diameters for given lengths, as is shown in Table 5-6A.

To be in a position to order the required number of common wood screws for a particular application under consideration, Table 5-6B is presented, indicating the approximate count per pound of steel wood screws. In the case of brass wood screws, 8 per cent can be added to the number of steel screws per pound.

As a result of this general availability of a wide variety of wood screws to choose from for given applications, a wood screw of optimum size may be used which is most effective and efficient. Bright and blued low-carbon-steel and brass wood screws of the common type were the generally stocked standard items in production. Today, hardened high-carbon-steel, aluminum, silicon-bronze, and stainless-steel wood screws are also available from screw manufacturers.

In its simplest standard form the wood screw has a continuous helical short-lead thread around a slightly tapered shank from which the thread projects. The thread-crest diameter is that of the plain-shank section between head and thread, with the thread-root diameter being relieved. The point of this wood screw is slightly blunt and eccentric.

TABLE 5-6A

Length In Inches	Nominal Sizes of Common Wood Screws, for Given Shank Diameters, In Inches																	
	.060	.073	.086	.099	.112	.125	.138	.151	.164	.177	.190	.203	.216	.242	.268	.294	.320	.372
1/4	0	1	2	3	----	----	----	----	----	----	----	----	----	----	----	----	----	----
3/8	----	----	2	3	4	5	6	7	----	----	----	----	----	----	----	----	----	----
1/2	----	----	2	3	4	5	6	7	8	----	----	----	----	----	----	----	----	----
5/8	----	----	----	3	4	5	6	7	8	9	10	----	----	----	----	----	----	----
3/4	----	----	----	----	4	5	6	7	8	9	10	11	----	----	----	----	----	----
7/8	----	----	----	----	----	----	6	7	8	9	10	11	12	----	----	----	----	----
1	----	----	----	----	----	----	6	7	8	9	10	11	12	14	----	----	----	----
1 1/4	----	----	----	----	----	----	----	7	8	9	10	11	12	14	16	----	----	----
1 1/2	----	----	----	----	----	----	6	7	8	9	10	11	12	14	16	18	----	----
1 3/4	----	----	----	----	----	----	----	----	8	9	10	11	12	14	16	18	20	----
2	----	----	----	----	----	----	----	----	8	9	10	11	12	14	16	18	20	----
2 1/4	----	----	----	----	----	----	----	----	----	9	----	11	12	14	16	18	20	----
2 1/2	----	----	----	----	----	----	----	----	----	----	----	----	12	14	16	18	20	----
2 3/4	----	----	----	----	----	----	----	----	----	----	----	----	----	14	16	18	20	----
3	----	----	----	----	----	----	----	----	----	----	----	----	----	----	16	18	20	----
3 1/2	----	----	----	----	----	----	----	----	----	----	----	----	----	----	----	18	20	24
4	----	----	----	----	----	----	----	----	----	----	----	----	----	----	----	18	20	24

TABLE 5-6B

Length	Approximate Numbers of Common Steel Wood Screws Per pound, for given Shank Diameters																	
In Inches	.060	.073	.086	.099	.112	.125	.138	.151	.164	.177	.190	.203	.216	.242	.268	.294	.320	.372
1/4	7100	4700	2800	2300	----	----	----	----	----	----	----	----	----	----	----	----	----	----
3/8	----	----	2200	1600	1200	970	750	610	----	----	----	----	----	----	----	----	----	----
1/2	----	----	1600	1300	970	780	600	510	406	----	----	----	----	----	----	----	----	----
5/8	----	----	----	1000	830	640	510	406	355	287	244	----	----	----	----	----	----	----
3/4	----	----	----	----	680	515	415	340	287	248	213	185	----	----	----	----	----	----
7/8	----	----	----	----	----	----	364	305	253	208	189	163	143	----	----	----	----	----
1	----	----	----	----	----	----	322	267	221	189	163	144	124	99	----	----	----	----
1 1/4	----	----	----	----	----	----	----	----	216	180	151	131	116	101	78	66	----	----
1 1/2	----	----	----	----	----	----	----	204	179	144	124	108	94	84	65	52	44	----
1 3/4	----	----	----	----	----	----	----	----	----	123	109	93	82	72	58	43	37	32
2	----	----	----	----	----	----	----	----	----	109	95	84	73	63	49	40	33	29
2 1/4	----	----	----	----	----	----	----	----	----	----	86	72	65	56	44	35	30	26
2 1/2	----	----	----	----	----	----	----	----	----	----	----	----	51	41	32	27	22	----
2 3/4	----	----	----	----	----	----	----	----	----	----	----	----	----	36	29	25	21	----
3	----	----	----	----	----	----	----	----	----	----	----	----	----	34	27	23	20	----
3 1/2	----	----	----	----	----	----	----	----	----	----	----	----	----	----	----	19	17	12
4	----	----	----	----	----	----	----	----	----	----	----	----	----	----	----	17	15	10

A basic design improvement of this standard form of the common wood screw is the use of parallel twin threads. This feature allows faster turning of the screw into wood without decreasing its holding power. Increase in thread-crest diameter (beyond that of the plain-shank portion under the head), in other words, relieved shank diameter between head and threads, is another important improvement in the design of wood screws, since such fasteners offer increased holding power. If manufactured with a centered point, by extending the twin threads to the tip of the point, the wood screw is more likely to turn straight into the wood and can be driven more easily. These features are incorporated in the *"Twinfast"* wood screw. Its use results in better wood assemblies and fewer rejections, and allows faster driving and tighter fastening. Despite these improvements, the cost of this screw is the same as that of the conventional screw of the same size.

Probably the most recent design improvement is incorporated in the hardened *"Self-Drilling"* wood screw, with a cylindrical rather than tapered shank and a centered point. An off-center slot is milled from the point part way along the threaded shank. The keen slot edge cuts mating threads in the wood during turning of the screw. Space for a limited amount of wood borings is provided by the slot. This screw is hardened to allow it to retain a sharp cutting edge as well as to provide greater torsional, shear, and bending strengths. The sharp gimlet point assures fast starting and helps the screw to pull itself into the wood. As a result of these features, it is not necessary to drill a pilot hole into the member into which the screw penetrates. Thus, "lining up" of shank and pilot holes is no longer necessary. Less driving torque, that is, less energy is required for driving this screw than a common wood screw. Because of the cylindrical straight shank and the fact that the threads start at the sharp point, this improved screw offers the same or a slightly greater withdrawal resistance near the point end than a common

TABLE 5-6C
Recommended Predrilled Holes For Common Wood Screws If Predrilling Is Required

Nominal Screw Size	Shank Clearance Hole		Pilot Hole				Head Countersink Auger Bit No.
	Twist Bit In Inches	Drill No.	Hard Woods		Soft Woods		
			Twist Bit In Inches	Drill No.	Twist Bit In Inches	Drill No.	
0	1/16	52	1/32	70	1/64	75	--
1	5/64	47	1/32	66	1/32	71	--
2	3/32	42	3/64	56	1/32	65	3
3	7/64	37	1/16	54	3/64	58	4
4	7/64	32	1/16	52	3/64	55	4
5	1/8	30	5/64	49	1/16	53	4
6	9/64	27	5/64	47	1/16	52	5
7	5/32	22	3/32	44	1/16	51	5
8	11/64	18	3/32	40	5/64	48	6
9	3/16	14	7/64	37	5/64	45	6
10	3/16	10	7/64	33	3/32	43	6
11	13/64	4	1/8	31	3/32	40	7
12	7/32	2	1/8	30	7/64	38	7
14	1/4	D	9/64	25	7/64	32	8
16	17/64	I	5/32	18	9/64	29	9
18	19/64	N	3/16	13	9/64	26	10
20	21/64	P	13/64	4	11/64	19	11
24	3/8	V	7/32	1	3/16	15	12

For *"Twinfast"* wood screws, the diameter of the pilot hole should be that of the root diameter of the screw, extending almost completely for the length of the screw.

wood screw. This is the case despite the fact that the threads along the slot are discontinuous.

The last described wood screw takes advantage of an improvement in manufacture, which has been employed by manufacturers of many other fasteners: The use of medium-carbon steel and the hardening, that is, heat-treating and tempering, of the finished product. If properly hardened, such a wood screw is unlikely to strip during driving into hard woods and, especially, into dry dense hardwoods. It provides greater torsional, shear, and bending strengths. Because of its effectiveness, the hardened wood screw can be smaller in size and provide satisfactory holding power. It requires less driving torque and is less likely to split the wood. Thus, it can be driven nearer the edge and end of a wood member than the heavier non-hardened screw.

Most wood screws are available today with either slotted or recessed heads. The recessed "Frearson" and similar "Phillip" screw heads offer an advantage over the slotted heads in as much as the screwdriver is less likely to slip and mar the screw head and the material through which the screw is driven. The screwdriver is centered automatically and stays aligned. Thus, assembly time can be reduced and workmanship improved by using a screw with a recessed head.

The heads of most wood screws are of the flat, round, oval, pan or truss design. However, a number of advantages can be gained by the use of wafer heads, hexagon, hexagon-washer, acorn-hexagon-washer, and "Nib" screw heads. The latter screw heads can be driven as well with a screwdriver as with a wrench and are easily and quickly located and held by the driver bit of hand, electric or pneumatic screwdrivers. The manufacture of wood screws with such improved heads may find an impetus with the more common use of self-feeding automatic, portable, screwdrivers, which feed the drive wood screws at any angle as fast as the operator can move from screw location to location.

Predrilling lead and pilot holes is often a necessity, especially if wood screws are to be driven into dry dense hardwoods, in order to allow driving of the screw and prevent splitting of the wood. The length of the pilot hole for the common wood screw should be slightly shorter than the screw length penetrating into the member. The drill diameter may be 10 per cent smaller than the thread-root diameter in hard woods, 30 per cent smaller than the root diameter in soft woods, and even as much as 50 per cent smaller than the root diameter in such soft woods as poplar. Recommended sizes of predrilled holes for common wood screws are presented in Table 5-6C.

STYLES OF WOODEN BOXES

There are several distinct types of nailed wood boxes made from sawn lumber. These types are known as Styles 1, 2, $2\frac{1}{2}$, 3, 4, 5, and 6. The principal difference in the construction of these boxes is in the design of the ends. In Fig. 5-4, these box styles are illustrated and a brief description of each is given below.

Style 1 box. This box, the simplest of all, consists of ends of a single thickness of lumber made of one or more pieces of wood. These are nailed to the sides, top, and bottom. Since the side boards are nailed to the end grain of the ends, this box is not very strong and has been replaced in most cases by such containers as fibreboard or cleated fibreboard boxes.

Style 2, $2\frac{1}{2}$, and 3 box. In these three styles, each end is strengthened with four cleats. The ends are thus reinforced against splitting and the sides are strengthened by the greater holding power of the nails driven into the side grain of the vertical cleats.

These boxes are generally recommended for use for both overseas and domestic shipment of contents having a weight not exceeding 1000 lb. The full cleated end with butt joints (Style 2) is most commonly used, since less labor and lumber are required in manufacturing this style of container.

NAILED WOODEN BOXES AND CRATES

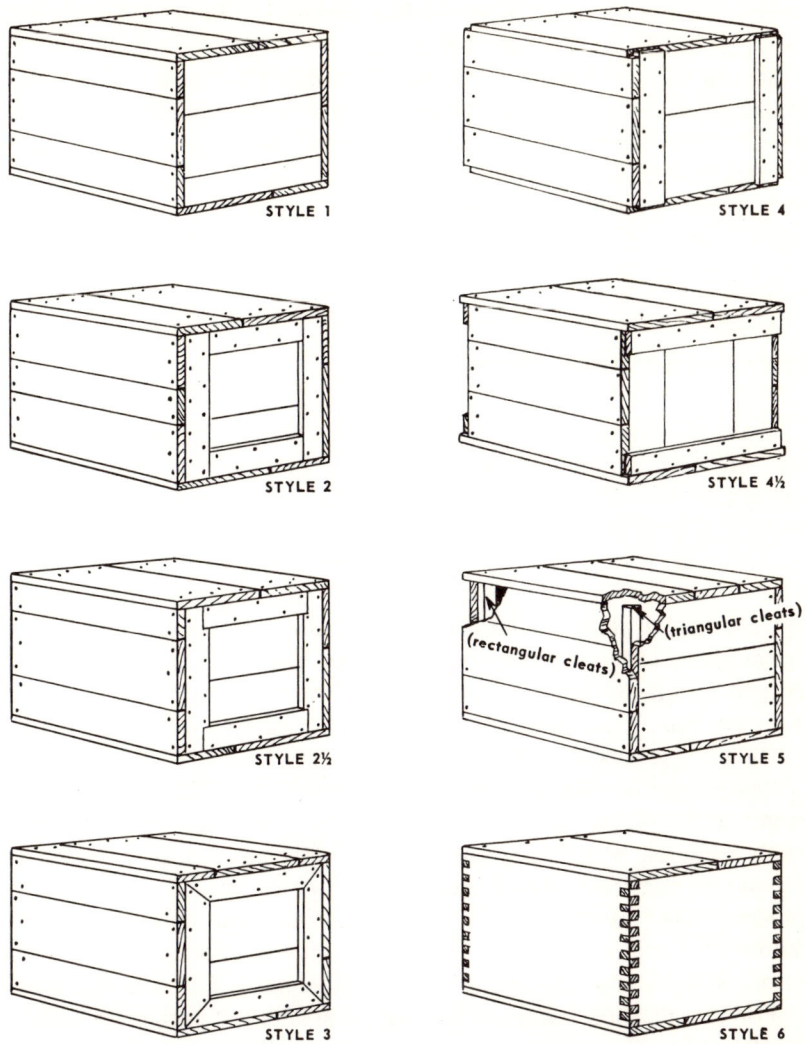

Fig. 5-4. Standard styles of nailed wooden boxes.

Style 4, 4½, and 5 box. Except for positioning of the cleats, the general requirements for the use of these styles of boxes are the same. They are recommended to be used for both overseas and domestic shipments for weight of contents not exceeding 400 lb and for all load types. The application of two cleats across the grain of each end increases the strength of the container, and end-grain nailing is reduced to a minimum. If the shape of the contents is such as to leave space in the corners of the box, the cleats can be placed inside of the box and thus reduce the outside dimensions of the container. In all other respects the three styles of boxes are similar.

Style 6 box. This box is commonly termed the "Lock Corner" box. It is constructed with ends and sides fitted together by tenons which are glued. This construction provides a package with tight corners, which is of particular value for those commodities that are subject to sifting or that require a particularly rigid box. It is frequently found as a reusable container for distributing of beverages and as a field box for fruits and vegetables.

STYLES OF WOODEN CRATES

From the foregoing paragraphs it will be noted that the top, bottom, and side sections of a nailed wooden box provide the main structural strength. The cleats or battens are supplementary structural members serving to reinforce the main assembly. As distinguished from a wooden box, a *crate* is defined as a structure in which *the frame members sustain the load and define the shape.* Sheathing is supplementary and is added primarily to enclose the structure. Generally the products are mounted to the base of the crate and the product is transported in its upright position. Thus the top and bottom of the crate take on a definite relationship pertaining to the contents, whereas in a box no such relationship generally exists.

Crates may be classified in two major ways: (1) Sheathed or closed crates and (2) unsheathed or open crates. Sheathed crates are generally required when complete puncture resistance on all parts of the container is desired. Open crates are frequently used if damage to the product is not likely to be encountered through outside hazards or the nature of distribution is such that the hazards can be minimized. The principal parts of sheathed and unsheathed crates is shown in Fig. 5-5.

For classification purposes, primarily for military packaging requirements, eight basic styles of crates have been classified. Five of these styles are sheathed with either lumber or plywood and the other three styles are open designs.

Sheathed crates.

STYLE *A*—STANDARD BOLT DEMOUNTABLE CRATE. This style of crate is for general use in crating aircraft and other large components, up to 10,000 lb. The crate is assembled with lag screws and bolts and can be disassembled for re-use. It is either sheathed with lumber or plywood.

STYLE *B*—STANDARD NAILED CRATE. This is a variation of Style *A* crate, except that it is more difficult to disassemble owing to the nailing.

NAILED WOODEN BOXES AND CRATES

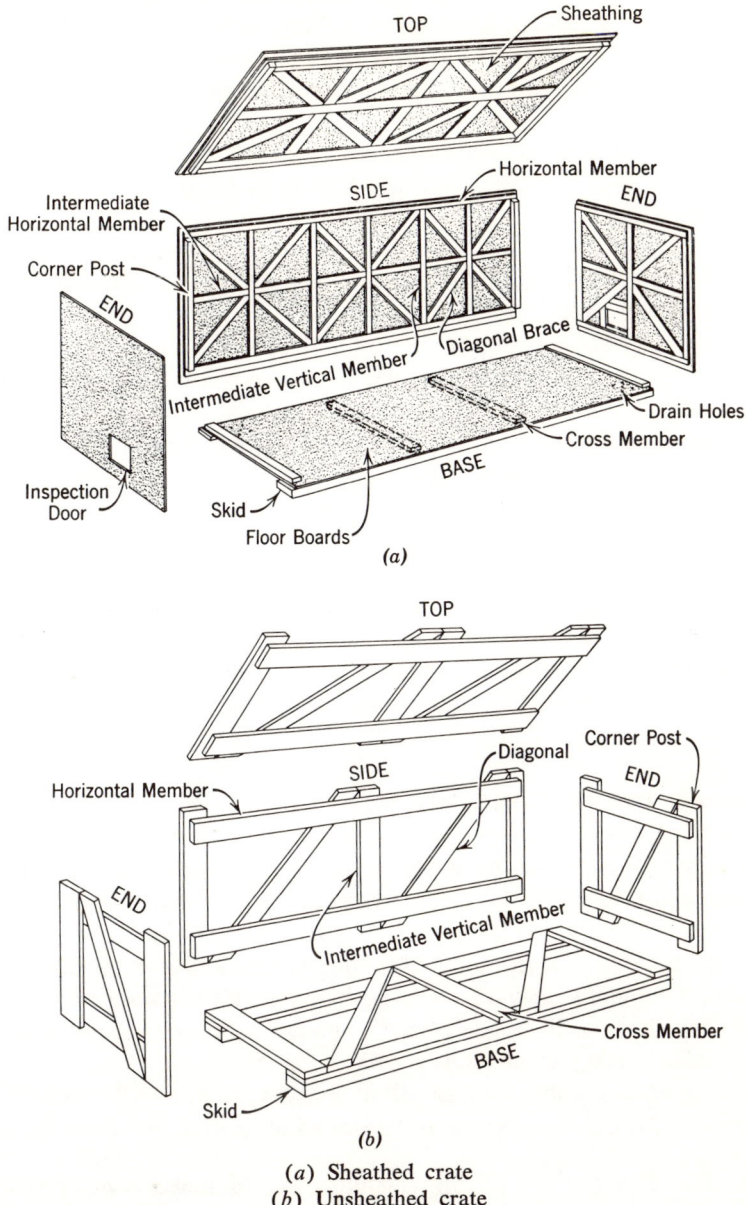

(a) Sheathed crate
(b) Unsheathed crate

Fig. 5-5. Principal parts of sheathed and unsheathed crates. Courtesy National Wooden Box Association.

STYLE *C*—LONG NARROW HIGH BOLTED DEMOUNTABLE CRATE. This style is used in crating items such as large airplane wing sections for which the complete crate is 10 feet or higher, and the height is greater than $2\frac{1}{2}$ times the width. The crate is assembled with lag screws and bolts.

STYLE *D*—LONG NARROW HIGH NAILED CRATE. This style of crate is of the same type as Style *C* but is not suitable for re-use because of its assembly with nails.

STYLE *E*—INTERMEDIATE CRATE. This style of crate is used for items which are too large to be packed in any of the styles of wooden boxes.

Unsheathed crates.

STYLE *A*. These crates are not more than 6 ft in length, 3 ft in width and 4 ft in depth. The net weight of contents is recommended to be kept under 250 lb.

STYLE *B*. These crates do not exceed 13 ft in length, 4 ft in width and $5\frac{1}{2}$ ft in depth, inside dimensions. The maximum weight of contents is 2,500 lb.

STYLE *C*. These crates are not more than 12 ft in length, 42 in. in width and 24 in. in depth. In no case may the length exceed 6 times the depth. The net weight of contents is recommended to be kept under 500 lb.

For further details on sheathed and unsheathed crates, refer to Federal Specification PPP-C-650, "Crates, Wood, Open and Covered."

BRACING AND BLOCKING MEMBERS

Frequently a fragile or irregular shaped product requires wooden members for internal blocking and bracing. Wooden containers provide excellent means to fasten these members to the sides, ends, and/or top and bottom. Bracing and blocking members may be held in place by direct nailing, with bolts or by positioning with vertical or horizontal cleats attached to the inside of the container. A typical illustration of an interior bracing technique is shown in Fig. 5-6.

The type of interior member will of course vary with each application. Important factors to consider in the design of bracing and blocking are:

1. The blocking or bracing member should make contact with a strong part of the product.

2. Contact point provisions for protection to the finish of the product should be made.

3. Boards are to be free of any defects and of sufficient strength to withstand internal stresses.

NAILED WOODEN BOXES AND CRATES

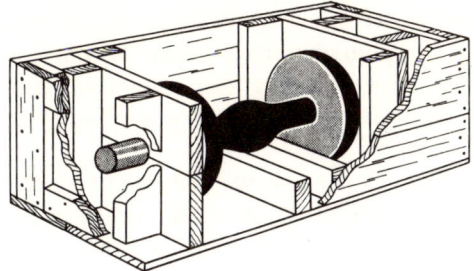

Fig. 5-6. Typical interior bracing technique. Courtesy National Wooden Box Association.

4. Fastening of members is to be adequate.

5. Outside container at points of fastenings should have adequate strength or be reinforced externally with steel strappings, cleats, or extra heavy lumber sections.

When products are bolted to the bottom of a box or crate, the bottom is frequently reinforced with an extra cleat, skid, or batten where the bolts pass through the base. This obviates the need to reinforce the complete base with heavier lumber sections and also creates a convenient skid platform for handling with such mechanical equipment as fork trucks. Fragile articles requiring a certain degree of shock isolation must be bolted with a shock mount, or other shock isolation material, between the mounting points and the base of the wooden container.

BASIC DESIGN AND CONSTRUCTION PRINCIPLES

The design of wooden containers is facilitated by reference to federal specifications such as PPP-B-621b and other similar data originally developed by the Forest Products Laboratory of the United States Department of Agriculture. The detailed information contained in the specifications permit the design of economically constructed containers. The empirical data relating to construction details have been developed as the result of extensive testing and field experience. However, individual judgment pertaining to such factors as internal blocking and bracing, finish protection, and provision for severe hazards anticipated must still be applied in the use of these basic data. Some of the important characteristics of wood and nails have already been covered in this chapter. A few additional design considerations are discussed below.

Types of loads. The prime factor influencing proper wooden container design is the physical attributes of the load. This includes the weight, size, fragility, shape, and capacity for supporting the container. For the purpose of classifying the contents which can be packed in wooden containers, three types of load categories have been defined. These are *Type 1, Easy Load, Type 2, Average Load,* and *Type 3, Difficult Load.*

TYPE 1, EASY LOAD. Type 1 loads consist of contents having low or moderate density and filling the inside of the container completely. The contents also consist of articles of sufficient strength to withstand the forces encountered in handling and transportation, and are of such shape as to fully contact all faces of the shipping container. Such items as boxed articles, chests or kits of tools, and wooden cabinets are examples of this type of load.

TYPE 2, AVERAGE LOAD. Type 2, average loads, consist of items which are moderately dense and which require a reasonable amount of protection. Items of this type may either be packed directly into the outer container or in an intermediate package which aids in supporting the faces of the outer container. The items themselves or their packages must provide a moderate amount of support for all faces of the

Fig. 5-7. Example of a Type 2, Average Load. Courtesy Kimberly-Clark Corp.

NAILED WOODEN BOXES AND CRATES

shipping container in order to be classified as a Type 2 load. In this group fall items in metal cans, bottles individually cushioned (Fig. 5-7), and numerous other items which are first packed in individual cartons.

TYPE 3, DIFFICULT LOAD. A difficult load consists of items which are highly concentrated or require a high degree of protection. Items in this category furnish no support to the faces of the shipping container but rather, in many instances, tend to apply concentrated forces to the containers' surfaces. Bolts, nuts, and other dense items which are free to shift or flow, as well as delicate instruments, assemblies, and others which do not completely fill the shipping container fall into this class.

In Table 5-7, recommended nailed wooden box styles, lumber thickness, and lumber grouping for domestic and export requirements for varying gross weights up to 1,000 lb are listed.

Thickness of sides, tops, and bottom. The thicknesses of sides, tops, and bottoms listed in Table 5-7 is also obtained by use of a formula developed by the Forest Products Laboratory. This formula may be used as a convenient guide in designing wooden boxes made of sawn lumber that are unstrapped and carry an average load in domestic shipment. The formula is as follows: $t = \frac{1}{8} \times W/b$, in which t = thickness of sides (top or bottom), W = gross weight of box and contents, and b = width across grain of side (top or bottom).

In Table 5-8 the thickness of sides, tops, and bottoms made of woods in Group I and II has been computed for several weights and sizes. When the computed thickness is not available, the nearest greater commercial thickness should be used. If Group III or IV wood species are specified, the thickness may be *reduced by 25 per cent* with a minimum thickness of ¼ in.

Dimensions of ends and cleats. The thickness of ends and cleats is determined by the thickness of the sides, top or bottom, whichever is the thicker. Field experience and laboratory tests have revealed that the relationship in thickness between sides, cleats, and ends of standard styles of unstrapped wooden boxes is as listed in Table 5-9.

Cleats of rectangular cross section should have a width of at least equal to twice their thickness plus ¾ in. As an example, a cleat which is ½ in. thick should have a width not less than 1¾ in.

Joints and number of pieces used. The economical use of lumber requires more than one piece for the sides, top, bottom, and ends of a box. Generally speaking, box sections made from one-piece are stronger than sections made from two or more pieces. There are several methods of joining pieces together. The most common are: (1) the *butt joint*, (2) the *ship-lap* or *rabbet joint*, (3) the *tongue and groove joint*, and (4) the *Linderman joint*. These joint constructions are shown in Fig. 5-8.

TABLE 5-7
SELECTION GUIDE TO THE USE OF WOODEN BOXES FOR EXPORT AND DOMESTIC SHIPMENT

Domestic Maximum Weight	Type Load	Export Maximum Weight	Style*	Group I and II Woods			Group III and IV Woods		
				S.T.B.‡	Ends	Cleats	S.T.B.‡	Ends	Cleats
50	1,2	...	1†	1/4	1/2	1/2 × 1 1/2	3/16	1/2	1/2 × 1 1/2
50	1,2	...	4, 4 1/2, 5	1/4	1/2	1/2 × 1 1/2	3/16	1/2	1/2 × 1 1/2
50	3	...	4, 4 1/2, 5	5/16	1/2		7/32	1/2	
...	1,2	60	1†	3/8	3/4		11/32	3/4	
...	1,2	60	4, 4 1/2, 5	3/8	5/8	5/8 × 1 3/4	5/16	5/8	5/8 × 1 3/4
85	1,2	...	1†	5/16	1/2		7/32	1/2	1/2 × 1 1/2
85	1,2	...	4, 4 1/2, 5	5/16	1/2	1/2 × 1 1/2	7/32	1/2	1/2 × 1 1/2
85	1,2	...	2	5/16	5/16	1/2 × 1 1/2	7/32	7/32	1/2 × 1 1/2
85	3	...	4, 4 1/2, 5	3/8	9/16	9/16 × 1 1/2	5/16	1/2	1/2 × 1 1/2
85	3	...	2, 2 1/2	3/8	3/8	5/8 × 1 1/2	5/16	5/16	1/2 × 1 1/2
...	1,2	100	4, 4 1/2, 5	7/16	3/4	3/4 × 2 1/4	3/8	5/8	5/8 × 1 3/4
...	3	100	4, 4 1/2, 5	1/2	3/4	3/4 × 2 1/4	7/16	5/8	5/8 × 1 3/4
...	3	100	2, 2 1/2	1/2	3/8	5/8 × 2 1/4	7/16	5/8	5/8 × 1 3/4
125	1,2	...	1†	3/8	5/8		5/16	1/2	9/16 × 1 1/2
125	1,2	...	4, 4 1/2, 5	3/8	5/8	5/8 × 1 1/2	5/16	9/16	9/16 × 1 1/2
125	1,2	...	2, 2 1/2	3/8	3/8	5/8 × 1 1/2	5/16	5/16	9/16 × 1 1/2
125	3	...	4, 4 1/2, 5	7/16	5/8	5/8 × 1 3/4	3/8	9/16	9/16 × 1 1/2
125	3	...	2, 2 1/2	7/16	7/16	5/8 × 1 3/4	3/8	3/8	9/16 × 1 1/2
225	1,2	...	4, 4 1/2, 5	3/8	5/8	5/8 × 1 3/4	5/16	9/16	9/16 × 1 3/4
225	1,2	...	2, 2 1/2	3/8	3/8	5/8 × 1 3/4	3/8	3/8	9/16 × 1 3/4
225	3	...	4, 4 1/2, 5	1/2	11/16	11/16 × 1 3/4	7/16	5/8	5/8 × 1 3/4
225	3	...	2, 2 1/2	1/2	1/2	3/4 × 1 3/4	7/16	7/16	9/16 × 1 3/4

234

Courtesy of the National Wooden Box Association.

* When the inside depth of a box is 5 in. or less, end cleats shall not be used. However, each side and each end shall be a one piece part, except when the end is approximately square, it shall be composed of two thicknesses with grain running at right angles to each other, with both pieces being of approximately equal thickness. Thickness of the end shall be not less than the combined thickness of the end and cleat as specified for style 4 boxes.

† Style 1 boxes shall be limited to 12 in. in height and the total dimensions (length, width and height added together) shall not exceed 50 in. Style 1 boxes may be used to carry a net weight exceeding 85 pounds and not exceeding 125 pounds if the boxes have single solid or built-up ends and sides.

‡ Sides, tops, and bottom.

TABLE 5-8
MINIMUM THICKNESS OF SIDES, TOP AND BOTTOM FOR BOXES
MADE OF GROUP I AND II WOODS

Maximum Gross Weight Pounds	Minimum Thickness of Sides, Top, or Bottom for Boxes of Different Inside Widths, Measured in Inches							
	8	9	10	12	14	18	24	30
25	1/4	1/4	1/4	1/4	1/4	1/4	1/4	1/4
35	1/4	1/4	1/4	1/4	1/4	1/4	1/4	1/4
45	5/16	5/16	5/16	1/4	1/4	1/4	1/4	1/4
55	5/16	5/16	5/16	5/16	1/4	1/4	1/4	1/4
65	3/8	3/8	5/16	5/16	5/16	1/4	1/4	1/4
75	3/8	3/8	3/8	5/16	5/16	1/4	1/4	1/4
85	7/16	3/8	3/8	3/8	5/16	5/16	1/4	1/4
100	7/16	7/16	7/16	3/8	3/8	5/16	5/16	1/4
125	1/2	1/2	1/2	7/16	3/8	3/8	5/16	5/16
150	9/16	1/2	1/2	1/2	7/16	3/8	5/16	5/16
175	5/8	9/16	9/16	1/2	1/2	3/8	3/8	5/16
200	5/8	5/8	9/16	1/2	1/2	7/16	3/8	3/8
250	11/16	11/16	5/8	9/16	9/16	1/2	7/16	3/8
300	3/4	3/4	11/16	5/8	5/8	9/16	1/2	7/16
350	13/16	13/16	3/4	11/16	5/8	9/16	1/2	7/16
400	....	....	13/16	3/4	11/16	5/8	9/16	1/2
450	....	....	....	13/16	3/4	5/8	9/16	1/2
500	....	....	....	13/16	3/4	11/16	5/8	9/16
600	....	....	....	7/8	13/16	3/4	5/8	9/16
800	....	....	....	1	15/16	13/16	3/4	11/16
1000	....	....	....	1 1/8	1 1/16	15/16	13/16	3/4

Courtesy of the Association of American Railroads.

Some of the general rules concerning number of pieces and joints are discussed subsequently. The width of box lumber, except when used for cleats and battens, should average not less than 3 in. The following may be considered as the equivalent to one piece stock: (1) When a glued Linderman joint is used, (2) when boards not less than 1/2 in. in thickness and not narrower than 1 1/2 in. are fastened together with three

NAILED WOODEN BOXES AND CRATES

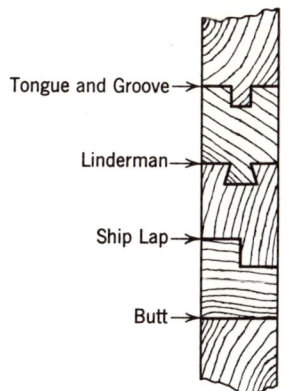

Fig. 5-8. Joints used in wood boxes.

or more corrugated fasteners spaced not more than 8 in. apart, (3) when pieces are tongued and grooved, glued and joined with corrugated fasteners. In Table 5-10 the maximum number of pieces allowed in sides, top and bottom, for different widths is shown. If the sides, tops, and bottoms are of one piece or equivalent construction, a reduction in thickness of $12\frac{1}{2}$ per cent is permitted from the thicknesses computed in the above formula. Ends of boxes must be of one piece or equivalent, or they must be cleated.

Corner constructions. As discussed previously, *end-grain nailing* provides considerably *lower* nail holding power than does *side-grain nailing*. Therefore, when boxes are joined by nails, the design should permit side-

TABLE 5-9
THICKNESS OF ENDS AND CLEATS BASED ON THICKNESS OF SIDE

Style of Box	Thickness of End	Thickness of Cleats
Style 1	2 times thickness side*	
Style 2, $2\frac{1}{2}$, 3	$1\frac{1}{4}$ " " " *	$1\frac{1}{4}$ times thickness side*†
Style 2, $2\frac{1}{2}$, 3	1 " " " *	$1\frac{1}{2}$ " " " *
Style 4, 5	$1\frac{1}{2}$ " " " *	$1\frac{1}{2}$ " " " *
Style 4, 5	$1\frac{3}{4}$ " " " *	1 " " "

Courtesy of the Association of American Railroads.
* End and cleat thickness should be based on thickness of top if latter is thicker than side.
† Minimum allowable thickness $\frac{7}{16}$ in.

TABLE 5-10
MAXIMUM NUMBER OF PIECES FOR PARTS FOR SIDES,
TOP AND BOTTOM OF DIFFERENT WIDTHS

Width of Side, Top or Bottom, in.	Maximum Allowable Number of Pieces
5	1
6	2
10	3
12	4
15	5
18	6
24	8

grain nailing whenever possible. In Fig. 5-9 the *three-way corner* is illustrated. This corner provides for the strongest and most rigid construction for a box or crate because (1) all nails are driven into side grain, (2) each member is nailed to a second member and has a third member nailed to it, and (3) if any member starts to work loose, one of the two sets of nails will counteract this tendency. Examples of corner constructions in which this is not accomplished are shown in Fig. 5-10. Corner *A* shows poor workmanship because the nails holding one member are driven in the end grain of an abutting member. In this case, the nails have comparatively low holding power. In Corner *B* a similar arrangement having the same fault as Corner *A* is shown. Corner *C* presents another poor corner construction, because the only nailing possible, except toe-nailing, is through one member into the end grains of

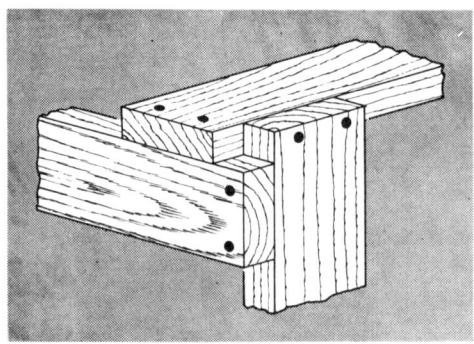

Fig. 5-9. Three-way corner construction.

NAILED WOODEN BOXES AND CRATES

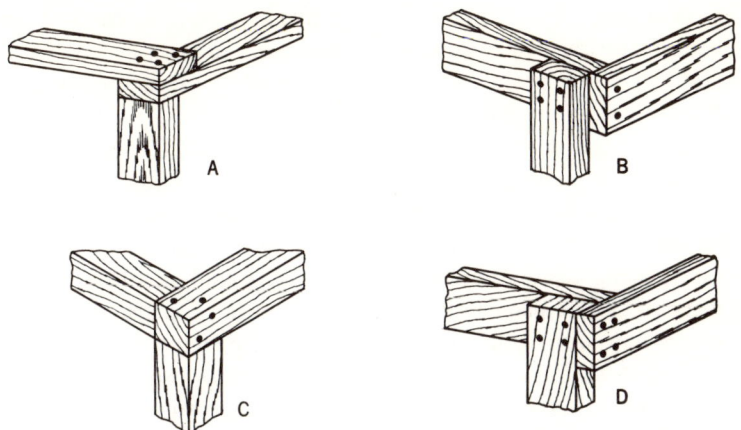

Fig. 5-10. Four examples of inadequate corner construction.

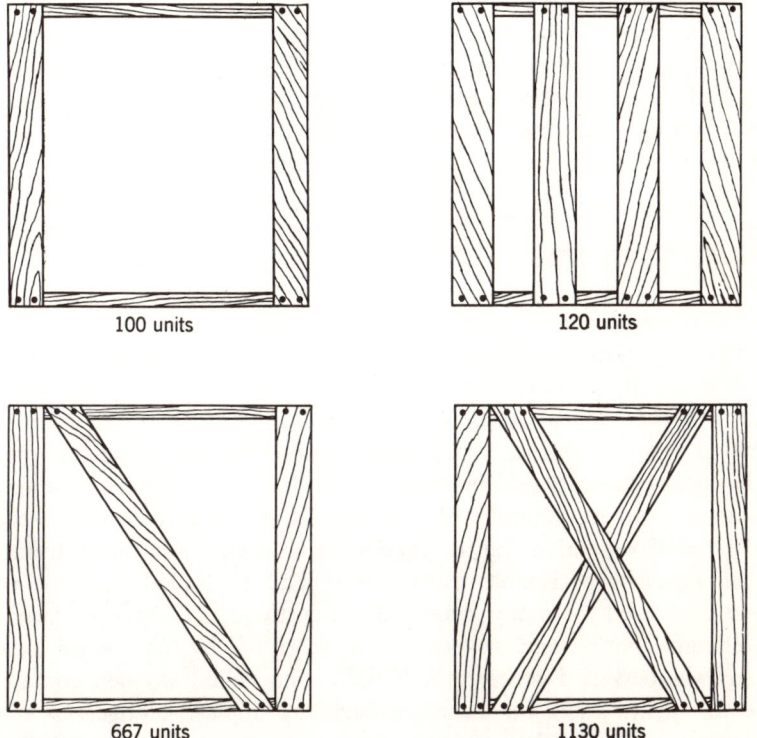

Fig. 5-11. Relative resistance to diagonal distortion provided by vertical and diagonal boards.

other members. In Corner *D* an improvement over Corners *A* and *B* has been made by placing horizontal members in such a way that nails penetrate into side grains of the various members, thus greatly increasing the nail-holding power.

It is contended that a three-way corner is difficult to open and to reclose. Thus, a container incorporating this construction is difficult to re-use. A nail puller is required to withdraw the nails without destroying the corner. If a crate or box need not be re-used the corner may be opened by using a saw through one of the horizontal members.

Diagonals. The use of diagonal bracing in the construction of boxes and crates has been widely accepted as another factor in increasing the strength of containers with the most economical use of lumber. In Fig. 5-11 the relative resistance to diagonal distortion of vertical and diagonal boards is illustrated. Diagonals may be employed singly, in a V or X construction, double X diagonals, or in other arrangements. The maximum benefit from a diagonal member is obtained if it is oriented in an approximate 45° angle.

The strength of wooden boxes and crates is often greatly enhanced by reinforcing the container with corner irons, flat or round steel banding, and special interior metal reinforcing irons which may form part of a supporting structure. Reinforcing materials are covered in Chapter 10, where their application to wooden containers is discussed.

SPECIAL CONTAINERS

There are numerous types of wooden containers which serve only the particular packing and distribution requirements of a specific commodity or a group of products. Some of these containers may also serve a dual purpose such as forming the outer crating case of a re-usable exhibition display. Other types serve as handling units such as tote or pallet boxes for inter-and intra-plant handling of parts and semi-finished products. A few examples of these specialties will now be discussed.

Fruit and vegetable nailed wooden containers. For the distribution of all types of perishable fruits and vegetables, nailed wooden boxes have been used for many years. Some of these have been replaced by wirebound boxes and others, more recently, by fibreboard boxes. Many shippers and receivers still prefer the nailed wooden containers for the distribution for such products as apples, oranges, peaches, berries, cantaloupes, lettuce, and celery.

Several basic styles of nailed wooden containers for perishables are illustrated in Fig. 5-12. It will be noted that all of these provide for ventilation and/or icing requirements essential to the shipment of perishable commodities. The construction details of most of these containers

NAILED WOODEN BOXES AND CRATES

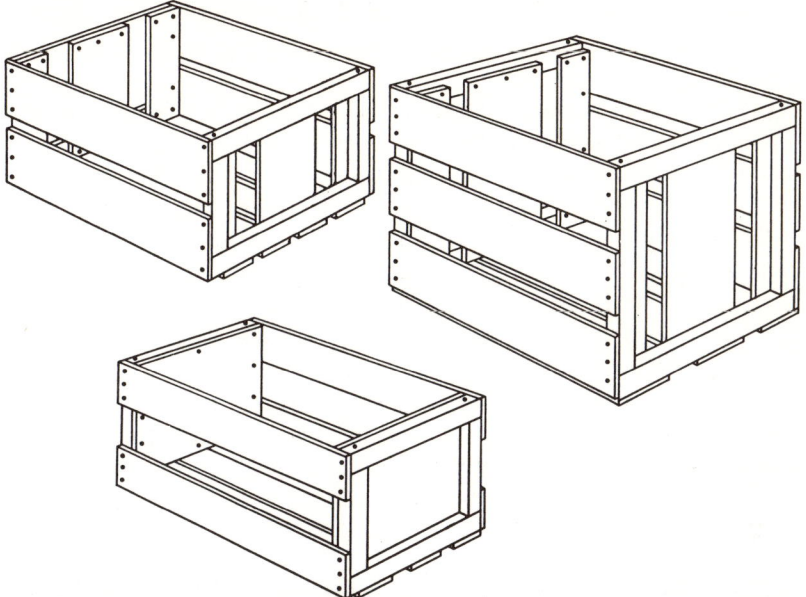

Fig. 5-12. Examples of typical fruit and vegetable nailed wooden containers.

is governed by industry associations and by applicable freight regulations pertaining to the shipment of fruits and vegetables. Special equipment for the assembly of these containers and their lidding has been developed which eliminates manual nailing for mass-production packing lines. For further reference to the packaging and shipping requirements of perishables, consult the *Freight Container Tariffs*. In addition, the United States Department of Agriculture has published considerable data on the packing, packaging, refrigeration, ventilation, deterioration, and shipping of foodstuff.

Tote and pallet boxes. An application of efficiencies attained in material handling through the use of pallet boxes is shown in Fig. 5-13. This is an example of wood used for expendable or re-usable bins and boxes. These boxes may be used either internally within a plant or for shipment. Containers of this type may follow the design principles of standard boxes or crates, as described previously, or may be constructed to satisfy specific needs. Special construction features include modified base design, special cleat arrangements for stacking, hinge or clip (Fig. 5-14) arrangements permitting collapse of the container, the incorporation of treatments to resist deterioration by the elements and fungi, and the use of reinforcements of various types.

One automotive company uses three sizes of pallet bins: 15 in. $\times$ 33 in. $\times$ 35 in.; 24 in. $\times$ 33 in. $\times$ 35 in.; and 24 in. $\times$ 40 in. $\times$ 48 in.

All are made with pallet bases designed for two-way entry. Skids or runners are $1\frac{1}{2}$ in. $\times$ $3\frac{3}{4}$ in., with the length equal to the length of the pallet base.

Cleats for sides and ends are made of $\frac{3}{4}$ in. $\times$ $2\frac{1}{2}$ in. material, whereas the sides and ends themselves are $\frac{3}{8}$ in. in thickness. All parts are made from Group III or IV hardwoods and are machine assembled. For additional strength the bins are banded with 14-gage wire. Two horizontal wires are used on each of the smaller size bins and three are used on the larger size. The pallet bins are designed to handle loads weighing up to 2,400 lb. They are used for a wide variety of bulk items such as standard parts, nuts and bolts, cold heading items, small stampings, brackets, rubber parts, and weather stripping.

Re-usable containers. Although tote and pallet boxes are generally re-usable, their primary function is to facilitate the economical handling of materials. Re-usable containers are so classified when they are primarily intended to either protect a special commodity that may have to

Fig. 5-13. Mechanical handling of a 2,400 lb load of potatoes in a pallet box. Courtesy Lane Container Co.

NAILED WOODEN BOXES AND CRATES 243

Fig. 5-14. Re-usable pallet boxes employing specially designed wire fasteners. Courtesy Navan Products, Inc.

be returned at some future date for repairs, or when they are intended to be returned because of high initial cost.

Containers of this type are generally outfitted with provisions for opening and reclosing, hardware for handling, reinforcing media to prevent wear on the corners, skids, and other vulnerable points of the containers, and with interior cushioning material such as felt, which may be permanently attached to the blocking or bracing components.

ECONOMIC FACTORS

In obtaining cost data on wooden boxes and crates the following basic costs must be included:

1. The purchase cost of a container, or (if fabricated in the box shop of a plant), the material, labor, and overhead costs of the box shop.

2. The shipping cost as influenced by the tare weight and/or the cube of the container.

3. The packing and/or unloading cost which includes lidding, reinforcing, and the expense of the materials used in this operation.

The cost of a style of wooden container can generally be estimated for *comparative purposes* by calculating the board footage requirements for each design analyzed. This will not always give an accurate indication for the purpose of evaluating the *final cost* of the container, since such items as waste, fabrication costs, delivery costs, and other indirect expenses are not known. However, unless special fabrication requirements enter the analysis, wooden containers utilizing the least amount of board footage of lumber will be the least expensive. Therefore this estimating procedure is adequate.

The unit of measurement of lumber is board feet. A board foot measures 1 ft long, 1 ft wide, and 1 in. thick, or its equivalent volume. In computing board footage, when the width and depth dimensions are given in in., the following formula is used:

$$\text{Board Footage} = \frac{\text{Length (in ft)} \times \text{Width (in.)} \times \text{Depth (in.)}}{144}$$

When dressed lumber is used, the nominal dimension and not the actual dimension must be used in board footage calculations.

In summation, there are four major approaches in designing and specifying economical wooden containers.

1. Selection of the proper style of container in consideration of the type of product and load, weight, handling, storage, and distribution requirements.

2. The use of sound design principles and materials, including attention to corner construction, nailing, bracing, defects in lumber, and other details discussed in this chapter.

3. Consideration of the use of Group III and IV hardwood species in lieu of Group I and II species, provided the fabrication can be carried out economically and the materials can be economically procured. Savings in board footage, tare weight, and cube can be obtained if this substitution is feasible.

4. Utilization of proper reinforcing media and techniques to permit reduction in board footage through reduction in lumber thickness. This topic will be discussed in Chapter 10.

BOX FABRICATION AND ASSEMBLY EQUIPMENT

When the decision has been made to use a nailed wooden box or crate for a distribution packaging application, consideration should be given to the following procurement alternates:

1. To purchase a fully assembled container requiring user closing of a prefabricated lid or crate element.

NAILED WOODEN BOXES AND CRATES

Fig. 5-15. Multiple head nailing machine.

2. To purchase pre-assembled box or crate elements (shook), and to perform the complete assembly operation prior or during the packing operation.

3. To fully fabricate and assemble all box or crate elements from a supply of lumber.

Some of the principal factors dictating the selection of one of the foregoing alternates have been discussed in the introduction to this chapter. Such factors as quantity, supply availability, freight costs, and size and style varieties influence procurement alternates for most shipping containers. User fabrication of nailed wooden boxes and crates, either partial or complete, is particularly motivated by the high cube of a pre-assembled container resulting in high freight, storage, and handling costs.

Wood-working machinery available to the user is essentially the same as that employed in suppliers' plants. The degree of mechanization of this equipment depends, however, on its economic justification, which is primarily based on the volume to be produced on this equipment.

Fundamentally, the following machines are employed:

1. Cutting machines to convert the lumber to desired size and form.
2. Fastening machines to combine individual lumber sections into sub-assemblies.
3. Assembly machines to form the container from either sub-assemblies (shook) or from precut lumber sections.
4. Miscellaneous wood-working machinery to drill, slot, plane, rout, etc.

Fig. 5-16. Open back nailing machine with double arm set up.

NAILED WOODEN BOXES AND CRATES

Fig. 5-17. Single head nailing machine.

Cutting machines include power saws of many types, which are available with varying degrees of infeed and exit mechanized provisions and are capable of performing single or multiple cutting.

Most wood box fastening is accomplished with nails, staples, or corrugated metal fasteners. Machines available to mechanically join wooden box elements are designed either for sub-assemblies (shook) or for the main assembly. Some machines are adjustable for both operations. In Fig. 5-15, a multiple head-nailing machine for the pre-assembly of crate sides is shown. Selective nail driving is feasible with equipment of this type. A similar machine illustrating the fastening of pre-assembled box elements is shown in Fig. 5-16. If volume is limited and rapid changeover from one box size to another is required, a single head-nailing machine can be used advantageously. (Fig. 5-17.)

Corrugated metal fasteners are inexpensive devices used to butt join two or more narrow boards to form a wide section. A typical machine with a double head, to drive such fasteners, is illustrated in Fig. 5-18.

Complete box assembly of a Style 1 box, where no prior pre-assembly is feasible, is accomplished on specialized equipment. In Fig. 5-19 a machine automatically nailing the sides and the bottom to the box ends is illustrated. The box elements are hopper-fed and the unit is turned for each multiple nailing operation.

Machines to position and secure lids to a filled box (Fig. 5-20) pro-

Fig. 5-18. Metal corrugated fastener driving machine with dual head.

Fig. 5-19. Automatic boxmaking machine. Courtesy FMC Corp.

Fig. 5-20. Lidding machine which automatically nails operator-positioned lid to box body. Courtesy The Wooden Box Institute.

vide economies in the closing operation. Centralized lidding of filled boxes feeding from one or more packing stations increases the operating rate and can justify high-speed lidding equipment.

Rating the economic feasibility of alternate fabrication equipment of the types surveyed can determine the true comparison of the total cost of nailed wooden containers as opposed to competitive types.

CHAPTER 6

Wirebound and Cleated Containers

In an attempt to reduce the lumber requirements of wooden containers and to facilitate ease of assembly of pre-fabricated sections, wirebound and cleated containers have been developed. Both of these types of container incorporate heavy lumber sections in the form of cleats to provide their basic structure, and both rely on relatively thin lumber or composition-material sections for faces or panels. Wirebound boxes in addition utilize the strength of wires as a reinforcement. Cleated boxes may or may not be reinforced with steel banding, but their use is not inherent in the container design.

Wirebound and cleated containers are primarily used when a substantial volume of the same size container is required. Such items as large appliances and industrial products such as transformers, water pumps, and pole-line hardware are packaged in these container styles. The degree of prefabrication obtainable with these containers usually provides economy through simplified packaging techniques.

WIREBOUND BOXES AND CRATES

Wirebound boxes were first developed in 1904 in conjunction with machinery which permitted the production of the box blank in a continuous process. Since the very crude beginnings, equipment for the manufacture of wirebound boxes and crates has been greatly refined and consists today of high-speed devices which turn out an accurately assembled product made in many styles and utilizing different types of raw materials in varying thicknesses. Since 1931 additional equipment placed in tandem with the blank fabrication machine was developed for the forming of a looped closure commonly referred to as *Rock Fastener*. This type of closure has largely replaced the *twist* wire closure formerly used, because it simplified and speeded up the closing operation.

For the manufacture of wirebound boxes rotary-cut and resawn lumber are primarily used. In addition, such other materials as plywood, paper-over-laid veneer, corrugated and solid fibreboard, and other similar materials are employed. Rotary cut lumber for box construction is relatively thin, usually ¼ in. or less in thickness and has almost the same strength properties as sawn lumber of the same species of wood, grade, and thickness. Rotary-cut lumber is comparatively free from defects, since it is usually used only from relatively smooth logs. As a result, the quality of the boxes made from rotary-cut lumber is frequently better than that from sawn lumber of equivalent thickness. However, its chief weakness, as in sawn lumber, is its comparatively low resistance to splitting and to shearing along the grain.

Lumber used for wirebound boxes and crates complies to the same classifications that were described in Chapter 5. Requirements pertaining to quality are similar to those for nailed wooden boxes and crates.

STYLES OF BOXES AND CRATES

Boxes. There are four basic styles of wirebound boxes: Styles 1, 2, 2*A*, and 3.

STYLE 1 BOX (FIG. 6-1). This box is made with a twist wire closure with nailed-in ends. The box with twist wire closure is low in cost and is readily adaptable to every type of shipment. It is pilfer-proof, for it cannot be opened without destroying the closure.

STYLE 2 BOX (FIG. 6-2). Style 2 has a Rock Fastener loop closure with nailed-in ends. This provides faster closing than the twist wire closure. It is safer in handling and assembling since no wire ends

WIREBOUND AND CLEATED CONTAINERS

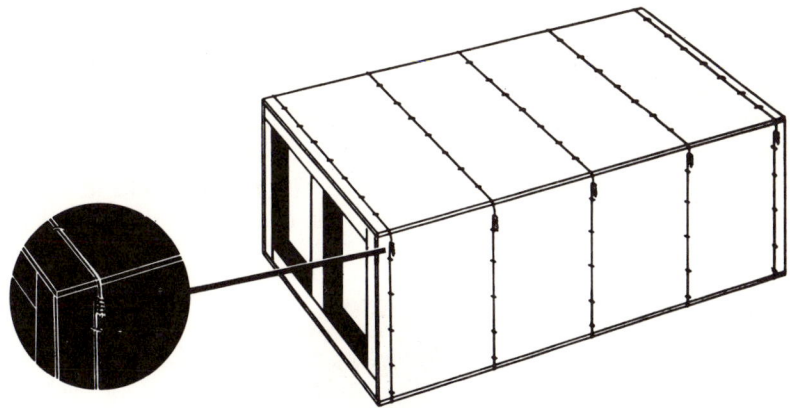

Fig. 6-1. Style 1 wirebound box.

are exposed. Also Rock Fastener wirebound boxes can be opened for inspection and easily reclosed for reshipping. For re-usable containers the Rock Fastener closure generally permits several opening and closing operations. Replacement Rock Fasteners have been developed which extend the usefulness of a re-usable container.

STYLE 2A BOX (FIG. 6-3). This container is a *twisted looped* wire closure with nail-in ends. This style offers similar advantages to the container with the Rock Fastener loop closure.

STYLE 3 BOX (FIG. 6-4). Style 3 box is the so-called *all bound* box with Rock Fastener loop closures and nail-less wired ends. It provides

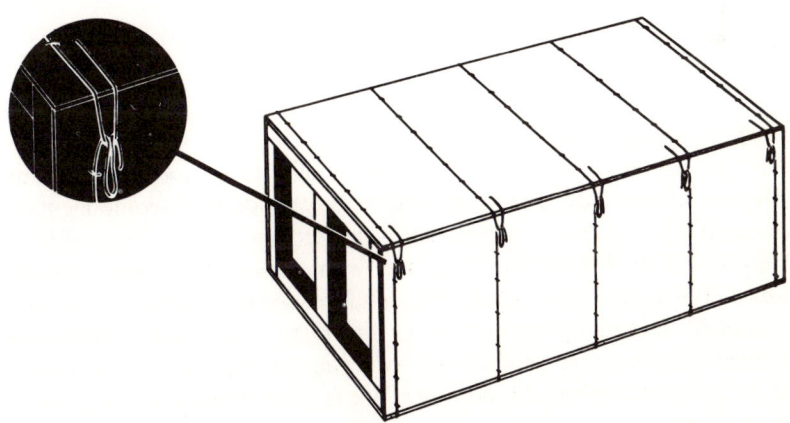

Fig. 6-2. Style 2 wirebound box.

pre-assembled ends which need not be stapled or nailed to the cleats of the box blank. Thus it offers the fastest assembly of all wirebound box styles.

Among Styles 1, 2, and 2A, which merely differ in the closure, there is no basic structural difference. They all require nailing or stapling of the ends into the cleats of the box blank. This nailing results in a more rigid end than for Style 3 box in which the ends are held to the cleats

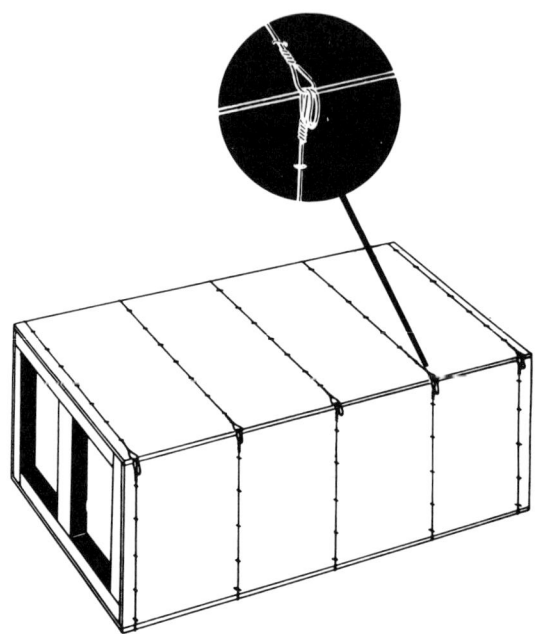

Fig. 6-3. Style 2A wirebound box.

with wire loops. This latter container is also faster to assemble, and only one knocked-down box instead of three individual components is delivered to the user and subsequently handled and stored.

Style 1, 2, and 2A boxes are made with *plain ends* or with *battened ends*. The battens are used as a reinforcement of the end and may be applied in many different ways as illustrated in Fig. 6-5. Plain ends (without battens) are only suitable when made of plywood. If rotary-cut or re-sawn lumber is used, *liners* are stapled to the inner face at the edges of the end boards with the grain of the liners perpendicular to the grain of the end boards. *Liners* are thin boards which prevent the

WIREBOUND AND CLEATED CONTAINERS

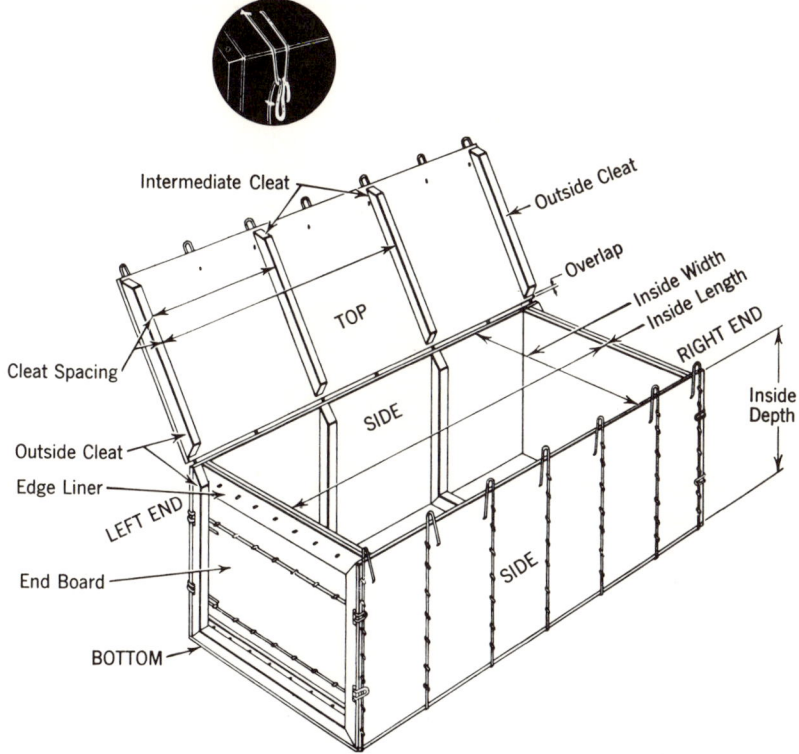

Fig. 6-4. Style 3 wirebound box.

end boards from splitting and the fastenings from pulling through the thin end boards. Also shearing of the end boards is prevented. In Fig. 6-6 three types of linered ends are shown. This illustration also includes a very wide end which has a center liner.

For Style 3 boxes, liners, and/or battens can be used. All wires are

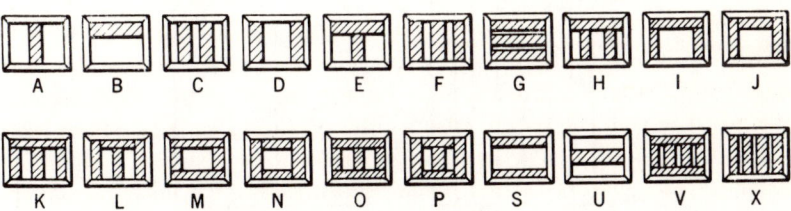

Fig. 6-5. Typical battened ends for Style 1, 2, 2A wirebound boxes.

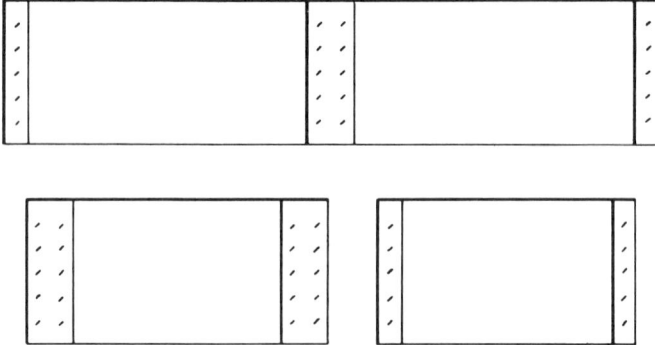

Fig. 6-6. Typical linered ends for Style 1, 2, and 2 *A* wirebound boxes.

stapled to the unreinforced area of the end boards. In Fig. 6-7 several typical ends for Style 3 boxes are illustrated.

The *cleats* of wirebound boxes are made with mitered ends or with mortise and tenon ends (tongue-and-groove). The mitered cleat is the more popular of the two. Generally, cleats are made of not less than $9/16$ in. $\times$ $13/16$ in. for boxes having a weight of contents not exceeding

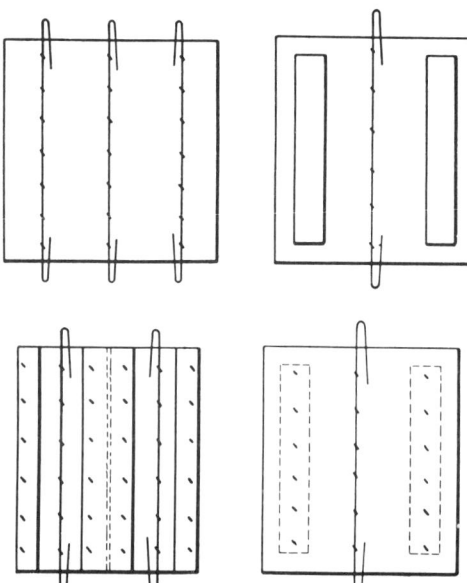

Fig. 6-7. Typical linered and battened ends for Style 3 wirebound boxes.

WIREBOUND AND CLEATED CONTAINERS

50 lb, and $1\tfrac{3}{16}$ in. $\times$ $1\tfrac{3}{16}$ in. for boxes having a weight of contents not exceeding 85 lb. When the contents weigh over 85 lb, cleats of at least $1\tfrac{3}{16}$ in. $\times$ $\tfrac{7}{8}$ in. should be used. When the inside length of the box exceeds certain limits, one or more intermediate rows of cleats should be added. Also double rows of cleats for the purpose of holding vertical partitions or for bracing and blocking members may be added along the length of the box blank. The minimum spacing of such cleats is $\tfrac{1}{2}$ in. (Fig. 6-8).

Crates. *Wirebound crates* have not been classified in the same way as wirebound boxes. However, the same types of closures, such as the twist-wire closure, the Rock Fastener, and a twist-loop closure may be

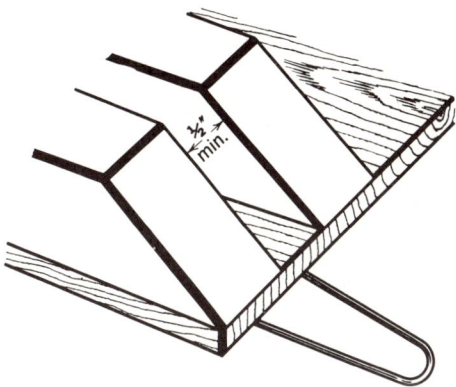

Fig. 6-8. Double row of cleats.

used on any type or style of crate. Basically, wirebound crates are defined as *containers with openings on one or more sides*. In Fig. 6-9 a wirebound crate similar in construction to the Style 3 box, except for partially open faces, is illustrated. This type of crate, in many variations, is used for the shipment of fruits and vegetables as well as for such consumer products as vitreous china lavatories and numerous heavy industrial products.

Another major style of crate, the upright crate, sometimes referred to as the *wrap-around* or *hood crate,* is illustrated in Fig. 6-10. This type is suited for appliances, motors, transformers, and a host of other similar products. In general, the item that is packaged is either bolted to the base or blocked in place by structural members fastened to one or more intermediate cleats of the crate blank. Many styles of *bases* may be designed. Several of these are illustrated in Fig. 6-11. The proper style of the base depends on the weight of the product, the

258   **DISTRIBUTION PACKAGING**

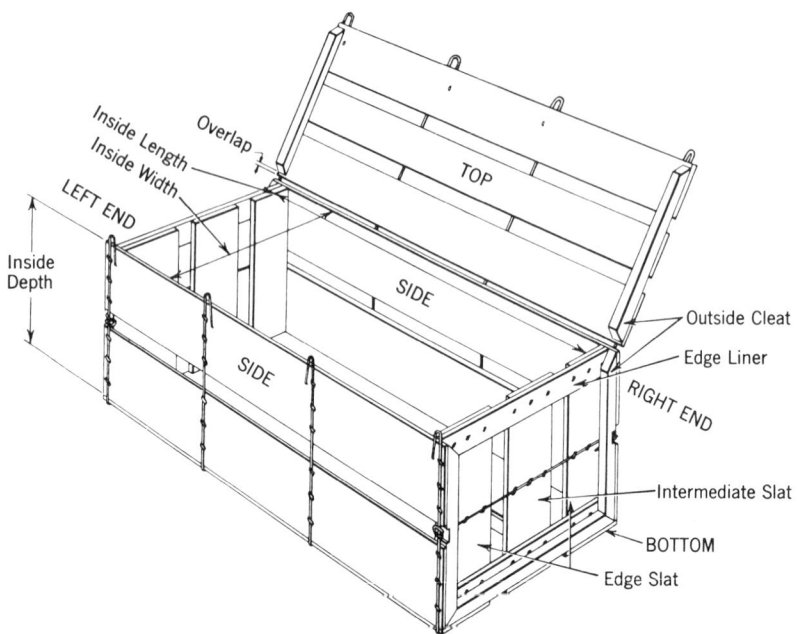

Fig. 6-9. Crate similar in construction to Style 3 box except for partially open faces.

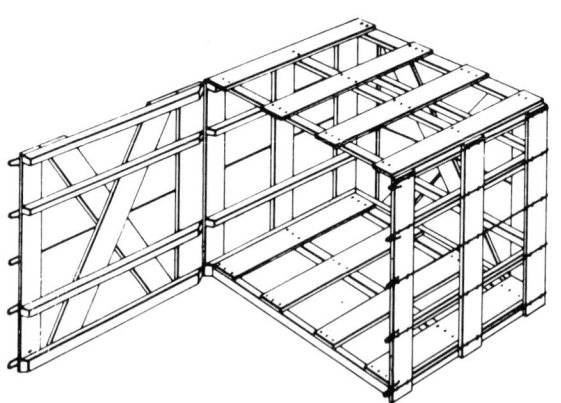

Fig. 6-10. Wrap-around crate.

WIREBOUND AND CLEATED CONTAINERS 259

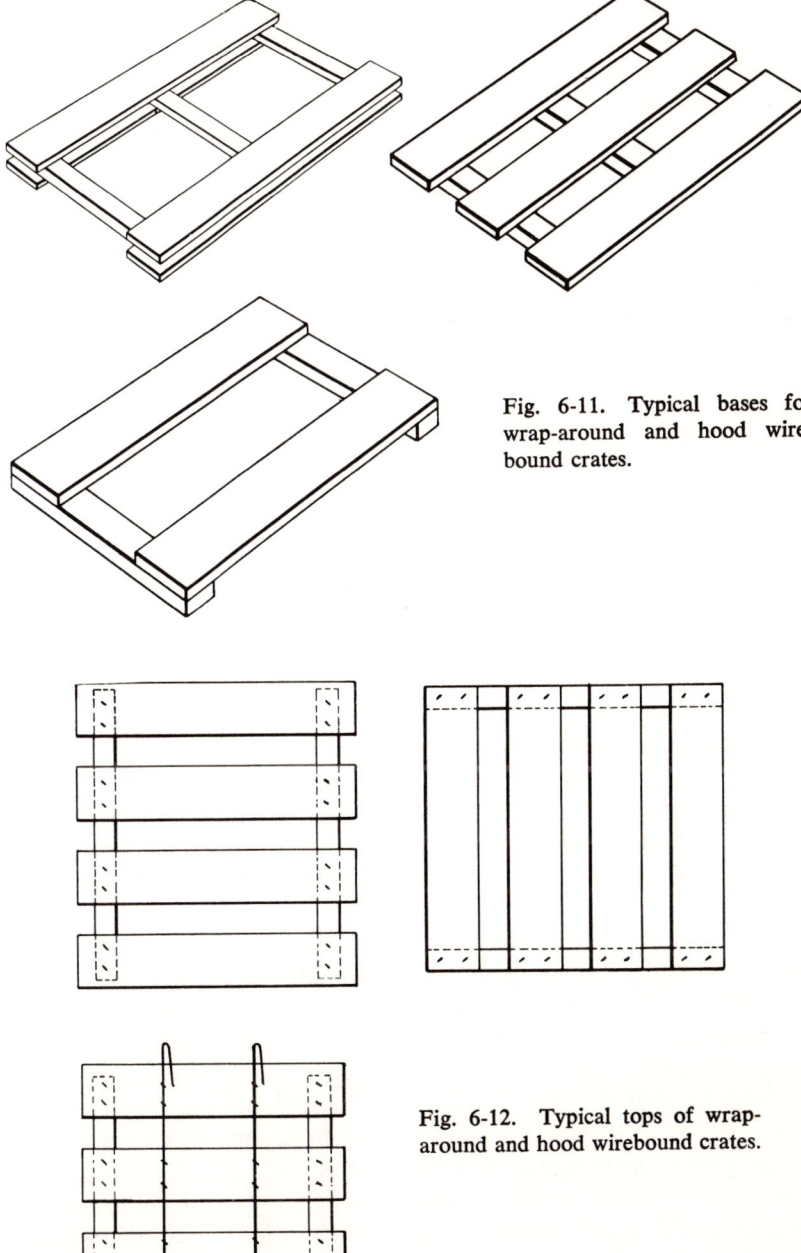

Fig. 6-11. Typical bases for wrap-around and hood wirebound crates.

Fig. 6-12. Typical tops of wrap-around and hood wirebound crates.

method of fastening or resting the product on the base, the method of holding the box blank to the base, and any special materials-handling considerations requiring a pallet or skid-base construction.

Most wirebound crates provide interlocking of the bottom and top cleats of the blank with the base and the top of the crate. Sometimes the tops are held in place with wires which are looped through the top cleats of the blank. A few typical top designs are illustrated in Fig. 6-12. Tops and bottoms are now made of stapled as well as nailed construction.

DESIGNING WIREBOUND BOXES AND CRATES

Through the efforts of the Forest Products Laboratory, United States Department of Agriculture, and the Package Research Laboratory, of Stapling Machines Co., a great deal of empirical information pertaining to the design principles of wirebound boxes and crates has become available. These data, developed through many tests simulating certain storage and shipping hazards, facilitate the design of properly engineered containers with a minimum of trial and error techniques. Yet, owing to the nature of the production equipment and to limitations that must be fully understood, it is frequently difficult to undertake a wirebound-container design without very close coordination with a representative of that industry.

Fundamentals of wirebound box design. The resistance of the various elements of a box to rough handling conditions is influenced by numerous design factors. Some of these factors will now be discussed.

FACE BOARDS. *Resistance varies as the square of the thickness;* thus, a board $\frac{1}{4}$ in. thick resists rough handling four times as well as a $\frac{1}{8}$ in. board.

BINDING WIRES. *Resistance varies as the square of the wire area.* For instance, 4-15 gauge wires having a relative cross-sectional area of 100 will withstand 100 units of rough handling, whereas, 4-14 gauge wires with a cross-sectional area of 123 will withstand 151 units of rough handling.

WEIGHT CARRIED. *Resistance varies inversely with the square of the gross weight.* Therefore, a box weighing 50 lb will withstand 4 times as much rough handling as a box weighing 100 lb.

BOX CROSS SECTION. *Resistance varies directly with the cross section;* thus, a box with a cross section of 300 sq in. (width times depth) will withstand *twice* as much rough handling as one with a cross section of only 150 sq in.

WIREBOUND AND CLEATED CONTAINERS

Staples in cleats. *Resistance varies directly with the staple holding power.* For staples which are *not clinched,* the *holding power varies directly with the length and diameter of the staple and with the species of wood used.* For instance, staples of identical length and diameter having a relative holding power of 100 in Douglas fir (Group II wood), will have a holding power of only 18 in white pine (Group I wood) and 168 in beech (Group IV wood).

Clinched staples have from 20 per cent to 100 per cent more resistance to pulling than unclinched staples. The increase results from the force required either to shear through the cleat material or to straighten out the wire.

Cleats. *Resistance varies with the type and cross section of the cleats.* Tests have indicated that mitered cleats resist approximately 60 per cent more rough handling than do tongue and groove cleats. Boxes made with cleats of $1\frac{3}{16}$ in. $\times$ $\frac{7}{8}$ in. cross section resist 140 per cent more rough handling than boxes made with cleats of $\frac{9}{16}$ in. $\times$ $1\frac{3}{16}$ in. cross section.

Species of face boards. *In general, harder woods withstand more rough handling than softer woods.* An approximate comparison of the strength of face materials for different groups of wood is shown in Table 6-1.

WIREBOUND BOX ECONOMICS

The cost of wirebound boxes, like the cost of other types of containers, depends, of course, on many factors including construction details and quantities purchased. As with fibre boxes, the shape of the container also has a bearing on the purchase price.

For resistance to rough handling a *cubical* package is most desirable. Long, flat, or deep packages with the *length more than twice the width or the depth* have the *least* resistance and should be avoided if possible. On the other hand, cubical boxes are not very economical to manufacture. The most economical shape for wirebound boxes has the following proportions:

$$L = 2W \quad \text{and} \quad W = D$$

$$\text{where } L = \text{Length}$$
$$W = \text{Width}$$
$$D = \text{Depth}$$

TABLE 6-1
Comparative Face Board Thicknesses for Equal Strength for Different Groups of Wood

Group I Soft Woods, in.	Group II and III Medium Woods, in.	Group IV Hard Woods, in.
3/16	1/7	1/8
7/32	1/6	1/7
1/4	3/16	1/6
5/16	7/32	3/16
3/8	1/4	7/32
...	5/16	1/4
...	3/8	5/16

Courtesy Package Research Laboratory, of Stapling Machines Co.

Relative factory costs of Style 3 boxes with the same capacity but designed with different shapes are illustrated in Table 6-2. These costs indicate a potential for savings when the product permits a choice of the shape of the box and uneconomical shapes are being used.

TABLE 6-2
Relative Costs of Style Three Boxes Designed with Identical Capacity but with Different Shapes

	Shape of Box		
Length	Width	Depth	Relative Cost
1	1	1	100
1½	1	1	96
2	1	1	94
2½	1	1	94.5
3	1	1	97
3½	1	1	101
2	2	1	107
3	2	1	106
4	2	1	109
2½	2½	1	111
4	2½	1	114
5	2½	1	116
1	2	1	113
1	1	2	109

Courtesy of the Package Research Laboratory, of Stapling Machines Co.

FUNDAMENTALS OF WIREBOUND UPRIGHT CRATE DESIGN

Most of the fundamentals discussed for wirebound boxes also apply to the design of wirebound upright crates. In addition, several other factors must be considered. Some of these are discussed.

Edge slats. The principal resistance to stacking is obtained from the edge slats of the blank. The stacking strength can therefore be increased by making the edge slats thick and wide. Additional strength in stacking is produced by intermediate rows of cleats and additional binding wires. In general, the contribution of intermediate slats in the side panels and the effect of the shape of the crate are considered insignificant in increasing stacking strength.

Intermediate slats. Intermediate slats contribute primarily to resistance to distortion and may be designed for this purpose or for other reasons such as most effective coverage or simplicity of manufacture. The most effective intermediate slat arrangement is accomplished through the use of diagonals, with the X-diagonal at or near a 45° angle, which provides maximum resistance to distortion. Single diagonals are found to be more economical than V-diagonals. The effectiveness of any diagonal depends on the proportions of the width to the length of the crate and the resultant angle of the diagonal.

Crate bases. Factors affecting the design of the base depend on the nature of the article mounted on the base and the method of attaching the blank to it. Several types of bases are illustrated in Fig. 6-11. In addition diagonal slats may also be incorporated into the design. X-diagonals may increase the resistance to base distortion by as much as 200 per cent. In general, the product is mounted to base slats, whose required width depends on the weight of the product, species of wood, the unsupported length, the number of weight bearing slats used, and the thickness of the slats. Similarly, the width of the batten or skid is affected by these factors.

If the product is very heavy, it should be bolted *through the slats and battens or skids,* the latter to be properly positioned for this purpose.

Crate tops. A few typical designs have been illustrated in Fig. 6-12. Tops are usually made of two or three levels. The *two-level top* consists of slats fastened to battens or liners and either nailed or stapled to the top cleats of the crate blank. The two-level top can also be fastened with binding wires. The *three-level top* has slats on top and bottom of the battens and forms a channel for insertion between the cleats of the blank.

The thickness of the slats depends on the species of wood and the gross weight of the crate. A minimum of 50 per cent coverage is generally essential. However, to protect against the stacking of heavy loads on top of the crate, workmen walking on it, or other anticipated hazards, greater coverage may be provided. Further, diagonals may also be incorporated into the top for resistance to crate distortion.

MECHANIZATION OF WIREBOUND BOX AND CRATE ASSEMBLY

One of the prime advantages of wirebound boxes and crates is the relative simplicity of assembly and closing facilitated by the inherent design features of these containers. Manual methods, aided by hand tools, are adequate for most assembly and closing requirements. Where high-volume packaging is practiced using wirebound boxes and crates, specialized equipment has been developed to increase productivity.

There are basically two types of machines used, (1) assembly machines and (2) closing machines. Assembling machines are designed to form

Fig. 6-13. Semi-automatic machine for closing wirebound boxes. Note wiper bar prior to its engagement with Rock Fastener loops. Courtesy Package Research Laboratory, of Stapling Machines Co.

WIREBOUND AND CLEATED CONTAINERS

the wirebound blank mechanically around a mandrel or the product. Closing machines, successfully used for many years for high-speed fruit and vegetable packing lines, have been adapted for distribution packaging purposes. In Fig. 6-13 a "flat pack closer" is shown. This is a semi-automatic machine in that an operator is required to fold the lid down and to align the Rock Fastener loops. The box is then inserted into the machine where clamps square up the box and a hold-down device brings the lid into position. An inclined wiping bar moves in a horizontal direction to bend the male loop and then wipes the closure shut by descending on the front face of the box. This position of the wiping bar is illustrated in Fig. 6-14 in which the closing of a crate of trapezoidal cross section has just been completed. This machine also incorporates a stapling head which automatically drives a staple through a slat on the top of the crate into an interior hold-down member to prevent this member from moving.

Closing machines for vertical crates (Fig. 6-15), to be used for such items as hot water heaters and transformers, which are mounted on a

Fig. 6-14. Closing machine for trapezoidal wirebound crate incorporating automatic stapling feature. Note the position of the wiper bar after completion of its cycle. Courtesy Package Research Laboratory, of Stapling Machines Co.

Fig. 6-15. Vertical wirebound closing machine incorporating automatic top and base stapling. Courtesy Package Research Laboratory, of Stapling Machine Co.

base have also been developed. After manual wrapping of the blank and top positioning, the machine automatically squares the crate, closes the Rock Fastener and drives staples into the base and top.

CLEATED BOXES

Cleated boxes use single pieces of stock of fibre, wood or combination materials, for the ends, sides, top and bottom, and edge cleats, which form the structure of the container. The stock for the faces consists of such materials as plywood, solid or corrugated fibreboard, paper-overlaid veneer and similar materials.

WIREBOUND AND CLEATED CONTAINERS

Some of the important features of cleated boxes are as follows:

1. Great strength in relation to tare weight of box.
2. Fabrication economically feasible on inexpensive equipment.
3. Smooth surface for printing.
4. Excellent resistance to infiltration of foreign matter.

STYLES OF CLEATED BOXES

There are eleven styles of cleated boxes in use. Originally the style designations of these from Style *A* to Style *K* were applied only to cleated plywood boxes. However, the same standard style designations are now also applied to cleated boxes using other facing materials than plywood.

In Fig. 6-16 these styles are illustrated. *Style A,* using cleats for the top and bottom which overlap all cleats of the sides, is easy to open and thus is recommended if re-use or in-transit or storage inspection is required. *Style B,* which incorporates a three-way corner construction, is more rigid than most of the other styles. On the other hand, this box is difficult to open, and a nail puller must be employed if the container is to have re-use applications.

The most economical style of cleated box is *Style D,* which is designed with the three-way corner and with only two cleats for each of the six box panels. Similarly constructed is the *Style G* container which has additional intermediate cleats for strengthening of the box panels.

All other styles are variations of *Styles A, B,* and *D.* Selection of a specific style of container depends on such items as strength, internal blocking, opening and re-use requirements, and the size and shape of box, as well as the panel material employed. Intermediate panel cleats may be placed in any number and desired spacing for support of internal members of the product or for additional panel strength on large-sized containers (Fig. 6-17).

Cleated plywood boxes. Cleated plywood boxes are widely used for export shipments as well as domestic application when a light and strong container is required. This box has high strength, resulting from the excellent resistance of plywood to splitting and the employment of a rigid single member for each face of the box.

The following items are some of the important features to consider in the design of a plywood container.

FASTENINGS. Either staples or nails are used to fasten the plywood to the cleats. Spacing and size requirements of the fastening are as important as for other styles of wooden containers and are described in

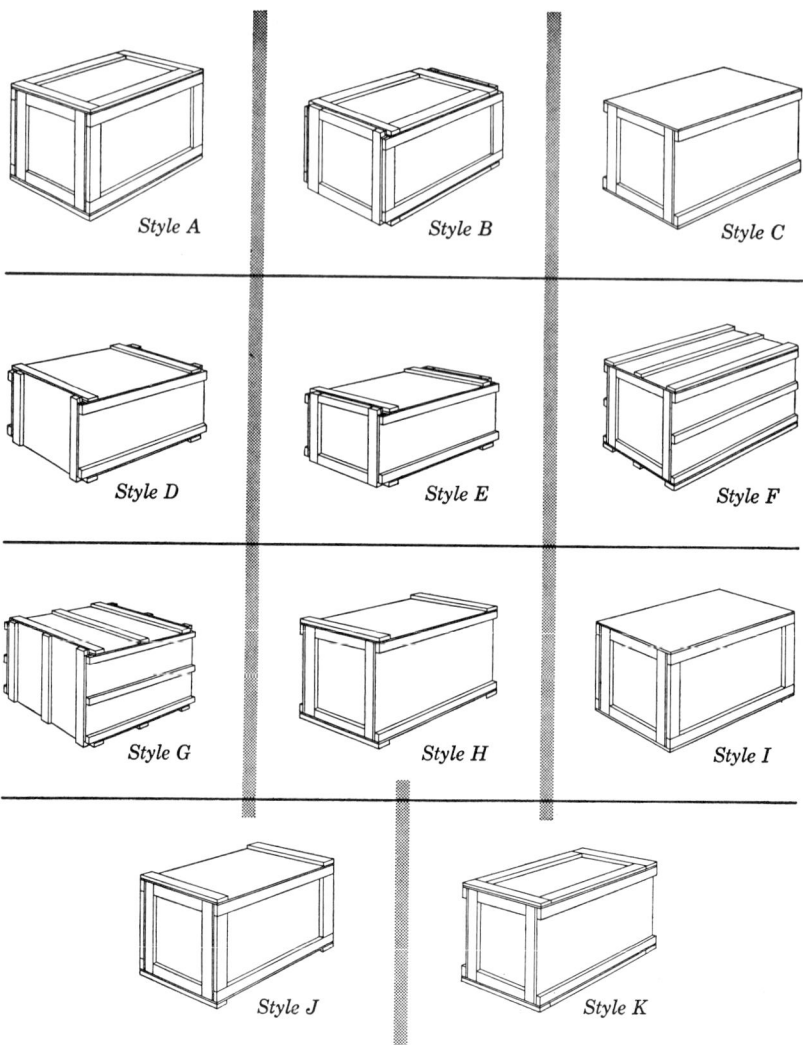

Fig. 6-16. Standard styles of cleated boxes.

Chapter 5. If nails are used, they should be large-headed so as not to pull through the plywood. A typical stapling machine, performing both stapling of the cleats to the plywood and cleat to cleat fastening is shown in Fig. 6-18.

CLEATS. The primary functions of the cleats are to provide means for securely fastening the box faces together and to reinforce the corners

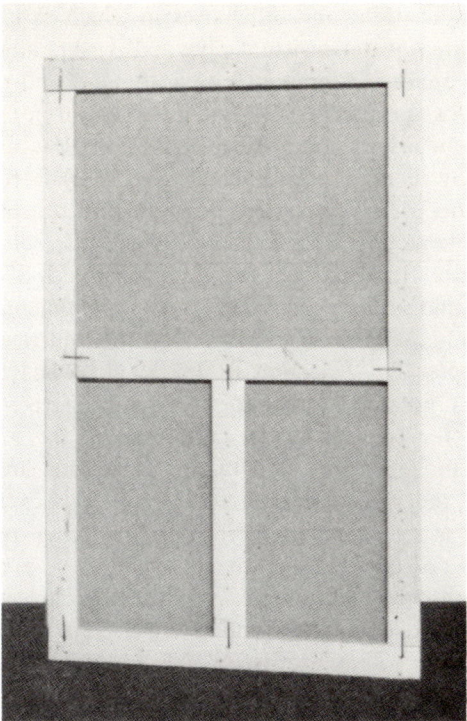

Fig. 6-17. Panel member with intermediate cleat.

against crushing. The size of the cleats depends primarily on the nailing necessary to hold the box parts together. Larger cleats are required when a single cleat is used along the edge of the box, as in *Style D,* than when two cleats are used along each edge, as in *Styles A* and *B*.

PLYWOOD. Tests have indicated that thicker plywood is required for single-cleated boxes than for double-cleated boxes. In the latter styles, failures in the plywood in rough handling are generally along the edge of the cleat, whereas, with single-cleated boxes the failures are localized around the nails or staples.

For best *cushioning* performance, plywood consisting of three plies of the same thickness with *the grain of the face plies of each box face running in the shortest dimension of the box face* is recommended. This arrangement permits maximum bending strength and thus good shock absorption in the container. If good *rigidity* of the container is of importance, *the grain of the face plies should be run parallel to the width of the box face* (longest dimension of box face).

270   DISTRIBUTION PACKAGING

For export applications and other uses in which exposure to the elements is anticipated, water-resistant plywood is generally specified.

Cleated fibre boxes. Fibreboard of both solid and corrugated construction provides an inexpensive facing material for cleated boxes. Corrugated fibreboard is principally used in single-wall constructions, although double-wall may also be employed. Grade ranges from 200 lb test and up are used. When greater puncture resistance is required than can be obtained with corrugated fibreboard, solid fibreboard is generally selected. In addition, weatherproof grades of corrugated and solid fibreboard find wide application for export shipments.

Details of the fibreboard materials incorporated in cleated boxes were covered in Chapter 2. Selection of the type of board (grade, flute, etc.) is generally not too critical for the compressive resistance of the container, since the wooden cleats take most of the load in stacking. On the other hand, puncture and tearing resistance are important for the durability of the box faces and retention of the box fastenings. Of importance also is dimensional stability of the facing material to avoid buckling, distortion, or stress on the fastening of the box panels. A tolerance for shrinkage of the material is therefore frequently specified.

Cleated fibreboard containers lend themselves to small runs of large ranges of box sizes. They are used for a multitude of items including

 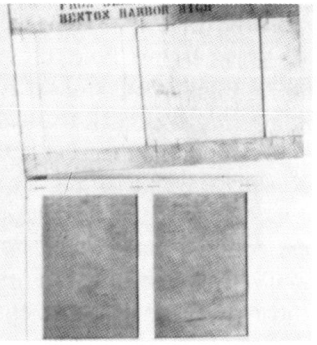

(a) Cleated box panel machine   (b) Stapling of cleats to plywood and cleat–to–cleat fastening

Fig. 6-18. Stapling of cleated plywood crate panels. Courtesy Saranac Machine Co.

WIREBOUND AND CLEATED CONTAINERS

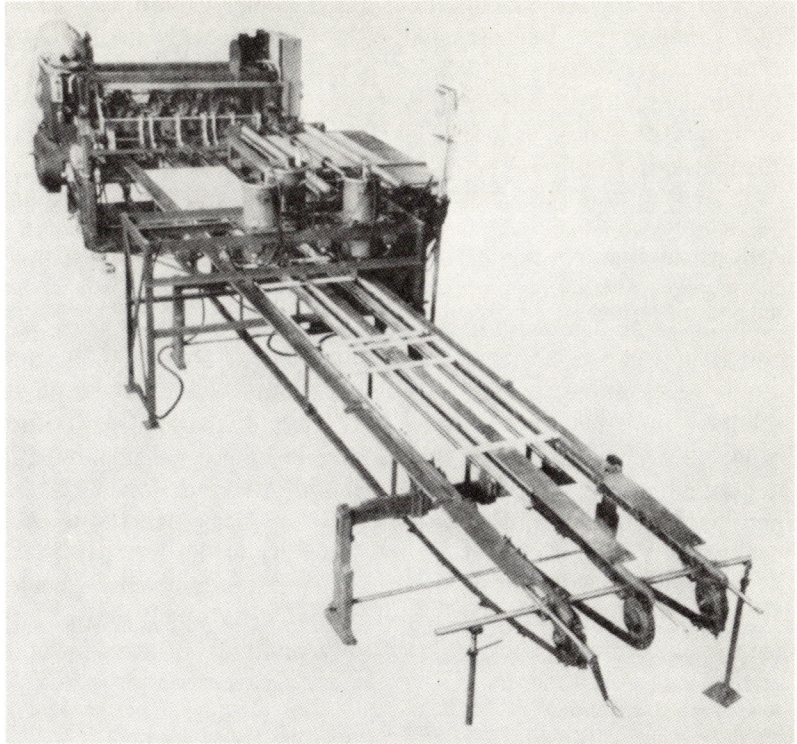

Fig. 6-19. Combined gluing and stitching cleated panel assembly machine. Courtesy Saranac Machine Co.

delicate machinery, long products, (extrusions, rugs, etc.), and appliances. A machine combining wire stitching and gluing of wood cleats to sheets of fibreboard, is shown in Fig. 6-19. The use of glue reinforces the fastening of the cleats to the panels.

Special cleated boxes and materials. An intermediate grade of facing material between plywood and fibreboard was provided through the development of *paper-overlaid veneer*. This is a three-ply material which consists generally of a hardwood core overlaid on each face with kraft paper. The thickness of veneer is varied for the required properties and can be $\frac{1}{16}$ in., $\frac{1}{12}$ in., $\frac{1}{10}$ in., $\frac{1}{8}$ in., $\frac{1}{7}$ in., and $\frac{3}{16}$ in., in thickness. The paper is either cylinder or Fourdrinier kraft; it is furnished untreated, is asphalt-impregnated, or is given other treatments depending upon any special requirements. The paper is adhered with an adhesive to the veneer, with its machine direction perpendicular to

DISTRIBUTION PACKAGING

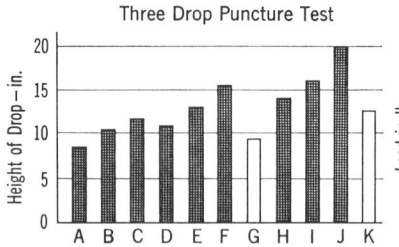

Height of drop from which 3 successive drops with a 5¼-lb spear will not cause puncture.

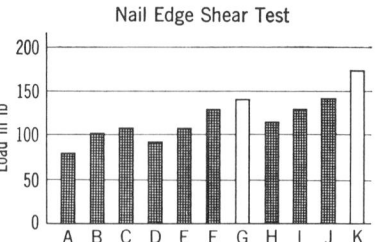

Pounds of pull parallel to grain of core required to cause shear failure. Pull applied on steel pin in hole ½ in. from edge.

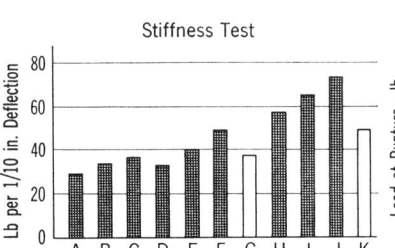

Represents deflection between loads of 33 and 53 pounds. Specimen placed on 12 in. x 12 in. frame with load applied on center of 12 in. x 12 in. specimen.

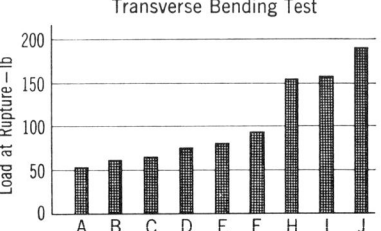

Maximum load in transverse bending required to produce rupture. 5 in. x 12 in. specimen tested on a 10 in. span. The grain of the core is across the supports. Load is applied at 1/16 in. travel of head per minute.

A. 1/10-in. Douglas fir veneer core with 42–lb kraft liners on both sides.

B. 1/10-in. Douglas fir veneer core with 42–lb kraft liner on one side, 90–lb on the other.

C. 1/10-in. Douglas fir veneer core with 90–lb kraft liners on both sides.

D. 1/8-in. Douglas fir veneer core with 42–lb kraft liners on both sides.

E. 1/8-in. Douglas fir veneer core with 42–lb kraft liner on one side, 90–lb on the other.

F. 1/8-in. Douglas fir veneer core with 90–lb kraft liners on both sides.

G. 1/8–in. hardwood plywood—1/24–in. poplar faces, 1/16–in. gum and poplar care.

H. 3/16–in. Douglas fir veneer core with 42–lb kraft liners on both sides.

I. 3/16–in. Douglas fir veneer core with 42–lb kraft liner on one side, 90–lb kraft on the other.

J. 3/16–in. Douglas fir veneer core with 90–lb kraft liners on both sides.

K. 3/16–in. hardwood plywood—1/20–in. Tupelo gum faces, 1/12–in. gum and poplar cores.

Fig. 6-20. Test data comparing various grades of veneer and plywood. Courtesy Silvatek Products Division, Weyerhaeuser Timber Co., and Elmendorf Research, Inc.

WIREBOUND AND CLEATED CONTAINERS 273

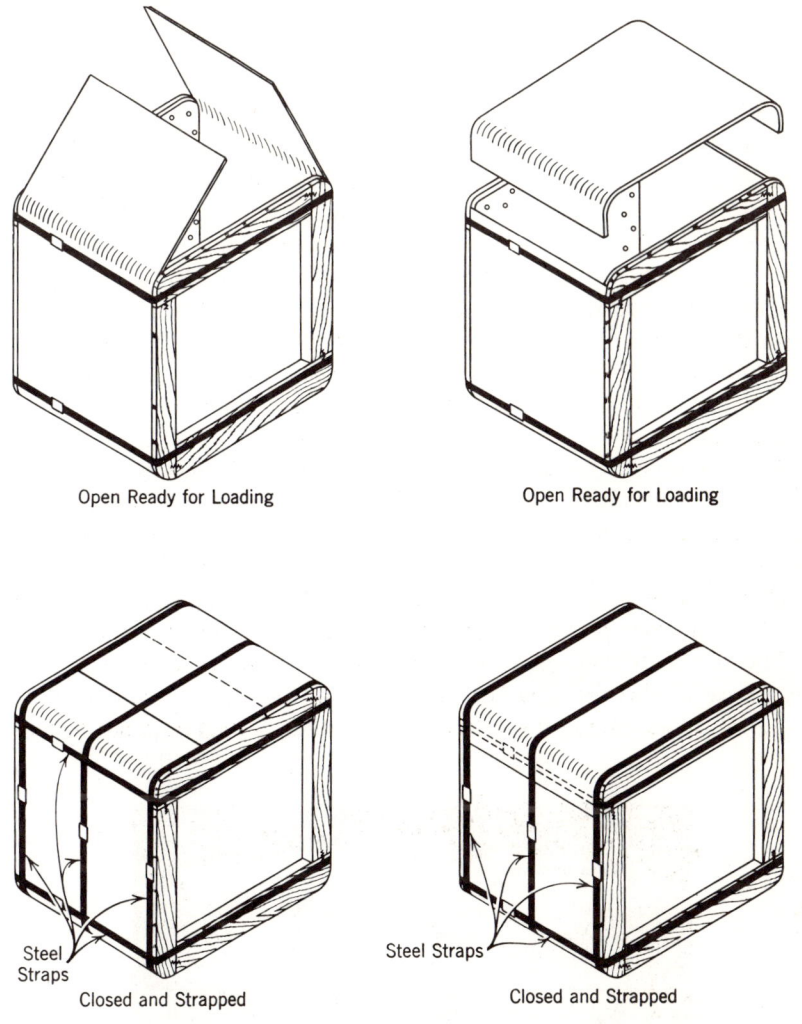

Fig. 6-21. "Straparound" boxes—two styles. Courtesy The Nelson Co.

the grain of the hardwood core. Various basis weights of papers can be used.

The strength properties of paper-overlaid veneer were extensively evaluated by the Forest Products Laboratory. In their tests it was found that all 19 combinations of paper and veneer utilized in varying manufacturing processes were suitable alternates for $3/20$ in. container grade plywood when fabricated into a certain style of cleated box.

Additional comparative information developed by another source appears in Fig. 6-20.

A special cleated container using paper-overlaid veneer is the "Straparound" box (Fig. 6-21) which is designed to be assembled and closed without the use of nails, although these may be used if needed in a specific application. The box consists of two cleated ends, a three-sided formed body section of paper-overlaid veneer and a formed top section which overlaps the sides. Both horizontal and girth straps are used for assembly. A jig is recommended to simplify this operation.

The manufacturer of the Straparound box claims that a 37 per cent and 40 per cent cube saving, respectively, over a Style 2 nailed-wood box and a Style A cleated box is realized. Savings in weight are claimed to be 158 per cent and 73 per cent, respectively, over these two containers.

Another special cleated box patented as the "Panel-Lox-Box" also assembled and closed only with steel strapping, is shown in Fig. 6-22. The container is so designed that the cleats interlock during box assembly. Girthwise straps applied in a simple packing jig, as shown, hold the container together for shipment. The facing material can be constructed from plywood, veneer lumber, and paper-overlaid veneer.

One of the advantages claimed for *nail-less* type cleated boxes is in the re-use of the container or in the ability to open the container for in-transit or warehouse inspection. A container which does not depend

Fig. 6-22. "Panel-Lox-Box," shown fully assembled in a packing jig. Courtesy Cardinal Containers.

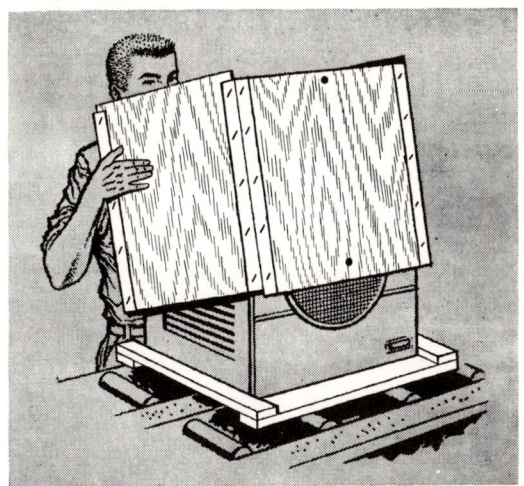

Fig. 6-23. Packing of a "Ply-Fold" box which utilizes a one-piece plywood tube. Courtesy Atlas Plywood Corp.

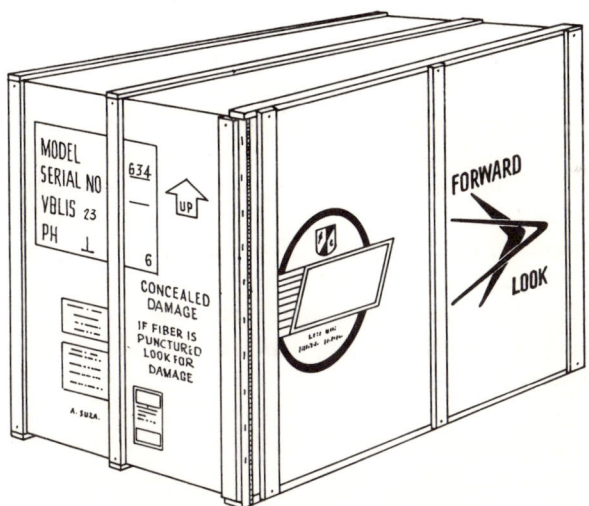

Fig. 6-24. "Watkins"–style box.

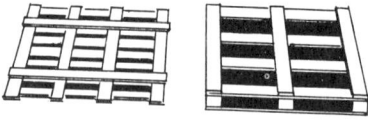 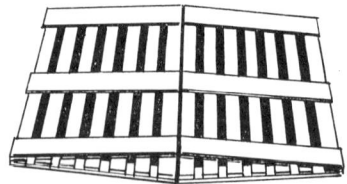

Fig. 6-25. "Hinged-Corner" crate. Courtesy Chicago Mill and Lumber Co.

on steel strapping for its assembly is claimed to have particular advantages for this purpose. This is the "Evans" container which uses specially designed flanged tubular sections which are inserted between two cleated face panels at the corners of the box. In addition, a certain amount of interlocking is accomplished with the cleats. The cleats are routed with a special tool for insertion of the tubular sections. Shipment and storage of household goods have been successfully made with this container. It is readily disassembled in a truck or at a home with the need of only a pair of pliers.

For the simplification of the cleated box assembly operation a partially pre-assembled container has been developed. This container, called by the trade name "Ply-Fold," consists of a plywood tube and a base and top. The tube, shipped in knocked-down form, is made up of four individual plywood panels which are held together with fibreboard seams stitched to the edges of the panel (Fig. 6-23). Assembly of the tube to the base and top is accomplished by heavy-duty staples, or by nailing. There are several other special or patented cleated containers available such as the "Watkins," "Hinged Corner" container and others. (Fig. 6-24 and 6-25). Selection of any of these will depend on strength, assembly, re-use, weight, cube, and other special requirements. The total container and labor cost of any of these special containers, together with special advantages they may offer, must be compared against standard cleated or other types of suitable containers, before their application is considered.

CHAPTER 7

Cylindrical
Shipping Containers

Cylindrically-shaped containers have played an important part in the field of distribution packaging and are one of the oldest forms of shipping containers suitable for the packaging of a wide variety of products. Cylindrical containers have been successful for the shipment of liquids, semi-solids, and solid products such as powders, flakes, and crystals. Moreover, cylindrical containers, by nature of construction and materials, have excelled in the packaging of dangerous and difficult products such as acids, inflammable or poisonous liquids and solids, and oxidizing materials.

In essence, a cylindrical container consists of a sturdy tube of metal, wood, fibreboard, or plastic, capped at one or both ends to provide an unsupported outer package that is usually shipped without further boxing or crating. Owing to a circular cross section providing a columnar construction, cylindrical containers provide excellent stacking strength and thus lend themselves to high storage conditions. Further, the columnar construction contributes to the uniformity of the side-wall strength. Most materials used in cylindrical containers also have a

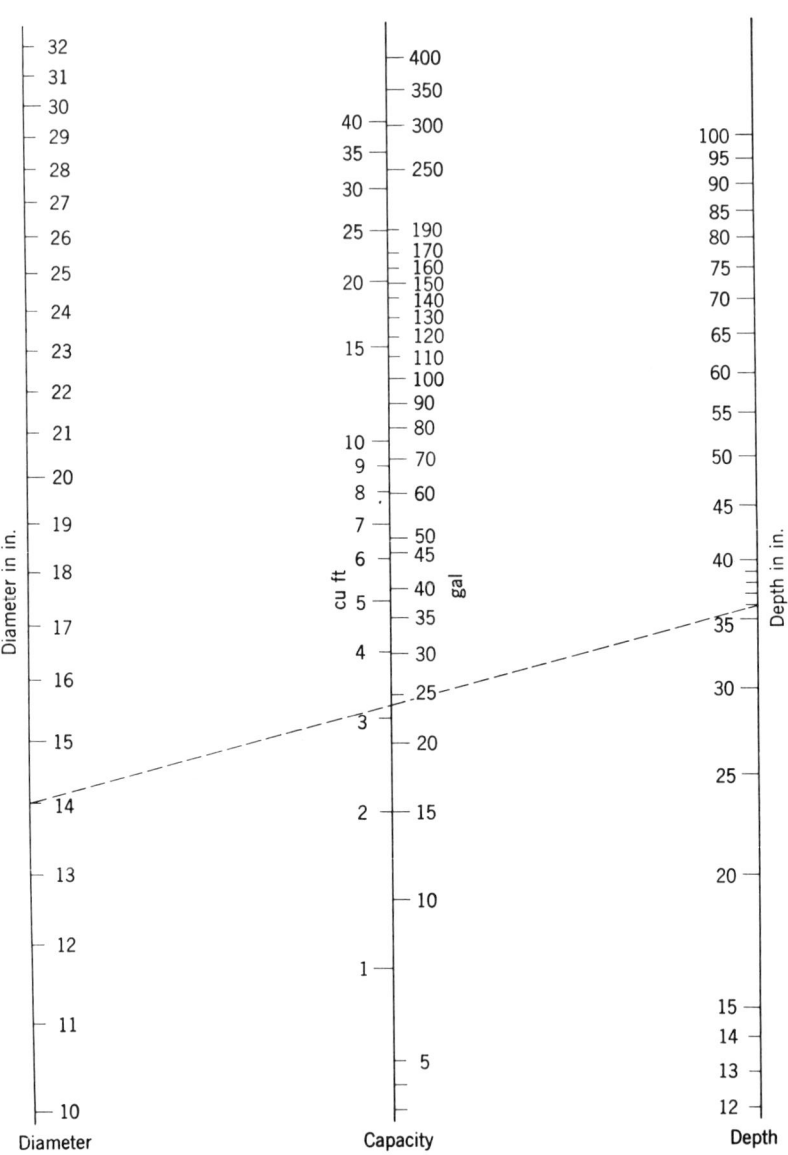

Fig. 7-1. Container capacity chart. Courtesy Rheem Manufacturing Co.

CYLINDRICAL SHIPPING CONTAINERS

high puncture resistance. These two factors impart to the container excellent resistance to external blows and impacts.

One of the primary advantages of cylindrical containers over square or rectangular containers is the *ease by which they can be rolled* in handling. This is particularly true with barrels in which the barrel bilge facilitates ease of turning. However, for industrial use, manual handling of shipping containers has been replaced by mechanical devices. Thus one of the former reasons for selecting a cylindrical container in preference to other styles is diminishing in importance. On the other hand, cylindrical containers require *more cubic capacity* for warehousing and shipping than do square and rectangular containers, even when some nesting is feasible. As an example, a *circular* container with an outside diameter of 24 in. has a cross-sectional area of approximately 452 sq. in. In storage it requires an area of 24 in. × 24 in. (assuming that the container cannot be nested), or 576 sq. in.

A *square* container having an equivalent cross-sectional area measures approximately 21¼ in. × 21¼ in. and requires only a storage area of approximately 452 sq in., or a saving of approximately 22 per cent.

The space requirements of a cylindrical container can be quickly determined by referring to the nomogram or alignment chart illustrated in Fig. 7-1. The diameter, and subsequently, the depth of container, can be varied by adjusting the line through the center column for a given capacity. Similarly, the capacity in gallons can be obtained from the inside container dimensions.

Cylindrical containers, which are fully prefabricated, require the same storage capacity for the handling, storing, and shipping of empty containers as the filled ones. To minimize wasted space in the shipment of various sizes of cylindrical containers, nesting or placement of small sizes within the larger containers is a general practice.

For the purposes of this chapter, cylindrical containers have been divided into five major categories by material, namely:

1. Wood.
2. Fibre.
3. Metal.
4. Plastic.
5. Glass and Earthenware.

DEFINITIONS

Prior to describing these groups and sub-groups in detail, several terms peculiar to cylindrically-shaped containers require definition.

Barrel. A barrel is a bilged (bulging) cylindical container of *greater length than breadth,* which has two flat ends or heads of equal diameter. Wooden barrels are made of staves bound together with hoops and may be either tight or slack (Fig. 7-2). Metal barrels are usually made of steel or aluminum. (Fibre and plywood drum containers are sometimes incorrectly called barrels.)

Drum. A drum is a cylindrical shipping container *having straight sides* and flat or bumped ends, designed for storage and shipment as an unsupported outer package that may be shipped without boxing or crating. The drum may be made of metal, plywood, or fibreboard with wooden, metal, or fibre ends. Drums are also made of rubber and polyethylene. Steel drums are available in capacities ranging from 13 to 110 gal.

Pail. A pail is defined as a container of circular cross section, either *cylindrical or a truncated cone* in shape, made of steel, black tin or terne plate, fibre or wood, fitted with a wire handle or bail, which has *a capacity of 12 gals or less* and is constructed of sheet steel of *28-gage or heavier.* Small pails for liquids, such as paint pails, usually have a part of the closure as an integral part of the pail in the form of a sealing ring into which the closure fits and is held in place by friction. Large steel pails are closed by lugs attached to the cover which are bent under the head or rim of the pail, or a ring seal that can be locked in place. In the industry, a container *without a handle or bail is a can,* whereas, a container *with a wire or bail handle is regarded as a pail.*

Can. A can is a relatively lightweight package, usually a consumer package, made of metal or paperboard or a combination of both.

Keg. A keg is a small barrel, technically of *10-gal. or less capacity.* It is usually any barrel of less than 30-gal. capacity.

Firkin. By definition, a firkin is a small wooden cask, usually used for butter. The capacity is not definite, but it generally holds about 56 lb of butter or 9 imperial gal.

Cask. A cask is a large, *tight, wooden barrel.*

Tierce. A tierce is a wooden barrel or cask of *about 42 gal capacity.*

Canister. A canister is defined as a rigid container, generally round or rectangular, made of fibre, metal, or combinations thereof, designed to pack or store dry products of not more than *5 lb in weight.*

Kit. By definition, a kit is a shipping container of metal or wood, of *truncated conical shape,* with fixed bottom in the large end and fixed or removable top in or on the smaller end. A kit usually holds *less than 5 gal.*

Tub. A tub is a small keg or cask holding *about 4 gal.*

CYLINDRICAL SHIPPING CONTAINERS

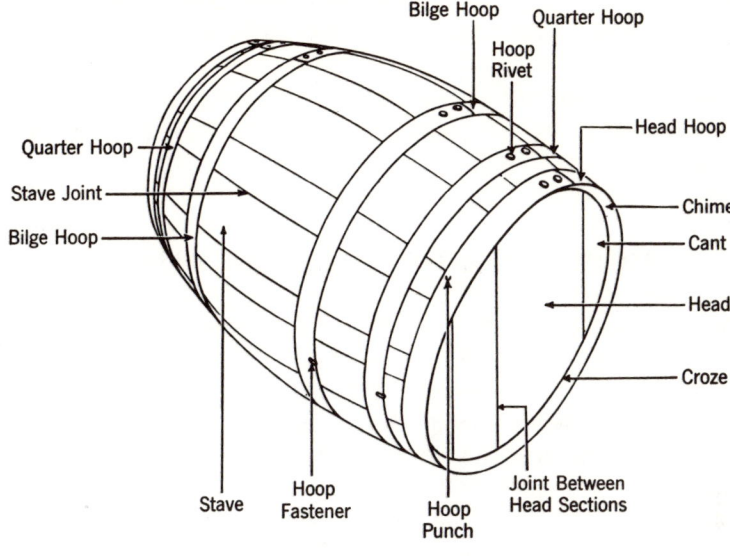

Fig. 7-2a. Slack barrel

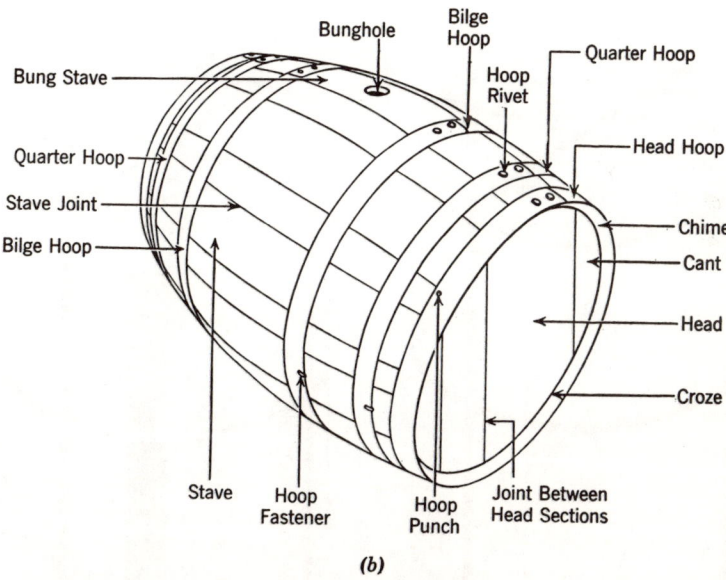

Fig. 7-2b. Tight barrel

Fig. 7-2. Slack and tight wood barrels and standard component nomenclature. Courtesy The Associated Cooperage Industries, Inc.

DISTRIBUTION PACKAGING

Earthenware. Earthenware containers are vessels made of coarse fired clay. Containers of this type are also generally referred to as crockery.

Carboy. A carboy is a large glass bottle enclosed in a box or wickerwork, used primarily for corrosives. Carboys may also be made of polyethylene.

WOOD CYLINDRICAL CONTAINERS

Wooden containers such as *slack and tight barrels* have gradually diminished in importance for distribution packaging purposes and have been replaced by fibre drums, lightweight metal containers and fibreboard boxes. However, wooden barrels have been retained for such important products as beer and whiskey, when aging and flavor are of prime consideration.

Construction details, specifications, and so forth, are contained in *The Wooden Barrel Manual* and in other publications issued by The Associated Cooperage Industries of America, Inc.

Another style of cylindrical wooden container is the plywood drum.

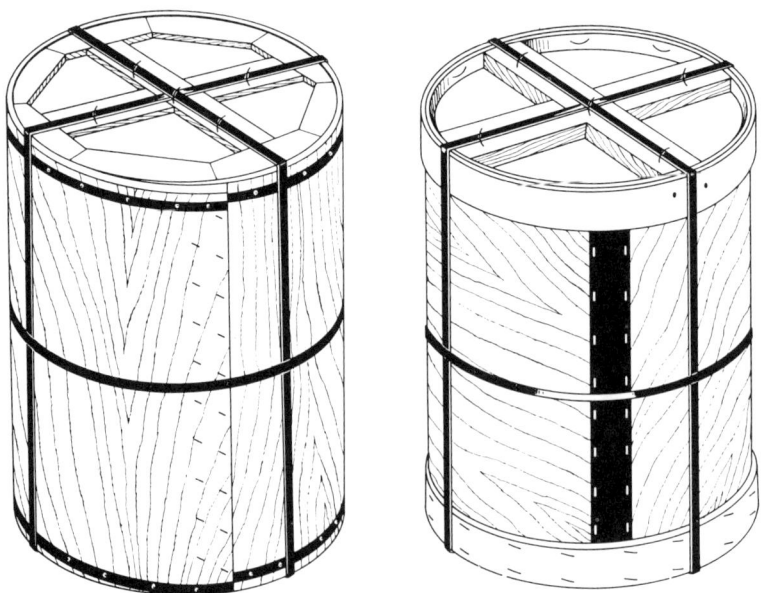

Fig. 7-3. Two basic types of plywood drum construction.

Essentially there are two basic styles of plywood drums: (1) those constructed with three-ply veneer which laps and is joined with staples and (2) those constructed with two pieces of two-ply veneer forming the cylinder with the edges butted together and stapled to metal strips. In Fig. 7-3 the two basic styles of drums including head construction and reinforcements are illustrated.

Plywood drums are used primarily to pack dry products and have an excellent weight-strength ratio.

GLASS AND EARTHENWARE CYLINDRICAL CONTAINERS

At one time, earthenware containers were used fairly extensively for the shipment of dangerous products. However, earthenware was largely replaced by glass for this purpose, and glass, in turn, has been substantially replaced by plastic and plastic-lined containers for the shipment of corrosives. Today, glass containers are fundamentally used for consumer packaging, but glass carboys are still used for industrial purposes as typified by the shipment of acids (Fig. 7-4). Complete details concerning glass containers can be obtained from the Glass Container Manufacturers Institute, Inc.

FIBRE CYLINDRICAL CONTAINERS

FIBRE CANS

The plain, all-fibre can which is almost obsolete today has been replaced by fibre cans with metal or other type ends. Its combinations are innumerable, considering the liners, coatings, outer plies, plastics, grades of paperboard and paper, and the pouring spouts, sifter tops, and a variety of closures which include slip-cover, friction plug, closed top, threaded top, and string openers.

The fibre tube today may be used for textiles, paper and paper converting, radio, radar, television, and electrical products, or it may be used for underground air ducts, for building and construction, and for thousands of other applications which depend on end-use requirements.

Recently added new characteristics to cans, tubes, etc., in the form of compositions, linings and special treatments, as well as better labels, have uncovered more uses for the can, such as for the baking industry and for packaging viscous materials, hardware, spare parts, chemicals, liquids, and precision instru-

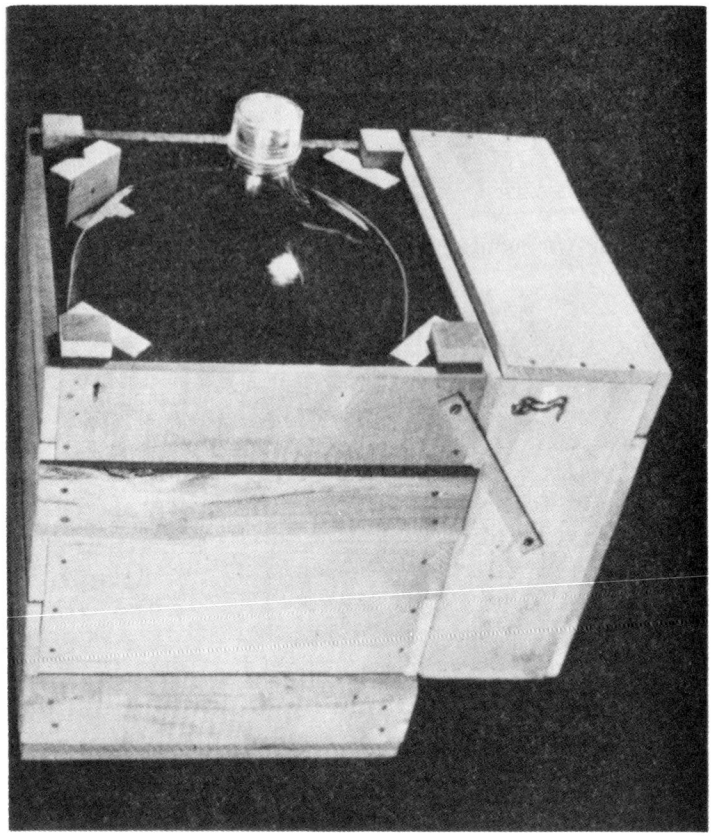

Fig. 7-4. Glass carboy in wooden shipping container with lid forming pouring platform. Courtesy Owens-Illinois Glass Co.

ments. The linings and treatments are designed to provide protection against the elements, against fumes, acids, salt air, and against extreme temperatures. As a consequence, fibre cans have wide application in military packaging, particularly for export shipments of spare parts for trucks, tanks, and aviation and electronic equipment. Spare parts containers (Fig. 7-5) are of single-bodied type with lacquered metal ends. One end is attached and one is supplied loose. A seaming machine is required to seam the loose end; therefore, this type is most adaptable to circumstances in which large volumes are required. (Fig. 7-6).

When volume is small, telescope style containers (Fig. 7-7) are frequently used, particularly if the purchase or rental of a seaming machine is not practical. Telescope containers are of three-piece construc-

CYLINDRICAL SHIPPING CONTAINERS

Fig. 7-5. Single-bodied fibre cans with metal ends. Courtesy The Cleveland Container Co.

tion. Closure at the butting seam is generally accomplished by means of a strong, waterproof, pressure-sensitive tape.

Fibre cans feature good strength characteristics because of laminated construction, ease of handling, simplicity in packing, and adaptability to high-grade labeling.

Additional details on fibre cans, cones, tubing, cores, ribbon blocks, and bobbins can be obtained from the Composite Can and Tube Institute.

FIBRE DRUMS

Certainly the best known and most widely used fibre cylindrical container is the fibre drum. Fibre drums are cylindrical shipping containers with straight sides, made of paperboard and used for the domestic and export shipping of dry powders, dry solids, semiliquids, and liquids. Fibre drums have been used to package such delicate apparatus as distribution transformers and electron tubes, heavy replacement parts for diesel locomotives such as cylinder heads and camshaft drive gears, and miscellaneous small metal products such as rivets, eyelets, stampings, and chains. Bulk shipments of shortening, lard, vegetable cooking compounds, oleomargarine, and related products formerly shipped almost exclusively in metal are now being packed in greaseproof fibre drums.

Specialty applications include the use of a patented drum with an interior core to package as well as dispense wire, which does away with a separate dispenser for this purpose. (Fig. 7-8).

The top and bottom may be made of paperboard, steel, plywood, solid wood, kraft-covered wood, kraft-covered steel, or other materials. Drum capacities range from ¾ *gal to 75 gal.* or from *173 cu in. to 10*

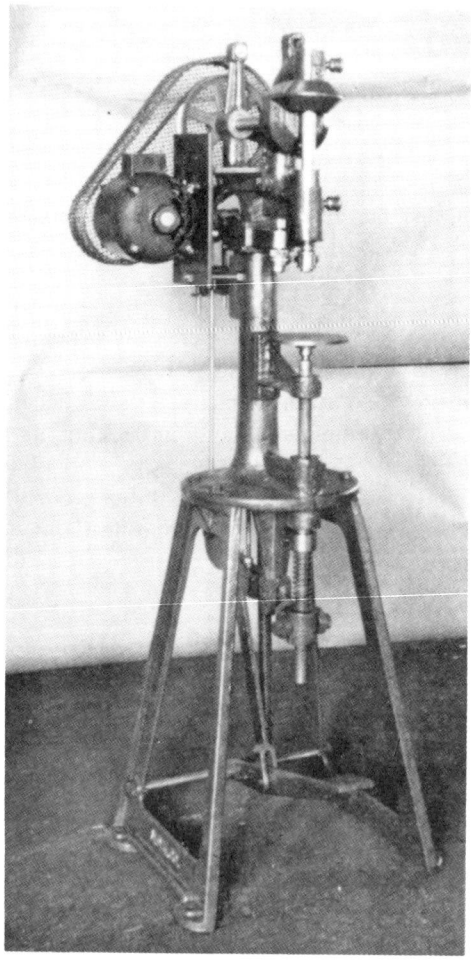

Fig. 7-6. Machine for securing metal bottoms and tops on bodies of round fibre cans. Courtesy M. D. Knowlton Co.

CYLINDRICAL SHIPPING CONTAINERS 287

cu ft. The reference to capacity in terms of gallons is in accord with general practices in the wood-cooperage and metal drum industries. However, fibre drums are used primarily for dry-bulk products measured in pounds rather than gallons.

Fibre drum diameters range from 8 in. to 23 in., and the depth ranges from about 3 in. to 42 in. or more. Weight limits range up to 400 lb, and higher weights are authorized for certain products and shipments. Fibre drums have great strength, are light in tare, and are relatively easy to open and close.

When the fibre drum was introduced in the first part of this century, it was used principally for bulk shipments of food products. The early drums were made of a single-ply laminated fibreboard sidewall with a glued, riveted, or stitched side seam or lap joint. However, the later development of the *convolute winder* permitted the winding of seamless multi-ply cylindrical sidewalls which greatly improved drum strength. Modifications in the winder design produced convolutely-wound drum sidewalls with a lining as the first ply, followed by paperboard plies as desired. Interiors may be lined with plastic materials or foils or may be given special coating treatments with wax, resins, etc., to impart desired vapor-barrier properties or resistance to chemical action. Side-

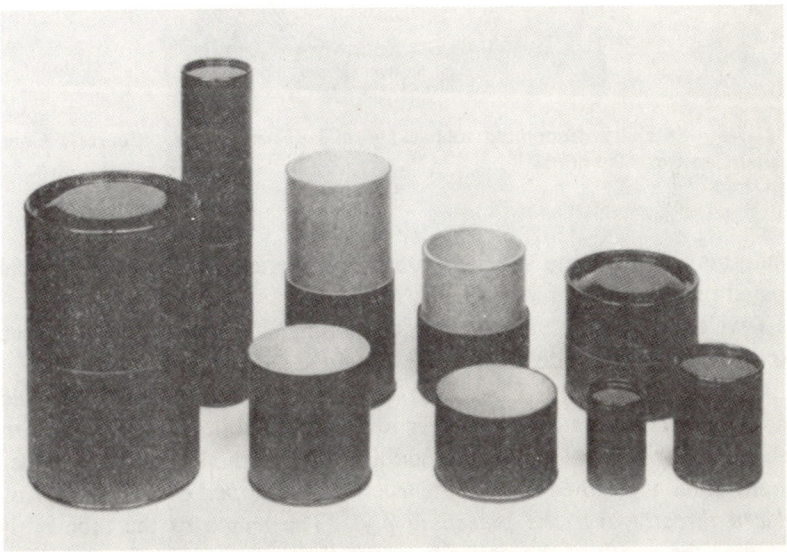

Fig. 7-7. Telescope-style fibre cans with metal ends. Courtesy The Cleveland Container Co.

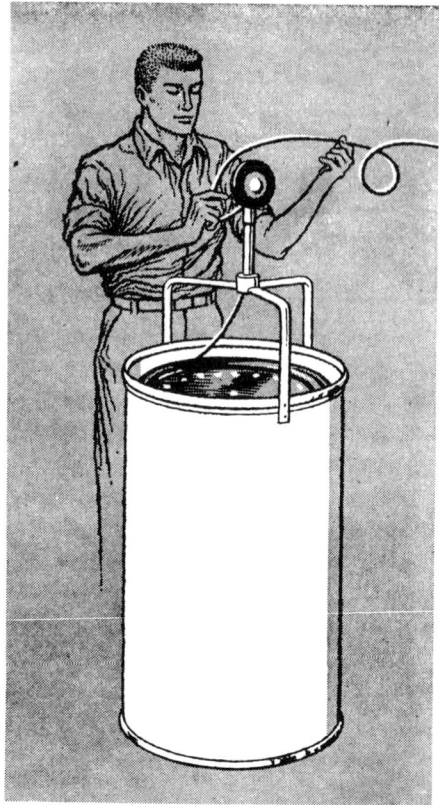

Fig. 7-8. Specialty drum used to package and dispense wire. Courtesy Continental Can Co., "Payoffpak."

walls may incorporate special waterproof barriers such as foil, asphalt coated plies or other materials.

Two terms, convolute and spiral winding, are frequently confused. Acceptable industry definitions are provided below:

Convolute winding. Convolute winding is a style of *straight winding,* in which the material advances to the mandrel in a direction perpendicular to its axis. Convolute winding is the method by which a paperboard tube is formed by having *each ply wrapped around itself and placed directly over the preceding ply.* The length of the tube is the same as the width of the paper from which it is wound. The tube may then be cut into individual fibre can or drum bodies.

Spiral winding. Spiral winding is a style of *continuous winding* in

CYLINDRICAL SHIPPING CONTAINERS

which the paper advances *at an angle* to the axis of the mandrel. To make a paperboard tube, the various plies *partially overlap preceding plies*. This process permits the production of continuous or endless tubing. The formed tube slides along the mandrel and is cut into lengths as required for fibre cans, mailing tubes, and other packages.

BASIC FIBRE DRUM TYPES

There are three fundamental fibre drum types:

1. *Plain drums* used for the packaging of dry bulk products which require no moisture protection.
2. Drums which *incorporate* a *water vapor barrier* in the sidewall and fibre headings. This construction is designed for the packaging of hygroscopic products and materials which become chemically reactive in contact with moisture.
3. Drums which have an inside *lining* or *coating* which prevents direct contact between the contents and the fibreboard. The lining or coating may be designed to provide moisture protection, chemical resistance, cleanliness, grease resistance, or to protect against contamination.

FIBRE DRUM BARRIERS

One of three types of constructions of drum barriers is ordinarily used. All have the basic characteristics of a laminated sandwich-type construction with the barrier positioned between two sheets of kraft so that a gluing surface is provided on both sides. This feature permits the barrier to be placed in any location in the sidewall or heading where it is supported by the rigid drum sidewall.

The most economical and popular type is an *asphalt barrier,* with two sheets of paper laminated by a heavy layer of hot-melt laminating asphalt.

The most efficient barrier, from a moisture-resistance standpoint, consists of *aluminum foil* laminated between two sheets of kraft with a special hot-melt laminant. Aluminum foil has a lower water vapor permeability rate than any organic film. The hot-melt laminant not only makes the foil adhere to the kraft, but also seals any *pinholes* which may be present in the foil. The laminant also functions to protect the foil from mechanical damage or corrosive fumes.

A third type of barrier consists of *polyethylene* extruded between two layers of kraft. Polyethylene is relatively inert, has no plasticizer to migrate, remains flexible indefinitely, has excellent low-temperature flexibility, and has good water vapor or water resistance.

FIBRE DRUM INTERIOR LININGS AND COATINGS

Interior linings and coatings can be grouped into two major categories. One consists of linings which *face the interior* of the drum and which are adaptable to being laminated, calendered, or coated onto the paperboard. These linings include glassine, parchment, polyethylene, pliofilm, cellulose acetate, aluminum foil, and vinyls. The other category is those which may be *flushed* or *sprayed* into the interior of a drum after it has been fabricated. Application may be made by flushing such substances as paraffin and microcrystalline wax hot or sprayed into the drum by means of a solvent which evaporates and leaves the residue as the coating.

In general, linings or coatings serve similar functions, although it is recognized that a lining, being more uniform, durable, and less apt to contribute to product contamination, is more effective than a coating.

FIBRE DRUM CLOSURES

Major types of drum closures in use are:

1. The *friction-type* or *telescopic slip-on* covers secured by gummed or pressure-sensitive tape.

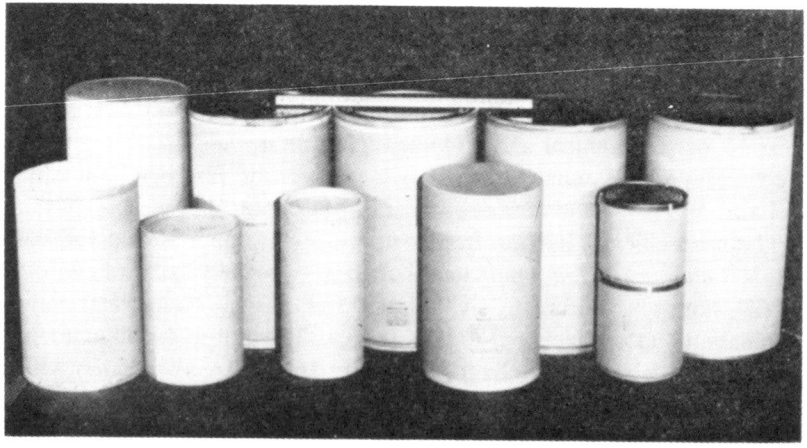

Fig. 7-9. Typical drum styles and different methods of closure.

CYLINDRICAL SHIPPING CONTAINERS

2. *Lever-activated locking* or similar type bands.
3. *Expanding* or *crimped* lids.
4. *Metal lugs* or *clips*.

Typical drum styles and different methods of closure are illustrated in Fig. 7-9. Less popular closures include *nailing, thread stitching,* and *metal stitching.*

Fibre drums can be efficiently opened and reclosed. Full open head drum construction aids in the filling as well as in the dispensing, especially when only part of the contents is removed and it is desired to reclose the drum.

DRUM SPECIFICATIONS AND REGULATIONS

Specifications and construction requirements pertinent to fibre drums are contained in Rule 51 of the *Uniform Freight Classification.*

As an example of the detailed construction requirements appearing in Rule 51 for fibre drums, Table 7-1 is reproduced. Additional Rule 51 fibre drum, pail, and tub references are contained in Sections 3, 4, 5, 6 and 7 of the ruling.

Fibre drums are being used successfully for the shipment of semi-liquid materials such as urea pastes and textile lubricants. For drum construction requirement purposes, semi-liquids are defined as products having a minimum viscosity of 5,000 centipoises at temperatures up to 100° F, exclusive of articles shipped under refrigeration, in which case the viscosity shall be measured at shipping temperatures. If the article contains a solid material susceptible to phase separation, the viscosity measurement is determined on the liquid component when this is greater than 10 per cent by weight. Construction of drums for such products must include treatments and processes or a plastic film bag liner which will prevent absorption of the contents by the sidewalls, tops, and bottom. Furthermore, the sidewall test per square inch as reproduced in Table 7-1 is increased. Pertinent details are contained in Rule 51, Section 5.

Although not all classes of liquids may be shipped in fibre drums, development along these lines has expanded the application of fibre drum packaging for liquids. Drums suitable for this purpose must have steel covers and locking bands and a bottom heading of fibre not less than 0.24 in. thick. The sidewalls and bottom of the drum must test not less than 1,000 lb per sq in. Liquid penetration into the drum components is prevented by interior lining or barrier forming an integral

TABLE 7-1
Minimum Requirements for Fibre Drums for Dry Products

Maximum Limit (See Note 5)		Side-Wall Test Per Square Inch, lb (See Note 1)	Fibreboard outer ply Waterproofed		Minimum Requirements Tops and Bottoms (each)	Wood Thickness	
Weight of Contents, lb	Capacity, gal		Thickness, in.	Steel Test, lb (U.S. Gage) (See Note 1)		Solid, in.	3 or more ply plywood, in.
Not over 60	Not over 30	400	0.120 or 0.090	300 or 600	30	½	3/11
Over 60 but not over 115	Not over 45	500	0.160 or 0.120	400 or 800	28	½	3/10
Over 115 but not over 150	Not over 55	600	0.160 or 0.120	400 or 800	28	½	3/8
Over 150 but not over 225	Not over 65	700	0.180 or 0.120	500 or 1000	26 (See Note 3)	½	3/8
Over 225 but not over 300	Not over 75	800	0.200 or 0.160	550 or 1100	26 (See Note 3)	25/32 (See Note 2)	7/16
Over 300 but not over 400	Not over 75	900	0.240 or 0.200	600 or 1200	24 (See Note 4)	25/32 (See Note 2)	7/16
Over 400 but not over 550	Not over 75	1000	0.220	1300	24 (See Note 4)	25/32 (See Note 2)	7/16

Note 1. Cady or Mullen Testing Method for Fibre Components: Either of the following test methods may be used. When more than single ply, test shall be determined from the summation of the tests of individual plies; *or,* when test is made on a completed drum, the punctures shall be made from the exterior to the interior surface, in which case the values for sidewall shall be not less than 80 per cent of the value in the above table. There shall be a minimum of 6 tests and the average shall be not less than the prescribed minimum requirements.

Note 2. Where the fibre shell of drum is provided with a formed inside bead to support the wood head and a steel rim holds the head in place and is locked in an external groove of the fibre shell, solid wood heads may be not less than ½ inch thick. Such ½ inch heads must be reinforced by covering both sides with kraft liner board not less than 0.012 in. thick securely glued throughout entire area of contact with a glue or adhesive which cannot be dissolved in water after the film application has dried.

Note 3. Bottom may be constructed of not less than 30 U. S. gage steel when combined with paperboard having a minimum thickness of 0.110 in. and Mullen test of not less than 400 lb.

Note 4. Bottom may be constructed of not less than 30 U. S. gage steel when combined with paperboard having a minimum thickness of 0.140 inch and Mullen test of not less than 550 lb.

Note 5. The minimum requirements for sidewall, top and bottom are governed by either the weight of contents in first column or by the capacity in the second column, whichever requirements are higher.

Requirements given here are as specified in Rule 51 Section 2 of the *Uniform Freight Classifications.*

CYLINDRICAL SHIPPING CONTAINERS

part of the drum and liquid-tight junctures. The permissible shipping weight must not exceed 400 lb and a capacity of 55 gal.

Certain types of drums, having special features or interior packaging materials, are authorized in the rail freight classification tariffs under "Package Description" designations. These packaging authorizations and exceptions are limited to certain specific commodities as described in the tariff. A few typical examples are reproduced below. These are included to illustrate special constructions and the manner of presentation. Since package number descriptions are constantly changing, the latest classification supplements should be consulted.

PACKAGE NO. 579. In fibre drums not exceeding 35 gallons capacity meeting requirements of Rule 51, Section 5, except viscosity test need not be complied with.

PACKAGE NO. 592. In fibre drums meeting requirements of Rule 51, Section 5, for drums testing not less than 1,000 pounds, except that the net weight must not exceed 550 pounds and product must have a viscosity of not less than 5,000 centipoises at shipping temperatures. Drums must be equipped with flexible plastic bag-type liner constructed of one or more plies, total thickness not less than 6 mils.

PACKAGE NO. 597. In fibre drums provided with a plastic interior lining so as to form an integral part of the drum meeting the requirements of Rule 51, Section 5, except sidewall must test not less than 600 pounds. Net capacity must not exceed 20 gallons, and net weight must not exceed 130 pounds.

PACKAGE NO. 599. In fibre drums having convolutely wound sidewall testing not less than 1,000 pounds and fibre bottom not less than .240 inch in thickness testing not less than 1,500 pounds, or bottom constructed of not less than 24 U.S. gauge steel combined with fibreboard testing not less than 1,000 pounds. Drums must be equipped with a flexible plastic bag-type liner not less than .004 inch thick, having a circular bottom without gussets or folds. The plastic liner must be protected at the bottom chime by one of the following methods:

1. By interposing a flexible creped Kraft paper liner having a basis weight of not less than 80 pounds between it and the drum bottom and extending not less than 4 inches up the sidewall.

2. By interposing a flexible 5-ply corrugated paper disk between it and the drum bottom.

3. By a plastic cuff not less than .004 inch thick permanently attached to the liner which must extend up the drum sidewall not less than 4 inches and extend under bottom of liner not less than 2 inches from the chime.

Covers must be made of steel not less than 24 gauge and must be equipped with a rubber or resilient plastic gasket to effect a liquid tight seal. Capacity of drum must not exceed 55 gallons and weight of contents must not exceed 500 pounds. Drum must withstand drop test prescribed in Rule 51, Section 6.

PACKAGE NO. 601. Resin plasticizers, in solid mass, in fibre drums meeting requirements of Rule 51, Section 2 for maximum weight limit of 125 lb, except that sidewall thickness must be not less than 0.098 in. and must test not less than 500 lb per sq in. Drum may have slip-cover fibre top. Gross weight must not exceed 75 lb.

In fibre drums meeting requirements of Rule 51, Section 2 for maximum weight of contents of 400 lb except that sidewall thickness must be not less than 0.126 in. and must test not less than 800 lb per sq in. Capacity must not exceed 44 gal.

PACKAGE NO. 605. In fibre drums meeting all requirements of Rule 51, Section 5, for drums having sidewall test not less than 1,100 pounds per square inch, except when sidewall tests not less than 900 pounds per square inch, net weight of contents may be increased to not exceeding 625 pounds, provided drums are integrally constructed with one ply of sheet steel not less than 34 gauge placed between plies.

PACKAGE NO. 606. In fibre drums meeting the construction requirements of Rule 51, Section 2, for 550 pounds net weight, except when gallonage capacity does not exceed 55 gallons, net weight may be increased to not exceeding 600 pounds. Interior sidewall and bottom of drum must be laminated with not less than .001 inch thick aluminum foil.

Since fibre drums are frequently used in the transportation of dangerous articles, applicable regulations may be obtained by consulting Department of Transportation, Specifications for Shipping Containers, Paragraph 178.224, Specification 21C.

PLASTIC CYLINDRICAL CONTAINERS

Improvements in materials and manufacturing methods have contributed to the development of commercially acceptable polyethylene drums and tanks. Of particular significance has been the development of the all molded polyethylene container. Due to the flexibility of the loaded polyethylene container it is necessary to use a rigid overpack for shipping purposes. The polyethylene units are designed in all size ranges to fit into standard steel drums or for smaller units, either into steel pails or corrugated boxes.

The polyethylene drums, including flanges, are molded in one piece to form smooth continuous units. Just as standard drums have two openings in the head, threaded to receive plugs, the polyethylene units have two heavy-duty flanges with standard threaded openings to accommodate the 2 in. and ¾ in. polyethylene drum plugs. The outer surfaces of the flanges are made with heavy-duty threads which accommodate standard metal, phenolic, or polyethylene screw caps. The screw caps can affect a primary or secondary seal and are generally expendable. The complete unit can be handled, filled, and emptied exactly like standard head fill drums by existing equipment. The polyethylene unit in its steel overpack has been classified as a drum by the Department of Transportation, thus affording low freight rates on shipments and return of empties.

Since the molded drums are interchangeable with any standard steel drum, the complete unit is available in two basic styles, with the steel

CYLINDRICAL SHIPPING CONTAINERS 295

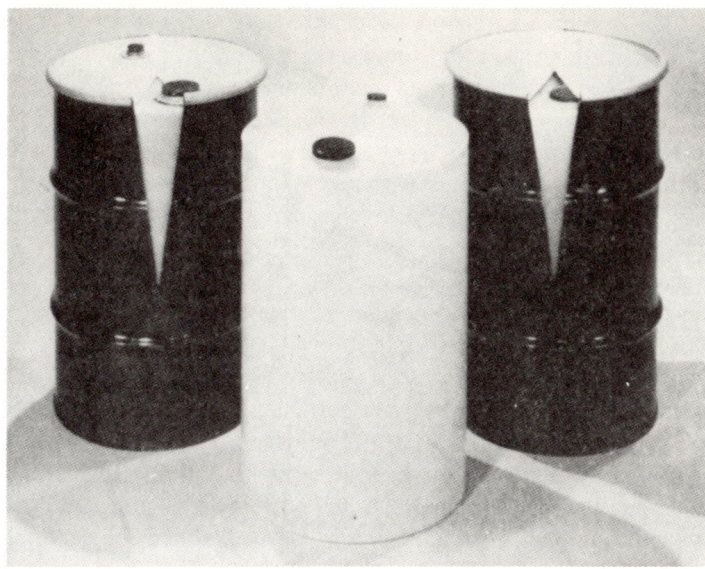

Fig. 7-10. (*a*) 55-gal polyethylene drum shown with two styles of steel drum overpacks. Courtesy Delaware Barrel and Drum Co., Inc.

(*b*) 55-gal full open head (FOH) liner used for either steel or fibre drum overpacks. Courtesy Hedwin Corp.

drum only changed slightly. The first is *totally enclosed* and the second features *protruding flanges* (Fig. 7-10).

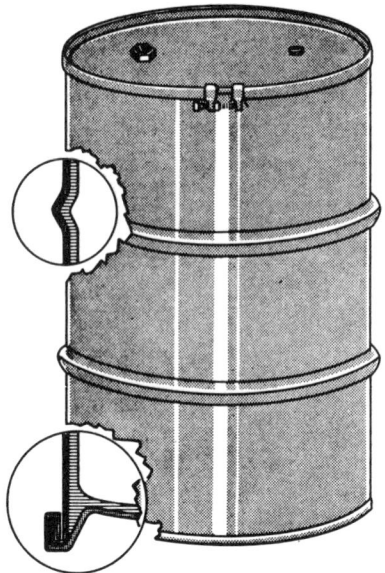

Fig. 7-11. Diagram of polyethylene-lined steel drum. Courtesy Delaware Barrel and Drum Co., Inc.

The advantages of molded polyethylene drums include flexibility, non-toxicity, light weight, durability, chemical resistance and re-usability. When overpacked in steel, these containers are ideally suited for the shipment of such items as food and pharmaceutical products and corrosives, and are available in standard drum capacities from 5 gal to 55 gal.

Polyethylene-coated or lined steel drums (Fig. 7-11) have had significant application for the bulk shipment of edibles, non-edibles, and corrosives. The coating is a continuous homogeneous film bonded firmly to the steel. The finished container presents a smooth, seamless, unbroken inner surface. An average thickness of 10 to 15 mils ensures a heavy barrier between metal and contents. Drum lids generally average 22 mils in thickness for added protection from vapors and fumes. *Hackney-type* drums which are frequently used for long term storage are lined with approximately 30 mils for added protection. The linings may be either clear or may be pigmented with various colors.

CYLINDRICAL STEEL SHIPPING CONTAINERS

Cylindrical steel shipping containers are not only widely used for liquid, granular, flaky, and powdered products, but also for the protection of delicate parts and accessories. Many difficult military export

CYLINDRICAL SHIPPING CONTAINERS

packaging problems have been successfully solved with a cylindrical metal container which affords a rigid structure with, if need be, reusable features.

A convenient classification of steel shipping containers, exclusive of such items as food or oil cans is as follows:

1. *Single trip containers* (Fig. 7-12) for storage and one-way shipment of oils, solvents, alcohol, anti-freeze, gasoline, food products, and various other liquids and semi-solids.

2. *Returnable containers* (Fig. 7-13) for storage and repeated shipment of the above items.

3. *Light weight containers* (Fig. 7-14) designed primarily for storage and shipment of dusts, flakes, powdered, granular, and solidifying materials.

These three classifications of steel shipping containers are furnished in two basic styles of (1) Fully Removable Cover for solids and viscous materials requiring a large opening for filling and dispensing and (2) Tight Head for liquids with mechanically inserted "fittings" to open and reclose for filling and dispensing.

Selection of any of these major classes of steel-shipping containers is governed by (1) the product which may be covered by transportation regulations and restrictions requiring a certain type of container, by (2) the end use of the product, and by (3) return or disposal considerations affecting the over-all economy of the container.

The basic capacity and size categories of steel shipping containers are *drums* 15, 30 and 55 gal., *and pails* 1-10 gal. which have been previously defined. A description of the major construction features of these styles follows.

STEEL DRUMS - Single Trip

Steel drums are made in various capacities generally containing 15, 16, 30, and 55 gal, although other capacities can be obtained. Most single trip containers (Fig. 7-12) have a U.S. Gauge Number of 18 and 19, but are also made of 20 and 22 gauge for such applications as oil and grease shipments.

Steel Drums - Returnable. Export and returnable containers (Fig. 7-13) have 14, 16, and 18 as the most common gauges with others also used for special uses.

Steel Drums - Light Gauge. Lightweight containers (Fig. 7-14) are made in 20, 22, 24, 26, and 28 gauges with the thinnest gauge recommended for asphalt and other solidifying products. The latter is called a strippable drum since the drum is stripped from the enclosed product in unpacking.

(a) SINGLE TRIP FULL OPEN HEAD LEVER LOCK RING CLOSURE 55-GAL-18 or 18/20 GAUGE WITH SWAGED ROLLING HOOPS AND DOUBLE SEAMED BOTTOM CHIME.

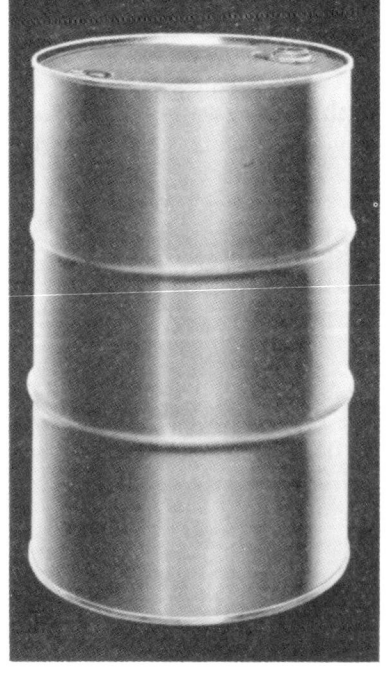

(b) SINGLE TRIP TIGHT HEAD 55-GAL-18 or 18/20 GAUGE DRUM WITH ¾" and 2" FLANGE AND PLUG FITTINGS. DOUBLE SEAMED CHIME AND SWAGED ROLLING HOOPS. FITTINGS MECHANICALLY INSERTED.

Fig. 7-12. 55-gal, 18-gauge single trip steel drums. Courtesy Rheem Manufacturing Co.

CYLINDRICAL SHIPPING CONTAINERS 299

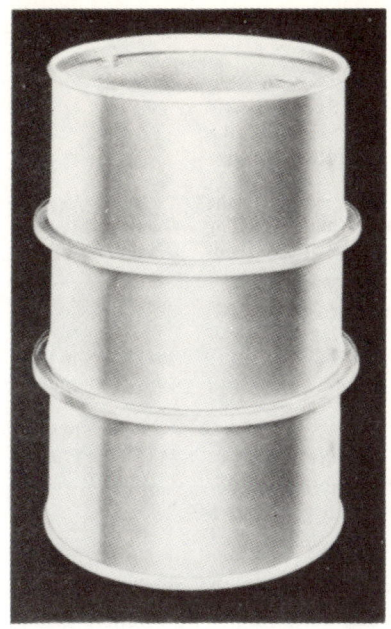

Fig. 7-13. Returnable tight head 55-gal-14-16 gauge drum with welded drop forge fittings in head with I-bar rolling hoops and reinforced chimes. Courtesy Rheem Manufacturing Co.

Fig. 7-14. 55-gal-24 gauge tight head light-gauge drum with 9-in. friction cover opening. Courtesy Rheem Manufacturing Co.

DISTRIBUTION PACKAGING

Dimensions and tare weights of a few standard drums are listed in Table 7-2.

TABLE 7-2
TYPICAL STANDARD STEEL DRUMS

Capacity	Name	Gage	Over-all Height	Chime Depth	Height Rolling Hoops	Over-all Diameter	Average Weight (Tare)
55 gal	Tight head universal	16	$34\tfrac{13}{16}$	$\tfrac{3}{4}$	$1\tfrac{3}{32}$	$23\tfrac{7}{16}$	62 lb
55 gal 400 lb	Full removable head universal	18	$38\tfrac{1}{16}$	$\tfrac{3}{4}$	$\tfrac{5}{8}$	$23\tfrac{27}{32}$	56 lb
30 gal	Tight head universal	19	$28\tfrac{7}{8}$	$\tfrac{3}{4}$	$2\tfrac{7}{64}$	$19\tfrac{3}{16}$	29 lb
16 gal 120 lb	Full removable head universal	20	$26\tfrac{3}{4}$	$\tfrac{1}{2}$	$\tfrac{7}{16}$	$14\tfrac{7}{8}$	16 lb
16 gal	Tight head universal	20	$26\tfrac{7}{8}$	$1\tfrac{1}{16}$	$\tfrac{7}{16}$	$14\tfrac{7}{8}$	17.5 lb

Drum constructions. Drum constructions vary in the manner in which (1) the drum head is fastened to the body (seaming), (2) whether or not rolling hoops are employed and how these are constructed, and (3) how drum opening and covering are accomplished.

The most common method of fastening a drum head to a body is the straight double seam shown in Fig. 7-15. At least 95 per cent of the 55 gallon

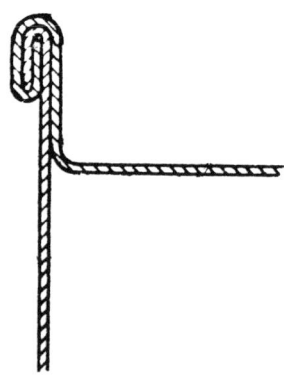

Fig. 7-15. Five-thickness double seam.

CYLINDRICAL SHIPPING CONTAINERS

drums in the U.S. are fabricated with this conventional five-thickness double seam.

Double-seaming consists of a process by which the chime is formed by rolling the body of the drum and the top head or bottom head together. Double-seamed chimes are furnished with or without reinforcement. A seam cement of a non-hardening type, which does not contaminate the product and withstands high temperature in baking of exterior and interior finishes, is used.

Rolling hoops not only provide a suitable contact surface when a drum is horizontally positioned for rolling, but serve to strengthen the sidewall of the drum. The number of hoops and their positioning, as well as the type of hoop used depends upon the size of the drum, the gauge of metal used and other factors including end-use requirements.

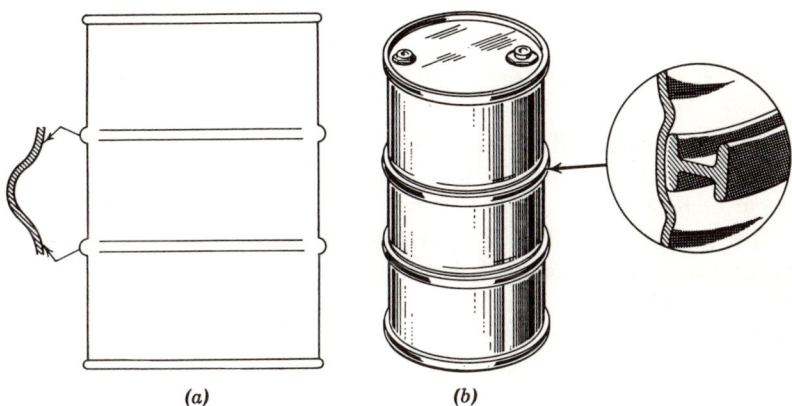

Fig. 7-16. Rolling hoop constructions. (a) Swaged (b) I-Bar.

Rolling hoops are swaged into the body of the drum (Fig. 7-16a). If damage to the sidewalls from abusive handling is anticipated, hoops are reinforced with circular steel I-sections (Fig. 7-16b), which are either attached by deforming the body of the drum to form continuous beads on each side of the hoop or by attaching welded metal lugs on each side of the I-section.

A recent development in rolling-hoop design permits, through an offset of the hoop, sidewall drum contact in shipping and handling. This offsetting facilitates unitizing and multiple mechanical handling of drums when suitable strapping is used (Fig. 7-17).

302   DISTRIBUTION PACKAGING

Fig. 7-17. Details of steel drums with offset rolling hoops and unitized load of four such drums. Courtesy Vulcan Containers, Inc.

LIGHT GAUGE DRUMS AND PAILS - Special Cover Features Used

Drum openings and covers. There are a variety of openings and covers which have been devised for these containers. The most popular types are described here.

The *friction lid* or friction cover type is preferred for products requiring a large opening for filling and dispensing and an easy method of reclosing. This feature is quite common for lightweight drums which are available with various sizes of friction covers (Fig. 7-18).

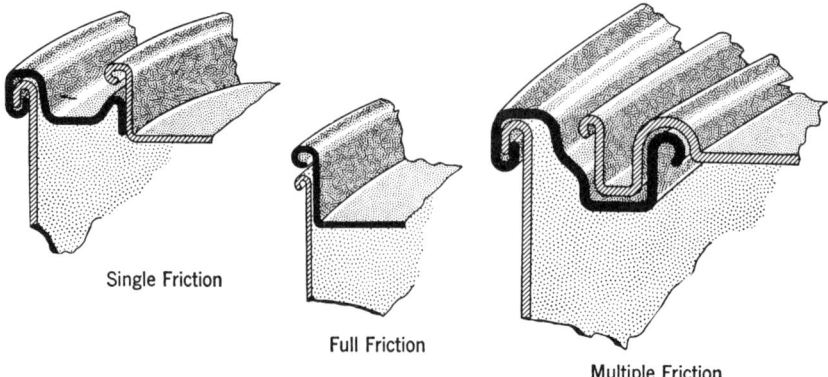

Single Friction

Full Friction

Multiple Friction

Fig. 7-18.   Three styles of friction covers.

CYLINDRICAL SHIPPING CONTAINERS 303

The *bolted cover* is usually specified for containers holding pastes, other semi-liquids, and powdered materials, when a large opening is required. These lids are secured to the drum with machine screws and utilize gaskets for leakproofness (Fig. 7-19).

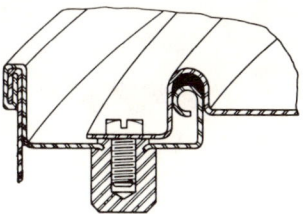

Fig. 7-19. Details of bolted cover type closure.

The *full-open head* is held in place with metal closing rings which are drawn tight with levers or bolts (Fig. 7-20). The heads of these drums are sometimes fabricated of heavier material than used in the bodies of the containers. Drums of this type are recommended when a complete opening is desirable for filling or dispensing purposes.

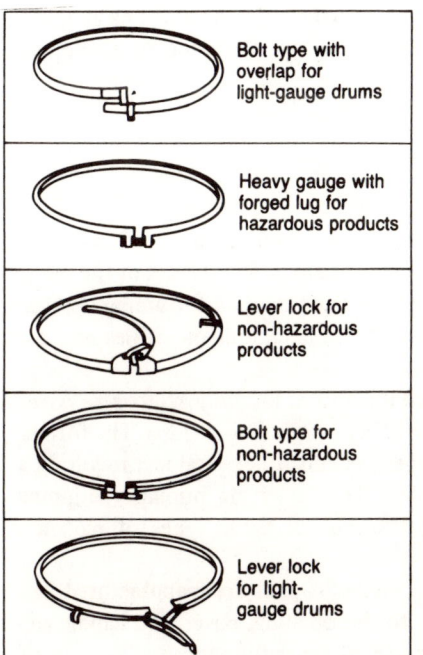

Fig. 7-20. Drum closure rings for full-removable head drums. Courtesy Steel Shipping Container Institute.

The *lug-cover* is most frequently used for grease drums. A sealing tool crimps the lugs under the curled portion of the head. Each lug must be pried open with a screw driver or other tool to open the drum (Fig. 7-21).

Fig. 7-21. Lug-type cover construction.

Lug Cover

Fittings (or closures) for *tight-head* drums used for liquids are equipped with pressed-in flanges and openings which can be obtained in ¾ in., 1½ in. and 2 in. sizes located on the head at any desirable position. A metal cap seal crimped over the openings affords a tamperproof and secondary seal. Openings are also furnished on the bilge of the drum either centered or located near one of the heads of the drum. These afford drainage when the drum is located on the bilge or standing in an upright position. A vent hole with a cast iron plug desirable for complete drainage is also often furnished. Since there are so many styles and types of flanges, fittings, seals, and vents, it is essential to rate the various feasible alternates in terms of filling and end-use requirements. Many of these features are proprietary and thus, supplier technical data should be consulted.

PAILS

Pails are manufactured in capacities ranging from 1 to 12 gal with the most common capacities being 1, 2, 5, and 6 gallons. The gauges of the metal used varies but generally is either 24 or 26. Pails are used for similar applications as steel drums, and are desirable for smaller unit quantities and for manual on-the-job handling.

Basically, two styles of pails are manufactured, (1) *fully removable cover type pails* (Fig. 7-22a) and (2) *tight head type pails* (Fig. 7-22b). The former are generally used for solids and semi-solids and other products requiring a large opening for filling and dispensing. The latter are primarily shipping containers for liquids. Additional illustrations of these types of pails are shown in Fig. 7-23.

Pail constructions. The above two styles of pails are manufactured in a variety of types differing with respect to the opening, cover, and handle employed. A discussion of some of the important types follows.

CYLINDRICAL SHIPPING CONTAINERS

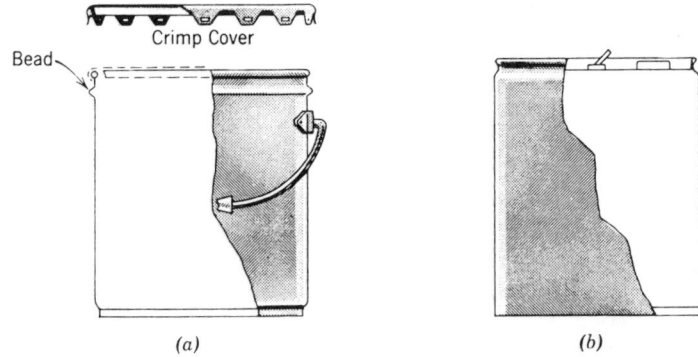

Fig. 7-22. (a) Crimp cover type pail (b) Tight head type pail.

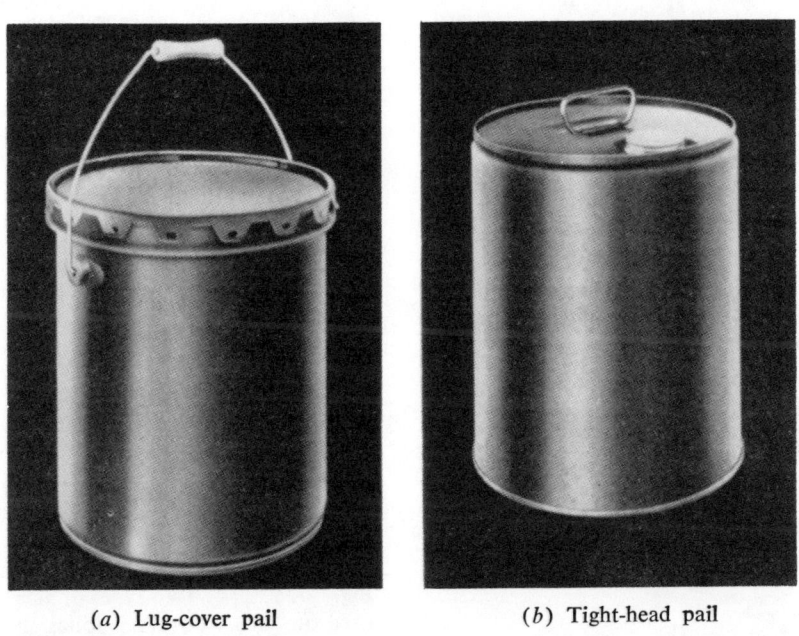

(a) Lug-cover pail   (b) Tight-head pail

Fig. 7-23. 5-gal-24 gauge lug-cover and tight-head pails. Courtesy Rheem Manufacturing Co.

The *crimp cover type* pail utilizes lugs on a full cover which is crimped around the perimeter of the opening with a special crimping tool (Fig. 7-24). This cover can also be furnished with a screw cap for such items as heavy fluids and gear oils.

DISTRIBUTION PACKAGING

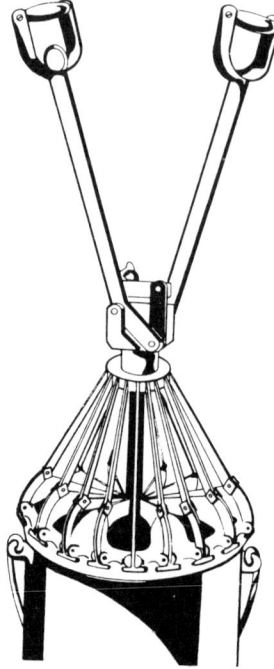

Fig. 7-24. Two types of manually operated crimping tools to seal lug-cover pails.

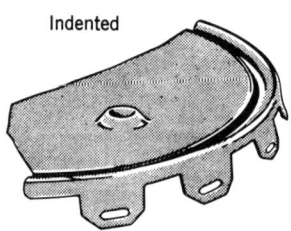

Indented

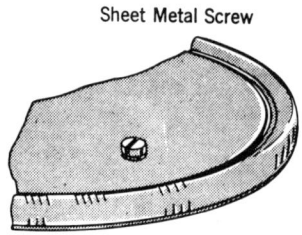

Sheet Metal Screw

Fig. 7-25. (a) Two types of vents used on pails. Courtesy Jones and Laughlin Steel Corp.

CYLINDRICAL SHIPPING CONTAINERS

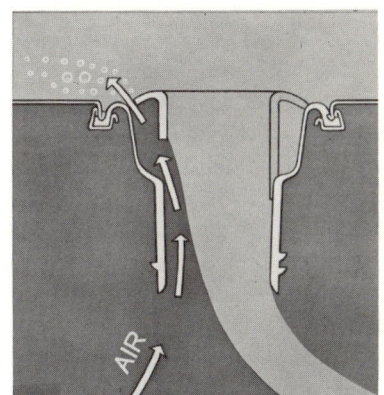

(b) Principle of operation of Flexspout® closure. Courtesy Rieke Corp.

The *tight head type* pail is furnished straight-sided as well as with expanded beads near the top and/or bottom. A variety of screw caps, pour spouts and filling devices can be furnished. These will vary with the filling and dispensing requirements of the product. One patented, self-venting system has been developed by one manufacturer to facilitate free flow during pouring. This fixture is a completely assembled, one-piece, polyethylene closure which provides leakproof, tamper-proof delivery of liquids. The principle of operation is shown in Fig. 7-25b.

Pail *handles* are attached to the body of the pail on cover type pails and to the top on the tight head type pails. The former is referred to as a *bail*, which is generally furnished with a wooden grip for ease of carrying. The latter are *drop handles* with or without wooden grips which do not interfere in stacking of the containers.

Vents may be incorporated as part of the screw cap or neck or separately as illustrated in Fig. 7-25.

STEEL CONTAINER LININGS

Generally speaking, linings are designed to protect the contents of a container from contamination or discoloration by the steel and not necessarily to protect the container from the action of the materials packaged. Specific problems frequently introduce special considerations and resistance tests may be necessary to determine whether or not linings meet particular requirements. In some instances, it has been found that two coats of linings are essential to provide protection for commercial periods of storage. Suitable linings have frequently enabled the use of steel containers for material which formerly had to be packed in containers made of wood, glass, and aluminum.

Baked lacquer linings provide excellent protection for food products, shortening, or any stock containing fatty acids, sulphonated oils, or other products which might otherwise be affected in taste or composition by contact with an unlined container wall. Lacquer-lined drums are being increasingly used for alkyds, resins, alcohols, castor oil, chocolate syrup, formaldehyde, glycerine, soaps and many other products.

Recently, lining materials of the vinyls, phenolics, and epons groups have been found to assist in extending the product types suitable for the shipment of steel shipping containers. Also separate interior polyethylene bags have permitted acids and other dangerous articles to be packed in steel shipping containers.

For products requiring extremely long service or for severe corrosion problems *stainless steel* containers are available. Essentially, these are furnished without linings or coatings (Fig. 7-26).

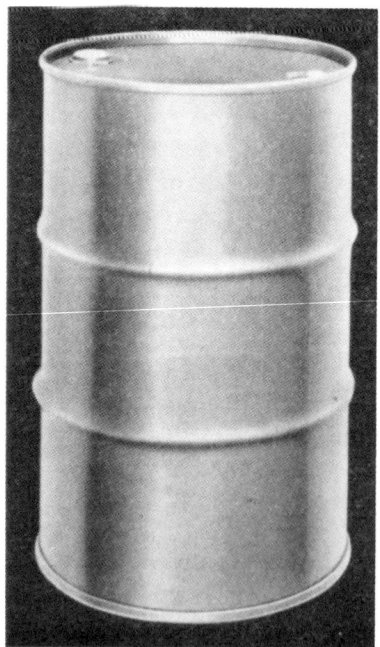

Fig. 7-26. 55-gal-16 gauge stainless steel drum for difficult corrosive material. Courtesy Rheem Manufacturing Co.

CYLINDRICAL SHIPPING CONTAINERS

EQUIPMENT FOR CYLINDRICAL CONTAINERS

To assist in the application of cylindrical containers various types of standardized and special equipment have been developed. Basically, such equipment can be classified as (1) packaging equipment, (2) sealing and shipping preparation equipment, and (3) unpackaging equipment. The type of equipment to be selected for the above functions in turn depends upon the following considerations:

1. Type of product and its physical and chemical form, that is, liquid, semi-liquid, and solids.

2. Degree of mechanization attainable in view of individual plant and product conditions.

3. Type of container selected for protection, cost, handling, and end-use requirements.

Obviously, the physical form of the product to be packed in a cylindrical container is of prime importance in the selection of equipment aids. In addition, the nature of the product influences the flow of material to the packaging station and into the container. For example, free-flowing liquids are exclusively handled through piping with gravity feed attainable into the shipping container. On the other hand, viscous materials require pressure feeding for filling.

Fig. 7-27. Automatic drum-filling equipment. Courtesy The Rucker Co.

Fig. 7-28. Automatic pail filling, weighing, and closing system. Courtesy MRM Co., Inc.

Because of product differences, various types of filling principles are being employed. For liquids, this includes gravity, vacuum, and volumetric filling. To avoid foaming or spilling due to entrapped air during the filling process, vacuum or subsurface filling is practiced. In vacuum filling most of the air in the container is removed by a partial vacuum either prior or simultaneously with filling. In subsurface filling the filling nozzle goes to the bottom of the container and rises with the liquid level, yet always remains just under the surface.

Free-flowing solids are filled quite similarly to liquids, although provisions for the removal of dust must be made. Certain viscous products may be heated prior and during the filling operation where no separation of the product will occur. This enables the employment of an essentially liquid filling system. Other viscous or non-flowing solid materials require a pressure-type filling system. Auger- and impeller-type equipment is frequently used.

CYLINDRICAL SHIPPING CONTAINERS

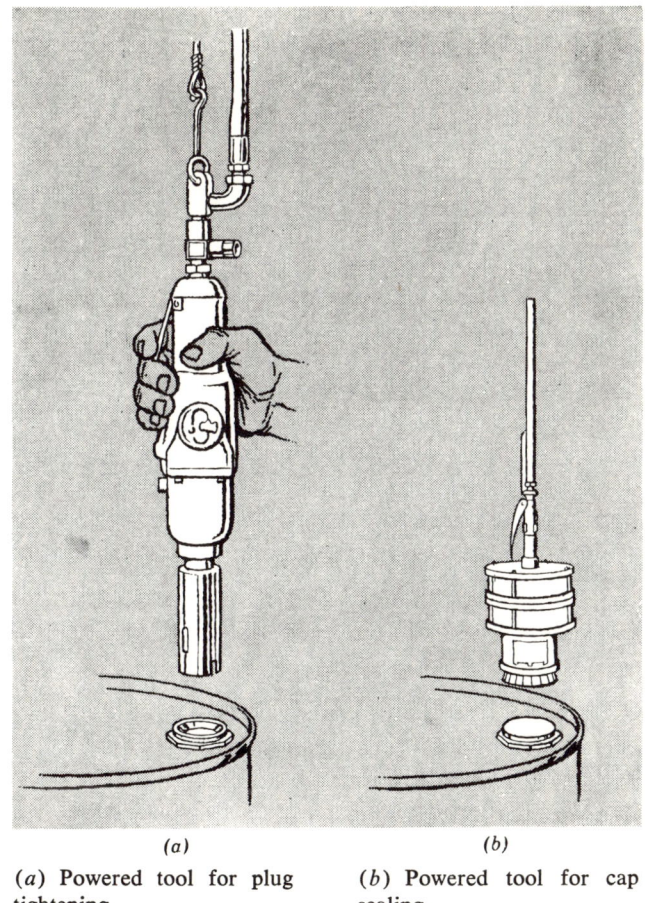

(a) Powered tool for plug tightening

(b) Powered tool for cap sealing

Fig. 7-29. Typical powered tools for plug tightening and cap sealing. Courtesy American Flange and Manufacturing Company, Inc.

Product variations and frequent difficulties in the filling process have led to the development of special equipment and to the modification of existing equipment for adaptation to a given packing line.

Although the type of product has the greatest influence in the selection of equipment, many other factors must be considered. These factors include process type (continuous or batch), materials-handling equipment employed, or potentially suitable processing rates and size and type of container. All of these factors determine the degree of mechanization attainable and thus the economics of the system. Consolidation of separate lines through higher-speed equipment or combination equipment (filling, weighing, sealing, etc.) usually permits highest labor productivity.

End-use requirements dictate to a large degree the type of container that is selected. These requirements include product protection, handling, unpacking and dispensing, re-use, return, and the cost to accomplish these objectives. Although all cylindrical shipping containers have great similarity in style, packaging equipment has been designed for particular container types when the demand for such specialization is justified. For example, in the oil industry, high-volume drum filling has created the need for equipment of the type illustrated in Fig. 7-27.

Sealing equipment is directly related to the style of cover or seal employed on the shipping container. The sealing operation may be accomplished separately or incorporated as part of a complete packing system. In Fig. 7-28 an automatic pail filling, lug-cover placement and lug-crimping machine demonstrates such an integrated operation. If sealing is performed separately, manual and powered tools are needed. Two different types of manual tools for lug-cover crimping have previously been shown in Fig. 7-24. Typical powered tools for plug-tightening and crimping of a secondary seal are illustrated in Figs. 7-29*a* and 7-29*b*.

Other shipping preparation practices pertinent to the various types of cylindrical containers such as marking and labeling and tape sealing are covered in subsequent chapters.

Unpacking a product shipped in a cylindrical container can be accomplished in many ways such as by means of special techniques adapted to cylindrical containers. Unpacking may simply consist of dumping or draining, stripping, pumping, or gradual dispensing. Vacuum removal processes are available for continuous or intermittent dust-free unpacking of many products.

CHAPTER 8

Wrapping, Barrier, and Cushioning Materials

The packaging materials discussed in the preceding chapters are ordinarily furnished in fabricated form and are designed with an established contour. However, there are other packaging materials, which either supplement another type of container or material, or in themselves serve as an external protective medium. These materials serve the following three major functions:

1. For wrapping purposes.
2. To provide a protective barrier.
3. For cushioning purposes.

Although there are several definite exceptions, most of the materials used for these functions are also considered as flexible packaging materials.

DEFINITIONS

Wrapping. Wrapping is broadly defined as the covering of a product by winding or enfolding with an unformed flexible packaging material. Wraps are further identified by function, method of application, and/or by their special usage. Owing to this concept, barrier materials may frequently be used for wrapping purposes. Similarly, cushioning materials may also serve the dual function of wrapping.

Barrier materials. These materials are designed to withstand, to a specified degree, the penetration of water, oils, water vapor, or certain gases. They may serve to exclude or retain such elements without or within a package. Like wrapping materials, barrier materials are also generally flexible and unformed.

Cushioning materials. Cushioning materials are primarily designed to absorb the shock or reactions caused by external forces acting upon a package. Cushioning materials vary widely with respect to resiliency and effectiveness and may be furnished unformed or preformed. A wide variety of materials are used for cushioning applications.

One or more of these three major packaging functions are accomplished by the following classes of materials: (1) *protective packaging papers*, (2) *protective packaging fabrics*, (3) *films*, (4) *foils*, and (5) *cushioning materials*. In view of the interrelationship of the functions of the above materials, this chapter is organized along material rather than functional lines.

PROTECTIVE PACKAGING PAPERS

The many types of packaging papers which are commercially available cannot be covered within the scope of this text. This treatment is principally concerned with papers, untreated, treated, coated, impregnated, refined, regenerated, reinforced, and laminated, which are used for *protective* purposes. Further and more detailed information on paper and papermaking is available in other literature.

KRAFT PAPER

For the wrapping of such products as textiles, paper products, furniture, and small commodities of a wide description, kraft paper is used (Fig. 8-1). For these applications, kraft paper is dispensed from a roll,

WRAPPING, BARRIER, AND CUSHIONING MATERIALS

 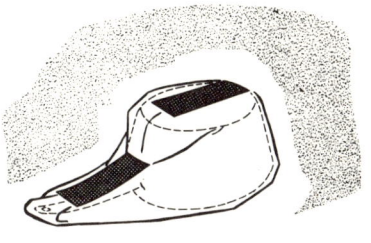

Fig. 8-1. Typical kraft paper wrapping applications.

and the wrap is formed around the product. When a large volume of the same size commodity is to be wrapped with kraft paper, it is frequently more economical to design preformed bags, tubes, or covers which can be applied to the product with greater ease and which will furnish a neater appearance. Wrapping machinery is also available to simplify this operation. In certain instances, wrapping from precut sheets may be advantageous, especially for saving in the use of paper.

Kraft paper used for wrapping purposes is a comparatively *coarse paper* made entirely from *wood pulp* produced by a modified *sulphite pulping process* and is particularly noted for its strength. It is usually manufactured on a Fourdrinier machine with a regular machine-finished or machine-glazed surface. It can be watermarked, striped, or calendered, and it has an acceptable surface for printing. Its natural unbleached color is light brown, but by use of bleaching it is produced in lighter shades or white. It may also be colored by the use of dyes. Kraft wrapping papers are produced in a large range of basis weights usually from 25 to 120 lb. on a ream weight basis (500 sheets, 24 in. $\times$ 36 in.).

The heavier grades are primarily used for exterior wraps, whereas the lighter grades serve for such functions as finish or contamination protection within an outer shipping container. Kraft wrapping paper may also be *treated* to impart greater wet strength characteristics if moisture is to be encountered. Laminated, reinforced, coated, and impregnated kraft papers are also available for specific purposes. Some of these papers are discussed later in this chapter. The properties of kraft papers are listed in Table 8-1.

It should be noted that papers other than kraft are also used in large quantities for wrapping purposes. These include jute paper, adulterated krafts, and other waste papers.

Basic papers selector chart

TABLE 8-1

	Kraft				Bleached sulphite	Pouch	Greaseproof	Glassine	Parchment
	Unbleached	Bleached	Creped	Extensible[1]					
General									**General**
Composition[2]	Sulphate pulp	Sulphate pulp	Sulphate pulp	Sulphate pulp	Generally a mixture of soft and hardwood	Usually virgin kraft	Hydrated pulp	Super-calendered hydrated pulp	Acid treated cellulose sheet
Ream weight range	20-90	10-90	Depends on % crepe	30-110	10-90	20-60	20-60	20-60	18-110
Yield, lb./sq. in.	4,800-21,600	4,800-43,200	Depends on % crepe	3,927-14,400	4,800-43,200	7,200-21,600	7,200-21,600	7,200-21,600	3,927-24,000
Color[3]	Tan to brown	White, colors	Tan to white	Tan to white	Opaque or colors	Opaque	White, amber, colors	Translucent colors	Translucent & opaque, white, colors
Mechanical									**Mechanical**
Tensile strength, CD lb./in. width/lb. basis weight (ASTM D 828)	0.35	Depends on functional characteristics desired	About same as flat kraft	Less than flat kraft	Depends on functional characteristics desired	0.4	0.3	0.4	0.4
MD	0.7					0.7	0.6	0.7	0.7
Elongation, CD (ASTM D 828)	2-7%	2-7%	Varies	8%	2-7%	Low	Low	Low	Varies
Burst (Mullen) (ASTM D 774)	90% basis weight	70% basis weight	—	More than flat kraft	70% basis weight	—	High	High	0.8 dry 0.4 wet

Tear, gms./lbs. basis weight (ASTM D 689)	CD	2.5	2.2	—	15% more than flat kraft	Depends on functional characteristics desired	0.9	0.8	0.7	0.85
	MD	2.3	2	—	5% more than flat kraft		0.8	0.7	0.6	0.85
Converting & use characteristics										**Converting & use characteristics**
Surface finish		—[3]	—[3]	Not calendered	Not calendered	—[3]	High	Matte	High	Matte to high
Printability		Excellent	Excellent	—	Excellent	Excellent	Excellent	Excellent	Excellent	Good
Coatability		Poor[4]	Poor[4]	Poor[4]	Poor[4]	Fair[4]	Excellent	Good	Excellent	Excellent
Grease, oil resistance		None	None	None	None	None	None	Excellent	Excellent	Excellent
Wet strength		No[5]	No[5]	No[5]	No[5]	No[5]	No	No[5]	No[5]	No[5]
Uses		Multi-wall bags, wrappers, liners, extrusion coating	Box wraps, bags, surgical packaging extrusion coating	Bags, surgical packaging	Multi-wall bags, wrapping, tape base, extrusion coating	Box wraps, waxing labels	Bags, wrappers	Bag liners, box liners	Bags, bag and box liners, wrappers	Butter wraps, meat wraps, release sheets, wet-strength wrappers, liners and bags

[1] Has a very high tensile energy absorption (toughness).
[2] Many papers are made from mixed pulps and recycled fibers and may be used for special properties or performance.
[3] Papers are generally available in a range of brightness (i.e. whiteness) with color added and also several degrees of surface finish.
[4] Depends on degree of calendering or use of surface primers.
[5] Special grades with wet strength treatment are available.

Reproduced by special permission of the publishers of Modern Packaging Encyclopedia and Planning Guide, December, 1974.

TABLE 8-1 (Continued)
Converted papers chart*

	Glassine			Pouch, lacquered	Polyethylene-coated kraft	PVDC-coated kraft	Waxed sulphite	Hot-melt coated paper
	Lacquered	Waxed	Laminated					
Ream weight range	23–45	23–65	45 and up	24–40	33 and up	35 and up	22 to 100	22 to 100
Total weight of coating	1–4 lbs. per side	2–5 lbs. per side	5–10 lbs. laminant	1–3 lbs. per side	8 lbs. and up	4 lbs. and up	5–20 lbs.	5–20 lbs.
Yield, sq. in./lb.	10,000–20,000	7,000–20,000	10,000 or less	11,000–18,000	13,000	12,000	4,000–20,000	4,000–20,000
Color[1]	Opaque, bleached natural	Opaque, bleached natural	Opaque, bleached natural	Opaque	Opaque or colors	Opaque or colors	Opaque	Opaque or colors
Use temperature Deg. F. High	140	120	120	140	160	180	120	130
Use temperature Deg. F. Low	−20[2]	−20	−20	−20[2]	−50	−[2]	−20	−30
WVTR[2], flat, ASTM E 96 gms./sq. m./100 deg. F., 95% R.H., 24 hrs.	4	4	4	8	30	8	8	8

Property								
Grease, oil resistance	Excellent	Excellent	Excellent	Fair	Fair	Excellent	Poor	Poor
Gas permeability	Low[2]	Low[2]	Low[2]	Fair	Poor	Fair[2]	Very high	Very high
Heat seal temp., deg. F.	200-350	150-200	No	200-350	200-350	250-350	150-200	150-250
Heat seal strength	High[3]	Weak	None	High[3]	High[3]	High[3]	Weak	Good to high[3]
Machineability	Excellent	Excellent	Excellent	Excellent	Good	Excellent	Good	Good
Printability	Good	No	Excellent	Good	Good if treated	Good	No	No
Uses	Bags, pouches, overwraps	Liners	Bags, liners	Pouches	Bags, pouches, wrappers	Bags, pouches, liners	Wraps, overwraps, liners	Bags, liners, overwraps

* The physical properties of converted paper do not significantly differ from the data shown under Basic Packaging Papers (or the equivalent total basis weight in the case of laminated glassine, etc.).
[1] Papers are generally available in a range of brightness (i.e. whiteness) with color added and also several degrees of surface finish.
[2] Depends on type and amount of lacquer or coating formulations.
[3] A high seal strength means the seal will pull paper fiber.

Reproduced by special permission of the Publishers of Modern Packaging Encyclopedia and Planning Guide, December, 1974.

TISSUE PAPER

Tissue papers have wide application in distribution packaging operations. They are used predominantly to protect against contamination, for the surface protection of a product against scratching and marring, and to fill voids in a package. Tissue is also employed for interleaving purposes, to separate products such as garments and sheeting. The use of tissue paper can frequently be eliminated by special treatment applied to the inside surfaces of the unit or shipping container. Such treatments as waxing eliminate the need for a separate packaging component and provide a simplified packaging method.

Tissue is a generic term for any type of lightweight paper. The term is generally restricted to paper weights of *less than 18 lb basis weight* (500 sheets—24 in. $\times$ 36 in.).

Tissue can be specially treated for the wrapping and packaging of such items as silverware, in which tarnishing may be encountered.

COATED, IMPREGNATED AND REFINED PAPERS

Ordinary untreated paper can be modified in its properties by altering the form of the basic fibrous structure of the paper or by additions to the paper of other materials. Included in this group of papers are *glassine, plain greaseproof papers, vegetable parchment, wet strength papers,* and *waxed papers*.

The major applications of these papers are for the packaging of food products such as liners, bags, and wraps in conjunction with other unit packages. However, they are also used for such industrial products as lubricating greases and oils, putty and caulking compounds, printing inks, paint products, and other materials containing petroleum derivatives, linseed oils, etc. For industrial applications, these papers are generally in the form of liners within other types of containers.

Glassine. Glassine is a supercalendered, smooth, dense, transparent or semi-transparent paper manufactured primarily from chemical wood pulp which has been beaten and refined to secure a high degree of hydration of the stock. It is grease resistant and has a high degree of resistance to the passage of air and many essential oil vapors used in food flavoring. When waxed, lacquered, or laminated, glassine paper is highly resistant to the transmission of water vapor. Glassine paper basis weight ranges from 12 to 90 lb (500 sheets—24 in. $\times$ 36 in.). Glassine paper is used as a protective wrapper for food, tobacco, chemicals, and metal parts. For the properties of plain, lacquered, waxed, and laminated glassine, refer to Table 8-1.

WRAPPING, BARRIER, AND CUSHIONING MATERIALS

Plain greaseproof. Plain greaseproof paper is a paper made from well-hydrated fibres possessing the ability to resist penetration of greases, oils, and fats, and is made by a process similar to that used for glassine. However, it is not supercalendered and therefore is less transparent than glassine. Basis weights range from 20 to 60 lb (500 sheets—24 in. $\times$ 36 in.). In Table 8-2, the packaging requirements of typical non-edible items are listed. These requirements are met by glassine and greaseproof papers either as a bag, an inner wrap, an outer wrap, a liner, or as a lamination to boxboard, paper, or other materials. In this table, in the column entitled *Moisture Control* the term "IN" means that moisture must be prevented from escaping from the product and the term "OUT" means that moisture must be excluded. The degree

TABLE 8-2
Packaging Requirements of Typical Non-edible Items Satisfied by Glassine and Greaseproof Papers

Items	Moisture Control	Grease Resistance	Other Functions
Soap—high moisture content	IN 2 to 3		Neutral pH, Odor retention
hard milled	IN and OUT 1 to 2		Mold inhibition
Dyes	OUT 2 to 3		
Sweeping compounds		3 to 4	
Metal products			
All manner of steel parts	OUT		Noncorrosive
hardware, ordnance, tools, etc.	1 to 4	3 to 4	Neutral pH
Photographic supplies	OUT		
Films, print papers, chemicals	2 to 4		Anti-fog
Razor blades	OUT 3 to 4	4	
Chemicals			
Drugs (pills & powders)	OUT		
Various dry chemicals	1 to 4		
Insect powders			
Putty products			
Paste putty, calking strips, calking compounds		4	
Waxes			
Floor wax, shoe polish, candles, paraffin		2 to 4	
Tobacco products			
Cigarettes, cigars, pipe tobacco, chewing	IN and OUT 2 to 3		Flavor retention

of moisture control required is indicated by codes 1 to 4, in which 1 is slight, 2 is moderate, 3 is considerable, and 4 is great.

Vegetable parchment. Vegetable parchment is a paper which is made by passing a waterleaf sheet prepared from rag or pure chemical wood pulp through a bath of sulphuric acid, after which the sheet is washed and dried. This treatment imparts grease resistance, wet strength, and freedom from loose fibres. It is also odorless and tasteless, is not disintegrated by water or salt solutions, either hot or cold, and is highly resistant to many other solutions. Vegetable parchment is made in weight ranges of 18 to 110 lb (500 sheets—24 in. $\times$ 36 in.) and its principal applications are for food products such as butter, margarine, meats, fish, and for the lining of containers for vegetables.

Parchment can be treated to impart mold-proofness, anti-rancidity, release towards sticky substances, and resistance to spread of oil and grease. In Table 8-1, the properties of vegetable parchment are listed.

Pouch paper. Pouch paper is made from virgin Kraft paper and is usually bleached, refined, supercalendered and highly plasticized. This paper is primarily produced for use in combination with film or foil to form pouches used for the packaging of food and pharmaceutical products, chemicals and photographic materials. Pouch papers are strong, pliable, have high density, but have no greaseproof properties. In Table 8-1 the properties of various types of pouch papers are listed.

Waxed papers. Waxed papers are papers that have been treated with wax to provide waterproof, vaporproof, and greaseproof qualities. They also afford low-temperature flexibility and scuff resistance. Industrial applications for waxed papers include wrapping, lining, or separating of greasy and oily metal parts, and to provide scuff resistance for appliances and furniture. Waxed paper is furnished in a wide range of basis weights and formulations, as listed in Table 8-1.

Coated papers. Coated papers include a large variety of specialty paper products which have been modified by the addition of coatings on one or both faces of the sheet. These coatings impart to the paper heat sealability, added decoration, scuff resistance, insect and rodent resistance, and other special properties. Economic considerations are the determining factors in selecting coatings over other methods that impart important performance characteristics to a wrapping material. Further consideration in this selection includes the type of closure required, the degree of flexibility for wrapping (machinability), and strength requirements, as well as specific physical and chemical product requirements.

WRAPPING, BARRIER, AND CUSHIONING MATERIALS

LAMINATED PAPERS

Frequently a single material is incapable of fulfilling all the *physical and chemical packaging requirements* for a given product. In other instances the properties could only be provided by the use of an excessive and uneconomical amount of a given material. For this reason many of the above materials and others discussed in this chapter are combined into laminated papers. Generally, these are defined as any *laminated structure* in which one ply is paper. Typical examples are asphalt, metal foil, and polyethylene laminated papers.

The combined structure often provides new properties to the laminated sheet through the type of laminant employed. This result is not achieved however in combining processes in which a separate laminant is not required (for example sheets having self-bonding characteristics under heat and/or pressure).

For distribution packaging uses laminated papers are designed to produce the function of a barrier, as defined previously. Selection of a given laminated material is governed by compatibility of materials, anticipated or required resistance to environmental conditions, workability, and cost. More specific applications and properties of laminated papers are covered below and in Chapter 9.

REINFORCED PAPERS

In many instances, the properties of a laminated sheet are insufficient to fulfill all the *strength requirements* of a specific packaging application. Therefore, it is necessary to introduce a reinforcing medium into the laminated structure. Typical reinforcing materials include *cotton, sisal, glass, rayon,* and *nylon fibres.* Ordinarily the fibres are embedded in the laminant itself, either at random (as is the case with sisal fibres), or in a distinct ordered pattern. Fig. 8-2 illustrates an asphalt-laminated reinforced fibre sheet. The reinforcing fibres themselves have different strength properties. For example, glass fibres have high tensile strength but relatively low elongation. On the other hand, rayon fibres have exceptional elongation but much lower tensile strength.

The selection of the particular reinforcement medium and its distribution in the sheet depends upon tensile and elongation requirements, puncture resistance, fibre compatability with anticipated or required environmental conditions, and cost.

324 DISTRIBUTION PACKAGING

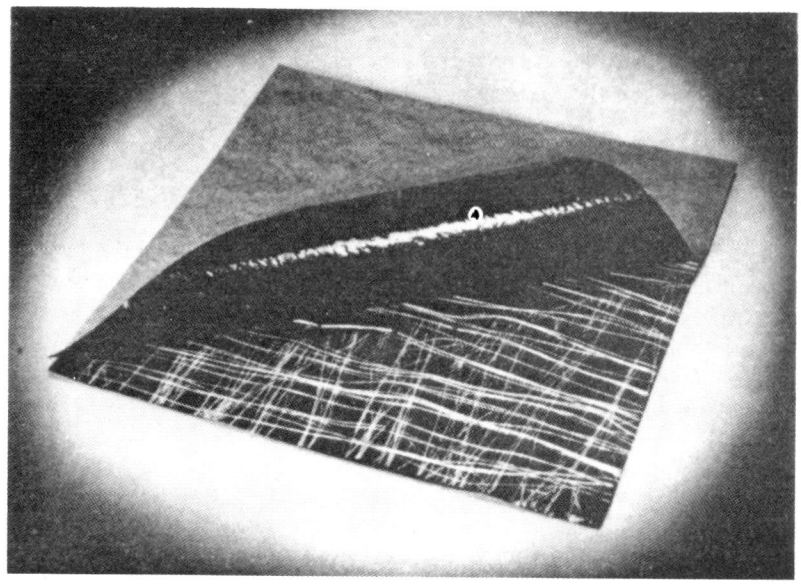

Fig. 8-2. Reinforced asphalt-laminated paper.

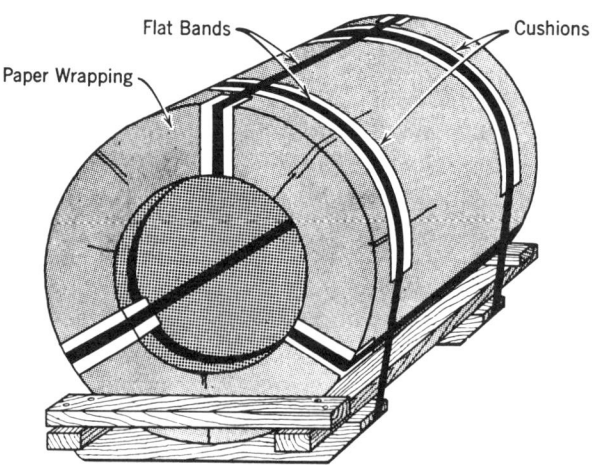

Fig. 8-3. Reinforced paper-wrapped skidded steel coil.

The paper stock of the outer and inner layers of the laminated reinforced sheet is usually of 30 lb stock (500 sheets—24 in. $\times$ 36 in.) but is furnished with other basis weights. Two principal laminants are used: (1) *asphalt* and (2) *latex-based* laminants. Papers using asphalt

WRAPPING, BARRIER, AND CUSHIONING MATERIALS

Fig. 8-4. Reinforced paper-wrapped textile roll.

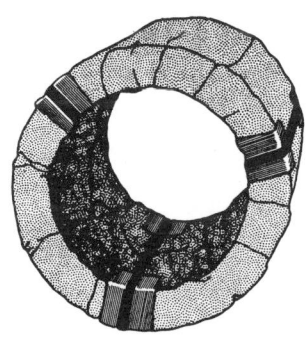

Fig. 8-5. Paper-wrapped coil of sheet steel.

may contaminate the product through bleeding upon exposure to excessive heat, and brittleness may result upon exposure to extreme cold. Latex-based laminated papers are usually reinforced with synthetic fibres, are generally of a lower over-all thickness, and are therefore more pliable. Both asphalt and latex cause problems in the reclaiming of wastepapers.

Reinforced papers are used in converted products such as tapes, bags,

326   **DISTRIBUTION PACKAGING**

Fig. 8-6. Paper-wrapped load of lumber in position on flat car. Courtesy Industrial Bag and Cover Association.

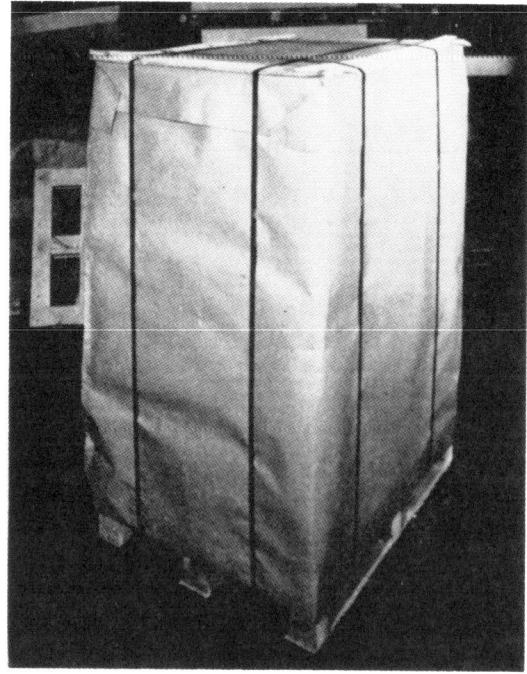

Fig. 8-7. Unit load of battery cases on an expendable pallet protected for shipment with reinforced paper. Courtesy Industrial Bag and Cover Association.

WRAPPING, BARRIER, AND CUSHIONING MATERIALS 327

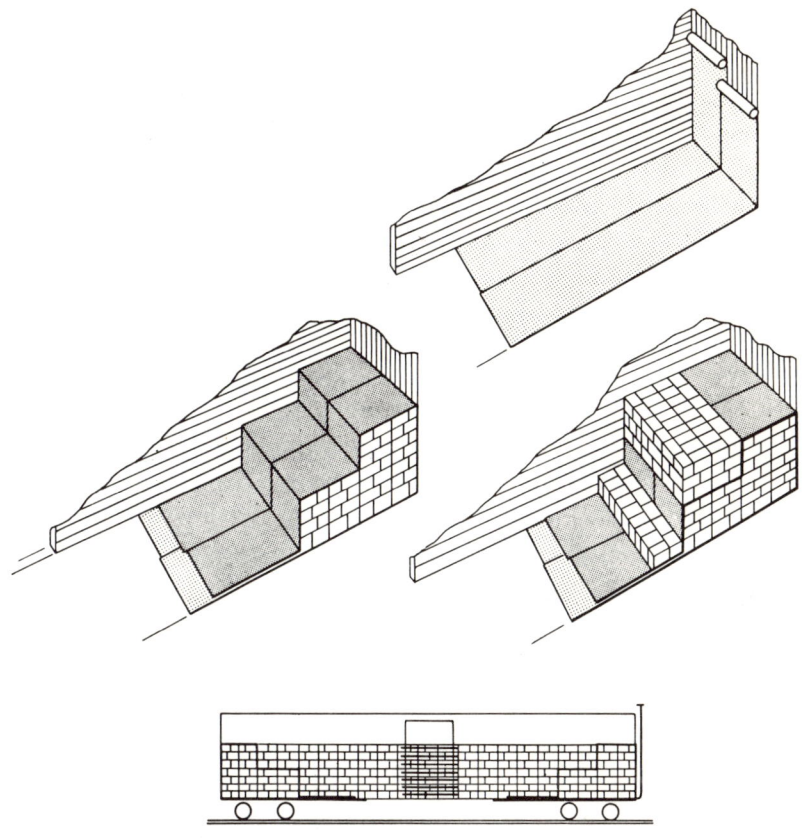

Fig. 8-8. One method of preventing shifting of solid loads of fibre shipping containers through the use of reinforced papers.

and covers. In sheet form they find extensive application in the bundling of lumber, steel, textiles, and other products. They serve for these purposes as exterior shipping protection. The combination of water resistance and strength that these papers provide makes them suitable for the packaging of products which are subjected to severe handling conditions, conditions in which an unreinforced sheet is liable to rupture and destroy the effectiveness of the water barriers. These papers have made possible shipment in open cars, outdoor storage of dressed lumber and other products, and have facilitated unit load handling procedures. In Fig. 8-3 through Fig. 8-7 a few such applications are illustrated.

DISTRIBUTION PACKAGING

Papers of this type are not suitable for products requiring cushioning. When strapping overwrapped bundles, edge protection must be provided under the straps to prevent strap indentation into the product.

Reinforced papers are also used for carloading applications, such as the lining of cars and for the separation of the load within the carrier that reduces shifting (Fig. 8-8).

Aside from important distribution packaging uses, reinforced papers are also extensively used in the construction and building industries.

PACKAGING FILMS

As defined by the American Society for Testing Materials, a film is "an unsupported, basically organic, non-fibrous, thin, flexible material of a thickness not exceeding 0.010 in. Such material in excess of 0.010 in. may be termed as sheet or sheeting." When separately formed films are combined with an adhesive, they are called a *laminate*. If films are extruded (co-extruded) and combined they are called composite films.

Packaging films have widest application in consumer packaging for such products as foods, cosmetics, pharmaceuticals, clothing, and household purposes. However, owing to the widespread extension of prepackaging techniques into industrial fields for such items as hardware, spare parts, and electronic components, it is necessary for the packaging engineer to be acquainted with these materials and their properties.

WRAPPING, BARRIER, AND CUSHIONING MATERIALS

Properties and accepted commercial standards of packaging films, are listed in Table 8-3. The properties of shrink films including representative temperature curves are reproduced in Table 8-4. The following section is a further discussion of these basic packaging films.

CELLOPHANE

Cellophane is the oldest and one of the most widely used of the packaging films. It has retained its popularity because of its low cost, transparency, packaging ease, and its ability to be coated for heat-sealing purposes and for improved moisture transfer properties. Cellophane is made of regenerated cellulose and is a *hydrophilic film, not a plastic.* Cellophane consists of a plasticized base sheet of essential greaseproofness.

For optimum package performance, especially on high-speed packaging equipment, control of the moisture content of cellophane is important. The percentage of moisture content influences flexibility, film strength, and sealability. To permit maximum flexural strength and to ensure a desired moisture content, plasticizers are added to cellophane. Changes in the moisture content produce dimensional changes in the film. This change is more pronounced in the cross-machine direction than in the machine direction.

Cellophane tensile strength is greater in the machine direction than in the cross direction. Conversely, its machine direction elongation is less than the transverse. The tear resistance of cellophane is greater across the machine direction than with the machine direction.

All cellophanes lend themselves to adhesive sealing, with no one particular recommended formulation suitable for all types. Because of its nonporous structure, cellophane is inherently grease-, dust-, and germ-proof. It is also impervious to infestation by insects of the non-boring type. When the film is dry, cellophane is quite gasproof; when the film is wet, gas transmission increases.

CELLULOSE ACETATE

This film differs from cellophane because the acetate base is water insensitive and can only be put in solution by certain solvents. Cellulose acetate has good dimensional stability, clarity, and general durability and has a high rate of water vapor and gas transmission. Cellulose acetate is a *thermoplastic material* made by the esterification of cellulose with acetic anhydride and acetic acid.

Cellulose acetate is more expensive than cellophane, since its manufacture is more complicated. Its applications include use as windows for folding cartons and for the packaging of certain perishables requiring vapor circulation provided by this materials' high rate of moisture transfer.

As a sheet, cellulose acetate is used for thermoforming of packages because of its economy, clarity, and formability. Its applications are for bubble and skin packs and for contoured containers.

RUBBER-HYDROCHLORIDE FILM

(Polyisoprene) rubber hydrochloride film (Pliofilm) is manufactured by treating a crepe rubber with hydrochloric acid. After neutralization in

TABLE 8-3

Properties of packaging films *

	Cellophane			Cellulose acetate[2]	Fluoro-halocarbon	Ionomer	Nylon (polyamide)	Nylon, cast, saran-type coating one side
	Lacquered	Polymer-coated	Polyethylene-coated[1]					
General								
Clarity	Transparent	Transparent	Transparent to translucent	Transparent	Transparent	Transparent	Transparent to translucent	Transparent to translucent
Yield (1 mil)	19,500	19,500	18,250 (1.1 mil)	22,000	13,000	29,200–29,500	23,500–24,500	21,200
Specific gravity	1.40–1.55	1.44	1.2	1.25–1.35	2.2	0.94–0.95	1.13–1.14	1.3
Mechanical								
Tensile strength, lb./sq. in. (ASTM D 882)	7,000–18,000	7,000–18,000	7,000–18,000	7,000–12,000	5,000–10,000	3,000–5,000	10,000–18,000	6,000–10,000
Stretch, % (ASTM D 882)	15–25	25–50	15–25	15–50	50–400	350–450	250–500	300–400
Impact strength, kg.-cm.	8–15	5–15	5–15	2–8	2–15	30–90	4–6	5
Tear strength (Elmendorf), gm./mil (ASTM D 1922)	2–10	7–15	2–10	2–15	10–40	15–25	20–150	20–60
Stiffness, gm. MD (Handle-O-Meter)[3] TD	37–65 / 18–31	37–65 / 18–31	40–60 / 20–40	25–40 / 25–45	20–45	5–10	5–35 / 5–40	5–30
Heat-seal range, deg. F.[4]	200–350	200–350	230–300	350–450[5]	350–400	190–400	350–500[5]	250–375
Chemical								
WVT, gm./24 hrs./1 sq. m. at 100 deg. F., 90% R.H. (ASTM E 96 Method E)[6]	3–15	6–14	18 and up	Very high	0.4–1.0	22–30	Very high	7–10
Gas trans., cc./mil/1 sq.m./24 hrs./1 atm. /73 deg. F., 0% R.H. (ASTM D 1434)	O_2 2–80[7] CO_2 15–95[7]	1–9[7] 6–9[7]	—[7]	1,800–3,100 7,700–52,000	100–300 250–750	3,500–7,500 9,700–17,800	30–110 150–390	6–15 20–30
Resistance to greases and oils	Impermeable	Impermeable	Impermeable	Excellent	Excellent	Good	Impermeable	Excellent
Permanence								
Maximum use temperature, deg. F.	Begins to char at 375	Begins to char at 375	180	250	250	190	350–475	250
Minimum use temperature, deg. F.	Depends on type and R.H.	About 0	Depends on cellophane	About 0	Minus 50	Minus 50	Minus 50	–40
Dimensional change at high R.H., %	3–5	2–3	3–5	0.2–0.6% at 80% R.H.	None	None	1–3	1–3
Flammability	Same as newsprint	Same as newsprint	Slow burning	Slow burning	Nonflam.	Slow burning	Self-extinguishing	Self-extinguishing
Converting characteristics								
Machine performance	Excellent	Excellent	Excellent	Good	Good	Fair to good	Good	Good
Printability	Excellent	Excellent	Excellent	Excellent	Good if treated	Good if treated	Good	Good
Sealing	Heat or adhesive	Heat or adhesive	Heat	Heat or solvent	All systems	Heat	Heat or adhesive	Heat or impulse
Heat shrinkable	No	No	No	No	No	No	No	No

* Data, unless otherwise specified, are for 1-mil (0.001-in. thickness). Properties depend on type and grade of polymer or resin used in base films or coatings. Properties of papers, shrink films, sheet plastics and laminations are covered elsewhere in this section.
[1] Various coating thicknesses available; also laminated combinations. Test data assume polyethylene side toward product.
[2] Modified acetates—cellulose acetate butyrate, cellulose propionate and cellulose triacetate—offer an additional range of properties.
[3] For 8-in. widths. 1/2-centi. slot, Handle-O-Meter.
[4] Heat-seal range, in deg. F., shows the lowest temperature at which effective seals can be made and the upper temperature at which seals begin to show degradation. For some films, special methods or equipment may be necessary.

WRAPPING, BARRIER, AND CUSHIONING MATERIALS

TABLE 8-3 (Continued)

Pliofilm (polyisoprene hydrochloride)	Polybutylene	Polycarbonate	Polyester, saran-type coating	Polyester (uncoated)	Polyethylene		
					Low density 0.910–0.925	Medium density 0.926–0.940	High density 0.941–0.965
Transparent to translucent	Transparent to translucent	Transparent	Transparent	Transparent	Transparent to translucent	Transparent to translucent	Transparent to opaque
24,000	30,000	23,100	20,500	20,000–22,000	30,000	29,500	29,000
1.11	0.91	1.2	1.4	1.35–1.39	0.910–0.925	0.926–0.940	0.941–0.965
3,000–4,100	4,500–5,000	8,000–9,000	17,000 and higher	17,000 and higher	1,000–3,500	2,000–5,000	3,000–10,000
Yield 10–20 ultimate 350–500	200-400	110	80–180	70–130	225–600	225–500	10–500
6–15	20–30	High	25–30	25–30	7–11	4–6	1–3
60–1,600	300-800	20–40	10–20	13–80	100–400	50–300	15–300
12–25 12–25	3–5 5	Stiff	40	40 40	2.5–4.5 3–7	5–10 6–14	8–16 10–20
240–300[5]	325-400	400–430	275–400	275–400[8]	250–350[5]	260–310[5]	275–310[5]
8 and higher	8–10	150	1–2	15–30	18	8–15	5–10
130–1,300 520–5,200	O_2—2,000 CO_2—10,000-15,000	4,000 12,000	9–15 20–35	52–130 180–390	3,900–13,000 7,700–77,000	2,600–5,200 7,700–13,000	520–3,900 3,900–10,000
Excellent	Good	Good	Excellent	Excellent	May swell slightly on long immersion	Good	Excellent
Softens at 200	180-230	265	Coating softens at 190	250	150 (softens at 230)	180–220	250
Depends on plasticizer	−30	Minus 100	Minus 60	Minus 80	Minus 60	Minus 60	Minus 60
None	None	None	None	None	None	None	None
Self-extinguishing	Slow burning	Slow burning	Slow burning	Slow burning	Slow burning	Slow burning	Slow burning
Fair to good	Fair to good	Good	Fair to good	Good	Fair to good	Fair to good	Good
Special inks	Good if treated	Good	Good	Good	Good if treated	Good if treated	Good if treated
Heat	Heat	Heat	Heat or adhesive	Heat[8] or adhesive	Heat	Heat	Heat
Some types	Some types	No	Some types	Some types	Special types	Some types	Some types

[5] Unsupported film cannot be sealed on all types of heat sealers. Special sealing may be necessary.
[6] To convert to gm. per 100 sq. in., divide gm. per sq. m. by 15.5.
[7] Depends on moisture content and plasticizer.
[8] Some types can be sealed only if coated.
[9] Coatings can be saran, acrylic or polyvinyl acetate, depending on end-use requirements.
[10] Saran is a registered trademark in certain foreign countries, but not in the U.S.

DISTRIBUTION PACKAGING

TABLE 8-3 *(Continued)*

Properties of packaging films (continued)

	Polypropylene (oriented)	Polypropylene (unoriented)	Polypropylene, balanced, oriented, stabilized, coated two sides [9]	Polystyrene (oriented)	Polyurethane	Polyvinyl alcohol	Polyvinyl alcohol (water soluble film[14])	Vinyl (PVC and plasticized)
General								
Clarity	Transparent	Transparent	Transparent	Transparent	Transparent to translucent	Translucent	Transparent	Transparent to translucent
Yield (1 mil)	30,600	30,800	29,500	26,300	23,000	21,600	16,300	19,000–22,000
Specific gravity	0.905	0.88–0.90	.93	1.05	1.12–1.2	1.21–1.31	1.59–1.71	1.23–1.37
Mechanical								
Tensile strength, lb./sq. in. (ASTM D 882)	25,000–30,000	3,000–6,000	15,000–30,000	9,000–12,000	7,000–12,000	5,000–9,000	8,000–20,000	2,000–19,000
Stretch, % (ASTM D 882)	70–100	200–500	100	10–60	300–700	400	40–80	5–500
Impact strength, kg.-cm.	5–15	1–3	10–25	1–5	40	Good	10–15	12–20
Tear strength (Elmendorf), gm./mil (ASTM D 1922)	4–6	40–330	5–8	4–20	High	300–500	10–20	Varies widely
Stiffness, gm. MD (Handle-O-Meter)[3] TD	5–40	11–27 11–27	20–30	50 50	Very soft	Soft	10 15	7.5–40 10–45
Heat-seal range, deg. F.[4]	Requires coating	325–400[11]	190–270	250–325[11]	300–375	375–490	280–300[11]	200–350[11]
Chemical								
WVT, gm./24 hrs./1 sq. m. at 100 deg. F., 90% R.H. (ASTM E 96 Method E)[6]	4	8–10	4–10[12]	100 and higher	Very high	None	1.5–5	8 and higher
Gas trans., cc./mil/1 sq. m./24 hrs./1 atm./73 deg. F., 0% R.H. (ASTM D 1434)	O_2 2,400 CO_2 8,400	1,300–6,400 7,700–21,000	—[12]	2,600–7,700 10,000–26,000	Very high Very high	Very low —	8–26 52–150	77–7,500 770–55,000
Resistance to greases and oils	Excellent	Excellent	Excellent	Good	Excellent	Excellent	Excellent	Excellent
Permanence								
Maximum use temperature, deg. F.	275	250	250	Shrinks at 185	220	240	Softens at approx. 290, melts at 310	Approx. 200 Depends on plasticizer
Minimum use temperature, deg. F.	Minus 60	—[13]	Minus 60	Subzero	Minus 40	15	Good flexibility at 0	Depends on plasticizer
Dimensional change at high R.H., %	None	None	None	Little or none	Little or none	3–5	None	None
Flammability	Slow burning	Slow burning	Slow burning	Slow burning	Self-extinguishing	Slow burning	Self-extinguishing	Self-extinguishing
Converting characteristics								
Machine performance	Good	Fair to good	Good	Good	Fair	Fair	Fair to good	Fair to good
Printability	Good if treated	Good if treated	Good	Special inks	Special inks	Good	Special inks	Special inks
Sealing	Adhesive	Heat	Heat	Heat or adhesive	Heat or adhesive	Heat or adhesive	Heat	Heat or adhesive
Heat shrinkable	Some types	No	No	Yes	No	No	Some types	Some types

[11] Unsupported film cannot be sealed on all types of heat sealers. Special sealing may be necessary.
[12] Varies according to composition or type and weight of coating.
[13] Not recommended where low-temperature durability is critical.
[14] Other types of water-soluble films are: methyl-cellulose, polyethylene oxide, starch and hydroxypropyl-cellulose. They provide a range of solubility temperatures and other properties.
Reproduced by special permission of the publishers of Modern Packaging Encyclopedia and Planning Guide, December, 1974.

WRAPPING, BARRIER, AND CUSHIONING MATERIALS

the acid, the rubber hydrochloride is put into solution in hydrocarbon solvents and then calendered into a film.

Polyisoprene hydrochloride film is transparent, water resistant, non-toxic, and essentially odor resistant. Whereas this film has greater resiliency than cellulose products, its stability is not as great. It becomes brittle upon exposure to heat, light, and excessive cold. Like cellophane, polyisoprene hydrochloride film can be printed and can be laminated to paper and other materials to improve physical or operating characteristics. Polyisoprene hydrochloride films are similar in price than cellulose films, but comparable lighter gauges frequently can be used with resultant economies.

POLYVINYLIDENE–CHLORIDE FILM

Polyvinylidene-chloride films (Saran) are produced by the extrusion process. They have the lowest water permeability of the packaging films. These films also have low gas permeability, have high specific gravity, and are among the heavier packaging films. The yield is relatively low and the base cost is relatively high. Polyvinylidene-chloride films are characterized by exceptionally high tensile strength and flexibility, complete transparency, and general toughness. They are also flexible at low temperature and can be heat sealed.

The characteristic properties of these films present difficulties on conventional packaging machinery. They are therefore not too widely used in high-volume packaging except when extreme clarity, chemical inertness, and the other foregoing properties are of importance.

POLYETHYLENE

Polyethylene is a straight chain hydrocarbon. Low-density polyethylene is made from ethylene gas under high pressure and temperature. High-density polyethylene is made at normal atmospheric pressure and room temperature using a catalyst of titanium tetrachloride and triethyl aluminum. In appearance it is transparent to opaque, and wax-like, and has good flexibility under extreme cold conditions. Polyethylene has excellent resistance to acids, alkalies, and inorganic chemicals, and has no known solvents at room temperatures. It has low moisture permeability, contains no plasticizers, and is of low cost. The resistance of polyethylene to organic substances such as fats and oils is relatively low, particularly in some essential oils used in perfumes, pharmaceuticals, etc.

Polyethylene film is being widely used for packaging applications because

TABLE 8-4

Shrink film selector chart*

Types	Typical gauge (mils)	Cost 1,000 sq. in. (cents)	Cost Per lb. ($)	Tensile (1,000 lb./sq. in.)	WVTR[1] (gm./mil)	Oxygen[2] (cc./mil)	Shrink maximum[3] %	Shrink tension range[4] (lb./sq. in.)	Film shrink temp.[3] range °F.	Tunnel air temp.[5] range °F.	Sealing temp. range °F.
Ionomer	1.0-3.0	3.8-11.4	1.10	3.8-4.9	20-30	6,000	20-40	150-250	195-270	250-350	250-400
Pliofilm	0.40	4.2	2.50	6-19	12-20	92	40-50	150-350	150-230	225-300	180-250
Polybutylene	0.5-2.0	2.0-7.1	1.10-1.5	17-21	8-10	1,500-3,000	40-80	100-350	190-350	250-400	300-400
Polyester	0.50-0.65	3.5-6.0	1.41-1.85	24-36	15	80-120	45-55	700-1,500	160-250	225-500	—
Polyethylene[6] Regular	1.0-2.0	1.6-4	0.50-0.60	1.6-3.1	12-18	6,000	20-70	50-100	190-300	250-375	250-400
Heavy duty[7]	2-10	4-16	0.50-0.60	1.6-3.1	12-18	6,000	20-70	50-100	190-300	250-475	250-400
EVA copolymers	1-10	1.6-16	0.49-0.63	2.-3.5	15	11,000	20-70	40-90	150-250	200-320	200-350
Polyethylene, oriented[8] Regular & modified	0.6-2.0	2.8-8.7	10.9-1.38	9	10-15	5,000	80	200-500	190-280	230-350	230-400
Cross-linked	0.6-1.5	2.8-6.5	1.29-1.38	8-19	5-10	5,000	50-80	250-500	160-290	225-600	300-500
Polypropylene[6] Regular & blends	0.50-1.5	2.6-4.3	1.33-1.74	15-27	4	2,000	50-80	300-600	200-350	300-450	350-400
Polystyrene	1.0	2.8	0.72	9-12	>60	3,500	40-70	100-600	210-270	270-320	250-300
Polyvinyl chloride Regular	0.5-1.5	2.9-5.0	0.90-1.50	10-19	>50	300-8,000	30-70	150-300	150-300	225-310	275-370
Heavy duty[7]	1.5-3.0	4.6-10.4	0.67-0.77	5-12	9	115	55	150-300	150-300	225-310	275-370
Polyvinylidene chloride-vinyl chloride copolymer	0.4-1.0	3.3-8.6	1.33-1.48	7-16	3-8	15-40	15-60	50-200	140-290	200-300	250-300

*The reported data for gauge, etc. are approximate values for the usual commercial grades, but the suppliers should be consulted for current prices and more complete information on other available gauges, types and the like.
[1] ASTM-Method #E96-66T. Units = gm./sq. m./24 hrs./mil at 100 deg F. at 90% R.H.
[2] ASTM-Method #D1434-66. Units = cc./sq. m./24 hrs./mil at room temperature and 1 atm.
[3] Shrink % (change from original) determined by 5 sec. immersion of marked film sample in water for temperature below 212 deg. F. or in silicone oil above 212 deg. F. (Much of the data in this category based on tests run in W. R. Grace & Co., Cryovac Div. laboratory.)

water used below 212 deg. F. and silicone oil used above 212 deg. F. (Tests run in laboratory of W. R. Grace & Co., Cryovac Div.) Shrink tension of cast polyethylene increases on cooling.

[5] Higher temperature may be used for faster shrink. But use care if temperature is above melting point.

[6] Shrink must be built into these films, by processing, such as stretching, during manufacture. Films employ modified polymers using copolymers and/or blends and in a range of densities. These films shrink after reaching their melting point.

[7] Intended for tray, case and pallet wrap, especially shipping or industrial uses. A major difference is greater durability and ability to hold heavier loads.

[8] Most of these films are biaxially oriented. The films are stretched and the stretch is locked in by holding the film during cooling. The films shrink rapidly at temperatures below their melting point. One type of film is cross-linked by irradiation or chemical means to provide additional properties such as improved heat sealing and resistance to burn-through.

Reproduced by special permission of the publishers of Modern Packaging Encyclopedia and Planning Guide, December, 1974.

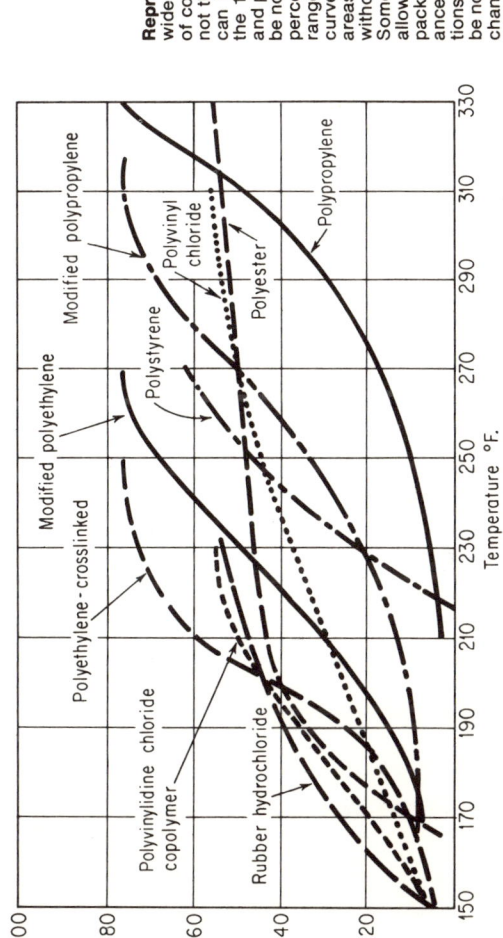

Representative shrink-temperature curves show wide range of shrink characteristics for nine types of commercial films. Temperatures are of the film, not the heating media. Shrinkage using hot water can be as high as 60%. Hot-air temperatures in the 160 to 280 deg. F. range are easily obtained and provide an excellent amount of shrink, as will be noted from a study of the individual curves, with percentages of 50 to 70 falling well within this range for eight of the films. Some films with steep curves (polyethylene—cross linked) can have film areas of high shrink adjacent to low-shrink areas without overheating the low-shrink portions. Some, like PVC and polyester, have flatter curves, allowing wider temperature variations during packaging. Similar shrink-temperature performance is obtained with the many different formulations offered by various film producers but, as will be noted, film modification can produce significant change in the shape of the curve.

of its excellent protective properties at low unit cost. Although its greatest popularity remains for the pre-packing of food and soft goods, its wide spread applications also include hard goods as hardware, silverware, parts, barriers for machinery, appliances, and for the shipment of liquids.

Polyethylene is used most extensively in combination with other materials such as paper in the production of multi-wall paper shipping sacks.

POLYESTER FILM

Polyester film (Mylar) is produced from a polymer of ethylene glycol and terephthalic acid and has gained acceptance due to its exceptional toughness and its impermeability to gases, acids, alkalids, greases, and solvents. In addition, polyester films offer clarity and good performance in a wide range of temperatures.

Polyester film is relatively costly on a weight and yield basis. However, because of its outstanding functional properties it finds application in the packaging of heavy, sharp-edged machine parts, such as window material, for folding cartons, and for articles subjected to excessive counter or shelf abuse.

The film has a high melting point making it suitable for boil-in-bag and bake-in-foods. Coated polyester can be run on packaging machines and heat-sealed. The uncoated film is not heat-sealable, but can be solvent-sealed.

POLYVINYL CHLORIDE FILMS (PVC)

Vinyl is made by the chlorination of ethylene or acetylene to produce vinyl chloride which is then polymerized to produce polyvinyl chloride. Two types of vinyl chloride are produced: (1) rigid PVC and (2) plasticized PVC. The characteristics are closely related to the additives used in the manufacture. The many possible formulations permit the production of a variety of films ranging from hard and brilliant to soft and flexible. Vinyl films have good dimensional stability, can be printed, and make strong heat seals. They are particularly well suited for stretch and shrink packaging and as a substrate for pressure sensitive tapes.

LAMINATED PACKAGING FILMS

In order to obtain the desired properties in a packaging material, combinations of two or more plies of similar or dissimilar materials are laminated

WRAPPING, BARRIER, AND CUSHIONING MATERIALS

into a single sheet. Some of these laminated structures have already been discussed in conjunction with paper and paperboard products. Other typical combinations involving film, foil, paper and paperboard are listed and their properties are detailed in Table 8-5. For example, the lamination of pouch paper with polyethylene is a very popular combination to produce packages for hygroscopic products such as dehydrated foods, detergents and chemicals, and in heavier form as a military barrier. In this combination, pouch paper features low cost, rigidity, strength and a good printing surface. Polyethylene also features low cost, and adds heat sealability, low permeability to gases and water vapor and excellent resistance to cold.

The proper selection of combinations of various films and other flexible and rigid materials can solve most protective packaging demands. The main limitation in the use of special materials is lack of volume for the fabrication of special combinations and formulations.

SHRINK FILMS

In principle, shrink films when subjected to heat, have the ability to draw or shrink down to the contour of an object being packaged. Orientation of the film for this purpose is accomplished in the manufacturing process and the amount of stretch can be balanced or unbalanced as desired.

Shrink packaging applications range from small units up to entire pallets loads immobilizing the individual components of the load to improve handling and shipping. Shrink packaging for single units provides excellent display advantages in addition to enhanced protection from dust, moisture and pilferage.

Properties of popular film used in shrink packaging are illustrated in Table 8-4 and representative shrink temperature curves are also shown for nine types of commercial films.

PACKAGING FOILS

A foil is an unsupported thin metal membrane less than 0.006 in. in thickness. Above 0.006 in. in thickness the thin metal is called a sheet. The principal foil materials for packaging are (1) *aluminum,* (2) *tin,* (3) *lead,* and (4) *tin and lead composition foil.* In Table 8-6 aluminum foil and laminations are discussed. Aluminum foil is by far the most commonly used metal foil.

TABLE 8-5

Laminations—types and properties*

Properties	Cellophane to cellophane with polyethylene or heat combined (also cellophane to glassine)	Cellophane, 250 to 195; polyethylene, ½-mil; foil, 0.00035; polyethylene, 1- to 2-mil	Foil, 0.0003; waxy adhesive; tissue, 17 lb.	Foil, 0.00035; wax, microcrystalline, 17-lb.; porous paper, 20-lb.[1]	Foil, 0.00035; casein adh., 2-lb. (or polyethylene, 7-lb.); backing paper[2]	Foil, 0.0035; polyethylene; kraft paper; hot melt
General						
Forms available	Rolls	Rolls	Rolls	Rolls	Rolls	Rolls
Thickness range, in.	.002 to .004	.003 to .004	.002	0.0025–0.003	0.0036	.0045
Area factor, sq. in./lb.	12,500 to 4,000	8,000 to 6,000	13,000	8,350	—	7,000
Mechanical						
Tensile strength, lb./in. (ASTM D 828) MD / TD	15,000 up / 6,000	15,000 up	15 / 8	21 / 11	19 / 12	30 / 20
Elongation, % (ASTM D 828) MD / TD	5–20	5–60	3	3 / 3	1 / 1	1
Bursting strength, lb./sq. in. (ASTM D 774)	50 and up	50 and up	15	21	17	40
Tearing strength, gm. (ASTM D 689) MD / TD	20	20–50	60	50 / 55	60 / 72	6
Folding endurance, 1 kg. load (ASTM D 643) MD / TD	—[7]	Good	50 / 20	48 / 7	110 / 13	>200
Chemical						
Permeability to gases[4] (flat sheets)	Very low	Approaches zero	Low	Very low	Very low	Low
Water-vapor permeability[4] (flat sheets)	—[7]	Approaches zero	4	Approaches zero	Very low	Very low
Resistance to greases[4] and oils	Excellent	Excellent	Fair	Poor	Fair	Poor
Permanence						
Resistance to heat[5]	Good	Good	Fair	Fair	Excellent	Excellent
Resistance to cold	Good[7]	Good	Excellent	Excellent	Excellent	Excellent
Dimensional change at high R.H.	2% plus some curl	Good, some curl	Good, slight curl	Good, slight curl	Good, slight curl	Good
Use performance						
Machine performance,[6] printability, sealing	Excellent	Excellent	Excellent	Excellent / Good dead fold	Excellent / Good dead fold	Excellent
Typical uses	Bags and form/fill/seal for snack foods because of fats and moistureproofness.	Form/fill/seal for nuts and snack foods because of barrier, greaseproofness and appearance.	Butter and margarine wrappers, generally embossed and printed.	Easy sealing overwrap or carton wrap for frozen foods, other items.	Label stock; also as liner or label for spiral-wound can.	Heat-sealed overwrapper on cartons.

*SOURCE: The values in this chart were obtained from various authorities. Grades may be obtained with properties different from the values shown. There are many specialized combinations. Manufacturers should be consulted for additional information or for new uses for their materials.
[1] An alternate construction employs an EVA blend rather than microcrystalline wax.
[2] Low density polyethylene extrusion applied as a laminating agent or coating can be used in a wide range of weights.
[3] Tyvek is Du Pont's spunbonded polyethylene.

TABLE 8-5 (Continued)

...oil, 0.00035; adhesive; paperboard	Glassine; wax, ½-mil; glassine	Kraft paper; polyethylene scrim; kraft; polyethylene	Paperboard; aqueous adhesive; white or fancy paper	Polyester film, ½-mil; adhesive; foil, 0.00035; polyethylene, 2½-mil²	Pouch paper (or glassine), 25-lb.; polyethylene, ½-mil; foil, 0.00035; polyethylene, 1-mil²	Pouch paper, 25-lb.; casein adh., 2-lb.; foil, 0.0007; heat seal coating, 3-lb.	Tyvek³ 10 Series 1 MIL polyethylene foil 0.00035 3 MIL polyethylene
							General
Rolls, sheets	Rolls	Rolls, sheets	Rolls, sheets	Rolls	Rolls	Rolls, sheets	Rolls, Sheets
—⁷	.004 up	—⁷	—⁷	0.003–0.0035	0.004–0.0045	0.0026–0.0032	0.009–.011
—⁷	9,000 up	—⁷	—⁷	7,000	7,225	7,450	3,875
							Mechanical
High⁷	35 / 20	Very high	Very high	15 / 20	23 / 14	24 / 24	90 lb.
—	3	2	1	24 / 45	3 / 10	2 / 4	—
High⁷	40	Very high	Very high	42	26	25	45
—⁷	50 / 60	Very high	High	71 / 98	48 / 52	37 / 37	408
—⁷	—⁷	Very high	—⁷	>800 / >800	431 / 92	250 / 17	—
							Chemical
—	Low	Poor	None	Approaches zero	Very low	Very low	Approaches zero
Very low	8	15	None	Approaches zero	Approaches zero	Approaches zero	Approaches zero
Poor	Excellent	Poor	None	Good to excellent	Good	Good to excellent	Good to excellent
							Permanence
Good	Fair	Excellent	Good	Good	Good	Excellent	Excellent
Good	Good	Excellent	Good	Excellent	Excellent	Excellent	Excellent
Good, slight curl	Poor	Good	Slight curl	None	Good, slight curl	Good, slight curl	None
							Use Performance
Excellent	Excellent	Good	Excellent	Excellent	Excellent	Excellent	Excellent
...ding cartons, ...s, oven-trays.	Inner liners in cartons.	A strong, tough wrap for industrial use.	Folding cartons, set-up boxes, trays.	Pouch, for items needing greater toughness or decoration—liquid, gas or vacuum packs. Drugs, etc.	Pouch stock for hygroscopic products, dehydrated soups, freeze-dried foods—and in heavier form as military barrier.	Cover stock for foil and plastic formed container unit packs. Excellent for pharmaceuticals, drink powders.	General packaging of military parts and equipment for storage or shipment.

...alues depend on thickness of foil and/or amount and type of laminating agent or coating.
...imited only by melting or softening temperature of components.
...urface slip, weight and components must be selected for specific requirements and surface primed for printing and coating.
...alues depend on grade and thickness of components or plies.
...produced by special permission of the publishers of Modern Packaging Encyclopedia and Planning Guide, December, 1974.

TABLE 8-6
PROPERTIES—ALUMINUM FOIL AND LAMINATIONS

Properties	Metal Foil — Aluminum	Specialized Aluminum-Foil Combinations—Heat-Sealing Types				
		Foil—paper—foil combination, heat-seal coating	Foil—water-resistant adhesive—paper—microcrystalline wax—porous tissue	Cellulose acetate—resinous adhesive—foil—heat-seal coating	Paper—waterproof adhesive—foil—heat-seal coating	Scrim—polyethylene—foil—vinyl film combination
General						
Forms available	Rolls, sheets	Rolls, sheets, bags	Rolls, sheets, bags	Rolls, envelopes	Rolls, envelopes	Rolls, pouches
Clarity	Opaque	Opaque	Opaque	Opaque	Opaque	Opaque
Thickness range, in.	0.0002–0.006	0.003	0.0044	0.0025	Approximately 0.005	Approximately 0.016
Maximum width, in.	54	—	—	—	36	36
Area factor, sq in./lb (approximations)	10,250 for 0.001; 29,200 for 0.00035	5,500	6,500	5,900	4,800	2,660
Approximate specific gravity	2.7	—	—	—	—	—
Mechanical						
Tensile strength, lb/in., TAPPI #T-404M	9–118	29	19	21	23	35
Elongation, %, TAPPI #T-457M	2.1–26	2.7	2.9	2.4	1.6	11.5
Bursting strength (Mullen), lb/sq in.	4–165	39	19	50	47	130
Tearing strength (Elmendorf), gm	5–425	60	15	40	135	880
Seal strength (Suter), lb/in.	—	3	1	2	4–5	5–6
Folding endurance (MIT)	4–18	154 at 1 kg	55 at 1 kg	240 at 1 kg	—	—

Chemical						
Water absorption in 24-hr immersion, %	Zero	—	—	—	Zero	—
Permeability to gases	Negligible	Negligible	Negligible	Negligible	Zero	Zero
Water-vapor permeability (Southwick-Rhoades), gm/24 hr/100 sq in.	Negligible	Negligible	Negligible	Negligible	Negligible	Zero
Resistance to acids	Poor to good	Poor to good	Poor to good	Poor to good	Good	Poor
Resistance to alkalies	Poor to good	Poor to good	Poor to good	Poor to good	Good	Poor
Resistance to greases and oils	Excellent	Excellent	Poor	Poor	Good	Good
Resistance to organic solvents	Excellent	Affected by some solvents	Affected by some solvents	Affected by some solvents	Affected by some solvents	Affected by some solvents
Permanence						
Resistance to heat, °F (limiting temp.)	700	120	120	120	120	160
Resistance to cold, °F (limiting temp.)	−120	Depends on laminating agent, etc.	Depends on laminating agent, etc.	Depends on laminating agent, etc.	−65	−65
Resistance to sunlight	Excellent	Excellent	Excellent	Good	Excellent	Excellent
Dimensional change at high rh, %	None	None	None	None	None	None
Resistance to storage	Unaffected	—	—	—	Good	Good
Flammability	Not combustible ordinarily	—	—	—	Partly combustible	Partly combustible

* SOURCE: The values in this chart were obtained from various sources and are believed to represent acceptable commercial standards. Special grades of some of these materials may be obtained whose properties differ from the values shown. A wide range of tempers and surface finishes is available. There are many specialized combinations of foil with other materials. Manufacturers should be consulted for additional information or for new uses for their materials.

ALUMINUM FOIL

One of the basic attributes of non-laminated aluminum foil is its appearance. For this reason, it is applied as a wrapper and used for other decorative purposes. It is non-toxic, non-absorptive, imparts no taste or odor, can be molded, crimped, and formed easily due to its flexibility, and has a varying degree of water vapor permeability depending on its thickness. Plain unsupported foil is used for semi-rigid containers for bakery and frozen food products and other consumer packaging applications. Coatings provide foils with heat-sealability, scuff resistance, and additional chemical resistance.

Foil laminations. For distribution packaging uses, aluminum foil is principally used in *combination* with other materials. Plastic, paper, and cellulose films laminated to foils enhance strength and utility on high speed packaging equipment. In combination, foil also imparts to the structure an extremely low water transmission rate. This is important for the protection of dry products and for the prevention of drying out and shrinking of moisture bearing products.

Some of the typical laminations are: (1) plain foil to paper; (2) heat-seal coat to foil to paper; (3) heat-seal coat to foil to paper to cotton sheeting or scrim; (4) heat-seal coat to foil to plastic film; (5) heat-seal coat to foil to plastic film to cotton sheeting or scrim.

Laminated foil barriers are used for sealed and unsealed wraps and as preformed bags for packing items as large as airplane motors. Foil-lined tubes and rigid containers are also used for protective packaging. Some of the foil-packaging applications include automotive parts, such as camshafts, heaters, gaskets, and valves. Foil-laminated bags and pouches are used for the packaging of electronic components, radios and radar components, spare machinery parts, pharmaceuticals, camera supplies, hypodermic syringes, scalpel knives, and sterilized bandages.

CUSHIONING MATERIALS

As previously defined, the principal purpose of package cushioning is to protect the article from the shocks and/or vibrations that are encountered in handling and transportation. This chapter covers only the principal types of materials which are used to provide cushioning. The basic materials used for cushioning and an evaluation of their properties are contained in Table 8-7.

The simplest and most appropriate way to classify the many available cushioning materials is by *material classifications and types*. These are:

WRAPPING, BARRIER, AND CUSHIONING MATERIALS 343

Non cellular, open cell plastics, and *closed cell plastics*. In addition a discussion of this subject must also include an overview of *mechanical cushioning devices* such as springs and shock mounts.

NON CELLULAR MATERIALS

This group includes such popular materials as *excelsior, cellulose wadding* and *paper*. These are inexpensive bulk cushioning agents which, in packing, conform to the contour of the article. The degree of compactness of the material is dependent upon the packer; thus, varying degrees of cushioning are achieved with these materials. The use of these and other bulk cushioning materials is being largely replaced by plastic preformed cushioning structures which provide greater uniformity, ease of packing and unpacking, and neater appearance.

Because of the limitations of bulk cushioning applications, many of these materials are fabricated into blocks, blankets, pads, and boards. They also provide the filler material for preformed cushioned structures such as bags. *Excelsior* typifies an inexpensive cellulosic material which is available in bulk, compressed into blocks, or covered with other materials in the form of pads and blankets. An application of macerated paper in blanket form is illustrated in Fig. 8-9. *Cellulose wadding,* available in sheet form as a wrapping as well as cushioning material, is made paper-backed and unbacked (Fig. 8-10).

A specialized non-cellular material, *cane fibreboard,* available in boards of varying thicknesses, is fabricated to exacting product contours by wood fabrication techniques or diecutting and serves frequently as a blocking, bracing, and cushioning medium. In Fig. 8-11 two contour-fitted cane fibreboard pieces are illustrated for the packaging of an aircraft engine cylinder. Each of these pieces is laminated from several layers of this board.

Fig. 8-9. Blanket of macerated paper used for furniture cushioning.

DISTRIBUTION PACKAGING

Fig. 8-10. Cellulose wadding for the wrapping of intravenous solutions packed in glass. Courtesy Kimberly-Clark Corp.

Fig. 8-11. Cane fibreboard interior packaging for airplane engine cylinder. Courtesy The Celotex Corp.

WRAPPING, BARRIER, AND CUSHIONING MATERIALS

Fibreboard cushioning materials include corrugated, creped, ridged, and molded structures including standard fibreboard materials and others which have previously been referred to in Chapter 2. Of these, *single-face corrugated fibreboard,* which performs a combined wrapping and cushioning function, has broadest commercial application. The wrapability is increased by prescoring the sheet in a criss-cross or other pattern. Most applications of single-face corrugated fibreboard are for odd shapes and sizes, and packaging is generally manually and individually performed from either a roll or pre-cut sheet. Typical uses are for glass and earthenware, luggage, shelving, pictures, and roll coverings. Solid molded pulp formed into the contour of the corrugated arch or nodule is a specialty for similar applications.

As discussed in Chapter 2, various other corrugated combinations and structures have been developed to enhance the cushioning properties of fibreboard. Included in this class are *multiple thickness of single-face corrugated, jumbo flutes, expanded paper cellular structures,* and other preformed shapes permitting vertical and horizontal suspension. These materials generally have combined uses of blocking, bracing, suspension, and cushioning, as well as the filling of voids within the

Fig. 8-12. Continuous strip of specially designed wrapping and cushioning material for automobile windshield glass. Courtesy Vanant Company, Inc.

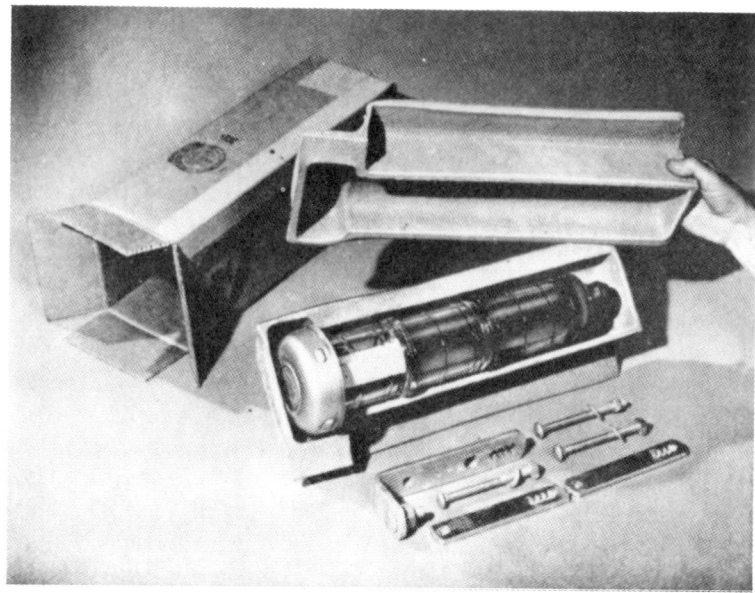
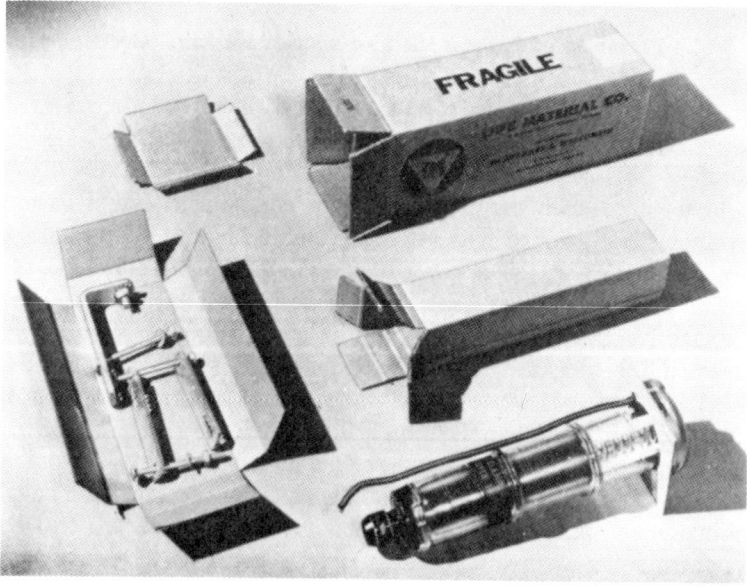

Fig. 8-13. Use of molded fibre sections shown in comparison with fibreboard interior components for a lightening arrester. Packaging was increased 35 to 80 units with this method. Courtesy Arvey Corp., Fibre Forming Division.

Fig. 8-14. Packaging of business machine with curled hair. Courtesy Blockson & Company.

shipping container. One development, of which an application is illustrated in Fig. 8-12, consists of a continuous strip of evenly spaced upstanding triangles which are slotted to a pattern to fit, suspend, and wrap such items as windshields, mirrors, clocks, and other fragile items.

Paperboard lends itself to other engineered structures which are suitable for cushioning and wrapping application. These include creping, embossing, and ridging.

Paper, papier mâché, and *wood pulp* are generally molded to fit the contour of irregular-shaped products. Owing to the relatively high die cost to produce such molded cushioning components, these materials are principally recommended for high-volume items. Where so applicable, a reduction in the number of interior packaging components for retention and cushioning of the item is usually possible however, and a less expensive exterior container may be more suitable. The pulp can be treated for moisture and oil resistance. The materials for such molded forms include waste news, kraft, cane, and other fibres. In Fig. 8-13 a typical industrial application is illustrated.

Non-cellular structures of *animal origin* include curled hair, wool, and felt. *Felt* is a fabric built up by the interlocking of fibres by a suitable combination of mechanical work, chemical action, moisture and heat, without spinning, weaving, or knitting. It consists of one or more classes of fibres: wool, reprocessed wool, and/or re-used wool, with or without admixture with animal, vegetable, and synthetic fibres. Felt finds application as a lining for re-usable' wooden containers in which it is adhered to the inner faces of the container or to interior bracing or blocking members.

Curled hair is supplied in sheet, roll, and die-cut stock and is produced by permanently bonding curled animal hairs with latex rubber, which is then vulcanized and cut into sheets. Curled hair is lightweight, highly resilient, moisture and fungus resistant, and free from dust. It is available in various densities and thicknesses.

OPEN CELL PLASTICS

As classified in Table 8-7 open cell plastics consist of *molded polyethylene film, flexible polyurethane* and *foam-in-place polyurethane*. All of these materials have low or low-medium density.

Molded Polyethylene Film. One of the simplest ways to provide surface protection and cushioning to light weight products is to mold open cells into polyethylene film in the shape of honeycombs, etc. Films, usually .002 to .004 in. in thickness, are thermoformed to a height such as 1/8 to ¼ in. The resulting flexible structure can be used as a wrap or an envelope for products requiring dust-free protection.

Polyurethane, flexible. This soft, sponge-like, easily compressible material is fabricated in a density range of 1½ to 6 lb./cu. ft., by polymerization and expansion of an isocyanate and hydroxil compound. Various stiffness characteristics are available. It can be molded or cut into shapes, produced in slabs and can be combined with other rigid materials such as polystyrene sheets for thermoforming specific shapes. One popular application is the use of a dimpled slab in corrugated fibreboard folders (Fig. 8-15) for the protection and shipment of odd shaped and delicate items. In other instances specifically molded or cut shapes are used for carrying cases such as photographic equipment.

Polyurethane, foam-in-place. This unique system eliminates special designs for shipping container interiors, especially for low volume products, requiring a high degree of protection. In essence, the product forms the mold. The system is excellent for odd shaped items and parts, business machines, instruments and many others where high unit cost is involved. In addition, the foam can serve as a thermal insulator.

WRAPPING, BARRIER, AND CUSHIONING MATERIALS 349

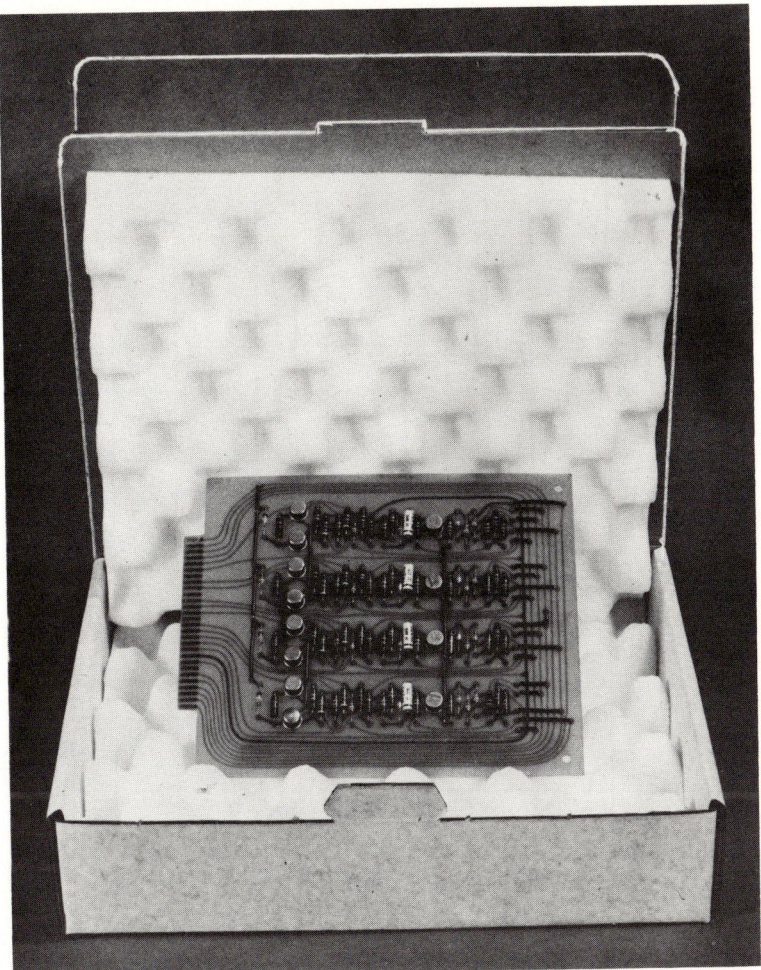

Fig. 8-15. Molded polyurethane foam utilized for positioning of fragile item in corrugated shipping container. Courtesy Lawrence Packaging Supply Corp.

The rigid foam is available in a density as low as ½ lb./cu. ft. which provides high compressive strength and optimum recovery characteristics. For heavy items, closed cell polyurethane is available with a density of 2 lb./cu. ft. and up.

The product is usually placed into polyethylene bags so that the foam does not adhere to it. The molded foam can even be reused when polyethylene sheets are placed between halves for easy separation. The foam application is simple and efficient.

TABLE 8-7

Cushioning materials chart*

Material classifications and types	Density[1]	Shock Absorption[2]			Resiliency[3]	Dampening[4]	Cleanliness[5]	Corrosivity	Liquid absorbency[6]
		Light loading 0.01 to 0.1 psi	Medium loading 0.1 to 0.4 psi	Heavy loading 0.4 to 1.0 psi					
Non cellular									
Bonded animal hair	Medium	Good	Good	Good	Good	Good	Fair	Low	Low
Cellulose wadding	Medium	Good	Fair	Poor	Fair	Good	Fair	Low	Low-High[6]
Excelsior fill	Medium	Fair	Fair	Fair	Poor	Fair	Poor	High	Medium
Excelsior pads	Medium	Fair	Fair	Fair	Poor	Fair	Fair	Low	Medium
Fibreboard Corrugated inserts	Medium	Poor	Fair	Fair	Poor	Poor	Fair	Low	Medium
Single face	Medium	Poor	Fair	Fair	Poor	Fair	Fair	Low	Medium
Paper Bogus and newsprint (crumpled)	Medium	Fair	Poor	Poor	Poor	Good	Poor	High	Medium
Indented kraft (multilayered)	Medium	Good	Good	Fair	Poor	Fair	Good	Low	Medium
Macerated pads	Medium	Good	Good	Good	Good	Good	Fair	Low	Medium
Shredded	Medium	Fair	Poor	Poor	Poor	Good	Poor	High	Medium
Open cell plastics									
Molded polyethylene film	Low	Good	Good	Fair	Good	Good	Good	None	None

Material									
Polyurethane, flexible	Low	Good	Fair	Poor	Good	Good	Good	None	Low
Polyurethane, foam-in-place	Low-Medium	Fair	Good	Good	Fair	Fair	Fair	None	Low
Closed cell plastics									
Air bubbles Barrier coated	Low	Good	Good	Good	Good	Good	Good	None	None
No barrier coating	Low	Good	Good	Good	Good	Good	Good	None	None
Polyethylene foam	Low / Medium	Fair / Poor	Good / Fair	Good / Good	Good / Good	Good / Good	Good / Good	None / None	None / None
Polypropylene foam	Low	Good	Good	Poor	Good	Good	Good	None	None
Polystyrene, expanded Molded sheets	Low	Poor	Fair	Fair	Poor	Poor	Fair	None	None
Loose fill	Low	Fair	Good	Good	Fair	Good	Fair	None	Low

*Based on information supplied by the producers and users of these materials. Data are intended *only* to provide a general guideline. Other characteristics to be considered are performance under high and low temperatures, material fatigue, thickness loss under static or increasing load, insulation, heat sealability, aesthetics. Form of use and objectives other than cushioning, such as protection of fine surfaces, ease of disposal, re-use and economy are also important. Ratings given here may vary considerably depending on manufacturer, thicknesses used, density, duration of use and conditions of use.

¹Low—less than 1.5 lb./cu. ft.; medium—1.5 to 3 lb./cu. ft.; high—3 lb./cu. ft.—as measured under a loading of 0.0243 psi.
²Capacity of material to absorb and not transmit shock.
³Ability of material to return to original shape following deformation after removal of load.
⁴Ability of material to minimize increase in shock resulting from a series of repeated impacts.
⁵Especially freedom from dusting.
⁶Available either as high or low absorbency.

Reproduced by special permission of the publishers of Modern Packaging Encyclopedia and Planning Guide, December, 1973.

DISTRIBUTION PACKAGING

CLOSED CELL PLASTICS

Air Bubbles—Barrier & Non-Barrier Coated. Entrapped air in flexible polyethylene film provides cushioning for light and heavy weight products and is made in two types.

One type consists of two heat-sealed layers of saran-coated polyethylene film, one with deep drawn pockets, the other to trap the air inside the pockets. This acts as a cushioning medium for products such as china, instruments, flowers and a host of others. The size of the pockets and the thickness of the film can vary depending on the requirements. The material is available in rolls, envelopes and pouches, and is suitable for mechanized wrapping operations.

The other type also incorporates the entrapment of air and the two-polyethylene sheet principle, but without the deep drawing and saran coating. Air is introduced between the two sheets and small, 1 in. square pillows are formed. Perforations in the heat-seal area permit tearing of the sheet into specific sizes.

In both types, the semi-transparent film allows product inspection without opening the wrap. The wrapping operation itself can be completed with or without heat-sealing.

In Fig. 8-16 an industrial shroud designed to protect a delicate industrial instrument is shown.

Fig. 8-16. Cushioned industrial shroud made of ¼ in. thick air bubble material. Courtesy Lawrence Packaging Supply Corp.

Polyethylene Foam. This is a closed-cell flexible and resilient material which resists temperature extremes and attack from chemicals. It provides water and dust protection plus the ability of reuse with little or no loss in performance. Polyethylene foam does not break or crumble. The material has a low density; 2 to 9/lb/cu ft. is the usual range of availability. Polyethylene foam is attractive, usually white, and it can greatly enhance product appearance. It can be molded or cut to the desired shape, therefore it is used extensively for blocking of heavy and light weight items as well. One of the unusual characteristics of polyethylene foam is that it can be reused without serious loss in cushioning characteristics.

Excellent surface protection is provided by enveloping delicate items with a premolded sheet, or by use of thin sheets or laminations to paperboard.

Polypropylene Foam. This material is a low density (approximately 0.7 lb/cu in.), closed-cell structure or oriented polypropylene containing approximately 50,000 bubbles per cubic inch. Since this material possesses a high coefficient of friction, flexibility, softness and resilience, it provides excellent surface protection and dunnage at low cost. Because of this high coefficient of friction, any internal shifting within a package usually takes place between the container and the foam instead of the foam and the product. For this same reason, the product furnishes excellent surface protection. The structure is non-dusting or fungus forming.

Foamed polypropylene has excellent heat insulating properties, remains flexible up to a temperature of -320°F, and can be used up to 250°F. The material has many useful properties which provide excellent applications for the wrapping of glassware, electronic parts, fruits, appliances, ceramics, etc. It is available in thicknesses ranging from 1/16 in. to 1/4 in.

Polystyrene, expanded, molded shapes and sheets. Polystyrene in the form of contour molded shapes has gained popularity for a number of reasons. In this form, it provides excellent product protection, low weight, a more precise control of foam density and affords increased packaging line productivity. Because of the predictability of performance, protection guidelines stating the number of square inches of loadbearing surface have been established for given product weight. The result is a scientifically controlled packaging system. Foam in the 1.5 to 2.5 lb/cu ft. range is customarily used in this application.

TV sets, small appliances, business machines and many other fragile products are packed in foam. For flat panels, slabs of the desired thickness can be used which do not require custom molds.

The most widely used forms of foamed polystyrene are end and corner pads for blocking and bracing, complete containers for the protection of glassware, cameras, etc. and platform trays which display and protect items such as cosmetics, housewares, etc. (Fig. 8-17).

354   DISTRIBUTION PACKAGING

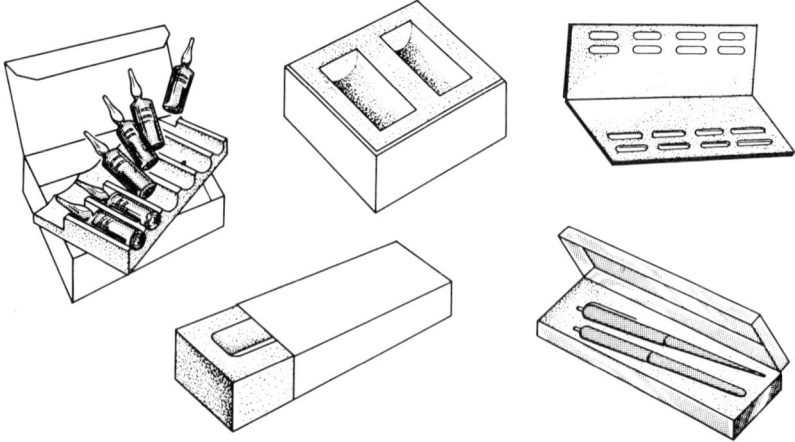

Fig. 8-17. Platforms of expanded polystyrene used with various types of unit containers. Courtesy Foampak Corp.

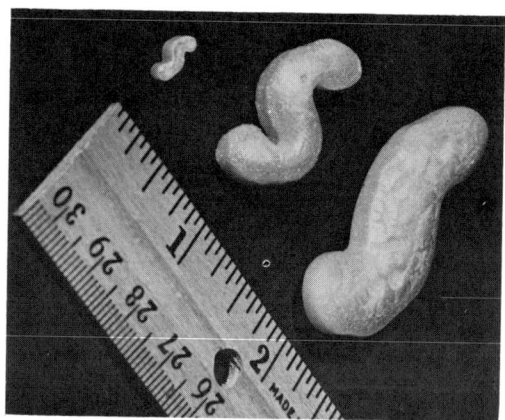

Fig. 8-18. Peanut shaped loosefill expanded polystyrene. Courtesy Norel Plastics Corporation.

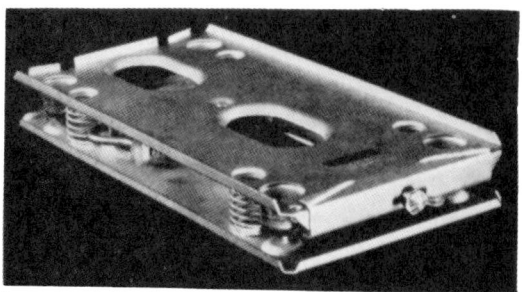

Fig. 8-19. Metal spring cushioning assembly. Courtesy T.R. Finn & Co., Inc.

WRAPPING, BARRIER, AND CUSHIONING MATERIALS

Loose Fill. Expanded polystyrene, in peanut (Fig. 8-18). or spaghetti form or other molded shapes, has become a popular form of dunnage and cushioning. The polystyrene consists of millions of tiny air cushions which provide shock and vibration isolating properties.

The most common density is ½ lb/cu ft. Because it flows freely into cartons and around product, packaging labor cost is reduced. This material is intended to replace such materials as excelsior, bogus or newspapers and its use is especially advantageous when replacing elaborate die cut inserts, cushioned interior pads or even molded shapes in shipping containers.

When packed under slight compression the strands or specific shapes interlock and resist settling of the product, an important consideration under excessive vibration. The material has a considerably higher ignition point than wood or paper. It can be purchased in bulk or expanded by the user for high volume operations.

SPECIAL CUSHIONING DEVICES

Special cushioning devices are employed when none of the standard cushioning materials provides the required isolation properties or is economical. This is particularly true with large, heavy, and fragile articles such as missiles, electronic instruments, and business machines. These devices generally provide a suspension system, in which the article is fastened to the cushioning device at convenient anchoring points and to the container. With most other types of cushioning materials contact is made with the product at as large a bearing area as possible.

Springs of the desired decelerating properties are examples of these devices. They are almost perfectly resilient and have no compression set. However, they are expensive and require considerable space in the container. (Fig. 8-19)

Fig. 8-20. Multi-directional cellular shock isolator incorporating rubber columns in the polyurethane foam to increase the load capacity.
Courtesy Hardigg Industries, Inc.

Multi-directional shock isolators (Fig. 8-20) are widely used in shipment as a mounting system particularly for the heavier fragile products. Both polyurethane and polyethylene foams are used in this application.

Another shock mount assembly system consists of the *tension mount*. This system is designed for very delicate articles where tension mounts are affixed between the corners of an outer container and an inner frame or container. Frequently solid rubber tension mounts are used. A modification of this type of system is shown in Fig. 8-21.

A specially designed corner pad based on the geometric shape, the *toroid*, is shown in Fig. 8-22.

For a comprehensive treatment of this subject see Military Standardization Handbook—Package Cushioning Design, Department of Defense, MIL-HDBK-304. Some of the equipment suitable for the distribution packaging functions of wrapping, cushioning and barrier application are included in Chapter 13.

Fig. 8-21. Tension mount utilizing polyurethane foam molded around tension coil springs for packaging of gyroscopes, computor elements and other delicate articles. Courtesy Hardigg Industries, Inc.

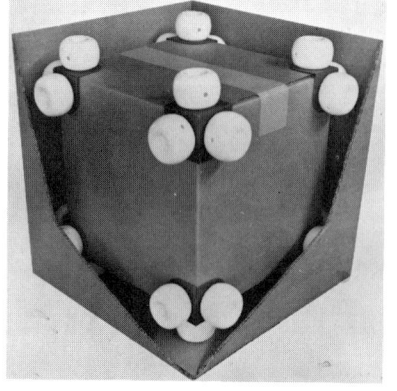

Fig. 8-22. Toroid-shaped expendable corner pads combine the elastic distortion of a polymer structure with the compression and discharge of air to achieve protective performance.
Courtesy Hardigg Industries, Inc.

Packaging Systems and Equipment

CHAPTER 9

Fastenings and Closures

In preceding chapters of this textbook the basic distribution container types and their materials were discussed. Most of these are fully prefabricated by the container manufacturer and are delivered to the user in a collapsed or knocked-down form in one piece, in several sections, or, such as with cylindrical containers, open for filling. In all instances the user of the container is required to perform one or more additional operations to prepare the container for shipment. In this chapter fastening and closing materials, systems, and equipment are covered. Subsequent chapters deal with other packaging operations frequently necessary to accomplish economies or to provide for full protection of the shipment.

The importance of adequate assembly and closure of distribution containers is not always fully recognized. A properly engineered container made of good quality materials frequently sustains damage in shipment due to spilling of contents, leakage, excessive distortion, or other typical failures which are attributable to a weak link in the engineering structure that a container represents, rather than in a basic weakness of the particular design that is used. In a transportation and packing survey jointly sponsored by the Railroads and the Fibre Box Association on 3,440 individual carload shipments of goods packed in fibre boxes, closure failures accounted for 4.5 per cent of all causes resulting in claims. Frequently the fastening and closing operations represent a major cost item in completing the packaging function. It is therefore important that the method of fastening and closure be fully analyzed and be included as part of an integrated evaluation to provide economical package protection.

DEFINITIONS

Fastenings are, broadly speaking, any devices which serve to secure one part against another. In packaging we refer primarily to *nails, screws, bolts, staples, special clips,* and other materials. *Strapping, twine, or certain tapes* may also be used as fastenings but are *primarily bundling and reinforcing materials* and as such are covered in Chapter 10.

Closure is defined as a sealing or covering device affixed to or on a container for the purpose of retaining the contents and preventing contamination thereof, or as the joint or seal formed in attaching two parts, such as the cover of a can to the body. There are many types of closures used which are specifically designed for a particular type of container. These include screw caps for various types of glass and plastic containers, and friction closures for metal, and fibre and plastic containers. Closures of this type are principally used for consumer packaging applications as defined in this book and are not covered here. This discussion is restricted to such closure materials and methods as *gluing, taping, stapling, and stitching* for paper and paperboard packaging. It is here that a choice is usually possible and an engineering analysis of the proper type is indicated. Most other closures for distribution packaging application are designed for the specific container, that is, lids for metal pails, covers for fibre drums, etc., and have been discussed in conjunction with these containers.

FASTENINGS

NAILS

The principal types and proper application of nails used for wooden containers have been discussed in Chapter 5.

SCREWS

When re-use of the container is anticipated or inspection of the contents in storage or transit is required, lids of wooden boxes should be closed

FASTENINGS AND CLOSURES

with *screws*. The countersunk-type is the most preferred. The *screw length* should be about 1¾ times the thickness of the piece holding the head. The required *gage of the screw* varies with the thickness of the board holding the point of the screw and the species grouping of the lumber. For example, for a ¾ in-thick board, a No. 9-gage screw is required for Group I wood, No. 8-gage for Group II wood, and No. 7-gage for Groups III and IV species. Similarly, 1¼ in.-thick lumber requires a No. 12-gage screw for Groups I and II, and No. 11-gage for Groups III and IV lumber, with the thicker board permitting a heavier screw and the harder species (Groups III and IV) not requiring as heavy a screw as the softer species (Groups I and II), owing to the greater holding power of the former.

Screw spacing. The screw spacing depends on the gage of the screw and the direction of the board grain into which the screw is driven. When screws are driven into the cleats of ends, the spacing for assembling sides, top, and bottom varies from 1¾ in. to 3½ in. Tops and bottoms are screwed to the sides with an 8 in. to 12 in. spacing, except that two screws must be used in the end of each piece. When the top of the box is fastened with screws *only*, the maximum spacing is reduced to 4½ in. with 12-gage screws and 3 in. with 7-gage or smaller screws.

For Group III and IV woods the drilling of shank clearing holes and pilot holes into the mating piece is recommended. Countersinking is only necessary for Group III and IV woods and not for the softer species of the Group I and II woods in which only drilling of the shank clearing holes is necessary.

Lag screws. On large crates, for the fastening of panels to frames and frames to skids, lag screws are frequently used. For details of the size, length, and number of screws required consult Federal Specification FF-B-561C which is the most suitable reference guide for this purpose.

BOLTS

Bolts find wide application in wooden shipping containers, primarily for the mounting of a product to a base. In addition, some large crates of the sheathed or unsheathed type are assembled with the use of bolts. For applications of this type, bolts should be standard machine, step, or carriage bolts. For mounting purposes any machinery handbook will furnish suitable strength data to evaluate the size of bolt required for a given weight. *Flat washers* under the bolt head and *flat* and *spring washers* under the nut should be used. Suitable locknuts can be substituted for a simplification of the mounting operation. Vibration in transit readily opens unsecured nuts, and crushing of the wood area under the head and nut may result in looseness unless washers are used.

STITCHING AND STAPLING

Wire stitches and staples are widely used for the fabrication of wooden containers but rarely for the fastening and closure of the container elements at the point of use. Stitching and stapling techniques for paper and paperboard products are discussed later in this chapter.

CLOSURES

ADHESIVES

Of the available closure types for paper and paperboard products, adhesives are the most economical to use, particularly if operating conditions permit the use of fully-automatic equipment.

Adhesives are materials in *liquid or soft plastic condition* for purposes of application, which are capable of changing to a semi-solid jelly or to a hard film of high-cohesive ability and suitable bonding power. When two surfaces are brought together they must be held together until a semi-solid condition has developed in the glue film that can successfully overcome the tension that may be imposed due to the resiliency of the flaps. In most applications the adhesive bond must have a tensile strength equal to or greater than that of either of the materials bonded. For certain easy-opening applications the adhesive is specially formulated to produce a *low tensile strength* which permits separation *within the film* of the set adhesive.

The most popular and extensive application for adhesives in the distribution packaging field at the user's plant is for the sealing of fibreboard shipping containers, and paper sacks and balers. Hot melt adhesives, silicate of soda (water glass), vegetable adhesives with a dextrin base, and synthetic resins of the emulsion type are the most widely used adhesives for this purpose. Hot melts used in packaging are thermoplastic, polymer-based compounds that become plastic when heated. Silicate of soda, although it has many excellent properties (including its relatively low cost), hardens to a glass-like consistency which presents problems of cleaning the more complex application equipment. For this reason, other types of adhesives have been developed which can more readily be washed off from moving machine parts and, the use of silicates has diminished.

The exact type of adhesive to be used depends on a number of variables which are considered in specific formulations of the adhesive manufacturer. Some of these variables are as follows:

1. The type of surface.
2. Rate of setting, which is a function of the speed of application as

FASTENINGS AND CLOSURES 363

well as the time under compression and tension factors (flaps, etc.) of the container.

3. Specific end-use requirements for water resistance, easy-opening, etc.

Adhesive sealing methods are divided into three general classes—manual, semi-automatic, and fully-automatic.

Manual sealing. This method is primarily performed by the shipper handling a relatively small quantity of containers or a wide variation in sizes. Sometimes, manual sealing is carried out in conjunction with another closure method such as taping. In this case the tape provides a means of compressing the box flaps as well as to seal any openings of the container through which foreign matter may enter in transit and storage.

Manual sealing is carried out with completely portable tools or with semi-portable equipment. Portable tools include brush, spray, or roller. With either one of these methods, full contact of the inner and outer flaps should be assured and proper *squaring* or alignment of the container should be carried out. Since these conditions depend on the operator, manual sealing is often of inadequate quality which results in subsequent closure failure. Compression to the flaps can be provided through (1) up-ending of the loaded container whose bottom has previously been sealed by other means, (2) placing another loaded container on top, or (3) suitable weights. Tape sealing as previously mentioned and stapling are also used. Semi-portable equipment is available for the application of both cold and hot melt adhesives. The former makes use of a dispensing head similar to that found on automatic equipment. Dispensing of adhesive is controlled by a hand operated valve and the adhesive is fed from a pressurized reservoir. For hot melts, there are two types of applicators or "guns" used in packaging applications: (1) using a solid cartridge of resin that is melted in the tool itself and (2) the resin in molten form is pumped from a heated reservoir and is dispensed through a trigger-controlled valve (Fig. 9-1).

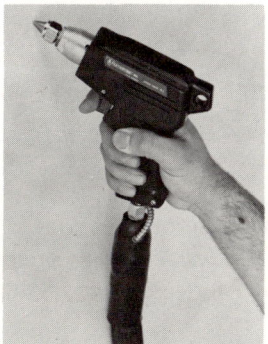

Fig. 9-1. Semi-portable hot melt adhesive applicator. Courtesy Spraymation, Inc.

Semi-automatic sealing. Semi-automatic sealing requires the involvement of an operator working in conjunction with mechanized equipment. Equipment of this type is available for either cold or hot-melt adhesives and the operator is usually required to fold the box flaps and push the cases manually through the sealing section, or to feed the boxes to a powered drive. Where cold adhesives are used, a moderately long compression unit is required. The compression unit (Fig. 9-2) consists of two endless belts, top and bottom, backed up by spring rollers and furnished with infeed and exit conveyors. The boxes are fed through the unit at a constant speed or are intermittently moved as boxes enter into the unit. Semi-automatic sealing is only recommended for infrequent or slow runs that do not justify the greater expenditure for fully automatic equipment or where space restrictions make the latter equipment impractical. In addition, the requirement for adjusting to different box depths does not offer any real advantages in operating when a multitude of case sizes is being packed.

Fig. 9-2. Compression unit for semi-automatic fibre box sealing. Courtesy ABC Packaging Machine Corp.

Fully automatic sealing methods. The full advantage of adhesives as a low-cost closure material is realized with *fully automatic sealing equipment*. The following conditions must be met in order for equipment of this type to be economically justified.

FASTENINGS AND CLOSURES

1. Direct infeed of the cases from packing stations.
2. Adequate volume.
3. Adequate floor and/or overhead space for the equipment and for accumulation lines where required.

Direct infeed of the containers is accomplished by connecting conveyors from packing stations. If top and bottom closures are simultaneously performed, long runs over curves, switches, and chutes must be properly engineered to avoid spilling of the contents during transit. If the output at one packing line does not justify a separate sealer, two or more lines can be combined into one infeed conveyor. If more than one size of container is packed on these lines, they must either first be accumulated in sufficient quantity before a manual or automatic adjustment on the sealer can be made economically, or self-adjusting equipment for random sizes must be utilized. The time element for clearing the machine of cases before it is feasible to adjust for another size, reduces the effective capacity of the machine. This is also a factor on self-adjusting machines. The length of the accumulation line, where used, must be designed for a balanced flow with minimum change-over delays. Frequently, the subsequent palletizing operation determines the number of cases to be accumulated. For instance, if a certain pallet pattern requires 36 containers, the accumulation conveyor ahead of the sealer will provide sufficient room for 36 containers. When these containers have been accumulated, the case sealer will either automatically or manually be changed to the size of this container. The line will be activated, and through switches or readers, it will feed the containers through the sealer to the palletizing station. In Fig. 9-3 a simple diagram illustrates such a system. Where self-adjusting case sealers are used for random box sizes, multiple pallet positions must be provided for palletizing by case or sorting and accumulation by size for automatic palletizing is necessary.

SELECTION FACTORS. The selection of the proper automatic sealer frequently presents an engineering problem in view of the many types that are available, which can be categorized as (1) sealers using cold adhesives best utilized for long, continuous runs of fixed size boxes (Fig. 9-4), and (2) hot melt sealers available for both fixed size boxes and for random size applications. Briefly stated, cold adhesive systems require more floor space and more time for clean-up and maintenance, but this is compensated due to the relatively low cost of cold adhesive. Hot melt adhesives, while more costly, do not require compression sections and thus these systems are more compact using less floor space. There are two basic types of random hot melt adhesive sealers available: those which accept a random flow of mixed size boxes fed from a common infeed conveyor (Fig. 9-5), and random-uniform systems which depend on prior accumulation by box size to obtain higher production speeds by minimizing the automatic adjustment of the

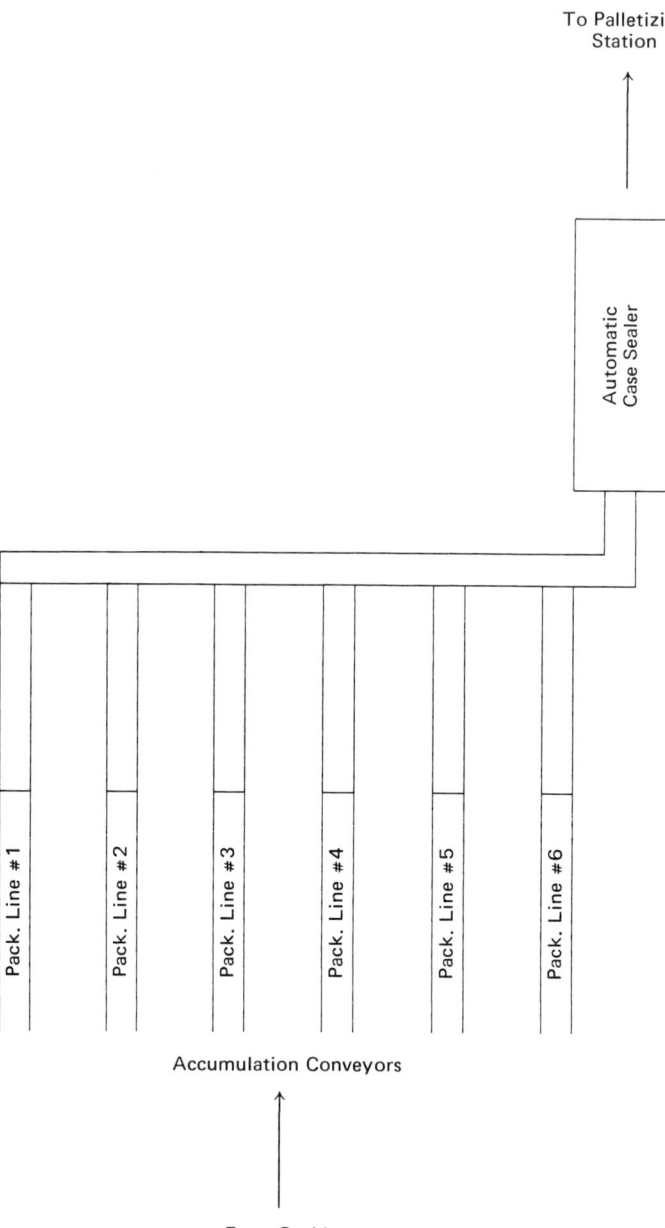

Fig. 9-3. Accumulation of different size cases for automatic sealing.

FASTENINGS AND CLOSURES

machine for a random mix of different box sizes. The following are some of the principal selection factors to consider:

1. Required floor space.
2. Container size ranges.
3. Methods of providing size adjustability.
4. Type of adhesive system and its cleaning and maintenance requirements.
5. Effective speeds in relation to its size.
6. Start-up time.
7. Initial purchase and installation cost.
8. Cost of labor to operate.

Fig. 9-4. Automatic sealer providing simultaneous top and bottom flap sealing. Courtesy Standard Knapp Division of Emhart Manufacturing Co.

TYPES OF SEALERS. Automatic sealers for cold adhesives basically consist of two units: (1) *the adhesive application unit* and (2) *the compression unit.* The application unit provides the infeed, flap opening, glue application, and flap closing mechanism, whereas the compression unit permits a prolonged tension-free contact between the freshly bonded surfaces to allow for the setting and to make no mechanical demands upon the adhesive until the preliminary setting is accomplished. The setting time is therefore a factor in the *length of the compression unit,* which varies with the speed of travel through the equipment. The application unit is of a *fixed size* and is not affected by the required sealing rate. These units are available as small as about 6 ft. To this must be added the compression unit which is made as

Fig. 9-5. Random size hot melt adhesive case sealing machine. Courtesy Padlocker Machinery Div., The Loveshaw Corp.

small as 4 ft for an output of approximately 4-12 in.-long cases per minute. In Table 9-1 the compression unit length and over-all length of one typical automatic sealer is illustrated.

Many attempts to reduce the overall length of sealers have been made by equipment and adhesive manufacturers to minimize the adhesive setting time.

TABLE 9-1
Typical Length of Automatic Sealer for Varying Production Rates

Rate per Minute*	Length of Compression Unit, ft	Overall Length of Sealer, ft
6	6	17
8	8	19
11	10	21
14	12	23
17	14	25
20	16	27

* For No. 2 can cases which are 13¾ in. long.

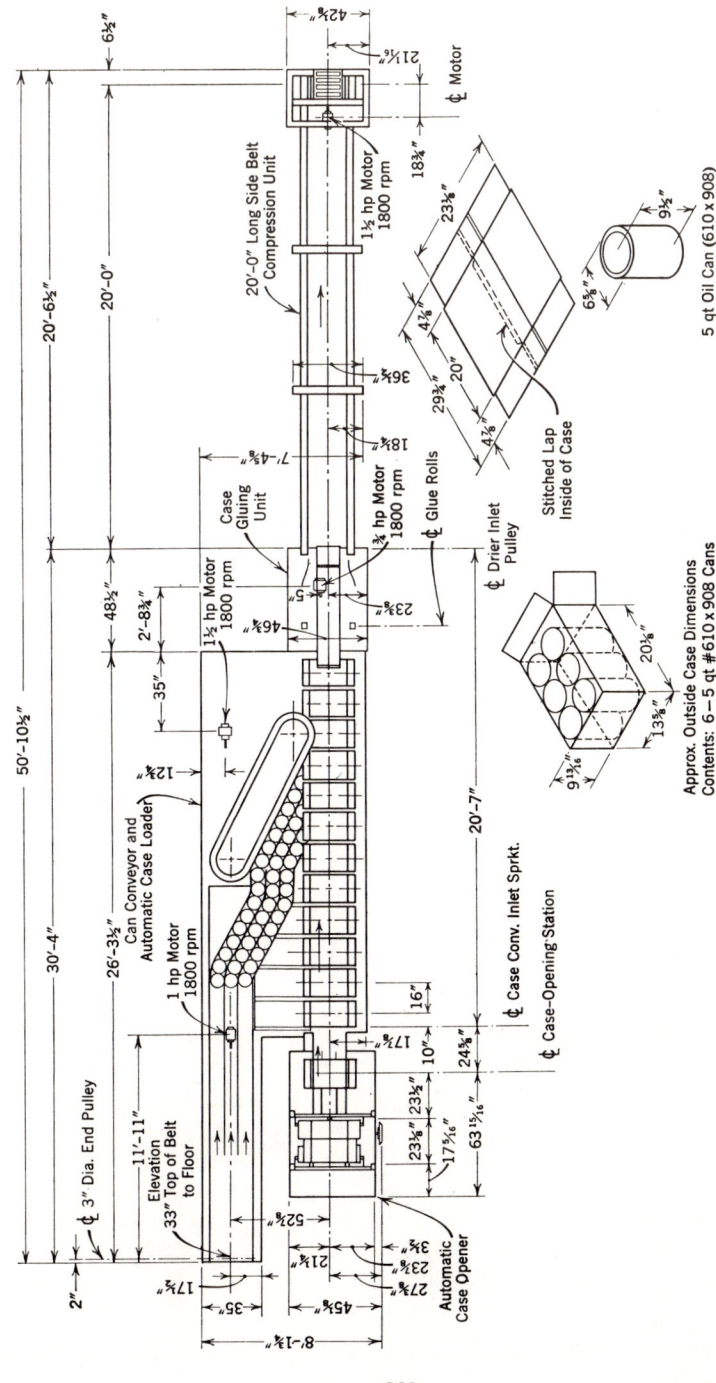

Fig. 9-6. Integrated loading and case-sealing system for *end-opening* boxes. Courtesy J. L. Ferguson Co.

Some of these include (1) the use of fast setting adhesives, (2) accurate control of the amount of adhesive placed on the container, (3) preheating of container flaps, and (4) an atomizer-like applicator head to spray adhesive on flaps accelerating the evaporation of water or solvent carrier.

However, the development and large-scale introduction of hot melt adhesives which feature virtual instantaneous setting have had a major impact on the design, size and cost of adhesive case sealing systems by entirely eliminating the need for comprehension sections. Hot melt adhesive generally costs more than cold setting adhesive and the application equipment is more complex and costly to operate due to the requirement of heaters for tanks, hoses and nozzles. This equipment has lower maintenance cost and simpler start-up procedures.

Sealers for *end-opening boxes* facilitate the economical integrated loading of such items as cans and the sealing of fibreboard shipping containers. Figure 9-6 shows such a system for the packaging of 5-qt cans packed six to an end-opening box at a rate of approximately 20 containers per minute. The more economical shape of the end-opening as compared to the previously used top-opening box, together with this packaging method, produced major economies for the oil company where this system was used.

The setting up of empty corrugated boxes and the sealing of the bottom flaps are often required if the packaging operation is performed at many different stations which are difficult to connect with a conveyor or when no

Fig. 9-7. Automatic corrugated box former and bottom flap gluer. Courtesy General Corrugated Machinery, Inc.

FASTENINGS AND CLOSURES

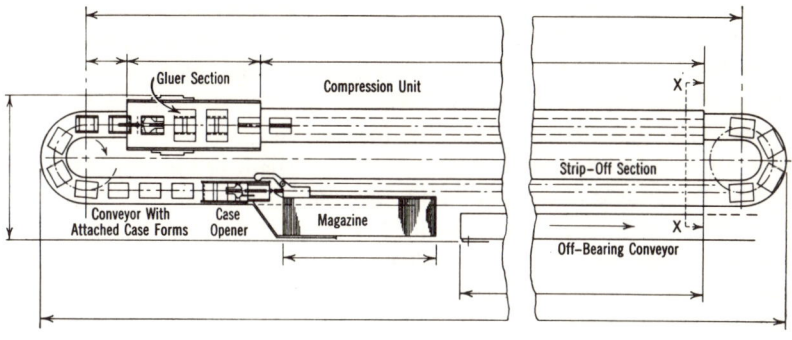

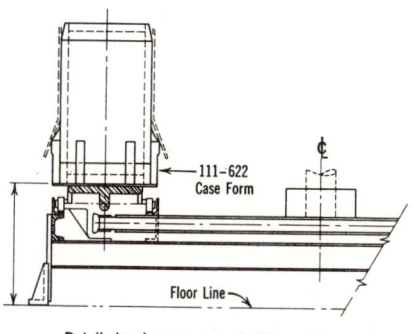

Detail showing conveyor, holding plate, form, and case in place on the form.

Fig. 9-8. Fully automatic case-bottom sealer of empty fibre boxes utilizing a magazine-fed container opener and a case removal mechanism. Courtesy Standard-Knapp, Division of Emhart Corporation.

top seal is needed as with empty glass bottles. The bottom closure can be effected by staples, stitches, tape, or with adhesive. For adhesive bottom sealing of empty boxes, equipment is available that can perform this operation automatically. In Fig. 9-7 one type of machine is shown. Knocked down boxes are opened by vacuum mechanism that contact the top and bottom of the flat blank. Once erect, the bottom inner flaps are closed and the box is squared. Adhesive is then applied to the inner flaps and the bottom outer flaps are closed.

Another machine uses a large "merry-go-round" for the compression and accumulation of sealed containers. This machine can be made fully automatic when combined with a magazine-fed case opener (Fig. 9-8) and a case removal mechanism (Fig. 9-9).

372 **DISTRIBUTION PACKAGING**

Adhesive patterns. As discussed later in this chapter, carrier regulations require not less than 50 per cent of the flap contact area of fibreboard boxes to be bonded together when cold setting adhesives are used. However, when hot melt adhesives are applied, at least 25% of the flap contact area must be bonded. The flap contact area is 100 per cent of the length by width dimension for a square box and 50 per cent if the length of the box is twice the width. To comply with regulations for cold setting adhesives, half of the flap contact area need only be bonded. Thus only ¼ of the length by width dimension must be covered with adhesive on a container that has the length twice the width. This arrangement permits the application of various glue patterns to the inside of the outer flaps where the adhesive is always applied by an automatic case sealing machine.

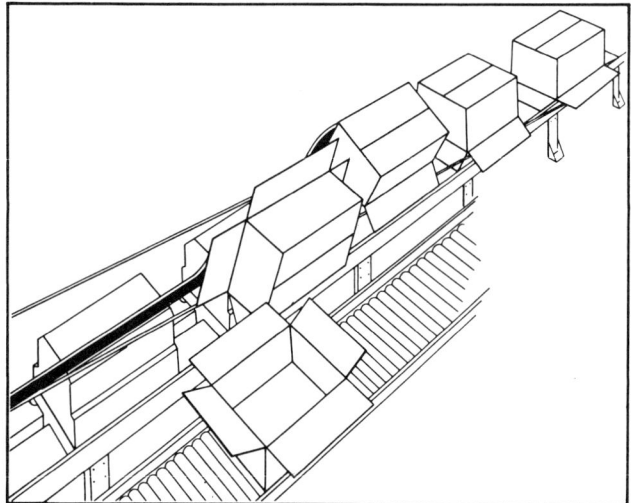

Fig. 9-9. Case removal mechanism showing guides which raise and strip case off the form on which the bottom flaps have been sealed. Courtesy Standard-Knapp, Division of Emhart Corporation.

In Fig. 9-10 various acceptable patterns are illustrated. The purpose of these is as follows:

1. Elimination of adhesive contact with product when the container has a rectangular shape.
2. Reduction of adhesive consumption.
3. Easier container opening and cleaner fibre or adhesive separation. (See Chapter 11.)

FASTENINGS AND CLOSURES

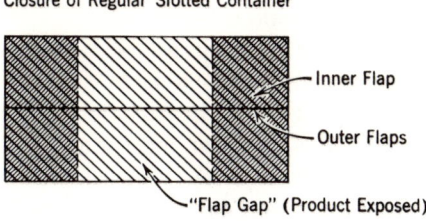

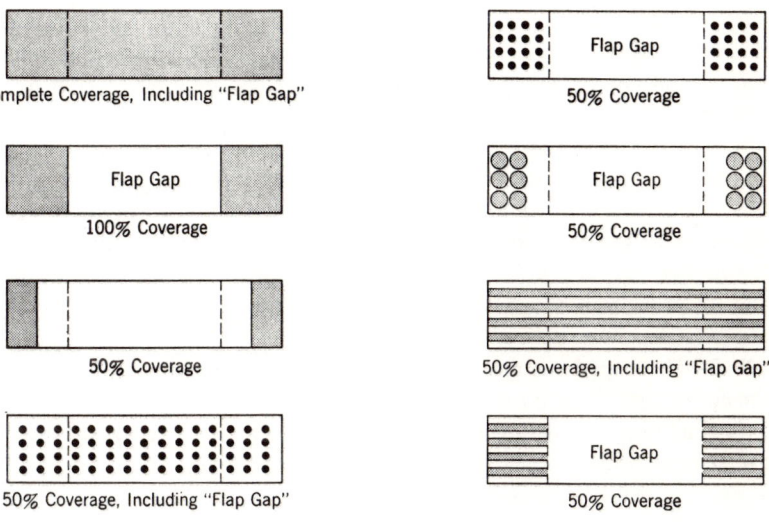

Fig. 9-10. Various glue patterns obtainable with automatic sealers.

TAPES

Various types of tapes are extensively used as a closure medium. Tapes for packaging purposes are classified by the type of adhesive used and by the material. The adhesive classification is as follows:

1. *Gummed tapes.* These tapes use a water activated adhesive.
2. *Pressure-sensitive tapes.* These use an adhesive which adheres under pressure and does not require moistening, heat, or solvent for activation.

Typical materials for fabrication into tapes are: (*a*) paper, (*b*) cloth, (*c*) film, and (*d*) laminated materials.

There are literally hundreds of various tapes available, many of which have been specially developed for certain applications. In this chapter, those tapes which are principally used to close distribution packages will be discussed. In Chapter 10, tapes designed for reinforcing and certain sealing and bundling operations will be covered.

Gummed tapes. There are two major types of gummed tapes used for the sealing of fibreboard containers and paper-wrapped bundles, i.e., *plain gummed sealing tape* and *gummed reinforced sealing tape*. Plain sealing tapes are made of kraft paper in a wide range of basis weights. The usual weights are 35, 60, and 90 lb. (24 in. x 36 in., 500 sheets). Gummed reinforced tapes consist of a laminated paper structure, reinforced with glass fibres which are embedded in an *asphaltic* or *nonasphaltic* laminant. Other reinforcing fibres such as *sisal*, *rayon* and *nylon*, while permitted, are rarely used. A full description of the construction details are covered later in this chapter.

In application, a single 3 in. wide strip of reinforced tape is applied over the center seam of the box flaps with minimum 2-½ in. end panel extensions. Paper tape on the other hand is applied over the center seam and over the end edges of the flaps and consequently three strips are required to effect closure for one set of box flaps. The great acceptance of gummed tapes stems from the following advantages:

1. When properly moistened and applied, they provide excellent protection against handling and shipping hazards.

2. If all open seams are covered with tape, as is the case with *plain* gummed sealing tape, protection against contamination by foreign matter is provided. Single strip sealing with reinforced tape is almost as effective since the critical flap gap area at the center seam is covered.

3. Imprinting of tapes enables the carrying of an inexpensive message for advertising, product identification, instructions, etc.

4. Small, compact dispensing equipment is available which permits flexibility of the taping operation in small packaging areas.

5. Re-usability of container due to ease of opening is facilitated.

The disadvantages of gummed tapes are as follows:

1. Water-activated tapes do not adhere well to inked, varnished or waxed surfaces.

2. Incorrect wetting or improper application can result in closure failure.

3. Labor cost involved in the application of three strips of plain gummed tape is generally higher than that of other types of closures.

FASTENINGS AND CLOSURES

The selection of a tape for a given operation depends on the above advantages and disadvantages in relation to the particular plant conditions, selection factors discussed later in this chapter, and individual preferences of the shipper.

Gumming adhesives. All water-activated gummed tapes use an adhesive which is applied to the backing of the tape during its manufacture. Both animal and dextrin adhesives make up the principal ingredients of the adhesive formulation. For reinforced tapes frequently subjected to heavy duty applications, a heavier film of adhesive is applied.

When gummed tape is applied to cold surfaces, (ice cream containers, for instance), problems of the water's forming ice crystals and thus preventing tackiness of the glue and proper fibre penetration are encountered. Special glue formulations are available for these conditions.

The reactivation of the gumming is provided through the formation of a sticky film when water makes contact with the dried adhesive. The rate of reactivation is dependent upon the *condition of the gumming*, the *temperature of the water*, the *mineral content of the water*, the *amount of water*, and the *degree of coverage*. In view of these reactivation requirements, tape storage conditions, as well as control of applying moisture to the tape, are very important. The latter condition is enhanced by heating the water, adding softening agents, and the condition and cleanliness of dispenser application devices.

Tape dispensers. There is a large variety of tape-dispensing equipment on the market which is basically classified by its method of performing the taping operation, that is, *manually, semi-automatically*, and *automatically*.

MANUAL DISPENSERS. Manually operated tape dispensers are made of the *pull-and-tear* type, and the *lever-dispenser* type. The latter type is furnished with *manual or automatic cut-off devices.*

The *pull-and-tear type* dispensers provide no pressure or water-level control and the quality of tape water-reactivation depends, therefore, to a large extent on the operator. Machines of this type are only recommended for very occasional taping operations and should be placed in an elevated position so that the natural tendency for the operator is to apply a downward pull over the brush or roller of the dispenser.

The *lever-dispenser type* uses a controlled water-level and pressure device above the tape brushes. Some machines provide for adjustability of the pressure device to be able to vary the water pick-up from the brushes for different speeds of operators or types of tapes. These machines are available with heating units, are suitable for incorporating printing and coding attachments, and are adjustable for various tape widths (Fig. 9-11). Usually smaller models are used for the lighter grades and narrower width tapes (35 lb, ½ in.-

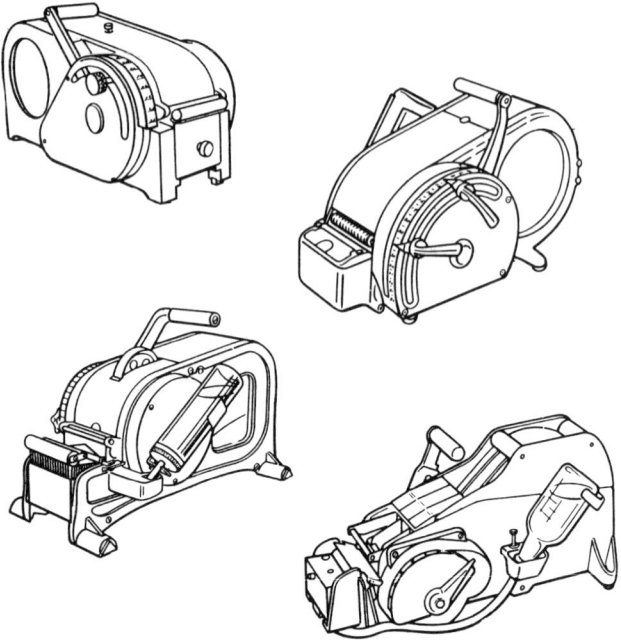

Fig. 9-11. Various types of lever-dispenser type gummed tape machines with automatic tape cut-off devices.

2 in.) than for the heavier paper, cloth, and reinforced tapes of wider width ranges. To facilitate the cut-off of reinforced tapes, heavy duty knives are required.

SEMI–AUTOMATIC DISPENSERS. Machines of this type have the tape-dispensing and cut-off operation *motorized* but depend on *manual tape application* to the container. The required tape length is either preset, selected by the operator, or premeasured electrically by the container.

In Fig. 9-12 a semi-automatic dispenser for the preselecting of two tape lengths is illustrated. Such machines can be operated either by hand buttons or foot pedals. Dual tape length dispensers are advantageously used when rectangular containers require different lengths for the center and end strips.

Other machines providing fully automatic pre-measuring of the center-strip of the tape are available. The controls for these machines can be built into the conveyor section ahead of the dispenser.

For the high-speed manual application of some of the heavier tape grades, frequently problems exist in properly obtaining water penetration into the thicker glue film of these tapes. This problem may be overcome by using a semi-automatic dispenser which prefeeds the tape and permits it to hang

FASTENINGS AND CLOSURES 377

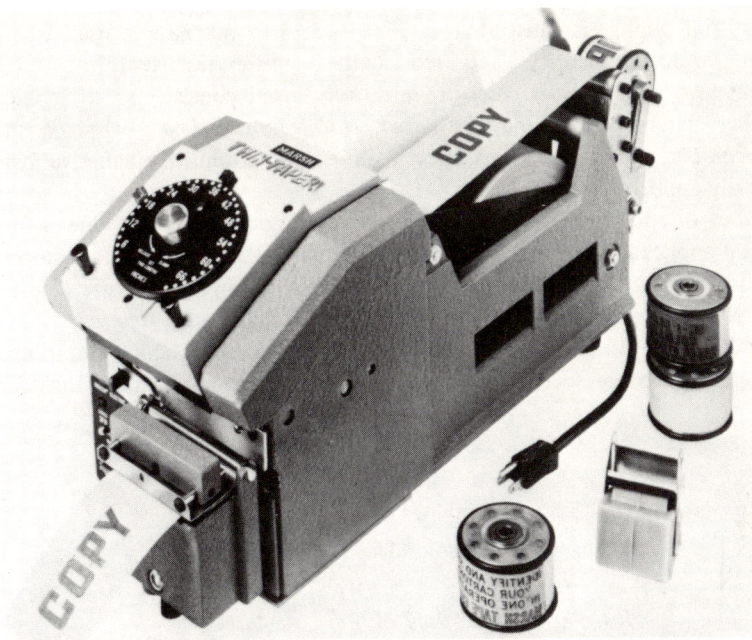

Fig. 9-12. Tape dispenser providing for preselecting of two tape lengths and mechanized tape feeding and cut-off. Courtesy Marsh Stencil Machine Co.

moist from the mouth of the tape dispenser while the previous piece of tape is applied to the container.

AUTOMATIC TAPE SEALERS. Fully automatic tape container sealers are available which apply the tape, after dispensing, to the top and/or bottom of the container. Tape can be applied only in one direction, which has precipitated the growth of reinforced sealing tape which does not require the use of end strips.

In Fig. 9-13 one type of automatic tape sealer for top and bottom application for uniform size boxes is shown. This particular machine provides for automatic length adjustment but requires mechanical height and width changes. The extended ends of the tape are pressed onto the end panels of the container by rubber roller devices.

In Fig. 9-14, a fully automatic random box taping machine is shown. This equipment self-adjusts to each different box size, irrespective of size or sequence, within its size capacity. The machine folds the top flaps of each box and then tape seals top and bottom flaps at speeds up to 16 mixed-sized boxes per minute, without an operator.

378   DISTRIBUTION PACKAGING

A specialized version of this random box sealing equipment, combining taping and gluing, is illustrated in Fig. 9-15. This machine is used where maximum closure security is required as with cigarettes, liquor, etc.

Pressure-sensitive tapes. Pressure-sensitive tapes consist of a flexible backing material which may be *paper, cloth, film, foil, polyvinylchloride* or a *combination* of these materials with a pressure-sensitive adhesive mass generally applied to one side.

Tapes of this type were initially used for surgical purposes. One of the original applications in the industrial field was to mask auto-bodies requiring a two-tone paint job. Since these rather humble beginnings, the use of pressure-sensitive tapes has grown into a multi-million dollar industry.

Today there are literally hundreds of pressure-sensitive tape types in use, many of which are specially designed for a particular application. Included

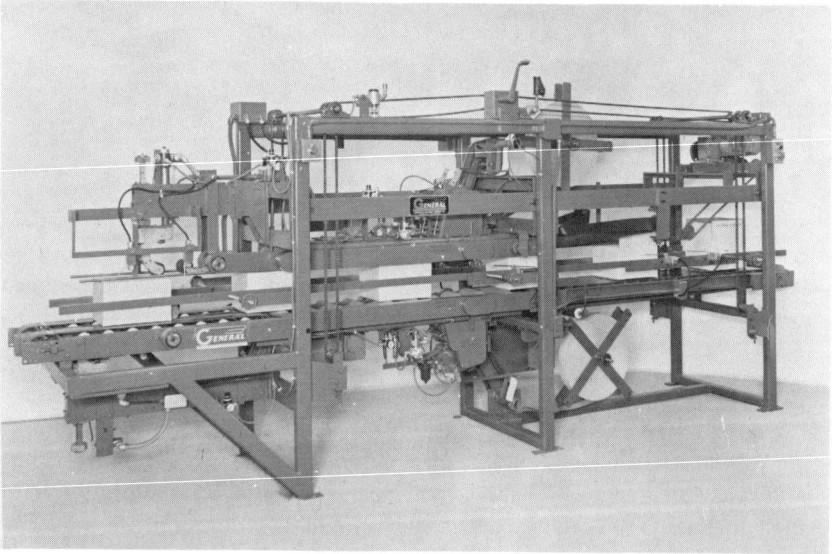

Fig. 9-13. Fully automatic tape sealer for top and bottom center-strip sealing of uniform size boxes. Courtesy General Corrugated Machinery Co., Inc.

are tapes for insulating, stenciling, and splicing. Although there are thousands of applications, these can be catalogued into ten basic categories. These are for (1) *holding*, (2) *masking*, (3) *sealing*, (4) *protecting*, (5) *reinforcing*, (6) *stenciling*, (7) *splicing*, (8) *identifying*, (9) *packaging*, and (10) *insulating*. For the purpose of this text three of these major categories are discussed, namely, *sealing, reinforcing,* and *packaging tapes*. Reinforcing tapes and their application are treated in Chapter 10.

FASTENINGS AND CLOSURES

Fig. 9-14. Fully automatic random container top and bottom tape sealing machine. Courtesy Padlocker Machinery Div., The Loveshaw Corp.

Fig. 9-15. Random top and bottom gluer-taper for applications requiring a high degree of sealing security. Courtesy Padlocker Machinery Div., The Loveshaw Corp.

Pressure sensitive tapes are advantageously used for packaging purposes for the following reasons:

1. Simplicity of application.
2. Adherance to all types of surfaces including metal, glass, plastic, and cloth.

3. Backing materials can be employed for merchandising appeal, clarity, strength, water, and/or vapor resistance.

4. Adhesive formulations can be modified to satisfy a wide range of application requirements.

As compared to water-activated tapes some of these additional advantages apply:

1. No moistening required.
2. Less complicated dispensing equipment.
3. Instantaneous adhesion without delays required for setting.
4. Easier to open.

The disadvantages of pressure sensitive tape include:

1. Relatively high cost of materials.
2. Limited acceptability for rail freight shipment.
3. Some plastic backings have a tendency to split in the lengthwise direction.

Types of pressure-sensitive packaging tapes. The principal types of tapes and the major fields of their application are as follows:

1. *Acetate tapes* consist of two types: acetate fibre and acetate film. Acetate fibre tapes are constructed of an adhesive-laminated structure with a ply of acetate film for waterproofness and a ply of rope fibre paper for strength with the pressure-sensitive mass applied to the rope paper side. Typical applications include sealing and reinforcing of fibreboard containers, bundling of containers, small parts, etc., bag and slip cover can sealing, holding lightweight items on display cards or on boxes in combination deals and attaching premiums. Acetate film is made of film of approximately 3 mils thickness and is used for mending, identification and reproduction applications. It features high clarity and dimensional stability.

2. *Cellophane* tapes are made from regenerated cellulose and are transparent. Their major applications, in addition to office and home use, include sealing regular kraft carry-home bags, treated bags, plastic bags and packages, small fibre tubes, slip cover cans, and holding light items to cards.

3. *Flat-back and creped-back paper* tapes utilize various paper stocks including rope papers in various basis weights. The paper may also be treated, saturated, or impregnated to impart moisture, grease, solvent, or stain resistance, low-release characteristics, etc. These tapes can be interchangeably used for most of the applications of acetate, fibre, and cellophane tapes. In addition, they are used for masking, banding lightweight wire coils, protecting finished surfaces in shipment, and holding moving parts of appliances such as stove doors and typewriter components.

4. *Cloth* tapes utilize cotton sheeting which is bleached, unbleached, or dyed and is treated to impart waterproof characteristics. Cloth tapes are used

FASTENINGS AND CLOSURES

for heavy duty packaging and bundling applications requiring greater strength than provided by paper tapes. Their major uses, however, are for military purposes in sealing weatherproof fibreboard containers, metal and fibre drums and cans, waterproofing and protecting tanks, aircraft, and automotive vehicles for overseas shipment.

5. *Filament* tapes are reinforced tapes constructed with a backing material of either film or paper to which is bonded tiny filaments of either synthetic or glass fibres embedded in a laminant over which is applied the mass of pressure-sensitive adhesive. These tapes are used primarily for reinforcing and bundling, and their applications are covered in Chapter 10. They are also finding application for the sealing of fibreboard containers such as telescope boxes and hardware containers.

6. *Filmic* tapes are either unplasticized polyvinyl chloride, polyesters or polypropylene. These materials provide adequate strength, without reinforcing filaments, for such packaging applications as carton sealing, horizontal banding of unit loads and for reinforcing. These tapes feature high abrasion and moisture resistance, and are usable at low temperatures.

The adhesion of all pressure-sensitive type tapes is measured in *ounces per inch of width*. This information is contained together with such other physical characteristics as *tensile strength, elongation*, and *thickness* in the specifications of the tape manufacturer. Average *adhesion* of pressure-sensitive tapes ranges from a low of approximately 5 oz per in. of width for low-tack masking tapes to approximately 70 oz per in. of width for high tack filament strapping tapes. The *tensile strength* of pressure-sensitive tapes ranges from a low of approximately 10 lb per in. of width to a high of approximately 80 lb for non-reinforced tapes, although there are many exceptions to this approximate range of tensile strengths. *Filament tapes* generally have a minimum tensile strength of 160 lb when rayon filaments are used, and as high as 600 lb with glass fibre reinforcement.

For many packaging applications the *stretch* or *elongation* properties of tapes are of importance. For flat paper tapes this property ranges from approximately 2 to 8 per cent, whereas, creped paper tapes have an elongation range from approximately 5 to 12 per cent. Filament tapes, reinforced with glass fibres, have an elongation of approximately 5 per cent and are therefore susceptible to breakage on impact. Rayon, although having lower tensile strength, provides an elongation of approximately 15 per cent allowing for energy absorption before breakage.

For many military applications pressure-sensitive tapes must meet applicable government specifications. Some of these specifications are mentioned later in this chapter.

Pressure-sensitive tape dispensers. Although one of the main advantages in the use of pressure-sensitive tape lies in its simplicity of application because preconditioning of the adhesive is not required, equipment has been developed to assist in accurate and rapid dispensing, cutting, and positioning. A large number of manual, semi-automatic, and automatic devices are available.

MANUAL DISPENSERS. The manually-operated tape dispensers are similar in construction and operation to gummed tape dispensers except that there are no provisions for moistening. They are made of the *pull-and-tear type* and the *lever-dispenser type*. The former is also made small and compact to be carried on the job by the packer. Two types of portable hand dispensers used for the application of carton sealing tape are illustrated in Fig. 9-16. Other pull-and-tear type dispensers for pressure-sensitive tape are the familiar office desk-type units.

 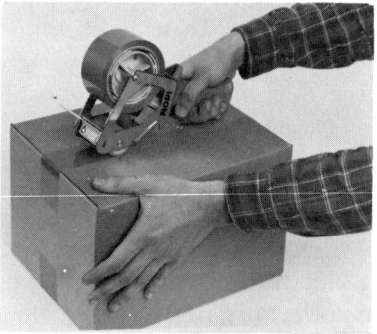

(a) using narrow width filament tape for sealing and reinforcing. Courtesy Permacel.

(b) using 2" wide filmic tape for center seam sealing. Courtesy Nopi, Inc.

Fig. 9-16. Portable hand dispenser for pressure-sensitive tape.

Lever-operated dispensers can accommodate tapes up to 4 in. wide and measure strips up to 21 in. long with one operating stroke. Wide rolls of tape can be slit into narrow widths on such machines with resultant economies. The tape is cut by an upward pull over a sharp knife which is attached to the housing of the dispenser. This principle of cutting pressure-sensitive tape is employed by most manually-operated pressure sensitive tape dispensers. For rapid closing of film, foil, and other types of bags used for the packaging of produce and other products, a specialized device has been developed, shown in Fig. 9-17. Other simple manually-operated dispensers are available. A sealer for small boxes and packages is illustrated in Fig. 9-18. The box or article is passed over a roller which applies a strip of 1½ in.-long tape over the side and bottom of the box. An automatic cut-off is also provided. Frequently fixtures or mechanical devices can be constructed to assist in the

FASTENINGS AND CLOSURES

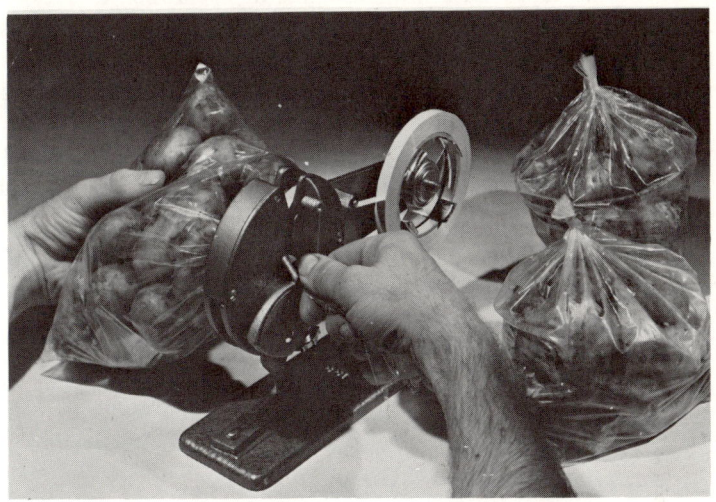

Fig. 9-17. Special bag sealer. Courtesy Nopi, Inc.

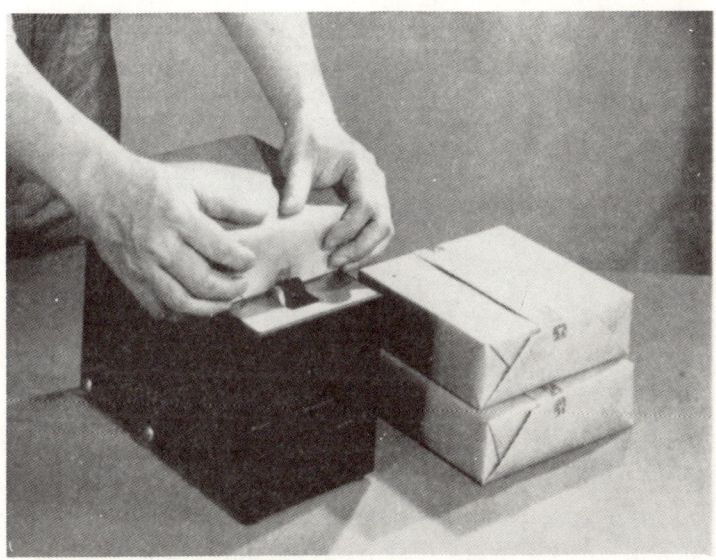

Fig. 9-18. Manual box sealer. Courtesy 3M Company.

384   **DISTRIBUTION PACKAGING**

application of the tape. Since no moistening or application of pressure is essential, such labor-saving mechanisms are usually simple and inexpensive.

SEMI-AUTOMATIC DISPENSERS. These machines usually have a power-assisted tape dispensing head with positioning of the package accomplished manually. In Fig. 9-19 and Fig. 9-20 two versions of these machines are shown. Machines of this type facilitate rapid and accurate tape sealing. Other semi-automatic machines are available for pipe wrapping, combination package bundling, and for several special purpose tasks.

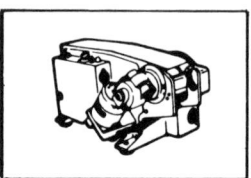

Fig. 9-19. Electronically controlled definate length dispenser. Courtesy Permacel.

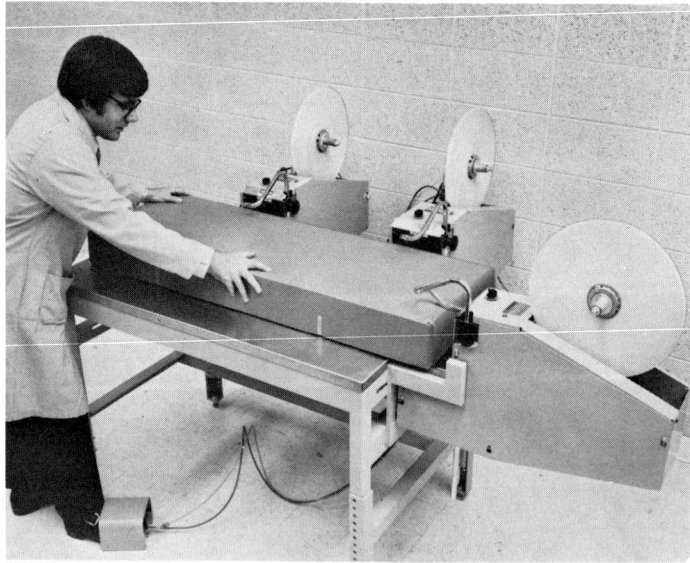

Fig. 9-20. Semi-automatic "L" clip taping system with multiple dispensing heads. Courtesy 3M Company.

SEMI- AND FULLY-AUTOMATIC SEALING MACHINES. The growing popularity of pressure-sensitive carton sealing tapes has spurred the development of semi- and fully-automatic machines for center strip sealing. These machines,

FASTENINGS AND CLOSURES

in addition to automatic *dispensing* of the tape, provide various additional functions in feeding, positioning and removing the package, and wipe down of the tape.

Models for uniform and random-size boxes are available with some designs featuring flap folding capability. In Fig. 9-21 a manually fed top and bottom

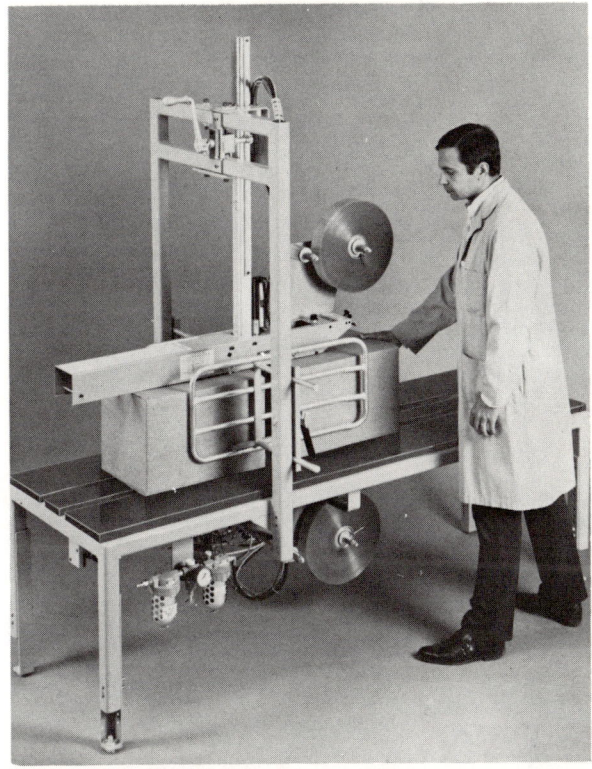

Fig. 9-21. Top and bottom semi-automatic tape sealing machine for uniform size boxes. Courtesy 3M Company.

box sealer for uniform sizes is shown. A similar version for sealing intermixed sizes is illustrated in Fig. 9-22 and a fully automatic machine including flap folding, requiring no operator, is featured in Fig. 9-23.

Manually fed machines of this type have an average productive capability ranging between 12 to 14 boxes per minute. This rate is influenced by the dexterity of the operator and box size and weight. The fully automatic version will seal mixed-sized boxes at a rate up to 16 boxes per minute, without an operator.

386   **DISTRIBUTION PACKAGING**

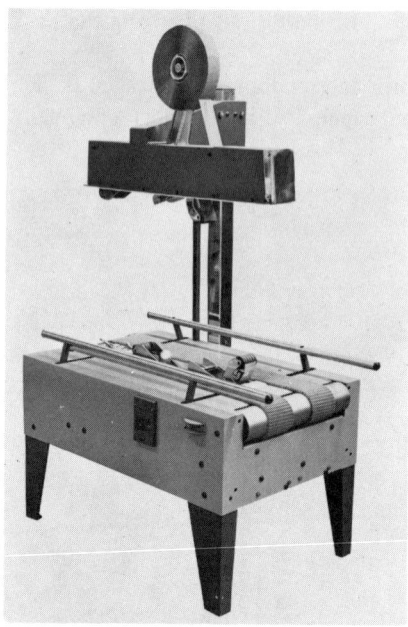

Fig. 9-22. Random box tape sealing machine without top box flap folding feature. Courtesy Little David Div., The Loveshaw Corp.

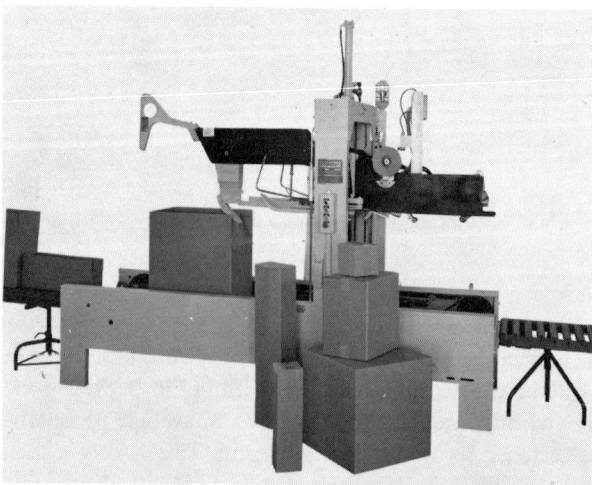

Fig. 9-23. Fully automatic random top and bottom box sealing equipment using pressure sensitive tape. Courtesy Padlocker Machinery Div., The Loveshaw Corp.

FASTENINGS AND CLOSURES

STAPLING AND STITCHING

Although these terms are frequently used interchangeably, the proper definitions are as follows. *Stitching* refers to the formation of a wire fastener from a continuous coil of wire during the fastening process. *Stapling* utilizes a preformed metal fastener which is dispensed, generally from a magazine-fed machine.

Stitching and stapling are used both for the manufacture of various types of shipping containers, that is, to form manufacturer's joints of fibre boxes and to fasten cleats of panel material to cleated and wire-bound boxes as well as to close containers. For manufacturing, high-speed production equipment is generally used. The discussion in this chapter covers stitching and stapling operations performed by the user to assemble and close the fabricated container. There are many other stitching and stapling applications of a non-packaging nature, such as in the fabrication of furniture, in the manufacture of automobiles, in construction work, in book binding, and in office and home use.

The major applications in packaging are as follows:

1. Assembly and closure of fibreboard boxes and drums and assembly of interior packing pieces.
2. Closing lids of wooden containers, fastening of straps and identification material to wooden containers, and securing linings within containers, box cars, etc.
3. To close bags and sacks and to attach a product or a small unit package to a display card.

The widespread use of stitching and stapling results from the following advantages as compared to other methods:

1. Low material cost.
2. Fast application with general or special purpose application equipment possessing flexibility for differences in container size.
3. Positive closure without subsequent compression or setting time.
4. Does not obscure printed matter.
5. Not radically affected by environmental conditions such as moisture and extremes in temperature.

The disadvantages of stitching and stapling include:

1. Potential damage hazard to merchandise by snagging, abrading, or puncturing by fastener, or during process of closing frequently necessitating the use of additional protective packaging materials.
2. Potential hazard to packers, handlers, and (upon opening) to ultimate consumer.
3. Provides no barrier against admission of foreign matter or moisture.

Types of stitches and staples. Whether the fastener is a *preformed staple or a stitch* formed in the stitching process, there are certain characteristics common to both. Stitches and staples are available from wire having different cross-section constructions, including *round, drawn flat, cut flat, oval,* and *arcuate*. The wire is made of high-quality carbon steel and is usually electrogalvanized for rust resistance. It may also be fabricated with a color coating.

TABLE 9-2
TYPICAL STANDARD STITCHING WIRE SIZES AND YIELD

Gage	Decimal Equivalent, in Inches	Feet per Pound
I. Standard Sizes Narrow Flat Wire		
No. 18 × 20	0.045 × 0.035	206
No. 19 × 21	0.040 × 0.032	248
No. 19 × 21½	0.040 × 0.029	290
No. 20 × 24	0.034 × 0.023	415
No. 20 × 25	0.034 × 0.020	478
No. 21 × 25	0.031 × 0.020	525
No. 22 × 26	0.028 × 0.018	644
II. Standard Sizes Round Wire		
No. 18	0.047	169
No. 19	0.041	226
No. 20	0.035	327
No. 21	0.032	370
No. 22	0.028	485
No. 23	0.025	608
No. 24	0.023	713
No. 25	0.021	903
No. 26	0.018	1172
No. 27	0.017	1320
No. 28	0.016	1483
No. 30	0.014	1940

Wire Size in Inches	Feet per Pound
III. Standard Sizes Wide Flat Wire	
0.103 × 0.023	129
0.103 × 0.020	148
0.103 × 0.017	175
0.103 × 0.014	211
0.103 × 0.011	270
0.103 × 0.008	370
0.057 × 0.024	223
0.057 × 0.020	268

FASTENINGS AND CLOSURES

Stitching wire is designated by gage and is available in gages from No. 18 to No. 30 for round wire, No. 18 X 20 to No. 22 X 26 for narrow flat wire, and 0.103 in. X 0.023 in. to 0.057 in. X 0.020 in. for wide, flat wire. Special sizes can also be obtained. In Table 9-2 standard sizes for these three types of stitching wire, including their footage per pound, is duplicated.

The wire sizes for staples are similar, although standards are not available since manufacturers are using many special staple types for the equipment they design.

Staple and stitch dimensions are identified by crown width and leg length. Both of these are inside dimensions. In fabricating stitches from continuous wire, the length of the leg is usually more readily adjustable than the width of the crown. Usually, the crown of the stitch can be made as small as $\frac{1}{4}$ in. and as large as $\frac{5}{8}$ in. Equipment for external flap stitching is designed for a crown width of 1 in. The leg length usually ranges from $\frac{7}{32}$ in. to $\frac{9}{16}$ in. Preformed staples are made with crowns as wide as $1\frac{5}{8}$ in. and the leg length generally varies from $\frac{3}{16}$ in. to $\frac{3}{4}$ in. For applications on nailed wooden boxes, larger size staples are obtainable.

TABLE 9-3
Stitch Yield for Various Types and Gages of Stitching Wire

Gage	Staple $\frac{7}{8}$ in.	Staple $\frac{15}{16}$ in.	Staple 1 in.	Staple $1\frac{1}{16}$ in.	Staple $1\frac{1}{8}$ in.	Staple $1\frac{1}{4}$ in.	Staple $1\frac{3}{8}$ in.	Staple $1\frac{1}{2}$ in.
Round Wire Staples per Pound								
No. 18	2320	2160	2030	1910	1820	1620	1480	1350
No. 19	3110	2890	2710	2550	2420	2170	1980	1810
No. 20	4500	4180	3930	3690	3500	3140	2860	2620
No. 21	5100	4740	4440	4180	3960	3550	3230	2960
No. 22	6660	6200	5810	5470	5180	4650	4240	3880
No. 23	8360	7800	7300	6870	6510	5840	5320	4870
No. 24	9860	9200	8620	8120	7700	6900	6290	5750
No. 25	12400	11580	10850	10200	9650	8660	7900	7220
No. 26	16110	15010	14080	13250	12550	11260	10280	9380
No. 27	18150	16900	15830	14900	14110	12670	11520	10550
No. 28	20400	19000	17800	16760	15880	14230	12980	11870
No. 30	26700	24850	23250	21900	20750	18600	16950	15530
Narrow Flat Wire Staples per Pound								
No. 18 X 20	2830	2640	2470	2330	2210	1980	1800	1650
No. 19 X 21	3400	3170	2970	2800	2640	2380	2160	1980
No. 19 X $21\frac{1}{2}$	3980	3710	3480	3280	3100	2780	2540	2320
No. 20 X 25	6580	6130	5740	5400	5120	4590	4180	3820
No. 20 X 24	5710	5320	4980	4690	4440	3980	3630	3320
No. 21 X 25	7220	6720	6300	5920	5610	5040	4590	4200
No. 22 X 26	8830	8170	7820	7270	6970	6180	5620	5150
Wide Flat Wire Staples per Pound								
0.103 X 0.023	1770	1650	1550	1460	1380	1240	1130	1030
0.103 X 0.020	2040	1890	1780	1670	1580	1420	1290	1180
0.103 X 0.017	2410	2240	2100	1980	1870	1680	1530	1400
0.103 X 0.014	2900	2700	2530	2380	2260	2020	1840	1690
0.103 X 0.011	3710	3450	3240	3050	2880	2590	2360	2160
0.103 X 0.008	5090	4740	4440	4180	3960	3550	3240	2960
0.057 X 0.024	3058	2854	2676	2519	2379	2141	1954	1784
0.057 X 0.020	3678	3430	3216	3027	2858	2573	2339	2144

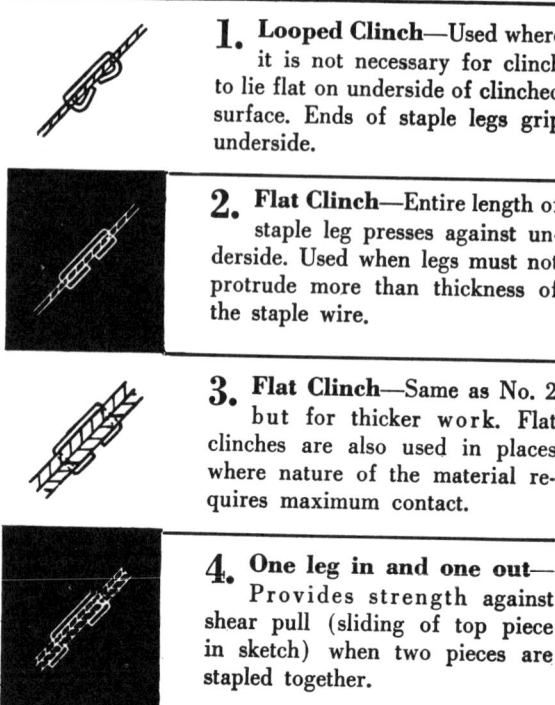

1. **Looped Clinch**—Used where it is not necessary for clinch to lie flat on underside of clinched surface. Ends of staple legs grip underside.

2. **Flat Clinch**—Entire length of staple leg presses against underside. Used when legs must not protrude more than thickness of the staple wire.

3. **Flat Clinch**—Same as No. 2 but for thicker work. Flat clinches are also used in places where nature of the material requires maximum contact.

4. **One leg in and one out**—Provides strength against shear pull (sliding of top piece in sketch) when two pieces are stapled together.

Fig. 9-24. Various types of stitch and staple clinching.

To determine the number of stitches produced from a pound of wire, consult Table 9-3. This table is based upon the *length of wire* which the *stitching machine draws per stitch* or the *length of the flattened stitch*.

In the process of completing the fastening of a stitch or staple, the type of clinch formed is extremely important and varies with the type of equipment used and with the application. In Fig. 9-24, eight styles of clinches are described. Improper clinching may result from several factors discussed in Fig. 9-25*a*. Additional information concerning efficiency of stitching is contained in Fig. 9-25*b*.

Stitching equipment. Stitching machines date back to 1875 when they were first used for the fabrication of books. Prior to the introduction of this equipment, the work was performed by preformed stapling machines.

Wire stitching machines draw wire from 5 to 50 lb coils, cut it to length,

FASTENINGS AND CLOSURES

5. Ends Turned sharply— Used for stapling compressible materials (foam rubber, etc.) Clinched ends bury themselves into undersurface.

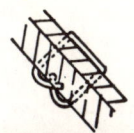

6. Rounded Staple (Hog Ring)—Points are turned in for anti-skid insurance when used for closure of bags made of slippery material etc.

7. Round Staple—Also for use in closure of bags. Points do not turn in, however. Used for jobs where it is not desirable to penetrate material.

8. Both legs driven straight in.—For tacking operations such as application of labels on wooden containers, installation of liners, cushioning, etc.

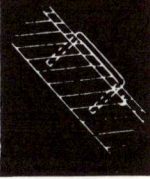

Courtesy Bostitch Div. of Textron, Inc.

form it, drive it, and clinch it in one continuous operation. The unit that performs this operation is termed a *stitching head*. This basic head can be mounted on different machines to perform various types of stitching operations including bottom stitching, side stitching, and top stitching. For special purposes multiple head machines are used. Stitching machines can be floor- or bench-mounted and can be adapted with ball bearing top tables, electric trips, tilt tables, and other fixtures to improve the efficiency of a given operation.

For bottom stitching to assemble regular slotted and overlap fibreboard boxes, a post type stitcher is used (Fig. 9-26). For telescope boxes and other styles of containers, including many types of interior packing components, the bottom post of the machine interferes with proper clinching and an arm-type stitcher must be used. Combination machines are available permitting the removal of either the post or arm to facilitate stitching.

A PERFECT STITCH

BASIC ELEMENTS OF PERFECT STITCH

Crown flat
Legs of equal length
Legs almost touching

Legs clinched back into underside of material
No damage to material

DEFECTIVE STITCHES

USUAL CAUSES FOR DEFECTIVE STITCHES

Stitch	Defect	Causes	More Causes
	One or both legs buckled	Stitch legs too long (1) Uneven leg length (3) Burred stitch legs (2) Clincher block out of line (4) Not enough compression (5)	Clincher block worn Worn anvil Wire size too small
	Buckled crown	Stitch legs too long (1) Uneven leg length (3) Burred stitch legs (2) Clincher block out of line (4) Not enough compression (5)	Clincher block worn Worn anvil Wire size too small
	Length of stitch legs varies		Dirty feed mechanism Worn feed mechanism Wire tangled on coil Too much tension on wire straightener Broken wire feed tension spring
	Corner of crown distorted or broken	Too much compression (5) Clincher block out of line (4)	Broken driver end Worn formers Worn anvil Worn clincher block Incorrect wire size
	Improper clinch	Clincher block out of line (4) Burred stitch legs (2) Incorrect leg length (3) Stitch legs too short (1)	Worn former legs Worn anvil Cut-off mechanism worn Material too thick
	Wire not formed into stitch Improper wire cut-off (2)		The following broken or inoperative stitch forming parts: Former plunger, wire gripper, anvil spring, gripper spring
	Stitch comes out in pieces		Wire not in line with grooves in stitch forming parts Cutter tube inserted incorrectly Wire size too large

(a)

FASTENINGS AND CLOSURES

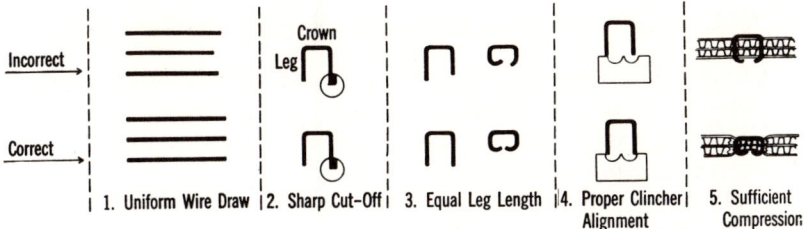

(b) Correct and incorrect stitcher adjustment for bottom stitching

Fig. 9-25. Stitching defects.

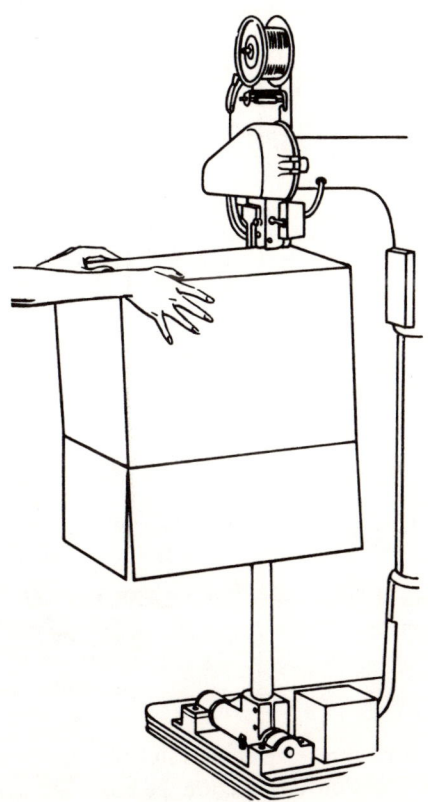

Fig. 9-26. Post-type bottom stitcher.

Stapling equipment. There are basically three types of available stapling machines, which vary in size and method of operation. These are:

1. Stapling equipment to fasten two surfaces together utilizing a clinching surface or bar under the surfaces to be bonded.

2. Stapling equipment incorporating a retractable anvil which provides the clinching surface under the staple in application.

3. Stapling equipment which drives unclinched staples into materials, such as wood, which have sufficient holding power to prevent withdrawal of the staples.

Equipment to accomplish clinching by providing a surface beneath the staple, ranges from simple desk-type staplers to elaborate multiple head bottom post stapling machines. A foot powered post type model is shown in Fig. 9-27. The hand machines of this type are desk- or floor-mounted or of the portable plier type. This equipment can be manually, electrically, or pneumatically operated.

Fig. 9-27. Foot powered post-type stapler. Courtesy International Staple & Machine Co.

The retractable anvil type stapler activates two anvils simultaneous to the entry of the staple. These anvils puncture the board ahead of the staple, form the surface for clinching, and are withdrawn upon completion of the cycle (Fig. 9-28). Staples for this equipment are available in staple rolls or cartridges.

FASTENINGS AND CLOSURES

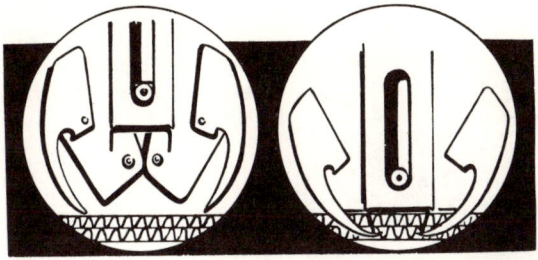

(a) Basic principle of operation. Courtesy Container Stapling Corp.

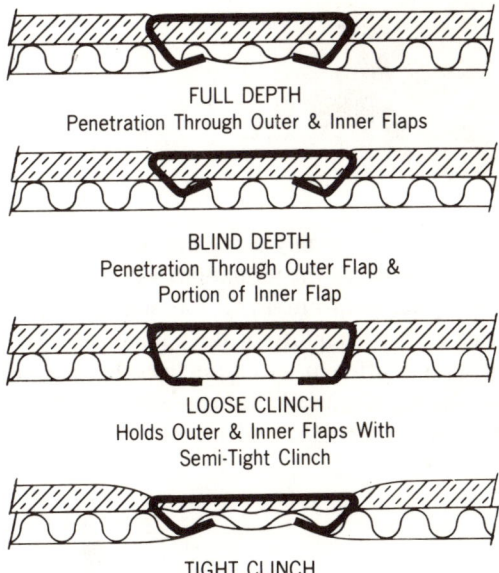

FULL DEPTH
Penetration Through Outer & Inner Flaps

BLIND DEPTH
Penetration Through Outer Flap &
Portion of Inner Flap

LOOSE CLINCH
Holds Outer & Inner Flaps With
Semi-Tight Clinch

TIGHT CLINCH
Outer & Inner Flaps Pulled Firmly Together

(b) Clinch adjustment variations. Courtesy International Staple & Machine Co.

Fig. 9-28. Retractable anvil type stapler.

This basic principle is employed on hand machines, either manually or pneumatically operated (Fig. 9-29). For high-speed closing operations, power-driven machines with staple heads for simultaneous top, bottom, and/or side stapling can be incorporated in packaging conveyor lines. Machines of this type are semi- or fully-automatic. Typical semi-automatic equipment features a centering device and a limited self-adjusting height range facilitating stapling of consecutive odd size boxes. Maximum efficiency on this equip-

Fig. 9-29. Pneumatically-operated, retractable anvil type hand stapler using staple rolls. Courtesy International Staple & Machine Co.

ment is obtained when stapling only along the center seam of regular slotted cartons is permissible. Refer to the regulations section of this chapter. Figure 9-30 illustrates a typical semi-automatic top and bottom retractable anvil type stapler. Owing to the flexibility of this equipment, a centralized closing station, servicing several different packaging lines, has become feasible. This flexibility of operation is responsible for the wide acceptance of stapling of loaded containers. Fully automatic stapling machines are designed for both

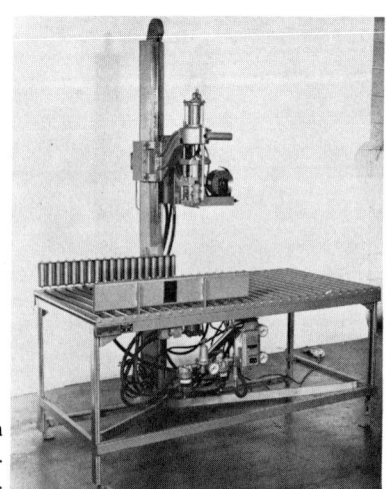

Fig. 9-30. Semi-automatic top and bottom retractable anvil type stapler with roll head. Courtesy International Staple & Machine Co.

FASTENINGS AND CLOSURES

uniform and random box size case sealing. The exit end of a top and bottom automatic stapler having a speed of up to 45 uniform size cases per minute is shown in Fig. 9-31. A fully automatic top and bottom random box size stapling machine featuring top flap folding capability is shown in Fig. 9-32. This machine requires no operator and has a productive capability of up to 14 mixed size cases per minute.

Fig. 9-31. Automatic top and bottom retractable anvil-type stapling machine featuring coil fed stapling heads. Courtesy Bostitch, Div. of Textron, Inc.

Equipment which does not require clinching of the staple is of the *hammer* or *tacker* type or *externally drawn*. The hammer type is activated by impact onto the surface to which the staple is applied (Fig. 9-33). The tacker type is operated by lever mechanism with the tacker placed against the surface to be stapled (Fig. 9-34). Externally-driven devices for unclinched staples usually utilize a hammer or mallet to provide the driving power. Machines of this type are also obtainable in air-driven models.

Tackers and hammer staples usually use a small staple and are therefore best suited for such applications as securing of linings, fastening of identifi-

Fig. 9-32. Fully automatic random top and bottom stapler. Courtesy Padlocker Machinery Div., The Loveshaw Corporation.

cation material, and for assembly purposes. Externally-driven machines are designed for heavier staples and therefore are best suited for strap and wire fastenings, lidding, and certain box assembly operations.

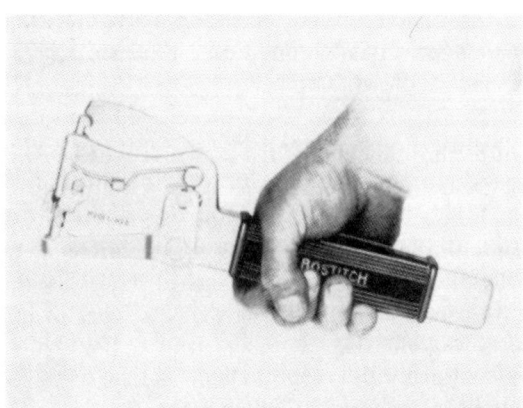

Fig. 9-33. Hammer-type staples (unclinched staples). Courtesy Bostitch Div. of Textron, Inc.

FASTENINGS AND CLOSURES

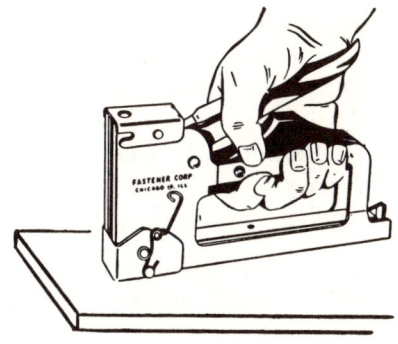

Fig. 9-34. Tacker type stapler (unclinched staples).

FACTORS INFLUENCING CLOSURE SELECTION

In the selection of the appropriate closure for a fibreboard box three important factors must be appraised:

1. Carrier regulations.
2. Protection requirements.
3. Economic considerations, labor and materials.

It is not always possible to effect a single closure completely satisfying all three of these selection factors. For example, a particular product packaging requirement may necessitate the use of waterproof tape over all box seams and edges to protect the product from moisture and contamination. Understandably, this is not the most economical closure available. Another example might be in the use of retractable anvil staples for box closure. If staples are applied only over the seam of the box flaps, only one pass through semi-automatic stapling equipment is required. However, since the rail freight regulations governing the use of these staples require the use of additional staples along the seams of the box, where the box width warrants their use, the box must be handled more than once in the stapling machine to correctly position the extra staples. This extra handling reduces the efficiency of the operation unless the volume warrants the use of multiple head machines. If the extra staples are not used, the closure may not be in compliance with the carrier regulations.

For further discussion on the selection requirements pertaining to labor efficiency under various conditions, refer to previous sections of this chapter.

Carrier regulations. Any closure type selected must first comply with minimum applicable carrier regulations. Closure or sealing regulations for slotted conventional boxes for rail shipment are contained in Section 7 of

Rule 41 of the *Uniform Freight Classification*. Section 8 of Rule 41 covers sealing requirements for other than conventional slotted boxes. Additional sealing requirements are referenced in Rule 5 of the classification regulations.

Other types of carriers have basically similar requirements which are listed in detail in appropriate regulations, that is, *National Motor Freight Classification, Hazardous Materials Regulations of the Department of Transportation*, etc.

Section 7 precedes a discussion of closure types by stating:

boxes must have both inner and outer flaps drawn together as closely as possible to insure tight pack; lengthwise flaps must meet or overlap; no flaps must project over edges, and boxes must be closed by one of the following methods or combinations thereof:

(a) All flaps must be firmly glued by one of the following methods:

(1) All flaps must be firmly glued not less than 50% of area of contact.

(2) All flaps must be firmly glued with a hot-melt adhesive of 100% solids contents of thermo-plastic materials which will maintain bond at temperatures ranging from 20°F. below zero to 165°F. above zero. Adhesive must be applied in not less than eight stripes on each inner flap, each stripe having a minimum width of 3/16 inch after compression. Stripes must be not more than 1-1/2 inches apart and not less than four stripes must be applied on each side of center seam on each inner flap for full length of flap overlap area with one stripe not more than ½ inch from each side of center seam. If less than eight such stripes are applied on each inner flap, adhesive must cover and securely bond not less than 25% of flap contact area with bonded areas extending to within ½ inch or less of center seam. Boxes must comply with all requirements of Rule 41 except gross weight must not exceed 65 pounds.

(b) By securely fastening all flaps with metal rivets, staples or stitches placed on each side of center seam in area where outer flaps overlay inner flaps and along outer edges of outer flaps, OR when lengthwise flaps overlap, with metal rivets, staples or stitches placed along entire length of seam and along outer edges of outer flaps. Fasteners must be spaced not more than 2½ inches apart (see Note 1), but allowing sufficient space to remove stitching device.

NOTE 1.-Staples or stitches made of wire of hardness not less than equivalent of Rockwell B-90 may be spaced not more than 5 inches apart along both the center seam and the sides of the outside flaps and may be used across the center seam where outside flaps meet in lieu of on both sides of center seam, but need only be used in area where outside flaps overlap inner flaps, OR where lengthwise flaps overlap, such staples or stitches may be spaced not more than 5 inches apart along the entire length of the seam and along outer edges of outer flaps. Such staples or stitches must meet the following minimum requirements:

(a) flat wire .037 inch thick, .074 inch wide, crown of 1¼ inch, or

(b) flat wire .028 inch thick, .103 inch wide, crown of 1¼ inch, or

(c) arcuate wire .027 inch thick, .095 inch wide, crown of 1 inch. Such stitches when placed over center seam must be spaced not more than 2½ inches apart, or

(d) arcuate wire .020 inch thick, .103 inch wide, crown of 1¼ inch.

FASTENINGS AND CLOSURES

(c) By securely sealing all outer seams full length with paper sealing strips. Paper sealing strips must be not less than 2 inches wide and must conform to one of the following sets of specifications:

(1)

Minimum basis weight of paper not gummed, 24 x 36 inches, 500 sheets	Sulphate Paper Minimum tear resistance		Minimum tensile strength long direction
	Long Direction	Short Direction	
Pounds 60	Grams 113	Grams 134	Pounds per inch width 45

(2) Rope stock paper described in Rule 40, Section 10(c), basis weight after sizing and coating not less than 85 pounds, testing not less than 65 pounds.

(3) Of two thicknesses of sulphate paper, total basis weight not less than 60 pounds, and testing not less than 60 pounds, combined with waterproof adhesive, and reinforced with glass fibres, distribution of fibre to give reinforcing in both cross and lengthwise direction, or reinforced with unspun sisal fibres not less than 13 to the inch running in lengthwise direction. Weight of glass fibres per ream of paper must be not less than 15 pounds.

(4) Rayon reinforced tape conforming to Paragraph (d) except tape may be not less than 2 inches wide.

(d) By securely sealing center seams only with reinforced tape conforming with the following and running full length of seam and extending over ends not less than 2½ inches. Sealing strip must be not less than 3 inches wide.

Tape must be made of two sheets of 100% sulphate Kraft, each not less than 30 lbs. basis weight, reinforced with glass, sisal or rayon fibre, combined with a laminant of asphalt or other material not affected by temperature extremes any more than would standard 180°F to 200°F softening point asphalt.

Tape must be reinforced by lengthwise fibres spaced not more than an average of ½ inch apart, and by crosswise fibres spaced not less than an average of 2 per inch except that when a diamond pattern is employed for crosswise reinforcement the spacing between the parallel sides of the diamond measured in the machine direction must be not more than 1 inch.

Glass or sisal reinforced tape must have a minimum tensile strength in the machine direction of 75 lbs. per inch of width and a minimum tensile strength in the cross direction of 45 lbs. per inch of width; rayon reinforced tape must have a minimum tensile strength in the machine direction of 57 lbs. per inch of width and a minimum tensile strength in the cross direction of 27 lbs. per inch of width with elongation not exceeding 15%. Tensile tests on the finished product shall be made on a 3 inch width sample.

Tape must have a performance test not less than 35% greater than paper sealing tape applied in accordance with Paragraph (c), when applied to 275 lb. test box 24 x 12 x 12¼ inches loaded with filled No. 2 cans to gross weight of 90 lbs. and tested in a standard 7 foot revolving drum.

Additional closure provisions are discussed in Section 7 of Rule 41 and will be discussed in Chapter 11.

Section 8 of Rule 41 covers sealing and construction requirements for other than conventional slotted boxes, including telescope boxes, boxes with covers, slide style boxes, folders, five panel folders, recessed end boxes, boxes with other than four sides and double thickness score line boxes.*

Protective requirements. There are three major interrelated functions that a closure is required to accomplish. These are not necessarily listed in the order of importance:

1. *Retention* of the flaps and direct assistance in the retention of the contour of the shipping container.

2. Adequate *durability* to resist the hazards of handling and distribution so that premature opening of box flaps and spilling of contents are prevented.

3. *Resistance to the admission of foreign or contaminating matter* when such protection is necessary.

When stitched, stapled or glued closures are used, the outer and inner box flaps are positively joined together thus restricting independent motion. In a taped closure, since the inner flaps are not secured, stress applied to the outer flaps is not necessarily transferred to the inner flaps. This fact is an important consideration in contour retention and, indirectly, in durability.

Extensive laboratory tests have disclosed that the type of closure utilized has some effect upon top-to-bottom box compressive rigidity. In this direction of compression, the securement or non-securement of the inner flaps influences performance with respect to the ultimate maximum load obtainable. Also, the type of closure does, at times, influence the amount of deflection required to achieve maximum load. With a taped closure, depression of the outer flaps against the nonresisting inner flaps may increase the deflection to maximum load.

In other directions of box compression, closure is of more critical issue. In the end-to-end and side-to-side directions of box compression, positive securement of the inner and outer flaps, by gluing or stapling, is desirable. This is particularly true in boxes having a very small gap or no gap between the inner box flaps. Positive flap joining provides a double thickness of board

---

*In addition to common carrier and postal regulations governing the type of closure permitted, some specific federal specifications are:
PPP-T-60(d), Tape: Packaging, Waterproof.
PPP-T-76(b), Tape, Pressure-Sensitive Adhesive Paper, (for carton sealing).
PPP-T-97(d), Tape, Pressure-Sensitive Adhesive, Filament Reinforced.
MIL-A-101A, Adhesive, Water Resistant, for Sealing Fibreboard Boxes.
MIL-T-5038F, Tape, Textile, Reinforcing, Nylon.
MIL-T-5237D, Tape, Textile, Rayon.
MMM-A-250, Adhesive, Water Resistant, for Sealing Fibreboard Boxes.
PPP-T-45C, Tape, Gummed, Paper, Reinforced and Plain, for Sealing and Securing.

FASTENINGS AND CLOSURES

under compressive duress as compared to a single thickness with a taped closure. Under compression the inner flaps have a tendency to depress inward, yielding to, rather than resisting the energy applied.

Other laboratory data stress the importance of container durability as a function of inner-outer flap securement. When the inner flaps are not secured, the container itself yields more easily to blows. The distortion of the container, which thus results limits the extent of container injuries such as scoreline rupturing. If the inner and outer flaps are securely bonded, there is less opportunity to yield with the shock, and container damage is more pronounced.

Economic considerations. The cost of the closure is influenced by the closure material cost and the labor required for the operation. Frequently a more expensive closure material in conjunction with labor saving equipment will lower over-all closure costs.

MATERIAL COSTS. The lowest unit material cost is provided by adhesives. The exact unit cost per container depends on the cost of the adhesive and the surface coverage yield. For example, the sealing of top and bottom flaps of a thousand 24-No. 2-can cases requires approximately 1 gal of case-sealing adhesive. Therefore, the material cost is the cost of one gallon of adhesive divided by one thousand.

Stitching-wire prices are based on the hundred weight price of the particular wire used and the yield for the size of stitch required. In Table 9-2 typical yield figures are presented which permit the computation of the cost per unit stitch or closure. Staples are priced in 1,000 units and direct cost is obtainable from a given price for a particular type of staple. Stitches formed from wire in the stitching operation are lower in cost than preformed staples of the same size, since no prior manufacture and packaging of the staples is required.

Gummed tape is furnished in rolls. A rol of the popular 60 lb (24 in. X 36 in.—500 sheets) grade contains 600 ft per roll. Reinforced tapes are put up in rolls varying in length from 300 ft. per roll to 600 ft. per roll. The cost per lineal foot of reinforced gummed tape is considerably higher than the per foot cost of plain gummed tape. However, this cost differential is minimized since a reinforced tape closure requires center strip sealing only, versus center and two end strips for gummed tape. In addition, savings in labor as the result of only a single strip application frequently favor the more expensive reinforced tape material.

Pressure-sensitive tapes are sold by the roll which contain usually 60 or 72 yd. There is wide variation in the price of the many types and grades of these tapes. Pressure-sensitive tapes are generally more expensive than gummed tapes. Frequently this unfavorable cost relationship is reduced by the permissible use of narrower widths of pressure-sensitive tape for specific carton sealing applications.

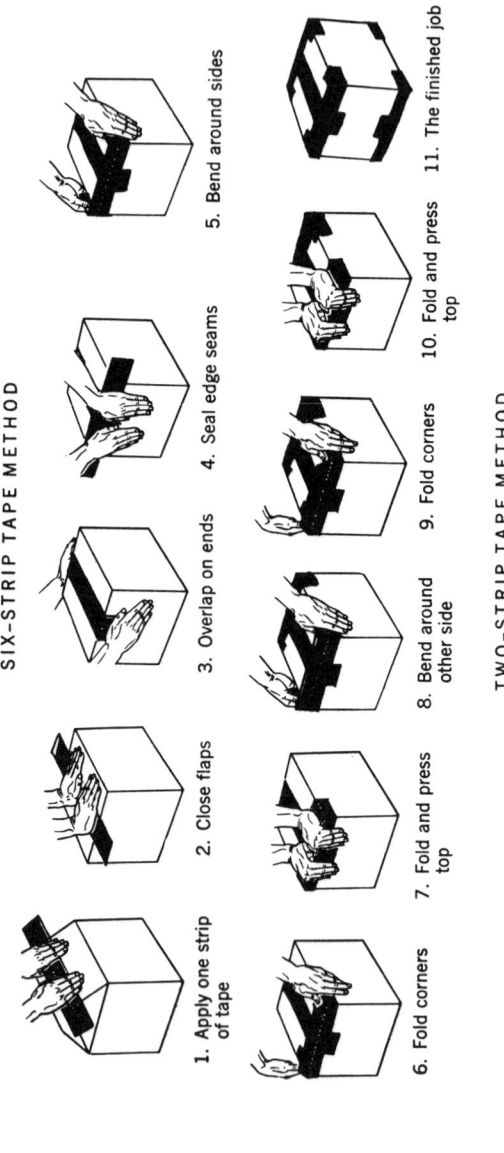

Fig. 9-35. Six-strip and two-strip taping methods.

FASTENINGS AND CLOSURES 405

LABOR COSTS. The application labor cost is greatly influenced by the type of equipment that is used. In fully automatic operations such as case sealing, only an indirect labor expense for maintaining, cleaning, and supplying the machine with materials is incurred. In the strictly manual operation, the closure cost may include setting up the container, packing, handling, labeling, and other related operations.

In an attempt to compare the labor expense of competitive closures utilizing equipment of similar efficiency, time studies on just the closure phase of a typical packaging operation revealed the following data:

A study of *hand application of two-strip taping* with a manually operated dispensing machine revealed:

1. Hand gluing was 85 per cent as efficient as the two-strip taping method (one strip, top, and bottom over the seam of the outer box flaps) in Fig. 9-35.
2. Hand stapling was 69 per cent as efficient as two strip taping.
3. Six-strip taping (3 strips, top, and bottom. In each set of three, one strip over the seam of the outer box flaps and one strip each over the ends of the flaps) was 34 per cent as efficient as two-strip taping (Fig. 9-35).

A further study of *motorized tape-dispenser application* disclosed the following relationships:

1. Hand gluing was 62 per cent as efficient as two-strip taping.
2. Hand stapling was 51 per cent as efficient as two-strip taping.
3. Six-strip taping was 35 per cent as efficient as two-strip taping.

CHAPTER 10

Reinforcing, Bundling and Unitizing Systems, Materials, and Equipment

In preparation for distribution, a shipping unit may require additional reinforcement to withstand the hazards of handling, storage and transportation. This reinforcement may supplement a previously performed closing or sealing operation or may accomplish both reinforcing and sealing as in such container styles as the telescope box. Reinforcement may also further facilitate ease of handling, as exemplified by a twine-tied package which provides a means of gripping.

When several units, within or without a shipping container are consolidated into a larger handling unit and held together with a securing medium, a *bundle* has been formed. The securing medium may add strength to the individual units in the bundle and also act as a reinforcing agent. When the bundle is designed for mechanical handling and is placed on a handling platform or is provided with other provisions for pickup and retrieval with mechanical equipment, the bundle is referred to as a *unit load*. The process of building up and securing a unit load is termed *unitizing*.

Frequently, the contribution that a reinforcing, bundling or unitizing method is making in strengthening the basic unit structure is difficult to determine. Some products are highly susceptible to damage when shipped individually and thus require considerable packaging protection. These same products, when combined into multiples and secured together, are less vulnerable owing to the support furnished by adjacent units. Therefore, when reinforcing, bundling or unitizing techniques are employed, a re-examination of product protective requirements should be made for possible economies.

BASIC PRINCIPLES AND APPLICATIONS

Reinforcing. The major applications of reinforcing are categorized for the following types of functions:

1. To *strengthen a shipping container* in anticipation of expected severe handling, that is, export shipment (Fig. 10-1).
2. To serve as the *closing or sealing* medium for a shipping container, that is, telescope boxes, partial overlap style boxes, etc. (Fig. 10-2).
3. To *secure and assemble individual components* of a shipping container, such as nail-less-type wooden or cleated boxes or crates and three-piece fibreboard boxes (Fig. 10-3).
4. To *serve as an integral part of a shipping assembly,* that is, to secure protective caps to a self-supporting unit (Fig.10-4).
5. To serve primarily as a *tying or holding agent* securing a product or component to the inside faces of a container, as in internal tying (Fig. 10-5).

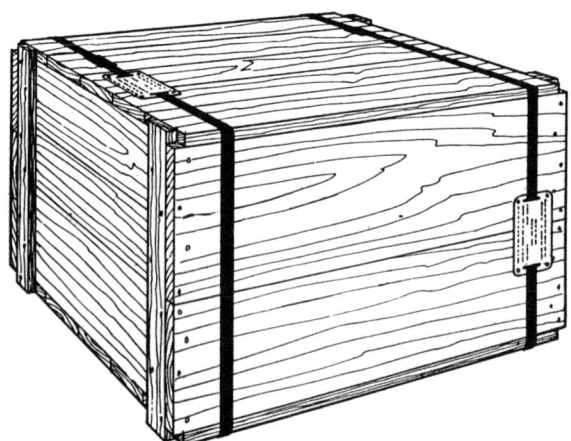

Fig. 10-1. Reinforcing used to strengthen shipping container.

REINFORCING, BUNDLING AND UNITIZING

Fig. 10-2. Reinforcing used to close shipping container. Courtesy 3M Company.

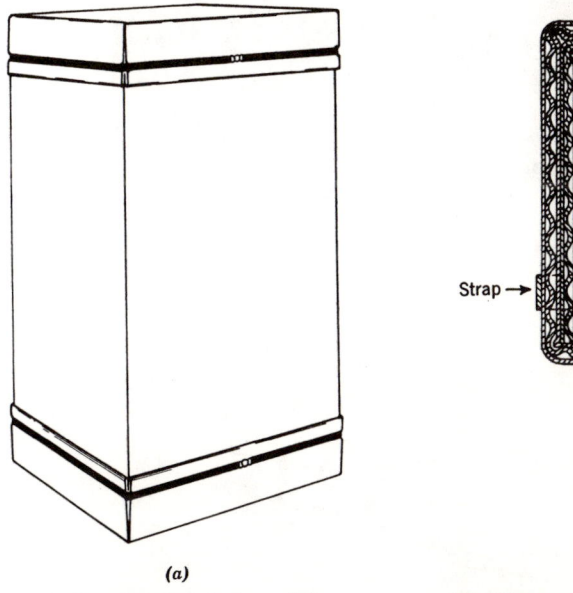

(a) General description of box

(b) Cross-sectional view of interlocking flanges held together with strapping

Fig. 10-3. Reinforcing used to assemble individual components of a shipping container. Courtesy Crown Zellerbach Corporation, Gaylord Container Division.

Fig. 10-4. Reinforcing serving as an integral part of a shipping assembly. Courtesy Signode Corporation.

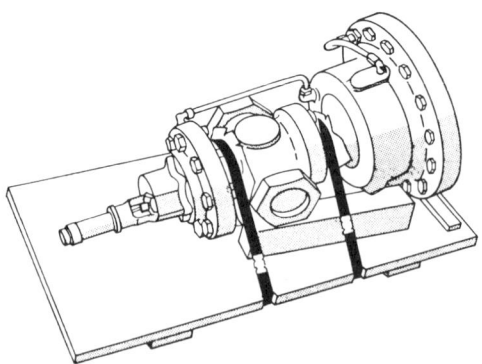

Fig. 10-5. Reinforcing serving as a tying or holding agent for internal bracing. Courtesy Interlake, Inc.

REINFORCING, BUNDLING AND UNITIZING

Fig. 10-6. Reinforcing serving to secure parts. Courtesy International Harvester Co.

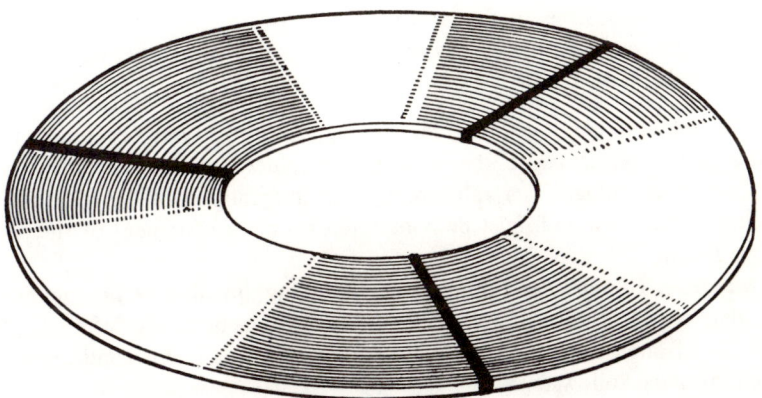

Fig. 10-7. Reinforcing and preventing unwinding of coil.

Fig. 10-8. Bundling of newspaper with non-metallic strapping using automatic strapping machine. Courtesy Signode Corporation.

6. To *secure smaller components* to the main body of the product, such as in agricultural implements (Fig. 10-6).

7. To *prevent unwinding or unrolling of wound units,* that is, steel rolls and coils, etc. (Fig. 10-7).

The method employed for reinforcing, securing, tying, or holding influences the selection of the shipping container, interior packaging, and method of handling. Thus, reinforcing is an integral part of the distribution packaging function and must be considered in the development of the packaging design.

Bundling. To a greater extent than reinforcing, bundling techniques apply to relatively small units as well as exceptionally large units. By definition, the bundles of large units is referred to as unitizing. The major applications of bundling are as follows:

1. To *combine several unpacked units* without supplementary support to provide a handling unit. Examples are bundling of newspapers, lumber, and pipes (Fig. 10-8).

2. To *unitize unpacked units* in conjunction with supplementary support,

REINFORCING, BUNDLING AND UNITIZING

(*a*) Unitized brick load with brick serving as support. Note strapping jig. Courtesy Interlake, Inc.

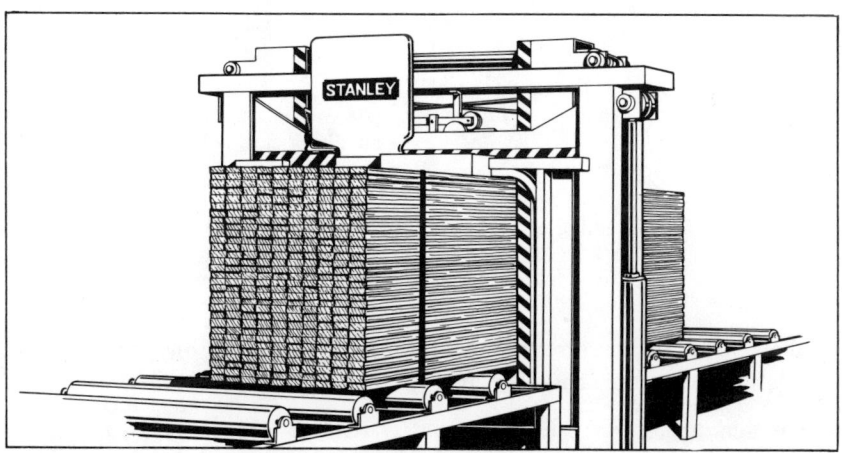

(*b*) Unitized load of lumber. Courtesy Stanley Strapping Systems, Division of the Stanley Works.

Fig. 10-9. Unpacked unitized loads.

414 **DISTRIBUTION PACKAGING**

either items to each other or to a base structure. Uses include brick, ingots, and lumber (Fig. 10-9).

3. To *tie several packaged units together without support* for ease of mechanical or manual handling, that is, small fibre boxes and large, paper-wrapped bundles of pipe, extrusions, coils, etc. (Fig. 10-10).

4. To *tie packaged units together with supplementary support,* either items to each other or to a base structure. Examples are fibre boxes on

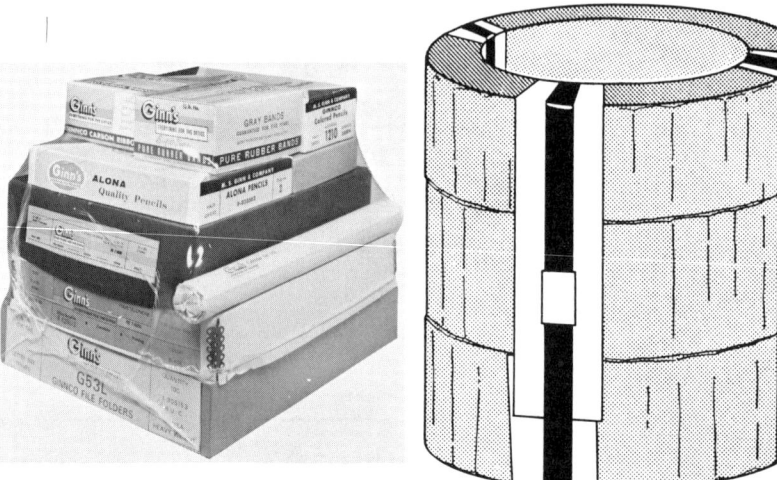

(*a*) Shrink wrapped bundle of multi-sized units for shipment. Courtesy Weldetron Corp.

(*b*) Steel-strapped and paper-wrapped bundle of steel coils

Fig. 10-10. Bundling of packaged units without supplementary support.

pallets, trays, separators, and other forms of partitions dividing layers of units and secured to a base (Fig. 10-11).

Within the above categories many variations are possible. The use of unitizing is becoming increasingly important because of the rapid expansion of mechanical handling of materials. As an illustration of the various bundling or unitizing systems that can be selected for the same type of product Fig. 10-12 has been included. This describes six acceptable methods of forming unit loads of fibreboard blanks.

REINFORCING, BUNDLING AND UNITIZING

(a) Shrink wrapped pallet load of small corrugated fibreboard boxes. Courtesy Weldetron Corp.

(b) Packaged brake drums utilizing separators on pallet. Courtesy Stanley Strapping Systems, Division of the Stanley Works.

Fig. 10-11. Bundling of packaged units with supplementary support.

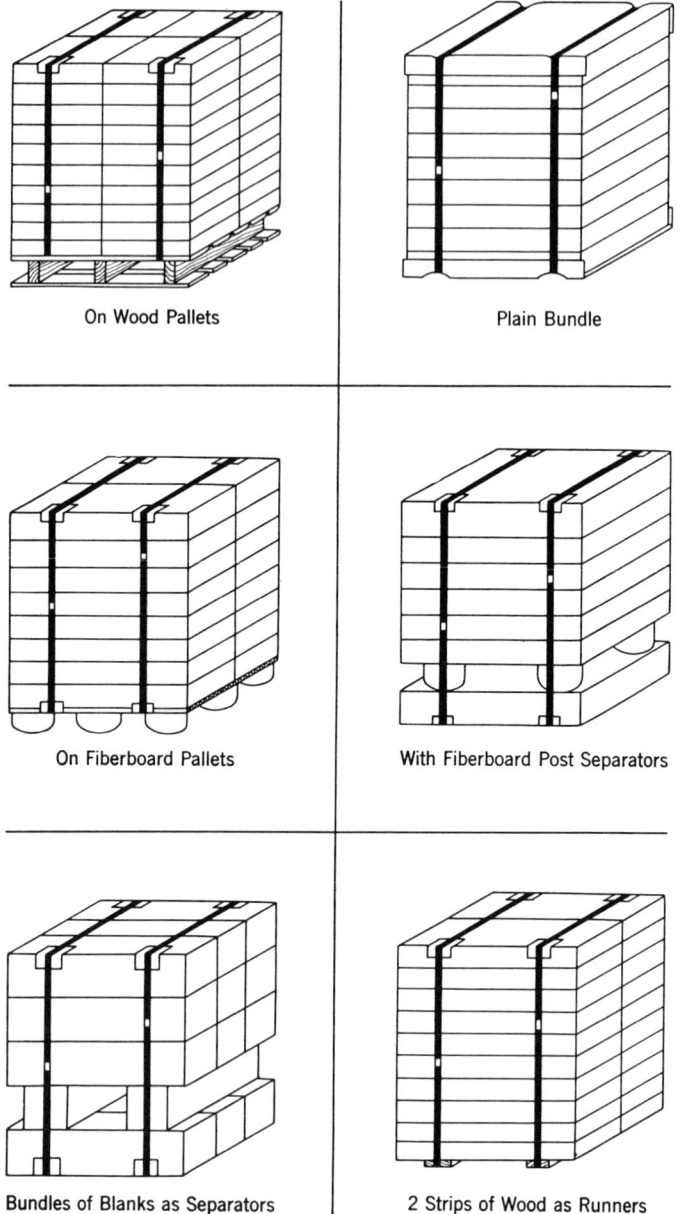

Fig. 10-12. Six methods of bundling and unitizing of fibreboard blanks. Courtesy Interlake, Inc.

REINFORCING, BUNDLING AND UNITIZING

REINFORCING AND BUNDLING MATERIALS AND EQUIPMENT

The seven basic applications of reinforcing and the four major areas of bundling all require the use of an additional element for assembly such as cordage, reinforcing and unitizing tapes, metallic and non-metallic strapping, and shrink and stretch film. These materials and their application equipment are discussed below.

Cordage. Cordage is the general name given to all *string, twine, cord, lines, rope, hawsers,* and *cables* made of twisted fibres. The principal raw materials are (1) *sisal,* (2) unpolished or polished *jute,* (3) unpolished or polished *cotton,* (4) *polypropylene,* (5) *rayon,* and (6) *nylon.* The primary product used for packaging is twine which is available in a wide range of constructions and put-up.

Cordage is primarily used for lightweight packages including retail carry-home bundles, but it also has wide applications for certain distribution packaging uses. Examples are bundling of knocked-down fibre boxes, set-up boxes, newspapers, and magazines.

Tying is accomplished either manually or with specialized tying equipment. In Fig. 10-13 a package-tying machine is illustrated. This machine wraps twine around the package to a predetermined tension, ties the twine into a non-slip, double-loop knot, and then cuts the twine. The operator positions the material to be tied and activates the tying cycle with a foot trip. Other specialized machines are available for bundling of set-up boxes, laundry packages, and trees and nursery products.

Fig. 10-13. Pedal operated twine-tying machine. Courtesy B. H. Bunn Company.

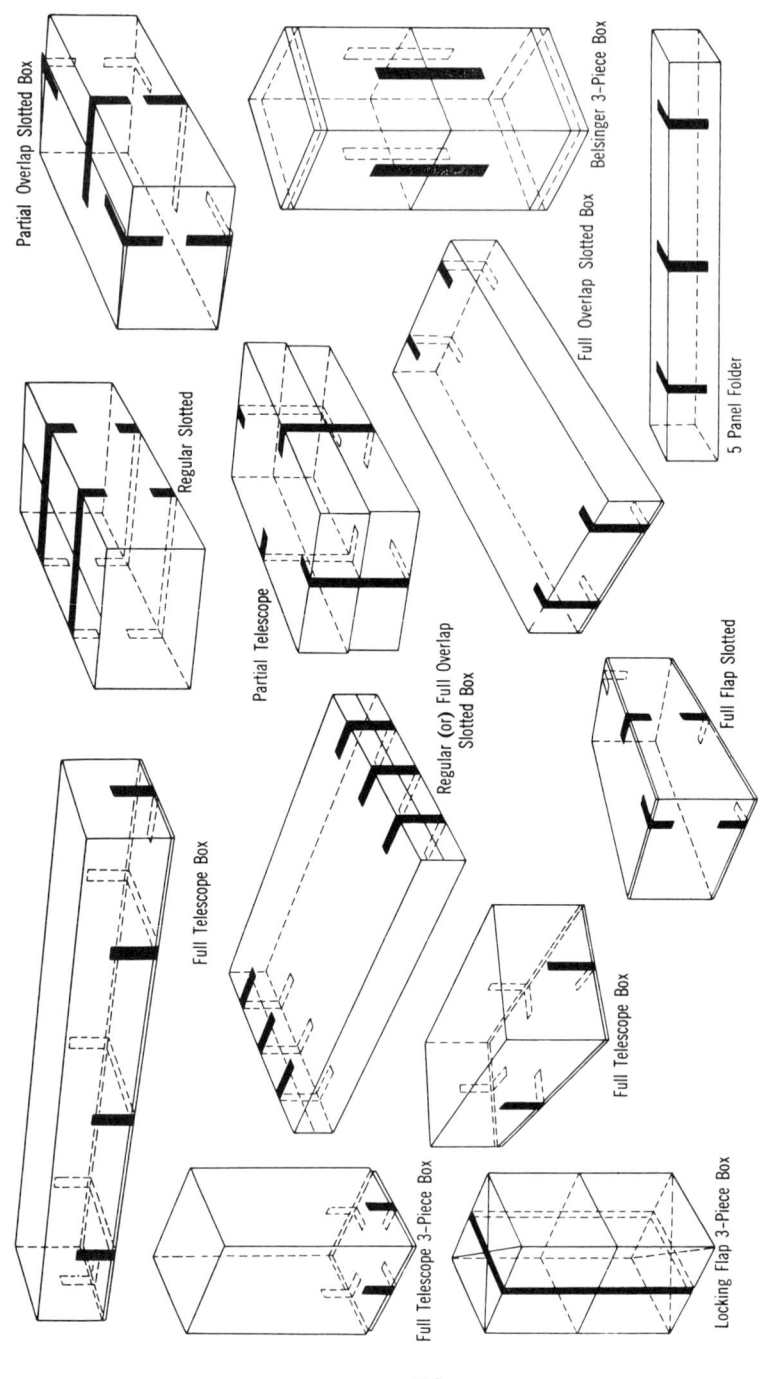

Fig. 10-14. Various types of closing and reinforcing methods for fibreboard boxes. Courtesy 3M Company.

REINFORCING, BUNDLING AND UNITIZING 419

Reinforcing and unitizing tapes. These are either filament-reinforced or of a single-ply construction (filmic). They are manufactured with a pressure-sensitive adhesive, and the filament types are available with various types of reinforcing fibres. Construction details of reinforcing and unitizing tapes were covered in Chapter 9 in reference to pressure-sensitive tapes.

MAJOR APPLICATIONS. There are four major reinforcing and bundling areas involving these tapes:

1. Closing and reinforcing of fibreboard containers (Fig.10-14).
2. Bundling and securing of multiple units such as piping and lumber (Fig. 10-15).
3. Bundling or unitizing of packaged products such as fibreboard boxes, pails, and barrels. (Fig. 10-16).
4. For the securing and holding of such items as coils, rolls, and accessories to the base product or the container.

Fig. 10-15. Bundling of pre-cut lumber with pressure-sensitive filmic reinforcing and unitizing tape. Courtesy Nopi, Inc.

Fig. 10-16. Horizontal unitizing of metal pails with filmic pressure-sensitive tape applied using portable dispenser. Courtesy Nopi, Inc.

There are also many other specialty applications for these tapes. These do not directly pertain to the reinforcing and bundling function. They are primarily of a securing and holding nature and apply to many nonpackaging operations.

For economy purposes, closing and reinforcing is effectively accomplished with the use of *short strips* rather than a continuous encircling band. Figure 10-14 illustrates typical approved applications for fibreboard boxes.

Owing to the relatively high cost of these tapes it is desirable to dispense no greater quantity than the required length of tape. Utilizing the pressure-sensitive tape roll without a dispenser or with a portable hand dispenser is generally wasteful, and this method is only recommended when reinforcing or bundling is carried out in widely dispersed areas in the plant. If centralized packaging is performed, dispensing and application equipment of various levels of mechanization can be utilized.

In Chapter 9 various types of dispensing equipment for sealing were detailed. Some of this equipment is also suitable for reinforcing and unitizing tapes.

Strapping. This technique has an extremely broad range of applications which varies from the securing of a distribution container to heavy duty carload bracing to prevent shifting in transit. It is the only packaging method which permits the application of tension to compress a load to reduce cube and enhance stability. Its versatility enables its application by varying types of equipment from manually operated hand tools to fully automated production systems.

TYPES OF STRAPPING. Strapping is usually classified as (1) *metallic* (steel), and (2) *non-metallic* (plastic). Steel strapping has been used since the beginning of the century and the first non-metallic strapping was introduced in the 1960's.

Steel strapping. The basic classification of steel strapping is by cross-sectional contour. These contours are: (1) *round* and (2) *flat*. Although each is intended for the same purpose of securing a load, the types of seals and application equipment for each contour are very different.

Round strapping has a circular cross section and is commonly called *wire strapping.*

Flat strapping is furnished either for *nail-on* or *nail-less* purposes. Nail-on strapping is fabricated from special hot-rolled or cold-rolled steel which can be readily pierced by the nails driven through the strap into a wooden container. Some nail-on straps have prepunched holes spaced at intervals less than 3 in. apart. By far, the largest proportion of flat strapping used is of the nail-less type.

A further classification or identification of steel strapping is based on the nature of the *strap finish,* the *tensile strength* of the steel, and the *size of the material.*

REINFORCING, BUNDLING AND UNITIZING

FINISHES OF STEEL STRAPPING. *Flat steel* strapping is furnished *bare, galvanized,* or with various coatings. Most grades of strapping are given a wax lubrication treatment of special importance for strapping tools that use a feed wheel-type tensioner. The lubrication makes it possible during the stretching operation for the overlapped faces to slide over each other.

Round wire strapping is furnished *bare* or *zinc-coated* and because of the type of application equipment used, there is no necessity for the lubricating wax treatment

Regular strapping finishes provide moderate corrosion protection. If strapping is subjected to corrosive conditions, a galvanized treatment is desirable. The surfaces and edges of galvanized strapping are completely covered with zinc. The weight of the coating ranges from 0.08 to 0.30 oz. per sq. ft. of surface. For very severe corrosion conditions, stainless steel strapping is used. Usually this material has mostly nonpackaging applications due to its high cost.

PHYSICAL CHARACTERISTICS OF STRAPPING. For the functional requirements of flat steel strapping, three basic types, having different physical characteristics, are in use: (1) *regular duty cold-rolled,* (2) *high tensile strength cold-rolled,* and (3) *heavy duty hot-rolled* strapping.

Regular duty grade is cold-rolled, low-carbon steel that is specially tempered to satisfy the requirements of tensional strapping. Produced under rigid specifications, this strapping grade is used in most general packaging applications and with powered equipment.

High tensile strapping is made from cold-rolled medium-carbon, high-manganese steel that is heat-treated to produce a product that combines the fine surface and controlled physical properties of cold-rolled strapping with the ductility and shock resistance of hot-rolled strapping. It is generally used where higher tensile strength is desired. Heavy duty strapping is a hot-rolled, normalized, high-carbon steel strapping made from special-analysis steel, tempered for high tensile strength and ductility, producing high resistance to impact and is designed specifically for such applications as car and ship-loading. In Table 10-1 typical ranges of sizes and thicknesses with corresponding physical characteristics are itemized.

Round wire strapping is furnished in three basic grades: (1) *high tensile,* (2) *standard,* and (3) *extra high tensile.* Typical physical characteristics are shown in Table 10-2.

Further information on the physical characteristics of strapping is contained in *Federal Specification* QQ-S-781h.

APPLICATION EQUIPMENT. Flat nail-less and round wire strapping are applied with special application equipment. The many standard and special machines which are available through the manufacturers of steel strapping are generally classified by the cross-sectional contour of the strapping for which they are designed. However, for comparison purposes and as an aid to

TABLE 10-1

TYPICAL SIZES, TENSILE STRENGTH AND YIELD OF FLAT METAL STRAPPING

Type	width	thickness	average strength lbs.	feet per lb.
REGULAR DUTY	3/8"	.010"	460	78.5
		.012"	540	65.4
		.015"	650	52.4
		.020"	880	39.3
	1/2"	.010"	590	58.9
		.012"	720	49.0
		.015"	870	39.3
		.018"	1,030	32.8
		.020"	1,170	29.4
		.023"	1,300	25.6
	5/8"	.012"	900	39.3
		.015"	1,100	31.4
		.018"	1,300	26.2
		.020"	1,460	23.6
		.023"	1,620	20.5
	3/4"	.015"	1,310	26.2
		.020"	1,750	19.6
		.023"	1,950	17.1
		.028"	2,300	14.0
		.035"	2,750	11.2
HIGH TENSILE STRENGTH	1/2"	.020"	1,450	29.4
		.031"	2,200	19.0
	5/8"	.020"	1,810	23.6
		.023"	2,000	20.5
	3/4"	.025"	2,620	15.7
		.031"	3,250	12.7
	1 1/4"	.031"	5,450	7.6
		.044"	7,610	5.3
	2"	.044"	12,300	3.3
HEAVY DUTY	3/4"	.035"	3,250	11.2
	1 1/4"	.035"	5,450	6.7
		.050"	7,400	4.6
		.065"	9,620	3.6
	2"	.050"	11,500	2.9
		.065"	15,720	2.2

Courtesy Signode Corporation

REINFORCING, BUNDLING AND UNITIZING

selecting the most suitable equipment, a more logical classification is achieved by the operational features of the equipment. Basically, there are three such major operational types: (1) *two-piece hand tools,* (2) *one-piece hand tools,* and (3) *stationary power-operated machines and machine systems.* The hand tools are either manually or power-activated, whereas, the stationary machines are all fully or semi-automatically power-operated. The following is a description of the basic features of these three classes of equipment.

Two-piece hand tools consist of a *stretcher* and a *sealer.* The stretcher tool (also called a *tensioner* or a *tightener*) is designed to draw the band tightly together. A sealer is then used to crimp or otherwise fasten a metal clip upon the overlapping ends of the flat strap. In Fig. 10-17 one of many avail-

TABLE 10-2

TYPICAL SIZES, TENSILE STRENGTH AND YIELD OF ROUND WIRE STRAPPING

Type	Gauge	Diameter, in.	Approximate Tensile Strength lbs.	Feet per Pound
HIGH TENSILE	8	0.162 in.	2,260	14.29
	10	0.135 in.	1,860	20.57
	12	0.1055 in.	1,130	33.69
	13	0.0915 in.	850	44.78
	14	0.080 in.	650	58.58
	15	0.072 in.	530	72.32
STANDARD	16	0.0625 in.	260	95.98
	16½	0.059 in.	225	111.30
	17	0.054 in.	195	128.60
	17½	0.051 in.	175	144.20
	18	0.0475 in.	150	166.20
	18½	0.0443 in.	130	191.94
EXTRA HIGH TENSILE	10	0.135 in.	2,220	20.57
	12	0.1055 in.	1,275	33.69
	13	0.0915 in.	1,050	44.78
	14	0.080 in.	800	58.58

Courtesy U.S. Steel Supply Division, United States Steel Corp.

424 **DISTRIBUTION PACKAGING**

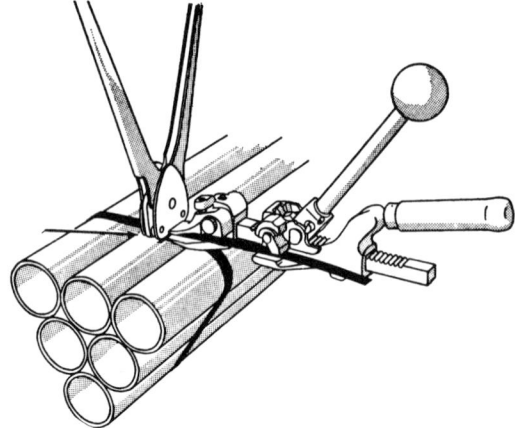

Fig. 10-17. Manually operated stretcher and sealer for flat metal strapping.

able variations of stretchers and sealers for flat metal strapping is illustrated. These two-piece tools are light-weight and portable and afford excellent service for low-volume strapping operations. Fundamentally, hand-tensioned tools are classified by the method of gripping the steel band. Three types of gripping methods are used (Fig. 10-18).

1. *Feed wheel take-up* of several serrated or fluted wheels under which the strap passes. Reciprocal motion of the stretcher handle turns the wheel and advances the strap. A ratchet drive provides a continuous unlimited take-up. This type of tool is most suited for flat surfaces on products providing a certain amount of compression where removal of the stretcher does not loosen the band.

2. *Rack gear take-up* type tool, also known as the *multiple gripping dog system*, consists of a ratchet arrangement that moves the gripping dog forward. The strap is grasped alternately by two gripping dogs that pull the strap forward when the handle is activated. At the end of the forward advance, the second gripping dog holds the strap while the moving gripper is repositioned. This basic type of tool (Fig. 10-17) is particularly recommended for non-compressible and irregular-shaped items in which no portion of the stretcher should come between the strap and the item to be strapped for maximum retention of strap tightness.

3. *Split drum with or without gripping dog* is generally found on heavy duty tension tools. To operate this tool, one end of a pre-cut strap is first locked in a gripping dog. The other end of the strap is placed around the item to be strapped and inserted into a slot in a split drum. As the tool

REINFORCING, BUNDLING AND UNITIZING 425

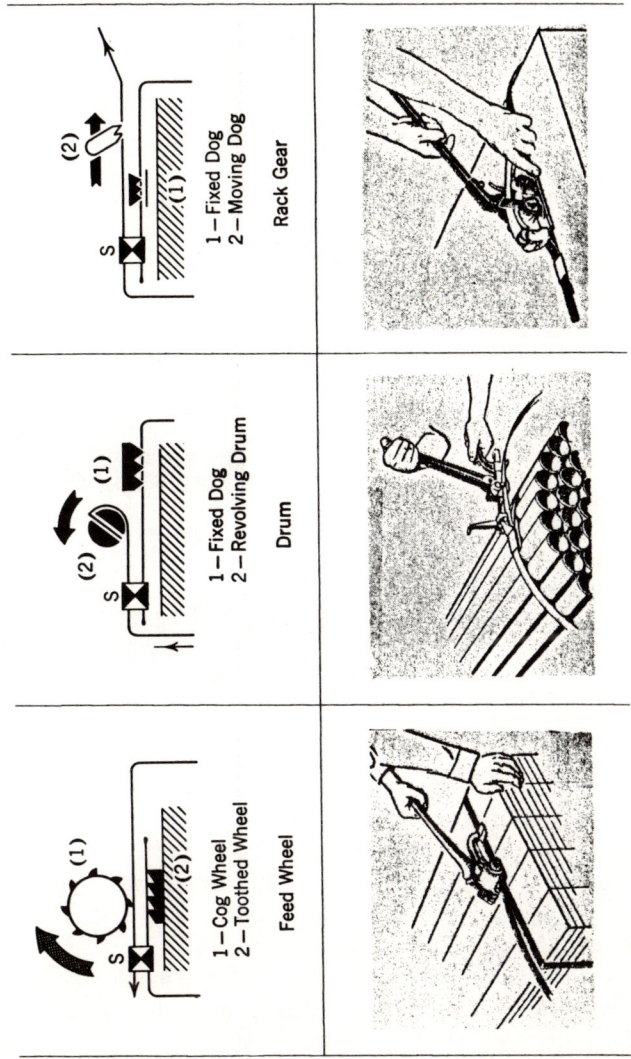

Fig. 10-18. Three types of gripping methods for steel strapping stretchers (courtesy Mr. Bernard S. Gaines) and application illustrations (courtesy Signode Corporation).

handle is reciprocated back and forth, the strap is wrapped around the drum and draws the strap tight. This tool requires precut strapping; thus, waste of strapping material is frequently encountered in operation.

With all two-piece tools, the seal is manually applied over the strapping either before or after the tensioning operation. With the rack gear take-up tool for noncompressible items, a *thread-on* seal is required. The strap is looped through the seal and is bent back under the seal. The stretching action is against the seal which is then crimped prior to removal of the stretcher.

Seals for two-piece tools are of the *thread-on* or *snap-on* type for single or double crimping.

Joining is performed with a lightweight tool of which there are a number of designs providing top, side, and angle activation of the levers. The joining action consists either of crimping the seal and strap in such a manner that a friction joint is formed, or by notching of the strap and a repositioning of the strap to bear against the seal (Fig. 10-19). The force required to crimp is

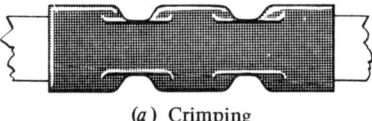

(a) Crimping

Fig. 10-19. Two basic types of seal formation.

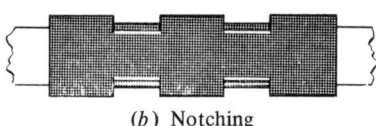

(b) Notching

considerably greater than by notching and most modern tools are therefore designed to use the latter technique. The notching operation reduces the effective tensile strength of the strapping by approximately 20 to 25 per cent.

Two-piece tools are primarily of the manually-activated type. However, pneumatic- and hydraulic-activated tensioners are also used. In Fig. 10-20 a pneumatically operated power tensioner is shown.

One-piece hand tools combine the tensioning and joining operation into a single device. Generally, a cut-off mechanism parts the applied strap from the reel. There are three basic on-piece hand tools:

1. *Round wire tools* which form the seal by twisting of the wire. These tools utilize the split drum tensioning mechanism described previously (Fig. 10-21). Discussion of the use of round wire with this and other tools in comparison to flat banding with or without seals is presented later in this chapter.

REINFORCING, BUNDLING AND UNITIZING

2. *Seal-less one-piece tools* are obtainable which utilize an interlocking die-cut joint. This eliminates the need for a separate metal seal. In Fig. 10-22 a pneumatically-operated tool with push button control is shown. The pneumatic feature assures uniform tension.

3 *One-piece seal or combination tools* are designed either with the feed wheel take-up or the rack-gear principle. These tools are available with or without a seal magazine, although the former is the more popular. The machine provides tensioning, placement of the seal, crimping, and cutting of the strap in one basic operation (Fig. 10-23).

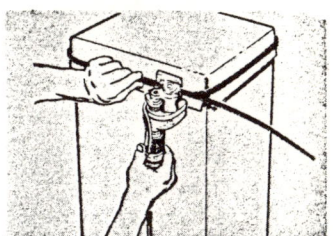

Fig. 10-20. Power tensioner particularly suitable for large or compressible loads. Courtesy Signode Corporation.

Seals for one-piece hand tools are either threaded on wire into stacks or are held together into stacks by interlocking nibs. Both types are of the *snap-on* design. If the wire type is used, it is removed from the stack of seals after they are inserted in the magazine of the machine. With the *wireless* type, it is possible to load partial stacks into the seal magazine.

Fig. 10-21. One-piece round wire hand tool. Courtesy U.S. Steel Supply Division, United States Steel Corp.

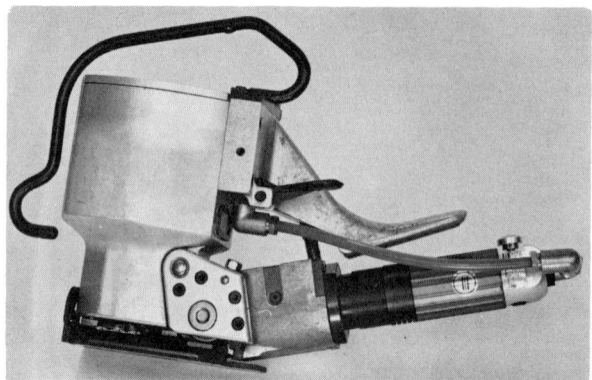

Fig. 10-22. Seal-less one-piece tool. Courtesy Gerrard & Company.

A comparison of the three basic one-piece tool types indicates that, potentially, the seal-less strapping tool can provide maximum material economy through the elimination of the seal. Assuming equivalent performance of the various available one-piece tools as to strap joining and operation of the tool, the elimination of the seal could result in a substantial material savings when high-volume strapping operations are performed.

Fig. 10-23. Fully automatic combination power hand strapping tool. Courtesy Signode Corporation.

Power strapping tools provide mechanical tensioning of the strapping and automatic fastening of the ends of the encircling band. These range from lightweight, hand-controlled tools that tension, seal, and cut off manually applied strapping, through to fully automated (operator-less) machines that

REINFORCING, BUNDLING AND UNITIZING

can compress the load, index it through to apply multiple straps, rotate it 90 degrees and index it through for multiple straps in a crosswise direction, then feed it to and outgoing conveyor.

Semi-automatic machines combine a strap feeding mechanism with a strapping head that is located in a stationary position (or on an adjustable platen). It is termed semi-automatic since the operator has to handle the strapping to thread it into the head before actuating the strapping functions (Fig. 10-24).

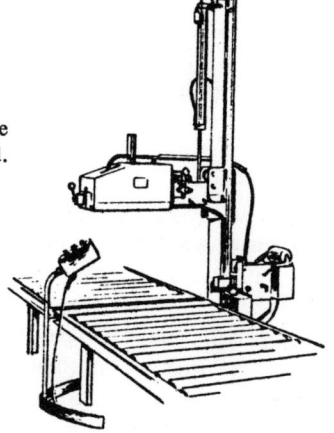

Fig. 10-24. Semi-automatic strapping machine requiring operator to drop strap around load. Courtesy Interlake, Inc.

An automatic machine eliminates any manual handling of the strapping (except for replenishment of the reels). However, an operator is required to position the packages, control the switches and pushbuttons and to monitor the operations. These machines usually have chutes to guide the strapping around the load and thread it into the strapping head. The chutes may be modular in design so they can be readily replaced to accommodate packages of different sizes. Other equipment under the control of the operator's panel could include lances to carry the strapping between the decks of a pallet, indexing conveyors, turntables, compression platens, (that also can be used with semi-automatic and fully automated machines), etc. (Fig. 10-25).

For high speed or high production strapping operations, even the operator can be eliminated, except for maintenance and for replenishment of the strapping reels and clips. Fully automated systems use sensors and computer-like operating controls for these strapping operations. They are usually designed for a specific product and are fully integrated into a production line operation (Fig. 10-26).

The power machines, like hand machines, provide variations in the securing of the ends of the strapping. Round wire is joined by twisting or braiding. Flat strapping is attached by metal seals or clips or by seal-less joining of the strap (Fig. 10-27).

Fig. 10-25. Automatic strapping machine for compression strapping of veneer. Courtesy Stanley Strapping Systems Division of The Stanley Works.

Production rates for manually attended power machines are affected by the complexity and sequence of the work elements performed by the operator. Factors such as machine feedout time, balanced design, control arrangement, tensioning speed, and other construction features of the machine itself, are of importance in attaining maximum machine output. For semi-automatic equipment the balance between the manual portion and the ma-

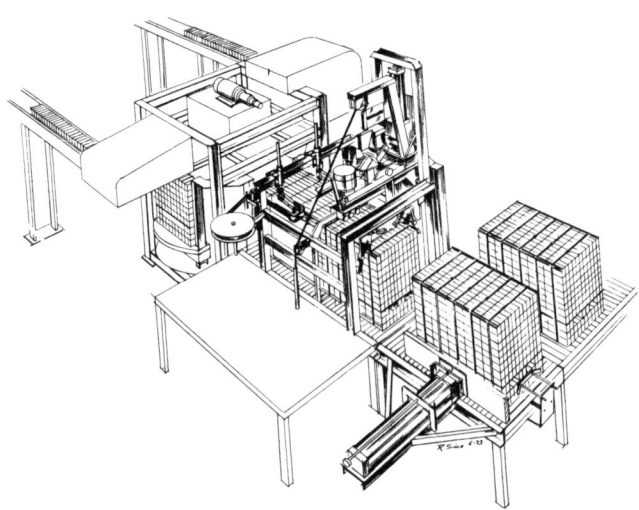

Fig. 10-26. Integrated automated strapping system with stackers for unit loads of brick. Courtesy Signode Corporation.

REINFORCING, BUNDLING AND UNITIZING

chine portion of the total strapping cycle is of importance in reducing the total strapping-cycle time. This is also affected by the speed of feed-out of the strapping, the speed of the tensioning gears, and the amount of slack material that is pulled back. Needless to say, the skill and aptitude of the operator play a very important role in the productivity of this type of semiautomatic equipment.

Output rates on fully automated systems depend on the flow rate of the product to the machine, the number of straps applied simultaneously, the rate of takeaway, and any supplementary operations performed at the same station.

PLASTIC STRAPPING. The first plastic strapping was basically woven rayon cord which was manually applied and held together with buckles or clips. This development was then followed by an entire series of materials including *polypropylene, nylon,* and *co-polymers* that can be manually or automatically applied and sealed. *Polyester* strapping has also been introduced having characteristics which overlap usages relegated to nylon and light-duty steel.

Fig. 10-27. Automatic strapping system using seal-less joining of straps. Courtesy Gerrard & Company.

PHYSICAL CHARACTERISTICS. Plastic strapping is generally made from four basic materials as summarized in Table 10-3. These consist of: (1) *co-polymers,* (2) *polypropylene,* (3) *nylon,* and (4) *polyester.* Co-polymers consist basically of polypropylene blended with other polymers like polyethylene or butyl rubber for specific characteristics such as elasticity, abrasion resistance, flexibility, etc. This material is generally used for light-to-medium duty applications.

TABLE 10-3
GUIDE TO PLASTIC STRAPPING

Material	Width (in.)	Thickness (mil)	Nominal Breaking Strength (lb)	Advantages	Disadvantages
Co-polymer	1/4	14	145	Light to medium-duty Low-cost High elasticity Good impact resistance High flexibility Abrasion-resistant Smooth Easily heat-sealed Stays tight on settling loads	Tends to "creep" (lose tension over extended time periods) High stretch Can split longitudinally
		18	200		
		22	255		
		27	310		
	3/8	18	300		
	7/16	25	500		
		30	600		
Polypropylene (PP)	1/4	15	225	Low-medium to heavy-duty Low cost High elasticity High impact resistance Stays tight on settling loads Maintains strength when nicked Easily heat-sealed Stronger size-for-size than co-polymer	Some creep Moderate stretch Can split longitudinally
	3/8	15	338		
		20	450		
	7/16	28*	550		
	1/2	15	450		
		17*	300		
		20	600		
		24*	500		
		25	750		
	5/8	20	750		
		30	1125		
	1-1/4	35	2500		
		50	3500		

Nylon	7/16	17 23 29	475 630 790	Good tension flow around corners Creep-resistant Heat or clip-sealed Stronger size-for-size than PP Medium-duty	Small amount of creep Small amount of stretch Weakened by moisture Loses strength when nicked
	1/2	15 20 25 30	475 630 790 950	High impact resistance Stays tight on settling loads Abrasion-resistant Resists splitting High tension capability with friction-wheel type tools	
Polyester	3/8	15 20	350 475	Medium-duty High impact resistance Holds in expanding loads Abrasion-resistant Will not tear at nicks or split lengthwise	Loosens on settling loads Difficult to heat-seal (gives off noxious fumes when melting)
	7/16	20	550	Unaffected by moisture High tension capability with minimum elongation No creep	
	1/2	15 20	475 630	Good tension flow around corners Stronger size-for-size than PP Clip-sealed	

*Waffled

Courtesy Modern Materials Handling.

Polypropylene strapping is made by extrusion and is generally used for the full range of strapping applications. Nylon strapping is produced exclusively in the United States by one company and is virtually the same when offered by different strapping suppliers. It is basically intended for medium duty applications. A new grade of nylon has been introduced which features a higher resistance to moisture, is less susceptible to splitting and is less affected by heat than polypropylene strapping.

Polyester has the highest resistance to elongation under tension of all of the plastic strapping materials and is, therefore, a lower cost alternative for many medium duty applications that previously specified steel strapping.

Further information on the physical characteristics of plastic strapping is contained in *Federal Specification* PPP-S-760a.

APPLICATION EQUIPMENT. By and large, the tools, equipment and equipment systems available for plastic strapping are very simmilar to those used for metal strapping applications. These were previously classified and described in this chapter. As a matter of interest, some equipment can be used interchangeably. However, there are a number of important differences which are discussed below.

For example, the feeding mechanisms on most automatic machines make use of the columnar stiffness of the strapping to push it through the chutes to encircle the loads. However, lighter-duty strapping of the co-polymer or polypropylene types may be too flexible to be handled this way. One solution is to use a device like a continuous bycycle chain on an automatic machine that grasps the strapping coming from the reel and pulls it around the bundle being wrapped, instead of being pushed. As the strapping head completes its functions and cuts off the strapping, the chain mechanism grasps the strapping still connected to the reel for the next cycle.

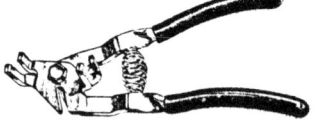

Fig. 10-28. Hand tool of the walking gripper type for plastic strapping. Courtesy Signode Corporation.

Other problems with plastic strapping can result from extrusion stresses that cause warping of camber preventing the strapping from properly passing through the chutes. Remedies, other than the rigorous care in extruding and coiling given by the strapping producers, include impressing a waffled pattern on the strapping when extruding it, using proper care in storage, and designing the chutes for each type of strapping used.

Hand tools also utilize the feed wheel take-up and rack gear take-up mechanism. In addition, a *walking gripper* tensioning tool (Fig. 10-28) is available

REINFORCING, BUNDLING AND UNITIZING 435

for portable applications for narrow-width strapping. A typical fully automatic strapping machine for pallet-size loads is illustrated in Fig. 10-29. This machine will apply, tension and heat seal plastic strapping around a pallet load.

Plastic strapping also differs from steel in the manner in which the ends are fastened. It can be machine sealed by crimped clips as with steel, or by friction-

Fig. 10-29. Automatic pallet strapping unit. Courtesy FMC Corp.

welding or heat-bonding the ends together. Manual methods (Fig. 10-30), include the use of buckles and hand-powered clip crimpers. Polyester strapping, because of the possibility of its producing noxious fumes when heated, is currently only held by a mechanical seal or clip.

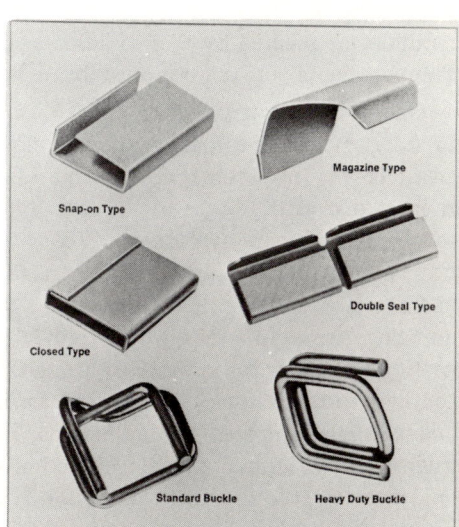

Fig. 10-30. Typical seals and buckles used for plastic strapping. Courtesy U.S. Steel Supply Division, United States Steel Corp.

The strapping ends are heat-sealed by one of three methods: (1) contact element, (2) air jet, or (3) friction. Contact element sealing consists of placing a heating element between the straps in the sealing area. melting the plastic, withdrawing the element, and compressing the straps together to form a bond before cutting the completed band free and releasing the tensioning mechanism in the strapping head or tool.

The air jet method uses a jet of hot air blown across a heating element and then between the strapping ends being joined. This method is claimed to eliminate any problems caused by a build-up of melted or burnt plastic on a contact heating element.

In place of a heating element, the third method of heat-sealing plastic strapping uses a friction shoe. After the band is pulled to the proper tension, the shoe clamps the two sections together and vibrates for a pre-set period. The heat built up by friction between the plastic surfaces is enough to melt the plastic and create the bond. This method requires the strapping to have certain friction characteristics, but eliminates the servicing of a heating element. On heavier plastic strapping, friction-welding takes about 1 to 4 seconds, on lightweight strapping it is almost instantaneous (1/50 sec).

Unitizing with Films. An entirely new concept in unitizing was made possible by the use of special films having the capability of conforming to the configuration of a bundle or unit load either by (1) *shrinking* or by (2) wrapping the unit under tension (*stretch* film). Shrink films, originally developed for consumer packaging, were later used for heavier duty applications involved in distribution packaging, offering particularly distinct advantages for the securing of palletized loads. A more recent application involves the securing of loose and irregularly shaped items within a shipping container minimizing the use of wrapping and cushioning materials. In such uses the products are secured by film to a fibreboard pad which when placed into the shipping container is positioned by a liner, creating a "floating" design.

One of the primary reasons for the use of films for unitizing is the ability to fully enclose the load, with compression applied over a broad surface area. This is in contrast with the use of strapping where localized pressure is exerted over the edges of the load and multiple straps may be required to adequately secure the load.

TYPES OF FILMS. The physical characteristics of shrink films were covered in Chapter 8 listing details of the films used in Table 8-4. Shrink films for unitizing are mostly heavy-duty polyethylene and require a high energy heating operation for activation. Stretch films are specially processed with additives to retain their elasticity and are slightly more expensive pound-per-pound when compared to shrink films. This film is under constant tension when wrapped around a load. This action serves to tighten loads that have a tendency to settle in transit, or are initially loosely stacked manually. These

REINFORCING, BUNDLING AND UNITIZING

films do not require heat for application and can therefore be utilized for products highly inflammable or potentially explosive, and for products (or their packages) which are heat sensitive. Stretch films are ideally suited for unitizing products in plastic bags, and shrink-wrapped trays or packages, since the films will not bond to each other.

Stretch films work best to unitize loads with uniform sides with a minimum of jutting corners or voids. By comparison, shrink film conforms to practically any configuration, and is therefore well suited for use with random height loads or mixed random shaped containers. Odd-shaped products, bicycles, furniture, and automotive parts can be virtually *skin-packed* onto a pallet.

APPLICATION EQUIPMENT. Unitizing with *shrink* film basically requires two operations: (1) bagging or wrapping, and (2) heat shrinking. Bagging or wrapping can be accomplished over four, five or six sides of the load. The simplest method of wrapping is with a preformed bag which is manually applied over the load. Equipment for bagging is offered ranging from semi- to fully-automatic, including manual and driven premade bag dispensers, premade bag positioners, automatic bag makers, and fully automatic unit load baggers.

Heat shrinking can be accomplished manually using (1) a hand-held hot air blower, (2) a shrink frame, (3) a cubicle, and (4) a heat tunnel. For moderate production needs of up to 30 loads per hour, the shrink cubicle is well suited for the intermittent flow of wrapped unit loads. A typical unit of this type is shown in Fig. 10-31. In operation, the unit load is automatically

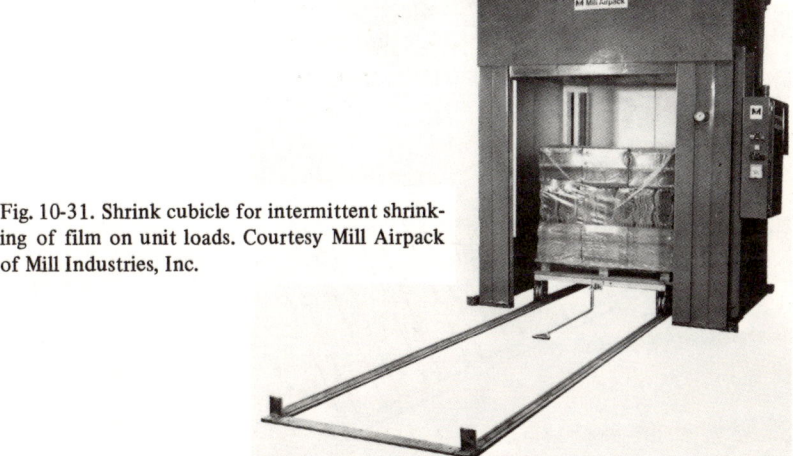

Fig. 10-31. Shrink cubicle for intermittent shrinking of film on unit loads. Courtesy Mill Airpack of Mill Industries, Inc.

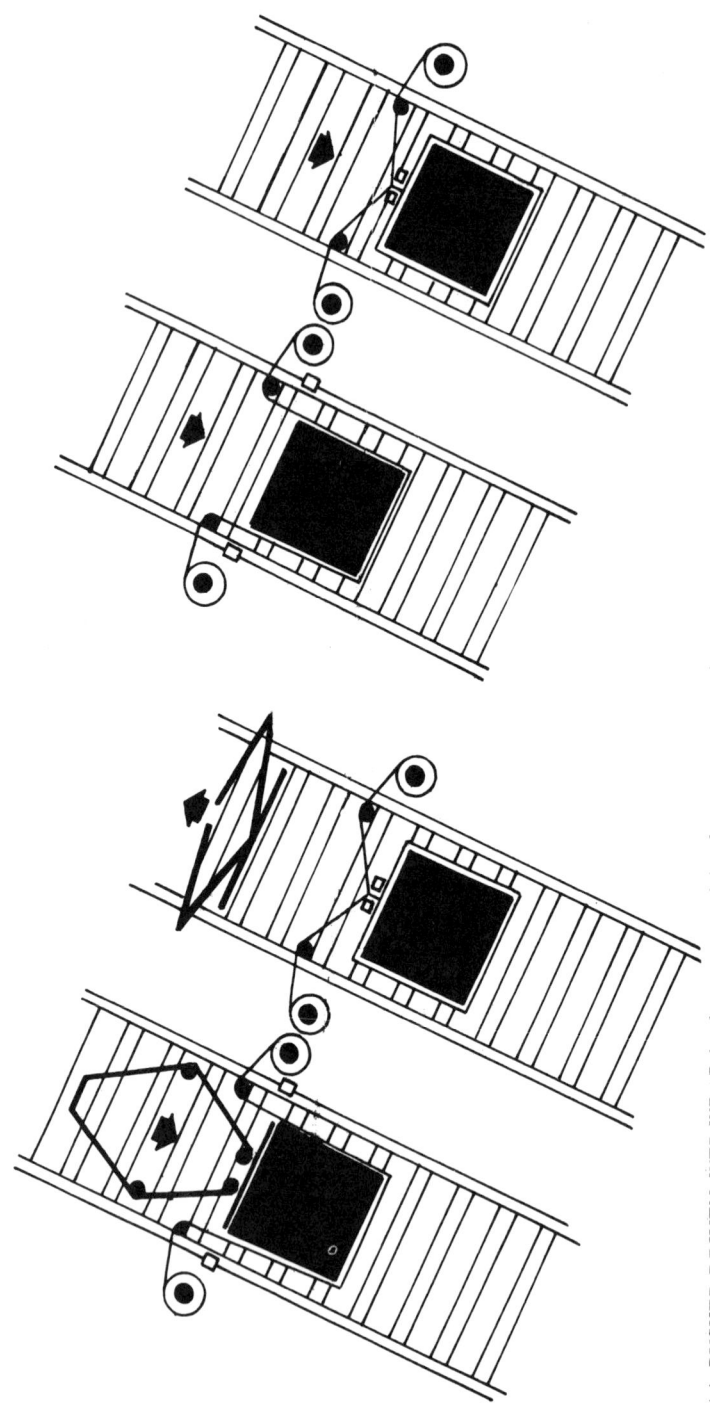

(a) PUSHER-DRIVEN WEB-WRAP loads are covered by 3 to 6-mil thick film in a single layer. A powered plate drives the load through the web formed by two vertically held rolls. Sealing bars complete the wrapping.

(b) CONVEYOR-DRIVEN WEB-WRAP LOADS are covered in a manner similar to pusher-driven. The powered conveyor drives the load through the web. The rolls are then locked so the sealing bars can tension the film.

Fig. 10-32. Two techniques of web-wrapping with stretch film. Courtesy MODERN MATERIALS HANDLING.

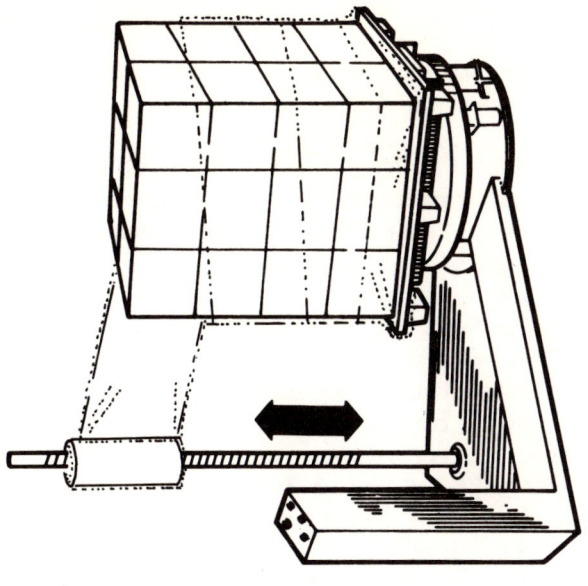

(b) SPIRAL ROTATIONAL-WRAP loads are covered by film pulled from a narrow roll. The roll, mounted on a helical shaft, can be moved up and down to wrap multiple layers of film anywhere desired on the load sides.

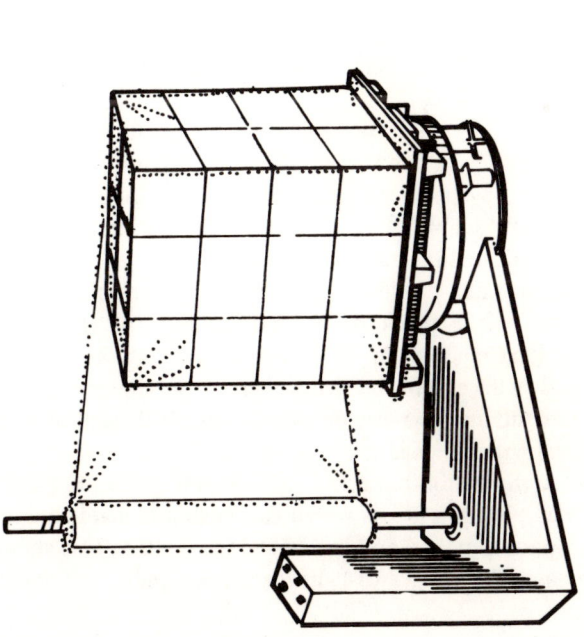

(a) FULL-HEIGHT ROLL ROTATIONAL-WRAP loads are covered by multiple layers of stretch film that overlap the top and bottom of the loads. The 1-mil thick film is pulled from a single, stationary-mounted roll.

Fig. 10-33. Two techniques of rotational wrapping. Courtesy MODERN MATERIALS HANDLING.

or manually covered with a bag while positioned on a dolly. The dolly is then either pushed or power-conveyed into the heated cubicle where shrinking occurs.

Shrink tunnels are designed for higher production rates of up to 200 unit loads or more per hour, providing continuous or indexed shrinking capability. Fully automatic systems use infra-red heat and such tunnels are usually a component of totally integrated handling systems.

Unitizing with *stretch* films involves two major methods: (1) *web-wrapping* and (2) *rotational-wrapping*. In the process of *web-wrapping*, the load is covered on all four vertical sides by the film web. The web is formed by two vertically held rolls of film, with the ends extending across a conveyor and joined over the conveyor center line. The wrap is accomplished by two techniques: (1) pusher-driven, or (2) conveyor-driven (Fig. 10-32). In the former, a large plate pushes the load through the web as brake pressure is applied to both rolls to maintain the desired tension. With the conveyor-driven technique, a minimal tension is applied to the film rolls while a powered conveyor drives the load into the web to drape the front and sides with film. Tension is applied by the sealing bars as they push the film across the back of the load and seal the two surfaces together.

In *rotational-wrapping*, a single roll of thin film (generally about 1-mil thick) is used. This is wound around all four vertical sides of the load in multiple layers to attain the thickness and strength desired. The first turn is usually applied without tension in the film to "catch" it onto the load. Subsequent turns pull up to a 30 percent stretch as the film winds up three, four, or more layers.

When the desired number of layers are applied, the end of the film is "tack-welded" to the load by a hand-held impulse sealer (or by an automatic unit on a fully automatic system), and the load is cut free from the film roll. Some films have enough "cling", like many household wrapping films, so that little or no sealing of the film end is required.

The simplest system holds the load on a stationary platform. A rotating arm extends from under the platform to hold the film roll (in a vertical positon) and the braking mechanism. The operator catches the end of the film on the load, pushes the end of the arm completely around the load, then sets the tension and continues to push the roll around on the end of the arm to stretch the film and wrap the load.

The more usual configuration has a film roll extending the full height of the load, and the load platform is a powered turntable. Standard equipment on some systems, optional on others, is a hold-down platen. This stabilizes the load and eliminates the danger of light loads being pulled off the turntable.

Spiral stretch-wrap systems use a narrow film roll mounted on a helical

REINFORCING, BUNDLING AND UNITIZING

shaft. This permits the operator to wrap several layers around the base of the load to lock it to the pallet, spiral-wrap the film up to the top of the load where he can again wrap several layers to lock in the top, and then wrap the entire load in overlapping layers in virtually any pattern desired.

Spiral-wrapping, although it involves more intricate equipment than the full-height roll-wrapping, has the advantages of being able to handle higher than normal loads, random height loads, and loads with less uniform stacking than those required by other stretch-film methods and equipment. In Fig. 10-33 two techniques of rotational wrapping are shown. Speeds of stretch film systems range from 15 to 180 loads per hour. The higher speeds are reached by the fully automated rotational-wrapping and web-wrapping equipment.

FACTORS INFLUENCING BUNDLING AND REINFORCING SELECTION

The major functions and applications of bundling and reinforcing have previously been outlined in the introduction to this chapter. In developing the final handling and shipping unit, a basic analysis needs to be made to determine whether or not bundling and/or reinforcing is a requirement at all. This determination is governed by many individual considerations pertaining to the product, the type of shipment, physical plant and warehouse conditions, handling systems used, and merchandising factors. There are no set rules which can be used to guide the engineer in this analysis toward the most economical system to handle and protect the merchandise. As an example, one metal parts manufacturer shipping his products in fibre boxes weighing not more than 50 lb may use no reinforcement on his shipping containers. Another manufacturer producing identical products has established a higher weight limit per fibre box and has found this system with the use of reinforcing more economical. Similarly, for handling and warehousing, including inventory control and customer convenience, the unitizing of individual units has proven to be advantageous. Basically, the major contributing factors motivating the use of bundling, unitizing and/or reinforcing are:

1. Protection requirements.
2. Economic considerations, labor and materials.
3. Carrier regulations.

After the decision is made to apply bundling, unitizing and/or reinforcing on the basis of a full analysis of the above factors, the selection of a particular material and/or system needs to be resolved.

It should be recognized that the materials discussed have certain inherent limitations which narrow their application, and consequently, selection is simplified. For example, pressure-sensitive filament tapes do not have comparable strength when compared to steel for heavy duty applications. However, the ease of application make it an attractive medium. But pressure-sensitive tapes are essentially limited to *smooth surfaces* to achieve adhesion. This limitation does not affect metallic and non-metallic strapping.

Further limiting factors influencing the selection of a particular material or system are:

1. The need to precompress the package, bundle, or unit load to obtain stability in handling and shipping.

2. Dynamic impact stresses encountered in shipment requiring special elongation and strength characteristics of the material.

3. The type of equipment system available for a given material.

Protection requirements. The contribution that reinforcing materials provide in shipment has been developed through continuing research by various agencies. An example of the value of metal strapping in reducing the cost of wooden containers is illustrated in Table 10-4. When boxes are provided with one or more metal straps, the sides, top, and bottom may be reduced in thickness by 20 per cent for one strap and by 35 per cent for two or more straps, except that no thickness below ¼ in. is to be specified. Table 10-4 gives the corresponding thickness for the box parts for unstrapped and strapped boxes, as well as for Group III and IV woods when substituted for Group I and II woods.

With fibre boxes no clear-cut reduction in the grade of board used is established through the use of a reinforcing medium. Frequently, the next lower test grade may be specified when a reinforcement is properly applied, providing that the lower grade also complies with the applicable weight and dimension limitations specified by carrier regulations and that the lower grade also satisfies cushioning, stacking, and puncture requirements.

The foregoing examples demonstrate the protective value that reinforcing materials may provide at the same time that they permit more economical packaging materials. In many instances, however, reinforcing is used as an *additive only*, without a corresponding reduction in grade or strength of packaging materials. The main purpose of this additive is to improve performance without resorting to the use of *higher* grade materials. A comprehensive analysis of oil can packaging for export disclosed that unreinforced weatherproof fibreboard boxes sustained an average of 30 falls in the revolving drum test prior to failure. Comparable boxes with a single girthwise strap sustained an average of 45 drum falls prior to failure. While strapping in this study provided a substantial improvement in performance, it should be recognized that with fibre boxes and with various types of products, this relationship can vary widely. To determine the value of reinforcing for a

REINFORCING, BUNDLING AND UNITIZING

given package, data comparing performance should be developed through field or laboratory testing.

As expressed previously, one of the main advantages of bundling or unitizing is the improvement of unit container performance through the added support provided by proximity with other or similar units in a bundle or load. Apart from the resulting handling savings, unitizing, by increasing the size and weight of the shipping unit, reduces the probability of such handling hazards as throwing and dropping, which are common with very small shipping units. If a bundled unit is dropped, fewer units are exposed to the direct impact, since the majority are protected by the adjacent units. A study of salt packed in fibre cans disclosed that damage to the unit containers subjected to handling hazards as simulated in the laboratory was 30 per cent more severe with individual shipping containers than with four units tied together into a bundle. The reduction in damage was primarily the result of the elimination of corners, that is, 32 corners for 4 individual shipping cases as contrasted to 8 corners for the bundle. Since the number of corners to be protected is reduced, a re-examination of the packaging system may reveal the possibility of redistributing packing protection to the vulnerable points with a resultant over-all saving in packaging materials. Similarly, with unitizing in conjuction with mechanical

TABLE 10-4
Reduction in Thickness of Wooden Boxes through the Use of Steel Strapping or When Groups III and IV Woods are Substituted for Groups I and II

Specified Thickness of Sides, Top, and Bottom for Groups I and II Woods, in.	Permissible Reduced Thickness When One or More Metal Straps Are Used, or When Groups III and IV Woods Are Used		
	20% Reduction When One Strap is Used, in.	35% Reduction When Two or More Straps Are Used, in.	25% Reduction When Groups III and IV Woods Are Used, in.
1/4	1/4	1/4	1/4
5/16	1/4	1/4	1/4
3/8	5/16	1/4	5/16
7/16	3/8	5/16	3/8
1/2	3/8	5/16	3/8
9/16	7/16	3/8	7/16
5/8	1/2	3/8	7/16
11/16	5/8	7/16	9/16
3/4	5/8	1/2	9/16
13/16	5/8	1/2	9/16

Courtesy of the Association of American Railroads.

handling, the need for unit protection is even further reduced because of the complete elimination of throwing and dropping, provided that the units are not subsequently individually distributed. With unit loads, the protection emphasis is therefore on stacking and resistance to impact. Unit load packaging has thus provided substantial economies in packaging materials.

On the other hand, unit loads have a tendency to lose dimensional integrity in handling and shipping which may create damage problems in confined storage areas. Where this is a problem, i.e., unit loads of charcoal, flour and pet foods in multi-wall shipping sacks, the use of shrink or stretch wrapping is recommended.

If a preliminary analysis has indicated that reinforcing and/or bundling is advantageous, then certain developed principles and techniques should be followed. Technical data relating to these principles are discussed below:

REINFORCING AND BUNDLING TAPES. Recommended procedures for the use of filament tape have been illustrated in Fig. 10-14. It will be noted that short strips suffice for fibre boxes. The minimum length of the strips, their number, width, and positioning have been specified in applicable carrier regulations to be discussed. The adhesive strength in shear on a paper surface should be equal to or greater than the tensile strength of the tape.

Filament tapes have been preferred over other reinforcing media by many shippers for the following reasons:

1. Application is easy and dispensing is less complicated.
2. Tape becomes an inherent part of package and does not loosen, snag, corrode, or displace in shipment.
3. Damage to product due to initial application pressure or subsequent weakening of container edges under stress does not occur, nor are edge protectors required to minimize this type of damage.
4. Safety to personnel in handling is greater.
5. Taping is generally a more portable system which requires no auxiliary tools.

On the other hand, there are certain limitations frequently encountered with the use of filament tapes which favor the use of other types of media:

1. Tape will not adhere properly to all surfaces.
2. The pull on the tape is no greater than the manual exertion of the operator during application; thus, unless precompression of the package is made, the tightness of the pack is limited.
3. Problems of tape with blocking, moistening, bleeding, curling, aging, and discoloration may be experienced.

REINFORCING, BUNDLING AND UNITIZING

4. Economic factors, from a material and application viewpoint, must be considered, especially when an encircling band is required because of the type of package.

When filament tapes are considered for the bundling of such items as piping, conduit, and extrusions, impact resistance is of prime importance in retention of a tight bundle and in prevention of tape breakage. Tapes having an elongation which will absorb shock should be selected for this purpose. This physical property is found in nylon and rayon reinforcements, whereas glass, although of greater tensile strength, will fail owing to low-impact strength and its lower elongation.

Single-ply pressure sensitive (filmic tapes) have replaced filament-reinforced strapping tapes for a variety of applications by providing comparable performance while offering important economies.

Carrier regulations currently require a minimum width of ½ in. for approved applications. For telescope boxes a width increase according to the gross weight of the package and the number of strips used is recommended. In Table

TABLE 10-5
Width of Filament Tape Recommended for Different Gross Weights of Telescope Boxes
(Tape Having Tensile Strength of 160 lb per in. of Width)

Gross Weight, lb	Number of Strips per Box		
	4 Strips	6 Strips	8 Strips
60 or less	¾ inch	¾ inch	¾ inch
70	⅞	¾	¾
80	1	¾	¾
90	1⅛	¾	¾
100	1¼	⅞	¾
110	1⅜	1	¾
120	1½	1	¾
130	1⅝	1⅛	⅞
140	1¾	1¼	⅞
150	1⅞	1¼	1
160	2	1⅜	1
170	2⅛	1½	1⅛
180	2¼	1½	1⅛
190	2⅜	1⅝	1¼
200	2½	1¾	1¼

446 **DISTRIBUTION PACKAGING**

10-5 the suggested width for 4, 6, and 8 strips of filament tape with a 160 lb per in. of width tensile strength is listed.

Application Equipment. Equipment suitable for the application of reinforcing and unitizing tapes has been described in Chapter 9.

The selection of the proper equipment depends on the volume of merchandise to be sealed or reinforced, the type and size of the merchandise, the method of packaging (centrally conveyorized or decentralized), the number and length of the tape strip, and the required method of application. When manual dispensers, either of the portable pull-and-tear type or the lever-dispenser type are used, the pull on the tape is no greater than the effort exerted by the packer. Therefore if the merchandise or container must be tightly compressed to retain its shape or retain the units within a bundle during handling or shipping, precompression of the product is desirable. Some of the semi-automatic and fully-automatic dispensing machines provide a controlled tension on the tape as well as precompression of the merchandise.

STRAPPING. The recommended method of applying nail-less steel strapping on nailed wooden boxes is shown in Fig. 10-34. Steel strapping should be applied immediately prior to shipment to avoid looseness of the strap due to shrinkage of lumber or progressive indentation of strapping into the edges of fibreboard boxes.

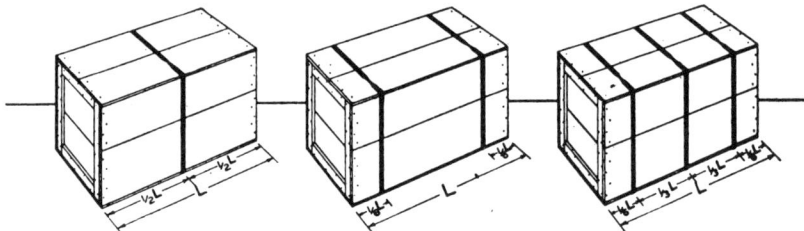

Fig. 10-34. Recommended spacing for nailless strapping.

A convenient way to determine strap requirements for boxes is to multiply the gross weight of the box by 10 and then divide by the number of straps used. The resultant value should then be equal or greater than the tensile strength of the steel strap to be selected. For example, a package weighs 200 lb is secured with two steel straps. A steel strap strength of 1,000 lb is required (200 lb times 10 equals 2,000 lb divided by 2 equals 1,000 lb). Referring to Table 10-1, a ½ in. × 0.018 in., and 5/8 × 0.015 in. strap has an equal or slightly greater tensile strength than the figure of 1,000 lb calculated as the requirement for the flat metal strapping in the above example. As a general rule, the widest possible steel strapping should be selected when

REINFORCING, BUNDLING AND UNITIZING

crushing at the edges of the container is objectionable. On the other hand, if the steel strapping may become excessively exposed thus introducing a snagging hazard, strapping of greater thickness rather than width should be selected. In the foregoing example, the ½ in. × 0.018 in. strap is most desirable for a wooden box, whereas the wider 5/8 in. × 0.015 in. strapping would tend to reduce edge crushing on fibre boxes. This would also minimize the need for supplementary edge protectors. Where crushing is a major problem, the use of non-metallic strapping should also be investigated.

As a further guide, Table 10-6 lists the gage requirements for round wire using different numbers of wires for the reinforcement of wooden boxes.

TABLE 10-6
Sizes, in Gage Number, of Round Wire for Strapping Wooden Boxes of Different Gross Weights

Maximum Gross Weight, lb	Gage of Wire When a Different Number of Wires Is Used		
	One	Two	Three
50	12 gage	14 gage	
100	11 gage	12 gage	
200	9 gage	10 gage	12 gage
300		10 gage	11 gage
400		10 gage	11 gage
500		9 gage	10 gage

Economic considerations. Since the cost of materials and the cost of application are interrelated in determining over-all expense, the selection of a bundling or reinforcing system should be made in consideration of both cost factors. A direct comparison of material cost should be based on material consumed per box to provide equivalent protection. As an illustration, a telescope box measuring 36 in. × 24 in. × 6 in. can be strapped with three encircling flat metal bands (one lengthwise and two girthwise straps) or can be secured with six 10 in. strips of ¾ in. filament tape. The total steel strap requirement, including some waste, is approximately 18½ ft. The tape consumption is 5 ft. If a ½ in. × 0.015 in. strap is used, which gives a yield of 39.3 ft per lb, approximately 0.47 lb of strapping is needed. The costs for either material can then be computed by multiplying the amount of material used by current unit prices. The determining factor in the selection of either of these two media is the cost of application and the protective advantages offered.

In a study made comparing the total equipment, labor, and material cost of various types of machines related to strap production rates per hr, the following data, at a rate of 750 straps per hr was compiled:

Type of Machine	Total Cost–Equipment, Labor and Materials (expressed in %)
Automatic wire tying machine	100
Automatic flat strapping machine	
Model A	104
Model B	115
Model C	135
Semi-automatic flat strapping machine	
Model A	151
Model B	178
One-piece magazine flat strapping tool	184
Two-piece hand flat strapping tool	250

The above comparisons include all cost items in performing the strapping operation.

The availability either of standard equipment or specially developed fixtures and accessories can greatly influence the production rate, and can consequently favor the selection of one or another of the many available reinforcing media.

Carrier regulations. As previously stated, the shipper frequently has the option of whether or not to reinforce or bundle his merchandise. Some notable exceptions are contained in military specifications, in some of the commercial export recommended shipping practices, in the separate package descriptions (special package numbers) contained in the *Uniform Freight Classification,* and in the *Department of Transportation Hazardous Materials Regulations.*

When reinforcing or bundling is contained in separate package descriptions, experience has indicated either the need for these media to provide adequate protection or has facilitated *more economical packaging* in comparison to compliance with *general carrier packaging regulations.* For example, in the description of articles in the *Uniform Freight Classification,* cabinets, air conditioning, without air conditioning equipment, may be packed set up, in boxes, crates or in *Package 1078.* Package 1078 consists of the following:

"In 3-piece fibreboard containers consisting of taped or stitched tubes with top and bottom flanged caps or in 2-piece fibreboard containers consisting of half-slotted containers with flanged cap fitting over open end. All fibreboard must be single-wall or double-wall, the fibreboard meeting require-

REINFORCING, BUNDLING AND UNITIZING 449

ments of Rule 41, Sections 2 and 3 for boxes testing not less than the test specified below.

Pallet load may consist of not more than five tiers, not more than four containers to a tier with individual covers or a common cover. Containers must be strapped to a wood pallet with not less than *two steel straps each not less than 5/8 x .015 inch.*

For gross weights not exceeding 300 pounds the fibreboard must test not less than 275 pounds. For gross weight exceeding 300 pounds, but not exceeding 500 pounds, bodies of containers must test not less than 350 pounds or not less than 275 pounds when lined on four sides with corrugated fibreboard testing not less than 200 pounds. Flanged caps must test not less than 275 pounds.

Surfaces of articles subject to damage by abrasion must be protected by non-abrasive materials. Sufficient vertical separators in the form of pads, partitions, or liners made of corrugated fibreboard testing not less than 200 pounds must be used to prevent crushing of the articles."

Referring to Chapter 2, it will be noted that the weight limitations now permitted exceed those allowed under Section 3 of Rule 41.

In another example under description of articles, sash or doors, glazed, or sash or doors, glazed and screens combined, iron or steel, or wood covered with iron or steel or tin plate, can be packed in boxes or crates and when shipped carload in package 419. Package 419 consists of the following:

"In bundles of one or two, four corners of each article must be protected by triangular corner caps made of single-wall corrugated fibreboard testing not less than 200 lbs. When two in bundle, articles must be further separated by wooden blocks at corners when necessary for clearance of projecting parts, and, when only one in bundle, article must be completely wrapped in single-wall corrugated fibreboard testing not less than 200 lbs., in addition to corner caps described. Bundles *must be metal strapped* at each end over fibreboard corner caps.

When loaded solidly from end to end of car in mixed CL with wooden doors or sash, aluminum doors protected by corner caps as described *need not be metal strapped* or wrapped."

This special packaging provides economy through bundling more than one door and through a packaging technique incorporating minimum packaging material requirements in conjunction with the use of steel strapping, except as noted.

The interchangeable use of filament tapes with metal straps, rope, or wire is permitted, provided the following conditions are complied with as stated in Rule 5, Section 3(e) of the *Uniform Freight Classification:*

"Unless otherwise provided, where the use of metal straps, rope or wire is required in separate package descriptions, extruded oriented nylon or polypropylene straps meeting the requirements of Rule 41, Section 8(a) (6), may be used as a substitute. Unless otherwise provided, where the use of metal straps, rope or wire is required in separate package descriptions, the filament reinforced, pressure-sensitive or gummed tape specified by Rule 41, Section 8(a) (3) is authorized. This tape must be used as a completed band with a 4-inch overlap of the tape on itself and in the manner prescribed by the package description. It must be in widths not less than ¾ inch or twice the required width of flat metal strapping, whichever is greater."

Section 8 of Rule 41 also contains pertinent details relating to the physical characteristics of filament tapes.

Economic factors should be considered in the selection of the desired reinforcing or bundling media.

CHAPTER 11

Easy-opening Devices

Preceding chapters have provided the basic framework for the proper selection of a packaging system with the primary considerations being to provide *adequate protection at lowest cost.* These factors alone do not necessarily satisfy the requirement of the ultimate consumer as to (1) ease of opening, (2) ease of handling, and (3) ease of disposal or re-use (including reclosing, etc.).

This chapter deals with the important subject of easy-opening devices as applied to distribution packages. Some of the devices and techniques have their origin in consumer packaging where the convenience of an easy-opening feature has played an extremely important role in merchandising. Features such as perforating, tear tapes, and special adhesives have found universal consumer acceptance. In distribution packaging applications easy-opening devices are incorporated to provide economy primarily through lower labor costs in the opening and unpacking operations, as well as for convenience. Additional functions frequently fulfilled by an easy-opening feature include display considerations of contents, reduction in damage to contents caused by opening tools, and ease of pricing.

452   DISTRIBUTION PACKAGING

Many shipping containers utilize a closure which facilitates easy opening. These closure types have been designed and are therefore selected to permit *reclosing the container* in the gradual dispensing of the product, or for *container re-use*. Other closure types provide for easy opening by nature of their construction. Examples of the former devices include the lever-activated fibre drum closure, tuck-type flap construction for fibreboard boxes and folding cartons, and screw-type or friction covers. Examples of closures facilitating easy-opening by nature of basic construction include the Rock Fastener closure for wire-bound boxes and nonreinforced gummed paper tape.

An easy-opening device is defined as a special process or material incorporated in the closure or in *another part of the package* to facilitate opening. The need for an easy-opening device exists mainly for expendable paper and paperboard packages and containers in which adhesives (either directly applied or in the form of gummed tapes), staples, or stitches are extensively used for closure purposes.

TYPES OF EASY-OPENING DEVICES

Easy-opening devices for paper and paperboard packages and containers are used for the following major distribution packaging applications:

1. Multi-wall paper shipping sacks (Fig. 11-1).
2. Paper-wrapped packages (Fig. 11-2).
3. Corrugated and solid fibreboard boxes (Fig. 11-3).

Shipping sacks with an easy-opening feature utilize a special stitching pattern known as the *chain stitch* which is formed with a single thread locking on one side of the bag material. The seam produced is strong enough to pro-

Fig. 11-1. Rip-cord easy-opening closure for shipping sacks.

EASY-OPENING DEVICES

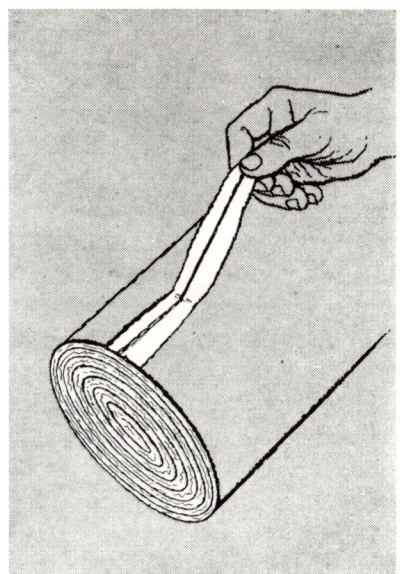

Fig. 11-2. Easy-opening device for paper-wrapped packages (roll of carpet).

vide ample security for average requirements, yet provides for easy-opening of the sack by the ultimate consumer. An illustration of the chain stitch is presented in Fig. 11-4.

Paper wrapped packages, as illustrated in Fig. 11-2, utilize a rip cord or tape which is inserted around the product before or simultaneously with the paper wrapping application. One end of the cord or tape is extended through the paper wrapping. In opening, this end is pulled with a swift lateral motion which in turn causes the cord or tape to slit the protective wrapping. In applications of this type the protective wrapping must be tightly fitted so that the opening cord is securely held, or the tape should be adhered to the inside of the wrapper. If the package is loosely wrapped the tearing medium has a tendency to pull out from underneath the wrapping or to effect an unsatisfactory tear. The latter is characterized by a build-up or bunching of paper or an irregular tear.

Fibreboard box easy-opening. For corrugated fibreboard boxes the provisions for opening can be effected by three major means: (1) *at the flap closure,* (2) *at the manufacturer's joint,* and (3) *peripherally located on the box panels.* The following paragraphs are a discussion of these techniques.

EASY-OPENING FLAP CLOSURE.—One of the most difficult flap closures to open is one that has been sealed with an adhesive. Therefore, considerable research has been devoted to develop techniques which would overcome this opening problem. Three methods are employed: (1) *printing* of the flaps, (2) *slotting and perforating* of the flaps, and (3) the use of a *low-tensile adhesive.*

454   DISTRIBUTION PACKAGING

Fig. 11-3. Tear strip tape easy-opening feature for a corrugated fibreboard box for nails. Note initiating tab. Courtesy 3M Company.

Printing of the outside of the inner box flaps is accomplished during the regular fabrication process and may consist of a tooth-like or similar pattern, usually applied in lines. Cold adhesives do not penetrate into the printed areas of the linerboard whose interstices have been filled with printing ink. Therefore, in opening, a break in the fibre tear pattern is accomplished where the printing appears. This reduces the effort required to open the box. This method is the least expensive of the flap opening techniques, but is not completely effective since no provision is incorporated to initiate the flap opening (that is, to lift the sealed edges of the flaps).

Slotting and perforating of the top flaps is an approved flap easy-opening feature described in Section 2 (*d*) of Rule 41 of the *Uniform Freight Classifi-*

Fig. 11-4. Easy-opening chain stitch used to secure shipping sacks.

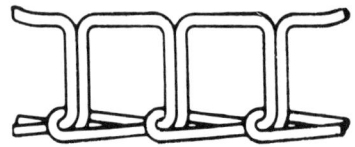

EASY-OPENING DEVICES

cation. The description is as follows: "Provided the carrying ability of the box is not materially impaired, boxes: ... (d) May have closing flaps perforated and inner flaps slotted with one slot in each flap to facilitate opening. Perforations must be at locations not closer to score lines than 1½ in. Slots must not extend beyond perforations."

This opening feature is illustrated in Fig. 11-5. Usually, some portion of the inner flaps is printed to facilitate removal of the flap section bordered by the perforations.

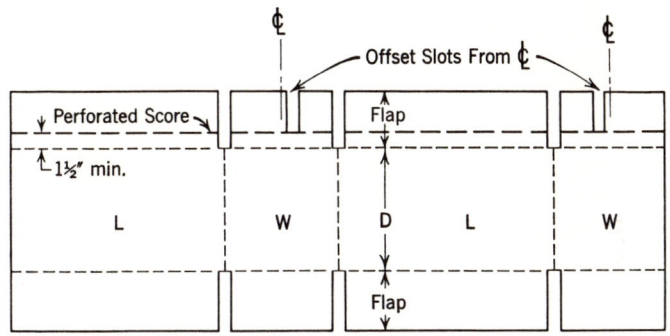

Fig. 11-5. Container with perforated and slotted flap for easy opening.

Opening is accomplished by depressing one center flap and lifting up the other flap which tears along the perforations. The other flap is then opened in a similar fashion. This method may necessitate a minor upcharge in the cost of the container because of the extra slotting and perforating operations. The limiting factor of this feature is in the initiation of the opening which is difficult with noncompressible contents.

Low tensile adhesives have been used fairly extensively to accomplish easy flap opening. It should be recognized that the adhesive must satisfy two major requirements in order to produce the desired results. First, the adhesive must insure adequate flap closure to prevent premature flap opening during handling and transportation, and second, the adhesive must separate readily when the flaps are lifted in opening. Adhesives of this type have *high shear* and *low tensile* strength.

Opening of these containers is usually facilitated by hand holes which are cut at the center of the outer flaps as shown in Fig. 11-6. The only significant cost factor in providing this easy-opening is in the extra cost of the special adhesive which is nominal per container.

As described in Chapter 8, hot melt adhesives are required to bond a smaller flap contact area than cold setting adhesives. This reduction in contact area provides comparable closure performance to cold setting adhesives. To enhance easy opening with hot melts, a stitch pattern of application is employed.

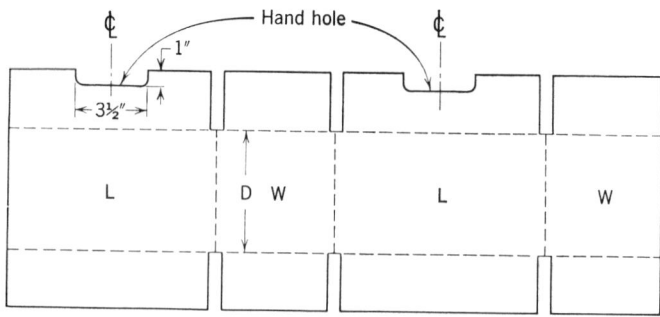

Fig. 11-6. Handhole in outer flaps, used in conjunction with low tensile adhesives for easy opening.

Flap closures accomplished with tape normally are easier to open than adhesive closures since no positive seal between the inner and outer flaps is made. It is therefore only necessary to cut or otherwise fracture the tape along the open box seams to effect opening. To further simplify tape closure opening, a special easy-opening gummed tape has been developed. This product, referred to as strippable reinforced gummed tape, is available in nonasphaltic grades only, is identical to regular reinforced tape, and is imprinted for identification. The gumming side kraft of this tape is made from a special type paper having a low internal fibre bond strength. When the tape is peeled from the box, the kraft sheet separates leaving a tissue-thin layer adhered to the box which can be readily separated. A lifting tab is formed by imprinting the end panels of the box with an oil base ink. The tape does not adhere in this area (Fig. 11-7).

EASY-OPENING AT MANUFACTURER'S JOINT. Easy-opening features may be incorporated at the manufacturer's joint of the box. One such method provides opening of one large side panel by a suitable combination of joint tape removal and slitting of the box panel adjacent to the joint. Tape removal is assisted by inked box areas at the ends of the manufacturer's joint which decrease the adhesion of the tape. Obliquely placed slits at the edge of the panel help initiate the tear as the side panel is pulled back. In Fig. 11-8 this type of easy-opening is illustrated. It will be noted that the box is a so-called "end-opener," and the panel which is removed exposes the tops of the unit containers for ease of price marking and removal.

Other modifications of easy-opening joints have been explored, including a joint incorporating a tear string strong enough to rupture the reinforcing fibres of the tape and spot glued manufacturer's joints. The success of these experimental and other easy-opening joints is limited by possible strength reduction. The factor of potential strength reduction is carefully considered by the carrier agencies prior to the approval of any new packaging material or method.

EASY-OPENING DEVICES

(a) Lifting-tab formed by imprinting end panels.

(b) Stripping of tape leaving tissue thin residue.

Fig. 11-7. Easy-opening strippable reinforced sealing tape. Courtesy Atlantic Gummed Paper Corp.

PERIPHERAL BOX EASY-OPENING. Peripheral easy-opening basically consists of separating the box into two sections. This separation can be performed parallel to the top and bottom flaps or perpendicular to the flaps. The former method is primarily used to remove a section of the container to expose the contents for point-of-sale display and for ease of marking. The latter method is a simple device to split the loaded container into two fractional units.

The following two basic techniques are used to accomplish peripheral box opening:

1. Opening by means of an additive, usually a tape, placed on the inside of the container.

2. Opening by means of *slitting,* or *perforating* the box panels.

Fig. 11-8. Easy opening through joint tape removal.

The techniques involving perforating and slitting are authorized in Section 2 of Rule 41 of the *Uniform Freight Classification* as follows:

Provided the carrying ability of the box is not materially impaired, boxes: ... (b) May be perforated once around to facilitate separation into two parts, OR may have not more than two lines of perforations provided the total lineal inches or perforation do not exceed the perimeter of the box (twice the sum of the inside length and width).

(c) May have inside facing slit to form not more than two lines or ribbons for easy opening or may have outside facing slit to form not more than one line, but not both, provided further the corrugating medium in either case is neither cut, crushed nor otherwise damaged.

The use of an additive such as a tape to facilitate opening, if properly applied does not adversely affect the strength of the box. Therefore, specific requirements pertaining to tape additives for opening are not contained in the rail freight regulations.

Peripheral box opening using tape is initiated by means of either of two systems: (1) a *tab* of the tape *extending through the box manufacturer's joint* on the outside of the panel adjacent to the joint, and (2) a *die-cut* in

EASY-OPENING DEVICES

the box panel *adjacent to the joint* in the same plane as the tape which is adhered to the inside of the box. The die-cut initiating device may be either an *"H" cut* or a *sunburst,* which also cuts the easy-opening tape. With both of these methods, identifying imprinting of at least two box panels is required to identify an easy-opening box and the position of the initiating device.

The tape used can measure as little as 3/32 in. in width and uses a pressure-sensitive or gummed adhesive or the adhesive may be applied in the application process. The principal tapes used are filament tapes, reinforced with either glass, rayon, or polyester. When the application is performed in a continuous process on a corrugator and box scoring is performed with the tape in place, the tape must have sufficient elongation to resist fracturing at the box vertical scorelines. The tab of the tape must have adequate tensile and impact resistance to avoid rupturing under the stresses of opening. For these reasons, rayon and polyester tapes are found to have the desired characteristics for tear tapes, and glass reinforced tapes are no longer too widely used.

The basic principle utilized in opening by means of slitting, or perforating is based on the inherent tensile strength of the fibreboard when one liner is cut and the tear is guided. A *ribbon of fibreboard,* formed by two parallel and adjacent cuts, is produced by slitting the inside liner of the box. After initiation of the tear, the ribbon which has been formed has sufficient strength to fracture the corrugating material and the outer liner around the periphery of the box. The resultant tear is reasonably straight and clean (Fig. 11-9).

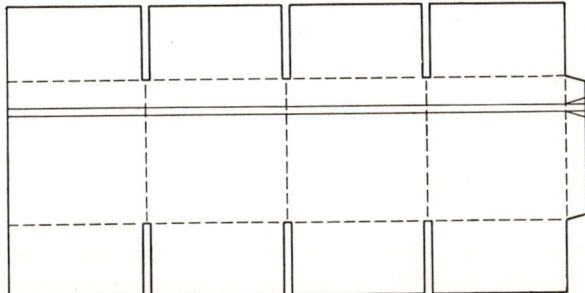

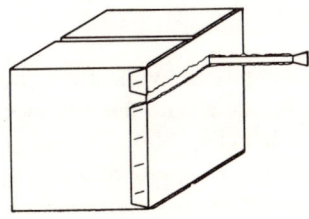

Fig. 11-9. Corrugated fiberboard box with easy-opening feature accomplished by slitting. Courtesy Weyerhauser Co.

Slitting is performed prior to the adhesion of the inner liner to the corrugating material. As a result, the corrugations are undamaged. The width of the ribbon formed by the two parallel slits is generally ¾ in. If slitting and subsequent combining are effectively accomplished, the loss in box compressive strength is negligible. Extensive laboratory tests have disclosed that a potential reduction of approximately 10 per cent in box compressive strength is possible depending on the positioning of the ribbon in relation to the horizontal box scores and the quality of workmanship. These data further indicated that there was no significant loss in bursting and puncture strength when tests were deliberately made at the slit area.

The position of the easy-opening tape or slit ribbon is dependent upon the type and arrangement of the contents as well as end-use requirements. With slitting, positioning too close to the horizontal box scores is not practiced because of potential reduction in box stacking strength. For heavy products such as business machines and furniture (Fig. 11-11), the position is generally near the bottom of the box. This position permits the top to be removed like a hood; thus, the merchandise is fully exposed obviating the need to lift the unit for installation. Obviously, when contents are arranged in two layers the opening device is preferred to be centrally positioned.

Perforating for easy opening is carried out *perpendicular* to the horizontal box flaps and usually is centrally located. One of the earliest applications of this method was for the packing of baby foods so that convenient break-up of the container into smaller units was permitted. For the merchandising of beer cans, breaking a case into two equal sections is facilitated by perforating. The primary function of perforating around the girth of a fibreboard box is in splitting the case into a convenient sales unit. Ease of opening and removal of the contents is frequently a secondary consideration.

APPLICATION EQUIPMENT

Easy-opening devices are incorporated into the container or package either during the packing, wrapping, closing, or sealing operation performed by the user or are an integral part of the boxmaking process in the manufacturer's plant.

The following is a discussion of user and boxmaker easy-opening application techniques and equipment.

Application in users' plant. Applications accomplished by the user in his own plant generally do not involve special purpose equipment. For instance, filled textile bags can be closed with a chain stitch (Fig. 11-4) on most of the conventional bag-closing machines equipped with an appropriate sewing head. For paper-wrapped bundles, rip cord attachments on wrapping equipment such as used for spiral wrapping of hoses, cable, etc. are available.

EASY-OPENING DEVICES

Adhesives of the easy-opening type are applied with any of the conventional application methods, including automatic case-sealing machines. As previously mentioned, strippable reinforced gummed tape requires no special dispenser or application equipment. However, the tape must be suitably identified and the box end panels should be imprinted as described.

Because of the ease with which any of these user easy-opening applications can be integrated into existing packaging lines, the application cost is nominal.

Application at box plant. Fibre box easy-opening devices, which consist of a modification of the basic box, are manufactured in the boxmaker's plant. The following methods may be used:

1. *On the combiner*, that is slitting or application of tear strip tape.
2. In conjunction with *subsequent finishing operations*, such as printing, perforating, die-cutting, and taping.

The inner liner is slit prior to the single face combining. Tear strip tape, when applied on the combiner, is positioned on the board at either of two stations: (1) just ahead of the hot plate section and (2) immediately before the slitter. Either pressure-sensitive, gummed, or ungummed tape can be applied on the combiner with suitable attachments. One typical device for applying easy-opening tape on the combiner is illustrated in Fig. 11-10. This equipment operates at board speeds up to 700 feet per minute.

Fig. 11-10. Application device for pressure sensitive easy-opening tape applied on combiner. Courtesy 3M Company.

Printing modification for flap opening or joint opening as well as special printed instructions for all other types of easy-opening devices is accomplished on the printer slotter in conjunction with regular imprinting of the box blank. Similarly, extra vertical scores, slots, perforations, and small die-cuts for initiating the opening are fabricated on this equipment. If excessive die-cutting is required, non-rotary equipment is needed which involves an extra operation.

Easy opening tapes can also be applied on joint taping equipment outfitted with a special attachment. The tape is applied in a separate operation on the inside of the box blank with an extension folded over onto the outside of the blank on one end. This extension forms a tab when the joint is subsequently applied.

SELECTION OF AN EASY-OPENING DEVICE

Prior to the selection of a specific easy-opening feature, an analysis should be made to determine the potential advantages as rated against the extra cost which will be incurred in the use of such a device.

The advantages resulting from the use of an easy-opening device can be summarized as follows:

1. *Tangible* benefits which can be measured in terms of potential dollar savings through faster speed of box opening, content removal (Fig. 11-11), price marking and coding, reduction in damage to contents in opening (since a knife or similar tool is not required), and the opportunity to develop convenient fractional sales units with resultant packing savings.

2. *Intangible* benefits which assist in maintaining and creating consumer good will and acceptance through ease of opening, possible display features, and all the tangible factors which make the user's job easier. Other intangibles include elimination of inconvenience through return of merchandise damaged in opening, reduction of injury to personnel in case opening, and the availability of usable box sections for store use (Fig. 11-12).

The tangible benefits are primarily obtained by retailers of merchandise such as supermarkets. However, the economic gains are potentially realized only if a substantial percentage of all incoming packages feature easy-opening devices. If these devices are widely different and potentially confusing to the unskilled labor used to unpack merchandise, the acceptance of an easy-opening device is limited. The general result is continued use of the conventional knife for the opening of *all* packages received.

Where unpacking of merchandise is directly under the control of the shipper, and adequate training can be provided, the full benefits of easy-opening can be attained. As an example, a manufacturer of electric type-

EASY-OPENING DEVICES

Fig. 11-11. Furniture box with easy-opening feature facilitates unpacking.

writers uses salesmen to install the machines in the purchaser's office. To assist the salesman in the unpacking of this heavy machine, an easy-opening device is positioned near the bottom of the box and thus eliminates the difficulties of flap opening and minimizes lifting.

It is therefore recommended that before an easy-opening device is incorporated, direct economic benefits should be rated. This will permit a sound decision when appraised in conjunction with the intangible benefits in creating consumer good will.

Economics. The cost of an easy-opening device is readily determined either by obtaining quotations from the supplier, if the device is applied entirely by the manufacturer, and/or through an analysis of material and labor costs when the easy-opening feature is applied during the packing and sealing operation.

The cost of an easy-opening feature represents only a fraction of the total cost of the package. However, since significantly wide variations in cost among different easy-opening devices exist, the user is given the opportunity to be selective. The following is a discussion of these economic factors.

Fig. 11-12. Availability of tray-like bottom box sections when box is opened with peripheral easy-opening device assists in store handling.

1. *Rip cords* for paper-wrapped packages increase the cost of the package almost directly in relation to the length of the cord, when attachments are available on the application equipment.

2. *Chain stitching* for shipping sack closure essentially involves no increase in material or application cost except for the capital investment for the special stitching head.

3. *Easy-opening box flap closures* vary in cost depending upon the method employed. Additional costs result from extra set-up time and/or die costs by the box manufacturer or special adhesives or tapes used by the box user. A one-time capital expenditure for special glue-application attachments may also be incurred. Since most easy-opening flap closures are relatively simple and involve only minor change, the additional cost is nominal.

4. *Easy-opening at the manufacturer's joint* requires no change in box

EASY-OPENING DEVICES

sealing techniques. The additional cost is therefore strictly an addition to the container cost. The upcharge is basically for increased set-up time by the box manufacturer, and in large volume the additional cost is very nominal.

5. *Peripheral box easy-opening* is also accomplished entirely by the box manufacturer with no change in the sealing method required. With this method, the widest price variations occur. Slitting and perforating do not involve additives and are performed during the normal box fabrication process. The additional cost is therefore reflected in extra set-up charges. When tape is used, the upcharge includes some of the above costs plus the cost of the material. The cost for the tape is dependent upon the actual box periphery and the type, width, and backing of the tape. When a tab of tape which extends through to the outside of the box is used, an additional fabrication cost is incurred due to this extra operation.

Economic considerations are further influenced by the required performance of the easy-opening device and the nature of the product. For instance, a flap opening device using a hand hole is very economical and provides good opening but should not be used for products requiring protection from foreign contaminants. Similarly, where the chance of acceptance is small owing to the nature of distribution, the investment for easy opening should be kept to a minimum or not at all considered.

Strength considerations. There has been considerable research conducted in the development and refinement of the various easy-opening devices. The objective of this work has been to incorporate a device with no appreciable reduction in box strength. In most instances, the present carrier regulations provide safeguards against methods which are likely to increase damage in transit.

Flap closures, embodying an easy-opening feature, must provide retention and must be sufficiently durable to withstand the normal hazards of transportation. Experience has indicated that presently accepted modifications of the flap construction through die-cutting, slotting, perforating, printing and/or the use of special adhesives does not materially influence minimum requirements for flap retention. Although a closure of this type may be adequate for shipment, an increased pilferage hazard may exist.

Peripheral box easy-opening has been carefully studied because of the possible influence on stacking strength. Essentially, it has been found that when tape additives are used, box stacking strength is not reduced. When slitting is employed, the critical consideration involves the positioning of the slit ribbon. For example, laboratory tests yielded the data listed in Table 11-1. It will be noted that in general, center slits were less deleterious to stacking than off-center slitting.

In selecting an easy-opening device, an analysis of any possible strength reduction should be conducted particularly in those circumstances in which minimum protection is currently employed.

TABLE 11-1
RELATIVE STACKING STRENGTH OF REGULAR SLOTTED FIBREBOARD
BOXES OF VARIOUS SIZE AND WITH DIFFERENT POSITIONING
OF PERIPHERAL SLITS*

	Top-to-Bottom	End-to-End	Side-to-Side
Size of Box (16¼ in. × 12⅛ in. × 11¼ in.), *Grade 200 lb, "A" Flute*			
Unslit box	100.0	100.0	100.0
Center slit box	107.0	100.0	99.0
Slit 1⅞ in. from top	99.0	95.0	93.0
Size of Box (21⅞ in. × 17⅝ in. × 23 in.), *Grade 200 lb, "A" Flute*			
Unslit box	100.0	100.0	100.0
Center slit box	101.5	98.5	93.0
Slit 5 in. from top	98.0	88.0	103.0

* Strength is expressed in percentage.

Carrier regulations. Pertinent carrier regulations have been included in the technical description of each of the easy-opening features. For the protection of the shipper and his customer, the carriers carefully review new applications for approval if the method does not comply with existing regulations and if there is a potential damage hazard involved.

Formal procedures exist to modify existing regulations when and if a promising new method is developed which provides the basic protective requirements in conjunction with a better and/or more economical easy-opening device. This same opportunity to modify or change regulations also exists for any other packaging change.

Development work in easy-opening devices continues and undoubtedly the user will have the opportunity to make his selection from an even greater variety of methods.

CHAPTER **12**

Marking, Labeling, and Coding Systems, Materials, and Equipment

A package is not complete until it is properly identified. The primary function of identification is to describe the characteristics of the contents within the package. However, it must be recognized that a proper identification system provides opportunities for the capturing of information which can be used for merchandising, warehousing, inventory management, material flow, and quality assurance. Increased sophistication of data processing and the accompanying development of scanning and reading devices permits the product code to serve as input to improved information management and material flow systems.

After product identification has been accomplished, the shipping container must be properly marked and identified to provide instructions to facilitate all functions in the distribution cycle. This includes addressing, routing, cautionary legends and handling and storage guidelines and may even include directions on the end use of the product.

This chapter is primarily concerned with the various options available to the shipper to achieve an effective distribution package identification system. It excludes esthetic considerations such as colors or graphics; neither does it deal with systems and equipment designed exclusively for consumer packaging such as high-speed labeling machinery, bread wrap seals, and can labeling.

PRODUCT IDENTIFICATION

Product identification, achieved through differences in colors, styles, designs, sizes, and capacities, has gained in importance because of the complexity and diversification of modern business and specifically because of the increase in point-of-sale merchandising. Another consideration involves the increased use of automated handling and information capturing systems. Product identification can be achieved by techniques utilizing alpha-numeric information that is visually discernible by a machine-readable code or by a combination of both.

Until quite recently every manufacturer utilized individual product identification number systems which were compatible with his own internal information and control requirements. This created many complications preventing standardization in the capturing of product data. With the advent of point-of-sales cash register equipment and the ability to automatically capture a product code through scanning, the need for industry-wide standardization became mandatory. The first major industry to adopt a *Universal Product Code* (UPC) was the food industry. Other industries are working towards a similar standard system of machine-readable product identification. At this point, such product identification codes are imprinted on the primary or consumer package and identification of the secondary package or shipping container is in the process of refinement.

Identification on the shipping container assists the distributor and retailer in the warehousing and stock control operations. However, since identification of expensive products on the shipping container may invite pilferage, a suitable recognizable code is preferred. Product identification on an interior container may even include an illustration of the product as well as assembly and use instructions. This technique assists in self-service retailing.

When sufficient volume of one item is produced, product identification and other instructions are usually preprinted by the packaging material supplier. No further product identification is therefore required. When the volume is small, or if large diversification of products would require an excessive packaging material inventory, complete or supplementary product identification is provided before, during or after packing. Sometimes combination runs of similar packaging materials are sufficiently large to permit fully preprinted product identification through the use of a printing pattern suitable for slug changing by the manufacturer.

The methods and equipment discussed in this chapter pertain to those situations in which the user is required to provide the product identification. These include the following:

1. *Printing* of unprinted or partially preprinted papers, boxes, drums, and sacks.

MARKING, LABELING, AND CODING

2. Printing and affixing of *labels*.
3. Printing and affixing of *tapes*.
4. Printing and affixing of *tags*.
5. Other product identification techniques, including stenciling, manual writing, stamping, and color coding.

The product identification system must be fully integrated with information management systems, equipment used for packaging, handling and warehousing, stock control, and sales. In addition, the identifying materials used must be scheduled with sufficient lead time so that the flow of production is not interrupted. In some companies, notably pharmaceutical plants, this function is so complex that a separate department to handle this activity is required. If materials management procedures utilize electonic data processing, product identification materials should be incorporated into this system.

DISTRIBUTION IDENTIFICATION

The primary function of distribution identification is twofold: (1) to provide specific handling, warehousing, opening and other instructions pertaining to a given product and (2) to direct the product or shipment to a particular consignee. The former identification can therefore be pre-planned and thus pre-indicated. This identification includes positioning arrows, precautionary legends, palletizing, stacking patterns, and sortation codes. The individual set of instructions which directs the package to the point of destination includes such information as name, address, routing, and order number, or in a captive delivery system, destination code, loading position and stop number. Documents relating to a complete shipment customarily accompany a shipment. These may be packed within a container or affixed on the outer surface of the container using a packing list envelope. (Fig. 12-1).

Precautionary legends should be used with discretion and should not be used to excuse inadequate packaging. Certain legends, required by regulation to appear for specific and hazardous products, will be covered later in this chapter. Typical cautionary markings in various languages are listed in Table 12-1. The most economical method to provide precautionary legends, if the container already requires printing, is preprinting by the manufacturer. If unprinted boxes are used, standard preprinted labels with many of the common cautionary markings are available.

The major cause of misdirection or loss of merchandise in transit is the result of unsatisfactory markings identifying routing and the consignee. When a used container is marked, all previous identification should be obliterated by taping, inking, or with special covering agents designed for this purpose. Whatever method is used to address the shipment, it should always be legible and indelible.

The methods employed to apply distribution identification are similar to those itemized for product identification. For this reason distribution and product identification techniques will be covered jointly. Another similarity is the need to integrate distribution identification with the materials management and distribution information system, including the preparation of invoices, picking labels, routing, etc. When merchandise is made strictly to order for a customer, product and shipping identification should be prepared simultaneously for efficiency in clerical procedures.

In advanced mechanized handling systems a destination such as a store can be optically scanned from a computer-generated, machine-readable label. This label serves also to locate the product in a picking slot, identifies the product (including price, where desired), and records other pertinent information such as the sequence of the shipment in a truck.

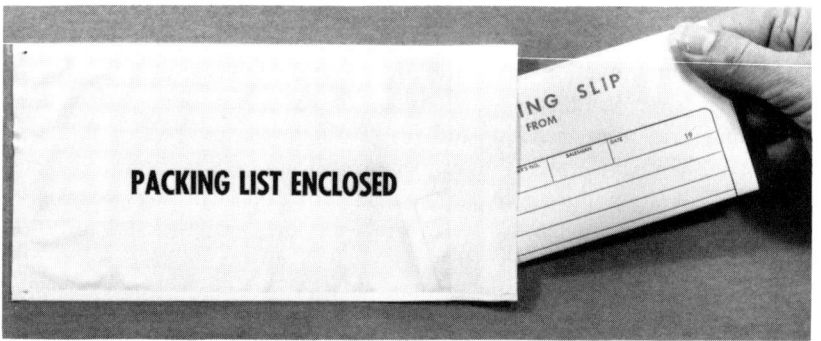

Fig. 12-1. Pressure sensitive packing list envelope. Courtesy Lawrence Packaging Supply Corp.

The recognition of this information serves to automatically sort and direct the shipping unit to a loading position. The label is further used for checking by both the shipper and consignee and for retail price marking. A typical label for this purpose is illustrated in Fig. 12-2.

Other types of computer-generated labels combining product identification, stock location and destination without automatic recognition are widely used for pick and pack operations where they are affixed to tote-boxes and shipping containers. These labels may also serve to identify and route a unit load to a preplanned warehouse location.

Where Universal Product Codes appear on the shipping container, it is possible to imprint the destination code automatically after the item has been identified by a reader and is matched against a specific order or shipment.

TABLE 12-1
Typical Cautionary Markings in Various Languages

English	French	German	Italian	Spanish	Portuguese	Swedish
Handle With Care	Attention	Vorsicht	Attenzione	Manejese Con Cuidado	Tratar Com Cuidado	Varsamt
Glass	Verre	Glas	Vetro	Vidrio	Vidro	Glas
Use No Hooks	Manier Sans Crampons	Ohne Haken handhaben	Manipolare senza graffi	No Se Usen Ganchos	Nao Empregue Ganchos	Begagna inga krokar
This Side Up	Cette Face En Haut	Diese Seite oben	Questo lato su	Este Lado Arriba	Este Lado Para Encima	Denna sida upp
Fragile	Fragile	Zerbrechlich	Fragile	Fragil	Fragil	Omtaligt
Keep in Cool Place	Garder En Lieu Frais	Kuehl aufbewahren	Conservare in luogo fresco	Mantengase En Lugar Fresco	Deve Ser Guardado Em Lugar Fresco	Forvaras kallt
Keep Dry	Proteger Contre Humidite	Vor Naesse schuetzen	Preservare dall umidita	Mantengase Seco	Nao Deve Ser Molhado	Forvaras torrt
Open Here	Ouvrir Ici	Hier oeffnen	Aprire da questa parte	Abrase Aqui	Abra Aqui	Oppnas har

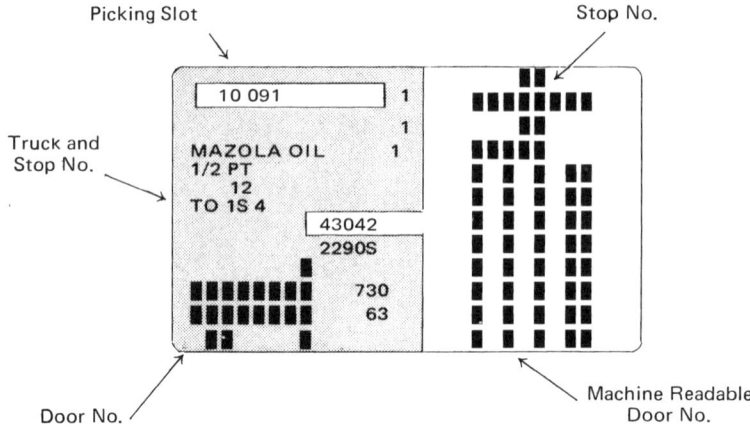

Fig. 12-2. Computer produced optically scannable identification label.

CODING

Additional identification in the form of a code is frequently assigned to the package. Coding is defined as the assignment of numerical, alphabetical, symbolic or scanner readable identifying marks to containers, packaging materials or articles to provide information concerning the characteristics of the container or the contents, or date, plant, or line in which it was manufactured, and information relating to its location or destination. The requirements for coding are dependent on the nature of the product, its manufacture, inspection procedures, distribution, and merchandising. Coding may be required in order to comply with carrier, customer, or governmental regulations. Coding with certain products facilitates immediate recall in the event that a defective lot has reached the market. Coding incorporating a date system permits proper rotation of stock and its replacement when the date has expired. Other code systems permit the positive control of various manufacturing steps and the automatic handling and sorting on conveyors and in storage systems.

Coding is accomplished by hand or mechanical embossing, printing, stamping, tagging, or labeling. The code may be applied directly to the container or formed package, to the label, the seal, overwrap, or to a tag.

This chapter is concerned with codes applied by the user during or after the process of packaging. Codes are also applied by the suppliers of packaging materials and containers in accordance with their own or customer requirements. For instance, codes on the bottoms of glass or plastic bottles identify molds, on fibre boxes a date code system is frequently employed, on labels the code may indicate the size of the run, and so forth.

MARKING, LABELING, AND CODING

Scannable codes. A *scannable code* allows automatic recognition of a symbol. Code recognition can be programmed to directly activate another function. Such a code is designated an *action code*. This may consist of simple verification or counting, or a more complex action. For example, it may involve a secondary piece of equipment such as a filling machine, a diverting mechanism, or automatically inspect a package. In the latter application, a scanner reads a codebar imprinted on a folding carton and activates an alarm or automatic reject mechanism when an improper carton is detected. (Fig. 12-3)

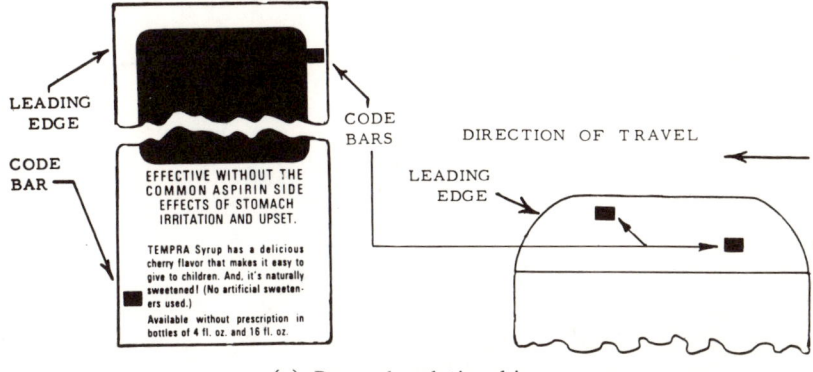

(*a*) Bar code relationship.

(*b*) Scanning unit installed on high speed cartoning equipment.

Fig. 12-3. Simple inspection system which verifies correct bar code on optical scanning unit. Courtesy Blen-Cal Electronics Industries, Inc.

474 **DISTRIBUTION PACKAGING**

If the code is intended to identify a specific product, it is termed an *information code*. This data is processed in a computer that initiates an automatic action, usually involving a piece of material handling equipment.

Scanning is based on the ability to sense variations in the reflectance of light from different surfaces. Reflection is of three types: (1) *diffused reflection* of light, (2) throwing a light beam on *retroreflective material* which gives a high level of response back to the area of the light source, and (3) *specular reflectance* in which the light strikes a glossy or mirror surface at an angle and bounces off a corresponding angle. The three types of reflectance are illustrated in Fig. 12-4.

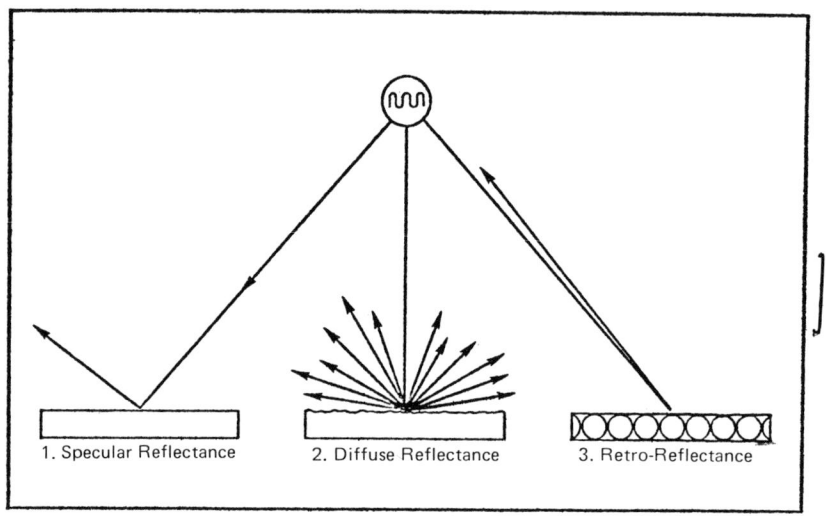

Fig. 12-4. Basic types of reflectances in scanning. Courtesy Scanmark Division, Markem Corp.

The simplest type of coding for distribution packages involves the use of photoreflective materials, such as inks and tapes. This type of code is recognized by a photo cell which in turn activates a simple function such as diverting, stopping or starting a conveyor, etc. Codes can incorporate complex information such as a product identification number. For example, the selective automatic diversion of containers or the automatic setting of stacking patterns on palletizing machines may be triggered by such a code.

CODE FORMATS. Code formats are basically collections of black bars, or bits, separated by "white" or unprinted spaces. The width of the bar determines whether or not it has a value. This can be clarified by comparing the number 107 in normal decimal form with computer binary values, which are 1 or 0 (see Fig. 12-5).

MARKING, LABELING, AND CODING 475

Decimal — Fields	100's	10's	1's				
Values	1	0	7				
Binary — Fields	64's	32's	16's	8's	4's	2's	1's
Values	1	1	0	1	0	1	1
Bar Code Format	▌	▌	▏	▌	▏	▏	▌

Thus, the wide bars or bits are the binary "1's" and the narrow are binary "0's". 107 is computed as:

Value	Times	Field	Extended Values
1	X	64	64
1	X	32	32
0	X	16	0
1	X	8	8
0	X	4	0
1	X	2	2
1	X	1	1
Total			107

Fig. 12-5. Example of how a binary code pattern develops the number 107. Courtesy Scanmark Division, Markem Corp.

A majority of the code formats in use today make provision for start bars at one end of the code to signal when the scanner should begin to read, and stop bars at the other end to stop the scanning. Such bars may also be direction-sensing codes to invert the numbers if read in reverse. A variation of this is provision in some codes for a "quiet area" of white or black in front of the first code bar, the function of which is to enable the scanner to stabilize for a period just before it moves into the coded area. While this quiet area may not be visible, it is an integral part of the format and must be provided for in laying out the mark.

Format dimensions may be affected by conveyor speeds. With a fixed rotating scanning speed, there will be as many "reads" of a 1 in. wide label traveling 100 feet per minute as of a 2 in. wide label at 200 feet per minute.

Also, it should be pointed out that, while usually the black bars are expected to indicate the binary 1's and 0's, some codes use the white spaces between fixed black bar widths for this purpose; further, recent codes are beginning to use adjacent black and white bars to increase the "packing density" of the format. In Fig. 12-6 various code formats are illustrated.

UNIVERSAL PRODUCT CODE (UPC). This code has been adopted by manufacturers who produce foods sold through supermarkets. An example of this code is shown in Fig. 12-7.

The symbol shows two sets of five digits each, separated by two tall center bars. The left series of five digits represents a number assigned to the food manufacturer by Distribution Codes, Inc., Washington, D.C., which is now charged with carrying out the program. The five digits to the right of the

476   **DISTRIBUTION PACKAGING**

Binary coded digit (BCD) in "ladder" format reads right to left.

The 2 of 5 is a variation of the BCD code.

Sunburst pattern normally permits printing of man-readable code in center.

Binary pattern uses wide and narrow lines of same length to code binary values.

Fig. 12-6. Various code formats for optical scanning.

MARKING, LABELING, AND CODING

Digital pattern uses variations of a horizontal bar to denote digits 0 to 9.

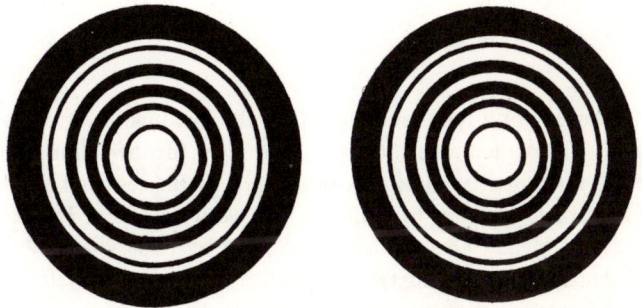

Bullseye, in which the ring widths correspond to BCD or 2 of 5 codes.

Double track pattern uses a straight binary code.

Fig. 12-6. (Cont.)

Fig. 12-7. Universal Product Code (UPC) format.

center bars represent a number assigned by the manufacturer to the individual product. These ten digits are only for item identification and do not include any reference to price. On both ends of the ten-digit symbol there are other characters added to facilitate and verify reading. To the extreme left is a white "dead" area, which is followed by two "guard bars" serving as a start signal. The following two bars indicate the number system character which, in the case of grocery products, is "0".

Following the right-hand set of five digits are two bars representing "the modulo check character" for verification. Following them are two bars representing the right-hand guard bars, then a required white "dead" area.

The pattern is infinitely variable as to size. The same digit will have a different bar configuration when appearing on the left-hand side of the center bars than it will on the right-hand side. Provision has been made in the symbol so that it may be read when placed on either the bottom or side of a package by a fixed scanner, or it may be read by a wand.

At the present time, this code is used on the primary package where the quality of printing is highly controllable and where point-of-sale scanning devices are gradually being introduced in food outlets. The printing technology to obtain desired quality levels on distribution packages is presently being explored. While every indication points to an ultimate success, a different type of symbology may have to be investigated. Some of the problems experienced include ink spreading, extent of coverage, distortion lines, etc.

MARKING, LABELING, AND CODING

Code selection factors. After the basic need to employ a coding system is established, several factors should be taken into consideration:

1. How the code is to be recognized.
2. Selection of the exact location for the code to appear either on the container, the seal, or on the identifying label or tag.
3. Determination at what station during the packing process coding will be most effectively applied.
4. Selection of the most economical and appropriate coding method and equipment for application and recognition.
5. Determination of whether or not coding can be integrated with another operation. An example would be tape dispensers with code attachments, conveyor-type code imprinters, etc.

The various pieces of equipment used for coding application by the user are similar in operation to those used for product and shipping identification. They will be covered in conjunction with these devices. Optical scanning equipment for distribution packaging applications only will be categorized later in this chapter.

SYSTEMS AND EQUIPMENT

The principal techniques for marking, addressing, identifying, and coding are (1) *printing,* (2) *labeling,* (3) *tagging,* (4) *stenciling,* and (5) *taping.* In addition, one of the most common, but least satisfactory techniques, consists of hand-writing with various writing implements. The following discussion is of the systems, materials, and equipment used for these principal techniques.

Printing. Printing is the primary method used for product identification. Printing is performed directly onto a label, tag, or tape, which is subsequently attached to the package. For economic reasons, the printing machines justified in the user's plant do not produce the same quality printing obtainable at the supplier's plant. Therefore, whenever possible, the basic information applicable to a complete company or product line should be preprinted by the supplier and the subsequent further identification operation by the user should consist of *imprinting* only.

Printing equipment available to the user consists of two basic types:

1. Equipment designed to provide the print job prior to the packing operation.
2. Equipment which prints after the package has been formed.

When the quality of the print job is of prime importance and in plants where proper scheduling and delivery of printed matter to the packaging line is feasible, printing prior to the packing operation is desirable. In Fig. 12-8,

Fig. 12-8. Magazine-fed rotary-type printing unit. Courtesy The Industrial Marking Equipment Co., Inc.

a magazine-fed rotary-type printing unit is illustrated, suitable for the printing of single thickness corrugated board, wrap around blanks and KDF boxes. For the imprinting of paper surfaces such as labels, any suitable duplicating process such as Addressograph, Multilith, or Multigraph can be used. In Fig. 12-9, a special tag and label printer is shown using either a rubber mat or stencil drum.

If the runs are small, the major importance in the duplicating process lies in the speed in changing plates or masters. The selection of the duplicating process is greatly influenced by the set-up and resultant down time of the equipment.

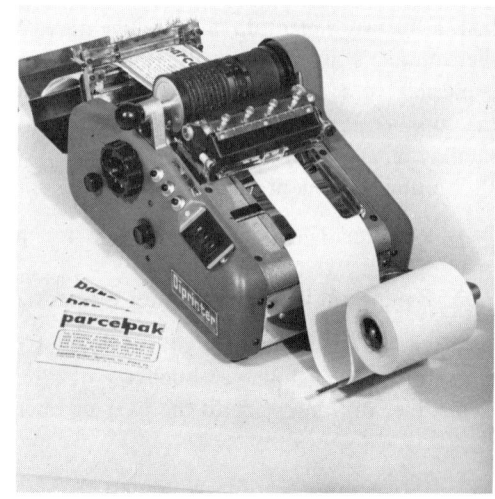

Fig. 12-9. Tag and label printer. Courtesy Diagraph-Bradley Industries, Inc.

MARKING, LABELING, AND CODING

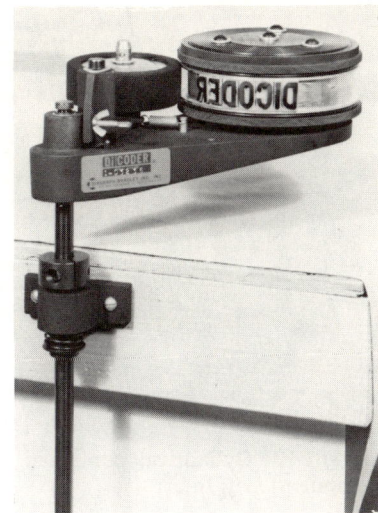

Fig. 12-10. Conveyor-line printing attachment. Courtesy Diagraph-Bradley Industries, Inc.

Printing onto the formed package is generally preferred, since no prior scheduling and handling of the packaging materials is necessary and separate inventories can be eliminated. The printing device may consist of a very simple conveyor line attachment as shown in Fig. 12-10 or may consist of an elaborate device which includes transfers for the printing of several box panels. Positioning of the conveyor line printers is determined by other printed matter already on the package or special identification requirements. Flexibility of positioning of conveyor line printing attachments is further illustrated in Fig. 12-11.

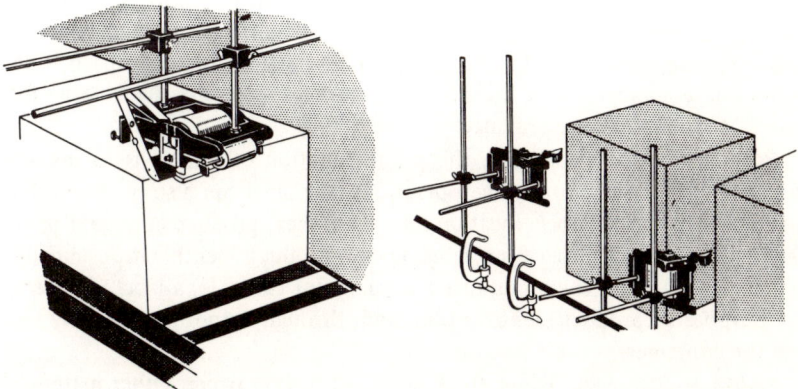

Fig. 12-11. Different positions of conveyor line printing attachments.

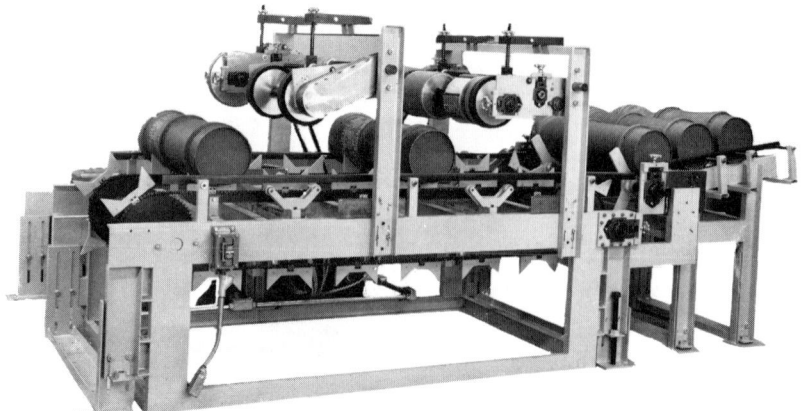

Fig. 12-12. Cylindrical container printer. Courtesy The Industrial Marking Equipment Co., Inc.

Imprinting devices provide identification as to codes, dates, types, sizes, colors, lot numbers, and content description. Equipment is available for printing on almost any surface, including cylindrical containers (Fig. 12-12), and for special purposes.

Labeling. A label is defined as a slip of paper or other material to be affixed to a container or article on which product or shipping information is printed. The direct printing of product information on a container or article rather than on a separate slip which is subsequently affixed, as on drums, bags, sacks, bottles, and glassware, is sometimes also referred to as a label; however, a label is rarely considered in this concept. For the purpose of this text, only separate slips, containing either product or shipping identification, are covered.

Labeling of consumer packages and products such as bottles, cans, and folding cartons is frequently an integral part of the filling, packing, closing, and weighing operations. Labels are affixed by specialized high-speed, fully automatic equipment, and the principal considerations are machine performance, economy, and appearance.

For distribution packaging, the prime function of a label is to provide shipping instructions. In captive operations, the label can also provide other instructions such as stock location, store number, product code and price. Overlapping of consumer-type labeling for product identification and distribution packaging functions often occurs. For example, a label may serve the purpose of product identification, and, through its positioning, may also seal the container.

LABEL MATERIALS. Label stock is almost always paper. Other materials, such as films and foil, are also used or used in conjunction with a paper

MARKING, LABELING, AND CODING

label. The label manufacturer selects the label stock for both manufacturing and end-use requirements. Papers used for labels may have a soft or hard finish, be coated or uncoated, and be selected for the type of adhesive subsequently used. Such physical characteristics as moisture content, absorbency, tensile strength, resistance to curl, ability to receive printing ink, and resistance to scuffing are prime considerations in label stock selection. After printing, most labels are coated with a varnish or lacquer.

The backing of the label depends upon a number of factors. These include (1) the surface to which the label is affixed, (2) the type of label construction and its positioning, (3) the type of equipment selected, and (4) special use requirements such as resistance to moisture and removal.

The following types of label backing are used.

1. Plain backing.
2. Gummed backing.
3. Pressure-sensitive backing.
4. Heat-seal coated backing.

LABEL APPLICATION EQUIPMENT. The equipment selected for the application of labels is influenced by (1) the type of label backing, (2) the type and size of container to which the label is to be affixed, and (3) production and/or shipping requirements.

The methods used for the various label backings previously recorded are as follows:

1. *Plain back labels* are used with high-speed labeling machines when economy is of utmost importance. Spot or lap gluing is accomplished on this equipment with minimum glue consumption. Frequently, portions of shipping documents which run through electronic data-processing equipment later become ungummed shipping labels. These labels are affixed either with a glue applicator or with a clear pressure-sensitive filmic tape (Fig. 12-13).

2. *Gummed labels* are more expensive than plain labels but require only simple moistening for application. The application of gummed shipping labels is usually a manual operation. Gummed labels for product identification can be applied with semi- or fully-automatic equipment.

3. *Pressure-sensitive labels* will adhere to nearly any type of smooth surface and do not require moistening or gluing for application. The labels are adhered to a low-release backing paper and removal of the label from the backing sheet can be accomplished either manually or by machine. In Fig. 12-14, a price marking machine which is actually a label dispenser is shown in a modern distribution center system. Individual units are price marked with a computer-generated label during the selection of merchandise from gravity flow racks. Machines are also available to apply pressure-sensitive labels or tape automatically to a container or product. These range from

484 DISTRIBUTION PACKAGING

Fig. 12-13. Affixing of shipping label with filmic pressure-sensitive tape. Courtesy Nopi, Inc.

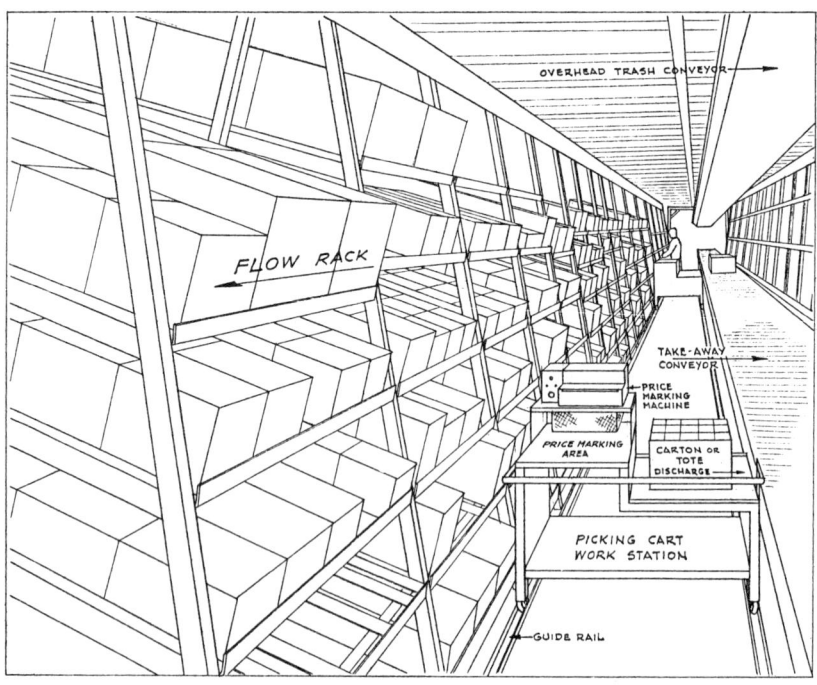

Fig. 12-14. Price marking label dispensing system in mechanized distribution center. Courtesy Walter Frederick Friedman and Co., Inc.

MARKING, LABELING, AND CODING

simple hand *labelling guns,* with or without price imprinting attachments, to high speed production line equipment essentially designed for consumer packages. Pressure-sensitive shipping labels, adhered to the backing, are suitable for addressing in typewriters and other business machines and can be integrated with clerical procedures (Fig. 12-15).

4. *Heat-seal coated labels* require a heating element to activate the thermoplastic adhesive. Pressure of the label on the desired surface can be accomplished manually or by mechanical means.

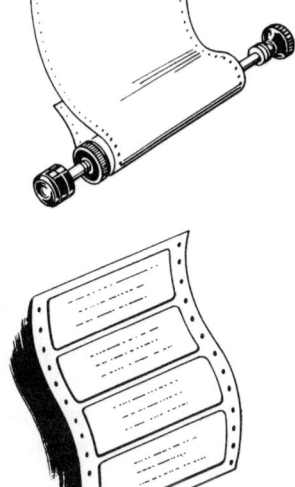

Fig. 12-15. Marginally punched pin feed labels on pressure-sensitive backing. Courtesy Kimbell Systems Div., Litton Industries, Inc.

Tagging. Tags are used primarily when no smooth surface is available for proper adhesion of a label or tape, or if direct imprinting, stenciling or writing is not feasible. Tags are used for product and shipping identification. Frequently, a tag may be the only identifying and addressing medium on a product which does not require packaging, such as, coils of cable, skids of ingots, and unit loads of lumber. Since tags are attention-catching, they are attached to many products which are subsequently packaged and further identified. Such tags may contain code numbers, serial numbers, warranty information, and prices. Since tags can be removed without damage or noticeable evidence of severance, they are selected for fabrics, meats, and furniture.

For shipping identification, tags are particularly suitable for wooden boxes and crates, since most wood surfaces are not preferred for printing or stencilling, particularly for printing of small legends, and wood surfaces present adhesion difficulties.

TAG TYPES. Tags can be made of metal, cloth, or paperboard, with the latter being most popularly used in packaging. The paperboard used for this purpose is referred to as *tag-board*. True tag-board is defined as a strong manila-type board with a high finish, suitably sized for writing or printing.

For heavy-duty purposes tags are reinforced at the eyelet with paper patch or metal reinforcing. Eyeletted tags are attached to the product or package with twine, cord, string, or wire. Cloth tags are usually sewn to the product or package, either in combination with a closure or separately. Tags are affixed to wooden containers with large-headed tacks, tag fasteners, or staples.

Tags are sometimes used in conjunction with tamper-proof seals which form the closure of the tying medium. Imprinting or writing on tags is performed prior to tagging. Any conventional printing or writing medium suitable for the tag surface may be used.

Tag application is manual, however, eyeletted tags are usually supplied with the tying material pre-threaded.

Stenciling. A *stencil* is a suitably perforated sheet, board, or plate through which ink or other coloring flows to make a printed impression. Stenciling may be directly applied to the package or to a tag or label which is subsequently affixed to the package.

Stencils are easily prepared by cutting through a reusable stencil board with a special machine, or through a single use master. No typesetting, dies, or plates are necessary in stenciling. Stencils can be cut to produce large, legible letters which are specially advantageous for shipping identification. For extra large letters, a thermal copying system is used which reproduces an expendable stencil from a master.

Integrated addressing systems, in conjunction with the preparation of other documents, have furthered new developments in stenciling. The equipment and methods of these and other methods is described below.

STENCILING METHODS. The conventional, reusable stencil is produced on a stencil-cutting machine which is illustrated in Fig. 12-16. Motorized machines of this type are now available. Stencil-cutting machines incorporate punches and dies which cut through the stencil board. The board is indexed after each impression thus permitting proper spacing between characters.

Stencil board, impregnated with linseed oil, does not absorb stencil ink. For single use stencils, a mimeograph-type stencil paper protected with pliofilm is used. Reusable stencils are inked by either spray, brush, or roller (Fig. 12-17). The ink flows into the cut-outs and forms the letters on the undersurface. Basically, this is an entirely manual method. Single use stencils are usually inked with an applicator which creates an image upon impact. These stencils can be made on special stencil cutting machines or can be processed by typing or handwriting (Fig. 12-18). The preparation of this stencil can be performed by any of the following three methods.

MARKING, LABELING, AND CODING

Fig. 12-16. Stencil-cutting machine. Courtesy Marsh Stencil Machine Co.

1. As a separate typing or handwritten operation as the need arises. This operation is usually performed in the shipping department after the shipment has been prepared.

2. As a by-product in preparing an invoice or bill of lading. In this operation the stencil may be routed separately or in combination with other papers to the shipping station.

3. On a continuous business form where the stencil is appropriately affixed (Fig. 12-19).

Stencils of these types lend themselves to preparation on manual or electric typewriters, automatic accounting machines, data processing printers, and other modern office machines. Procedures of this type eliminate a separate stencil preparation step.

Mimeograph-type stencils are applied by two basic methods:

1. Directly to the *container surface* with a special applicator containing stencil ink (Fig. 12-20). Either the stencil contains the basic precut information, or a positioning imprint is provided on the container surface during the container manufacturing process as shown in Fig. 12-20.

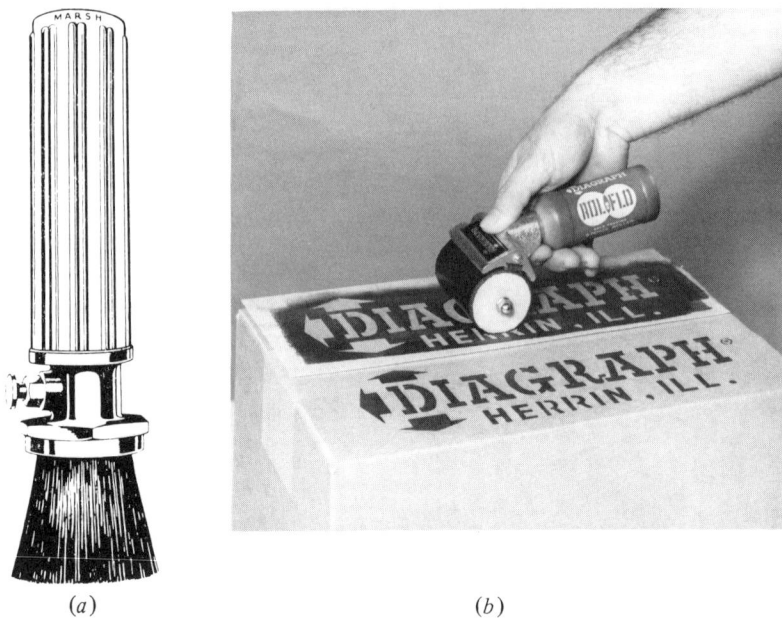

(*a*)   (*b*)

(*a*) Brush applicator. Courtesy Marsh Stencil Machine Co.
(*b*) Roller applicator. Courtesy Diagraph-Bradley Industries, Inc.

Fig. 12-17. Brush and roller, stencil ink applicators.

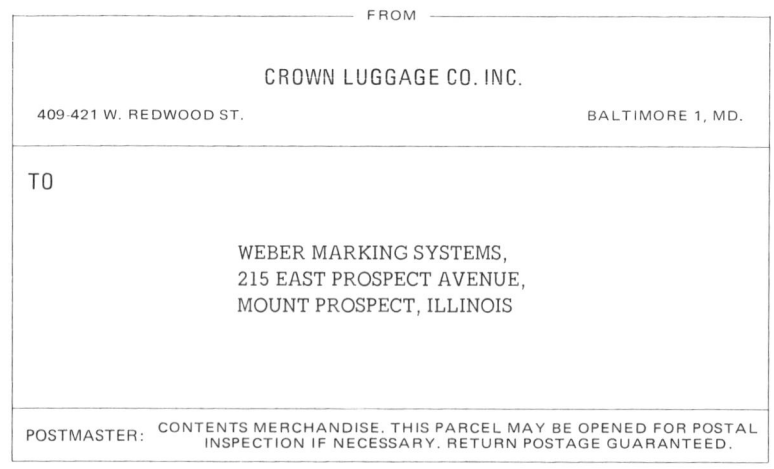

Fig. 12-18. Typewritten stencil with basic information precut. Courtesy Weber Marking Systems, Inc.

MARKING, LABELING, AND CODING

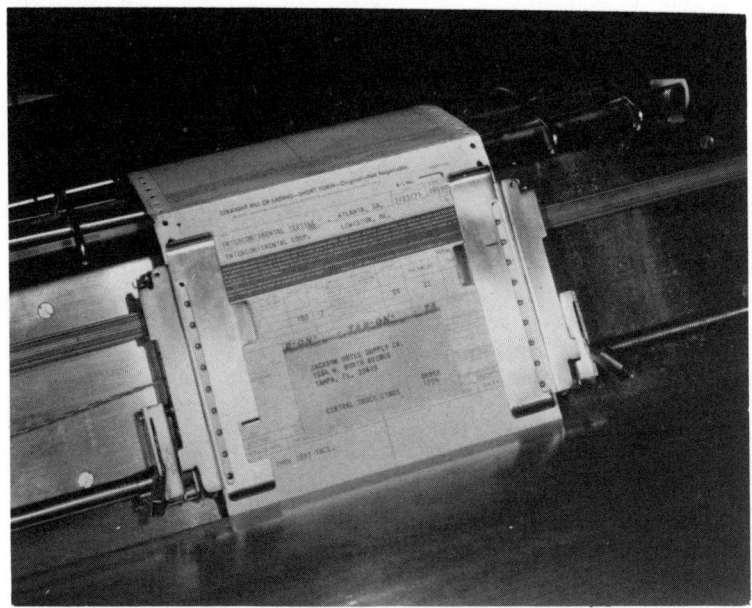

Fig. 12-19. Data processing printer cuts stencil as a by-product of bill of lading preparation. Courtesy Weber Marking Systems, Inc.

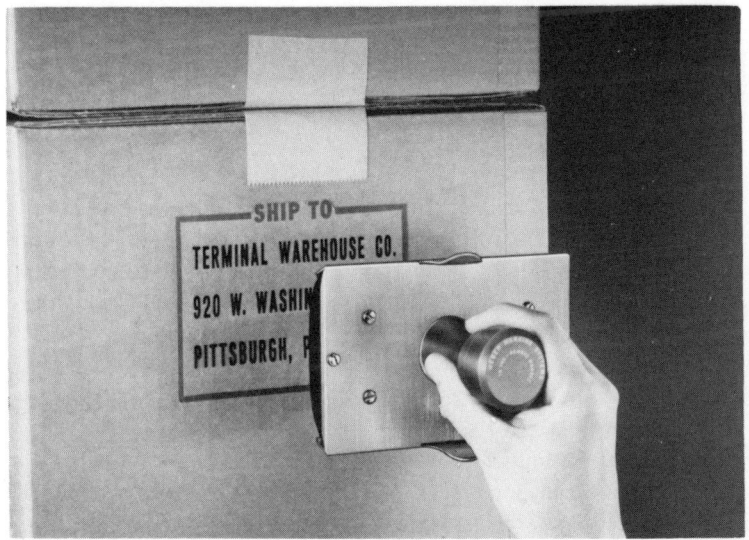

Fig. 12-20. Hand stencil applicator. Courtesy Weber Marking Systems, Inc.

2. Transfer to a label or tag which is subsequently affixed to the product or container. This transfer is performed either by a hand applicator or by a machine (Fig. 12-21).

Hand applicators are used when a relatively small quantity of containers are identified with the same information. This is particularly true for addressing. When a large number of identical reproductions is required, machine duplication onto labels or tags provides greater efficiency. Separate labels usually provide greater clarity of character reproduction than direct transfer to the container. Labels also enhance product or company recognition of the container and are therefore preferred when unprinted containers are employed.

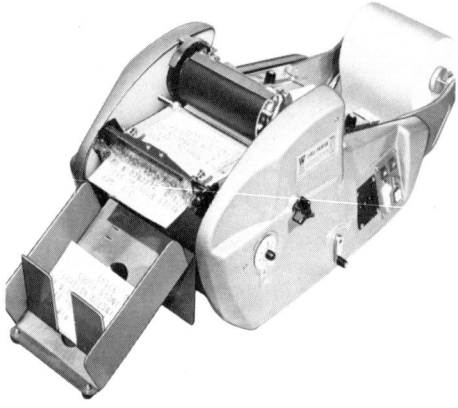

Fig. 12-21. Stencil imprinting on roll label stock. Courtesy Weber Marking Systems, Inc.

Taping. The use of printed tape as a means of identification is primarily practiced for product identification or coding when repetitive information is needed. Tape may serve the function of a label if the information or code is indexed by the dispenser or may be used as a continuous strip with repetitive information. The latter usually serves a combined function of sealing the container, as well as identifying the product, manufacturer, or trade mark. Imprinted tapes for closure also serve as a *pilferage deterrent.*

The main advantages of tape for the purposes described are: (1) low cost of printing in a continuous form and (2) inexpensive application from a roll. From an esthetic viewpoint, it is difficult for printed tapes to approach the quality and freedom of layout attained with individual labels.

TYPES OF TAPES. Tapes used for identification purposes are identical to those used for container closure. These were covered in Chapter 9. To review,

MARKING, LABELING, AND CODING

there are two basic tape types: (1) gummed tapes, and (2) pressure-sensitive tapes. Both types are available in imprinted form from the tape supplier with some minor exceptions. Gummed tapes can be imprinted by the user with simple dispensing machine attachments prior to the tape application. Pressure-sensitive tapes require special printing processing not readily accomplished on the user's premises.

Selection of the particular type of tape depends upon several factors:

1. The strength requirements if identification is combined with closing.
2. The backing surface, that is, gummed or pressure-sensitive, in consideration of the surface to be adhered or economy.
3. Esthetic considerations such as color and texture, to obtain the desired appearance.

TAPE APPLICATION EQUIPMENT. Factory preprinted tape is dispensed with any of the available tape dispensers discussed in Chapter 9. These machines do not provide for precise registering, and the information printed on the tape should be spaced in such a manner that the smallest length of tape dispensed will contain at least one complete message.

Unprinted gummed tape is imprinted during the dispensing operation, with an attachment on a standard dispenser or with a special device. The attachment basically consists of a print roller to which type is affixed. Either the type or the roller can be quickly changed to alter the message. The ink is transferred to the tape on a transfer roller during the dispensing process (Fig. 12-22).

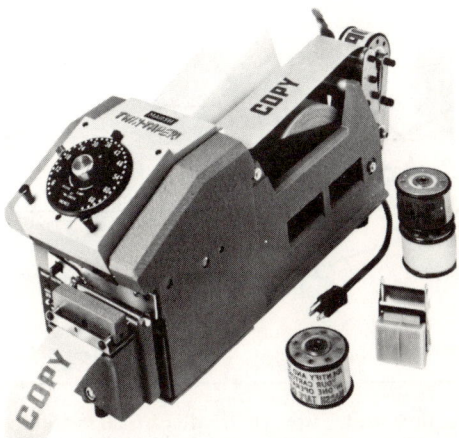

Fig. 12-22. Electric dispenser with attachment to imprint reinforced or regular gummed tape. Courtesy Marsh Stencil Machine Company.

The main advantage of imprinting attachments for tape is in their ability to make quick product information or coding changes and thus obviate the need to stock separate imprinted tapes or labels.

Scanning. Scanning devices are an integral part of automatic identification systems. This section relates only to the technology having application to the field of distribution packaging.

There are three basic types of scanners or readers: (1) fixed-beam, (2) moving-beam, and (3) hand-held.

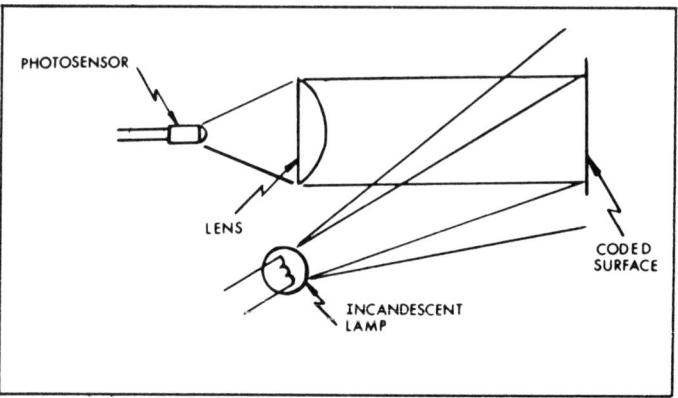

(a) The reflective reader uses a light source mounted above, below or to the side of its light sensor.

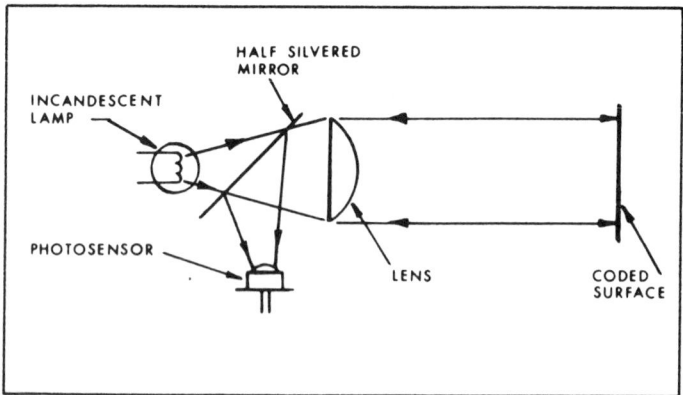

(b) The reflex reader uses the same optical system to illuminate and detect bar code patterns.

Fig. 12-23. Basic types of fixed-beam readers. Courtesy © The Material Handling Institute, Inc.

MARKING, LABELING, AND CODING

FIXED-BEAM READERS. The simplest type, the fixed-beam reader utilizes individual scanner heads with stationary light sources and stationary light sensors. The light source is normally an incandescent lamp or light emitting diode (LED). The sensor can be a silicon solar cell or a photo transistor.

The name "Fixed-Beam" is used because the light beam and its optics as well as the optics associated with the light sensor do not move. Accordingly, the code moves into the reader's field of view.

There are two basic types of fixed-beam readers: (1) *reflective* and (2) *reflex,* also known as *coaxial* readers. Their methods of reading are illustrated in Fig. 12-23.

MOVING-BEAM READERS. In recent years an entirely new approach to optical reading has been developed that increases the practical range of industrial applications. Utilizing a moving-beam to scan, this technology is based on the developments first used for ACI (Automatic Car Identification) in which multiple-digit, permanent retro-reflective railcar labels are read at speeds up to 80 miles per hour. Industrial scanners today will locate, read and decode miniature code patterns with high data content at speeds in excess of 500 feet per minute.

High-speed scanners normally utilize a low-power helium/neon laser beam, mercury vapor lamp or an incandescent light source. The light is usually reflected off a rotating mirror-faceted head which projects the beam through the scanning field.

Moving-beam readers operate by detecting variations in contrast between code marks and label background color or carton surface. These variations in reflected light are analyzed and counted by decoding electronics which, in effect, compare the response to a standard. If the code read is validated, including a correct count of code marks, a command signal will be transmitted to trigger a diverter, actuate a counter or transmit the product's identity to a computer for processing. In some applications, the digitized signals are transmitted to a computer or processor for decoding.

Scanners will read a one-inch label on a product as many as 18 times at a conveyor speed of 100 feet per minute. If the label width is doubled, the number of scans per label at a given speed will also be doubled. In other words, a moving-beam reader along with the code pattern and, as appropriate, label design can be selected to meet the technical and economic requirements of any data acquisition system. Typical scanning rates vary from 180 to 360 sweeps per second. Other methods of generating the moving beam may also be used to achieve approximately the same scanning rates.

Normally, the horizontal speed of the product is such that the moving beam can completely sweep across the label many times before the product moves out of reading range. Each scan is compared to a standard or checked against previous scans so that the final product identification is based upon several readings of each code pattern. Most manufacturers require at least

494　　　　　　　　　　　　　　　　　　　　　　**DISTRIBUTION PACKAGING**

three readings, two of which must be redundant before developing a valid code signal.

Most moving-beam readers can accommodate a wide variation in code alignment and orientation. For example, skew can range up to 45°, tilt to as much as ±25° from the vertical, pitch ±20° toward or away from the reader and distance to thirty inches (Fig. 12-24).

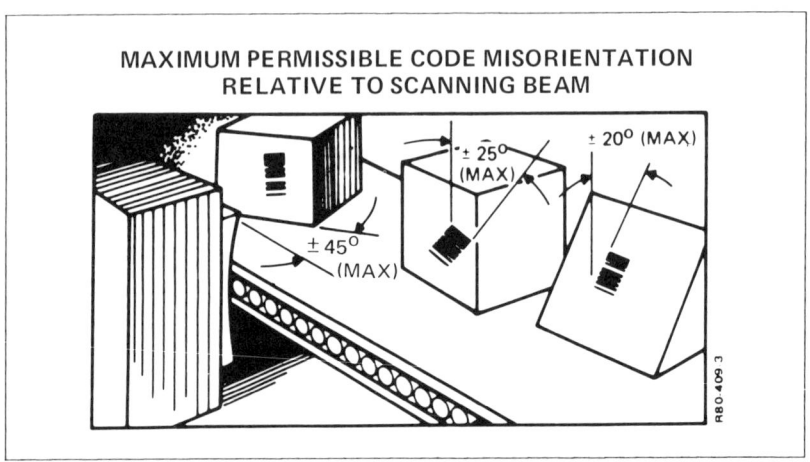

Fig. 12-24. Wide tolerances for code pattern tilt, skew and pitch can be accommodated by the moving-beam readers. Courtesy © The Material Handling Institute.

Tilt is defined as the amount of angular displacement of the code pattern from the vertical axis of the scanning curtain.

Skew is defined as product misalignment with respect to the perpendicular to the scanning curtain. Since skewing shortens apparent width of code marks, wider labels obviously permit greater allowable tilt or skew at a given conveying speed. In all applications, the label should be wide enough to allow at least 3 to 6 readings of the code pattern so that the advantages of redundancy are utilized.

Pitch can be defined as the degree of code displacement from the vertical scanning curtain. Accommodation of pitch must be considered when the application calls for identification of products whose shape or method of transport; e.g., power and free conveyors, increase its likelihood. Since the degree of pitch also varies the apparent width and the space between bars, greater pitch can be accommodated by increasing the size of the bars and spaces.

In Fig. 12-25 two representative systems which utilize optical scanning are illustrated.

MARKING, LABELING, AND CODING   495

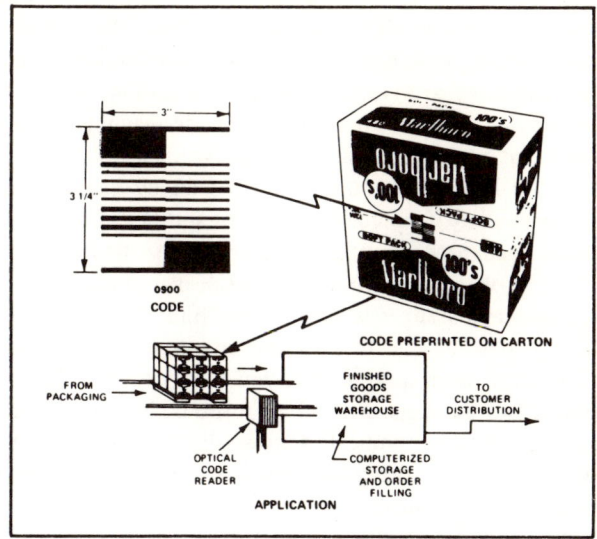

(a) Vertically mounted reader identifying preprinted bar code on corrugated cartons.

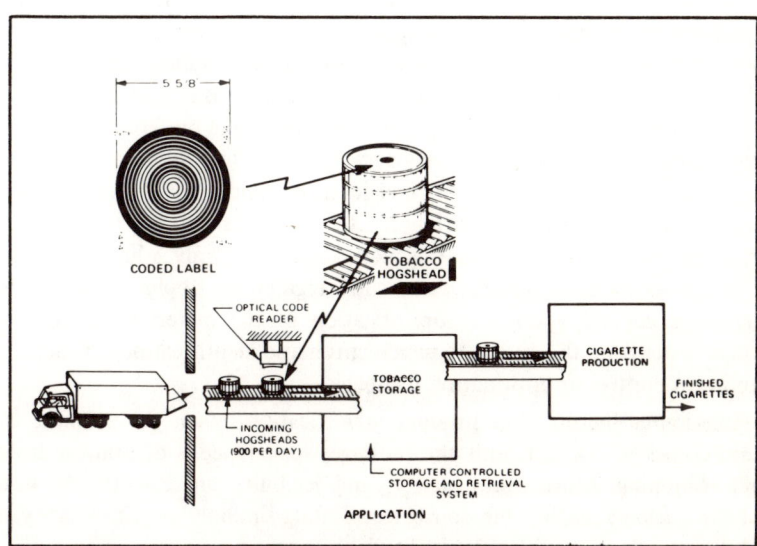

(b) Overhead mounted reader identifying and counting hogsheads of tobacco.

Fig. 12-25: Two optical scanning systems using moving-beam readers.

HAND-HELD READERS. These take the form of a pen or a wand, or the shape of a two-cell flashlight. These scanners are the newest development in the field and have adaptability for recording scanned information direct, or into portable cassettes that may accumulate information, subsequently to be fed into a computer. Typical applications include scanning, indoors or out-of-doors, items that are not adaptable to conveyorizing—such as small articles which have to be handled manually rather than by case or pallet; loose merchandise, particularly if stored one on top of another; and stationary items.

There is no fixed path here for the merchandise to follow; rather, the scanner is either stroked across individual bars or placed over the entire mark. Usually, the hand-held scanner is used for the finer code patterns and small formats. Also, it may be able to read and interpret a code bi-directionally. Some have a built-in time limit factor, which may tend to limit the length of the code or affect the speed with which it should be stroked. They usually provide an audible signal to indicate that a valid scan has taken place. With the stroking type, there is the possibility that repeated stroking may wear off the mark, leading to nonreads.

SELECTION FACTORS

The selection of a particular system for product, shipping, or coding identification depends upon (1) appearance requirements, (2) economics, (3) regulations, and (4) the interface with material handling and information processing equipment. The evaluation of *appearance* is frequently subjective and the selection is influenced, therefore, by individual preference. *Economic factors* are influenced by the over-all system of handling the physical product, the paperwork, volume and size considerations, and many other extraneous circumstances. *Regulations* pertaining to identification are prescribed by certain governmental and carrier agencies. Therefore only a limited freedom of choice exists in instances when such regulations apply. *Interface* considerations depend upon the sophistication of the materials handling and storage system and the extent to which automatic identification is utilized for equipment control or information feedback.

Appearance factors. For product identification, when the ultimate consumer comes in contact with the message, appearance is of primary importance. Shipping information, clarity, and legibility are essential to insure that the package reaches the consignee. Coding similarly requires clarity and legibility, particulary with machine readable systems.

Printing by the supplier permits the greatest latitude in colors, type variety, materials, and layout and is therefore recommended if the volume justifies separate imprints. The selection of the medium on which the message is to appear depends upon the container material and upon volume considera-

MARKING, LABELING, AND CODING

tions. For example, label stock provides an excellent surface for high quality printing, whereas with fibreboard, similar quality is not readily attainable. For product identification, when appearance is of prime importance, a fully supplier-printed label may therefore be selected to be affixed to a fibreboard container. If the volume is small, the basic information on a label would be imprinted by the supplier and the variable information would be subsequently imprinted by the user. Since only the variable information is furnished by the user, the over-all appearance of the label is not necessarily impaired.

Poor placement or adhesion of a label seriously detracts from appearance despite the quality of the label itself. The over-all appearance of the package is influenced by a number of individual elements such as the container, closure, and identification, which must provide an integral unit. The selection of the identification system for appearance reasons cannot be isolated to provide the desired effect.

Economic factors. Many of the economic factors influencing the selection of a given system have already been presented in the discussion of each basic identification method. The economic factors can be summarized as follows:

1. Preparation cost such as set-up time of message and equipment.
2. Material cost. Examples are labels, tags, tape, and ink.
3. System cost, including labor and equipment of application, and the interface costs.
4. Inventory, purchase and handling expense.

PREPARATION COST. This cost includes all work elements incurred in preparing the message before application. This cost may be part of the initial material cost when the identification is entirely accomplished by the supplier of the container, label, tag, etc. If the message is directly written by hand onto the package, no preparation cost is involved. However, the application cost will be comparatively high. When stenciling is performed as part of other paper work procedures, the stencil preparation cost is minimized, and only insertion of the stencil in the applicator or machine is required. As a general rule, the larger the run of identical messages, the lower the preparation cost per unit and the greater the permissible set-up time. Therefore, the preparation of a separate imprinting machine which is used for a run of several thousand boxes and takes several hours, is fully justified.

MATERIAL COST. This includes such items as labels, tags, stencils, ink, and rubber dies. In most instances the materials required for an identification system are secondary to the preparation and application labor costs. Reusable materials such as stencils lower the material as well as the preparation cost of an identification system. However, the expense of filing and relocating stencils may diminish or exceed these savings and, therefore, single use stencils may be preferred.

If the identification system requires rubber dies, an analysis of individual

Fig. 12-26. Authorized warning labels for hazardous materials

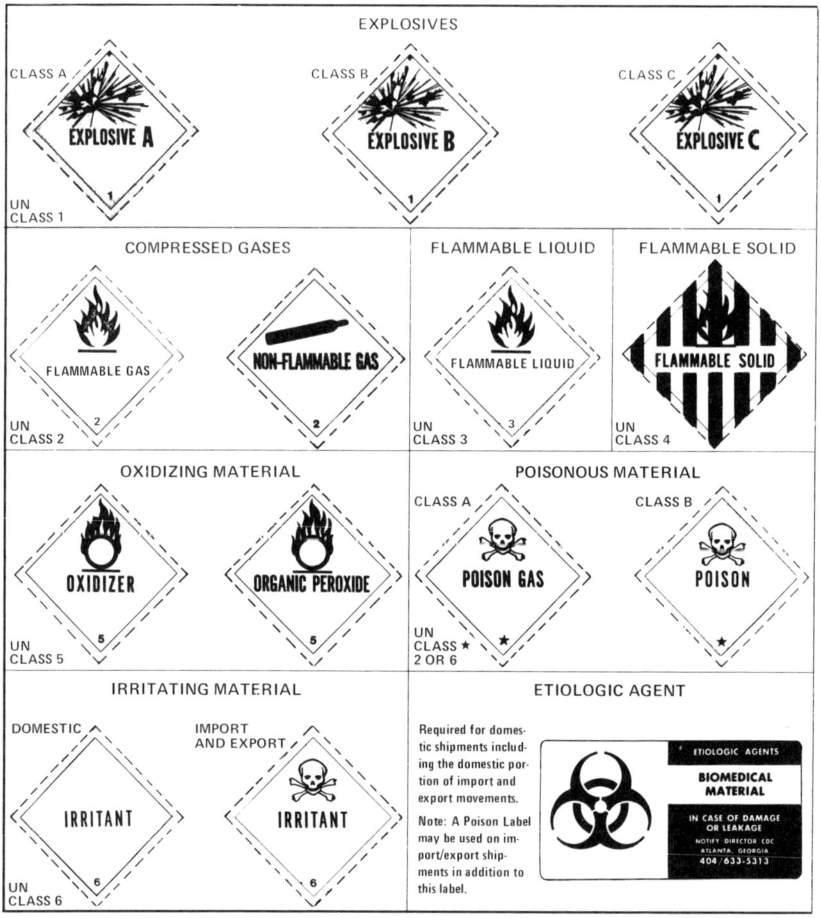

dies containing the complete message versus type setting should be made in order to compare initial purchase price with the set-up cost with these two systems.

SYSTEM COST. This includes the labor to apply the identification as well as fixed and operating expenses of the equipment. It also includes interface costs of operations and equipment required to interpret and utilize machine readable information. If possible, fully automatic devices should be investigated. These may be quickly amortized if labor savings are feasible.

Product identification usually permits a greater freedom in selecting mechanical application devices, since greater uniformity of containers and much longer runs are encountered. Shipping identification is an individualized

MARKING, LABELING, AND CODING

(color not indicated). Courtesy Lawrence Packaging Supply Corp.

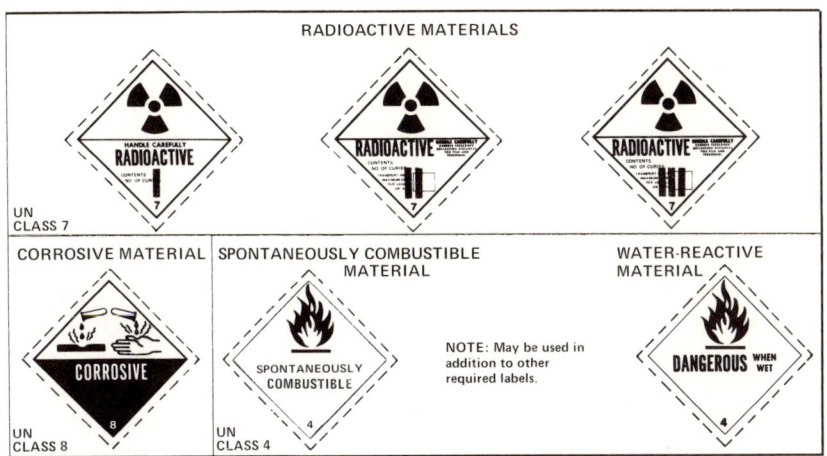

operation lending itself better to manual application techniques. Even within a given system, reduction of application cost is frequently feasible by proper work lay-out, mechanical aids, and proper scheduling.

INVENTORY COST. This expense consists of the working capital tied up in stocks of material especially preprinted for a given product and in the space allocated for this material. When each of these products requires a special size container, no savings in inventory cost are possible through a user-imprinting or other identifying system. On the other hand, if several of the same size containers can be consolidated in purchase orders and inventory through subsequent identification, the working capital in such containers may be reduced substantially. In addition, higher quantity purchases of containers will reduce the per unit container cost.

Similarly the cost of handling, ordering, and scheduling of preprinted stock can become a major expenditure if only small runs of a given product are packaged. In analyzing an identification system, a comparison of the preparation, material, and application costs versus inventory, purchasing, handling, and paper work and control expenses should be made for a realistic appraisal of the most economical system.

Regulations. Carrier and governmental agencies prescribe regulations for certain products and shipments to provide either safeguards in handling the product or uniformity in marking. In addition, industry itself, through trade associations, has established important guidelines and standards.

The following are some of the important regulations governing marking and labeling of articles.

1. Uniform Freight Classification. Rule 6 establishes basic principles of marking for rail shipment with accepted procedures fully specified. This rule also specifies marking for fragile items. In other rules, sections, and package descriptions, additional marking and labeling requirements are detailed.

2. United States Postal Regulations cover marking, labeling, and addressing of all mailable merchandise. General as well as specific product regulations apply.

3. For hazardous materials, The Department of Transportation has established different Codes of Federal Regulations (CFRS) which describe the rules which shippers must follow for the marking, labeling, packaging, handling and transporting of such products. They are: (1) CFR-49 Surface Transportation, (2) CFR-46 Water Transportation, and (3) CFR-14 Air Transportation. For rail and truck shipments, R. M. Graziano's Tariff also details specifications including marking and labeling of hazardous materials. Shipments via United Parcel Service must comply with United Parcel Service Guide for handling hazardous materials. For export of hazardous materials via water, two Codes of Federal Regulations specify that materials be shipped in accordance with the regulations of the country of destination. These regulations are known as the Maritime Consultive Organization Regulations (IMCO).

Export shipments via air are covered by the regulations of the International Air Transit Association (IATA).

Typical Department of Transportation authorized warning labels for hazardous materials are shown in Fig. 12-26. These are based on the United Nations Label System and are authorized for domestic and export shipments.

4. Federal and Military Specifications contain provisions for marking requirements which must be complied with on government contracts.

Other carrier and governmental regulations contain reference to marking, labeling, addressing, or otherwise identifying. The appropriate sections should be consulted when a shipment according to such regulations must be prepared.

CHAPTER 13

Packaging Equipment

The subject of packaging is not complete without a discussion of the equipment used to accomplish the packaging function at lowest possible cost. Equipment for this purpose may be defined as any device or contrivance which assists in the accomplishment of a task. In accordance with this definition, equipment may consist of a simple aid such as a jig, or a complex, high-speed fully automatic equipment system. The objective of all of this equipment is to increase productivity or to permit the performance of a task.

In the broadest sense, two classes of packaging equipment are employed. First, equipment to *fabricate* packaging materials and containers, and second, equipment employed by the user to *utilize* packaging materials and containers. Under certain conditions, the user may find it economical to incorporate in his operations, complete or partial, packaging material or container fabrication equipment. In these instances, the user performs the function of the packaging supplier. This text does not cover packaging supplier equipment and systems except in those circumstances in which the user elects to carry out supplier functions for economic reasons. Thus, equipment for the fabrication of corrugated boxes, set-up boxes, and crates has been surveyed in other chapters.

Packaging equipment employed by the user is often designed for a specific material or container, and as such, cannot be isolated in the discussion of a particular packaging practice. Therefore, many of the aids utilized in such operations as closure forming, reinforcing, bundling, labeling, and easy-opening have been included in previous chapters in conjunction with a particular material or container type.

A complete equipment survey would provide voluminous data because of the many hundreds of standard and specialized devices employed in the packaging field. It is simply the purpose of this chapter to broadly classify all user packaging equipment with emphasis on distribution packaging applications previously not included in this text and to assist in its selection.

CLASSIFICATION OF USER PACKAGING EQUIPMENT

For simplicity, packaging equipment is best classified by function. The following items represent the major functions that can be performed by packaging equipment:

1. Forming and assembly.
2. Filling, loading, and overwrapping.
3. Weighing and counting.
4. Closing and sealing.
5. Bundling, unitizing and reinforcing.
6. Identifying.
7. Miscellaneous.

Although a functional classification serves to distinguish between the available packaging equipment types, it does not delineate equipment on the basis of distribution or consumer packaging. Further subdivision is required because of the inherent differences in such equipment, despite the similarity of function. For example, filling equipment designed for vials is very different from filling equipment for 55-gal drums. Therefore, since this text is primarily concerned with distribution packaging, the equipment discussion will cover only those types used for distribution packaging materials and containers. It should be recognized, however, that some equipment is adaptable for both consumer and distribution packaging functions and are, as such, included.

In Table 13-1 a listing of the major consumer and distribution packaging machine types is presented. Devices and equipment which were not included in the individual chapters in conjunction with specific distribution packaging material and container types will now be discussed. This equipment is important in the attainment of economy and efficiency in distribution packaging operations.

PACKAGING EQUIPMENT

TABLE 13-1
Major Packaging Machine Types

Accumulating and collating
Adhesive systems
Aerosol machines
Ampule machines

Bag closing and sealing
Bag filling
Bag forming
Bag making
Bag opening
Bag sealing
Bagging
Baling
Banding
Blister packaging
Blow molding
Bottling
Box making
Bundling, sleeve wrapping

Capping
Cap sorting
Capsule filling
Carton forming
Cartoning machines
Case forming
Case loading, packing
Case opening, unloading
Case sealing, compression units
Check-weighing
Classifying
Cleaning, washing-container
Closing machines, carton
Closing machines, container
Coating machines
Coding, dating, marking, stamping
Computerized systems
Conveying, converging and dividing
Core cutting
Core winding
Corking
Cottoning
Counting
Coupon and leaflet inserting
Creasing
Crimping
Cutting and trimming

Detecting and inspecting
Die cutting
Dispensing, tape, label

Embossing

Feeding, orienting
Filling dry products
Filling liquids
Filling and closing
Filling viscous products
Folding
Form, fill and seal, bag or pouch

Gas flush packaging
Gas and vacuum equipment, packages
Granulators, dry, wet

Hooding machines

Imprinting, printing
Injection molding

Labeling
Laminating

Metal detectors, systems
Multipacking

Overwrapping, wrapping

Palletizing
Paper bag, feeding, opening, filling, weighing and closing
Paper web, printing
Paper winding
Partitioning
Pasters
Presses, flexography and gravure
Pumps, ink and adhesive

Rewinding
Roll stands

Scales
Scoring
Shrink packaging
Skin packaging
Slitting
Slitting and roll winding
Stamping machines
Stapling
Stenciling
Sterile packaging

TABLE 13-1 *(Continued)*

Strapping	Unit packaging
Stretch packaging	Unscrambling
Strip packaging	
	Vacuum forming
Tablet counting	Vial packaging
Tape dispensing	Vibrators
Tension control equipment	
Testing	Warmers, bottle
Thermoforming	Waxing machinery
Tray making	Web equipment
Trim removal, trim blowers	Weighers, scales
Trimming	Weight computers
Tube fillers and sealers	Weight statistical analyzers
Tying	Weight tabulators

MISCELLANEOUS EQUIPMENT AND DEVICES

Forming and assembly. The earliest attempt to form preformed blanks of fibreboard around a commodity involved book packaging. This method of packaging was economical because it combined forming, assembly, loading and sealing in one operation. This was later expanded to accommodate multiple items including caseloads of merchandise. In Fig. 13-1, the basic steps involved in *wrap-around* caseforming and loading are shown. This is a highly efficient technique of packaging which is limited to mass production operations of long runs of identical sizes.

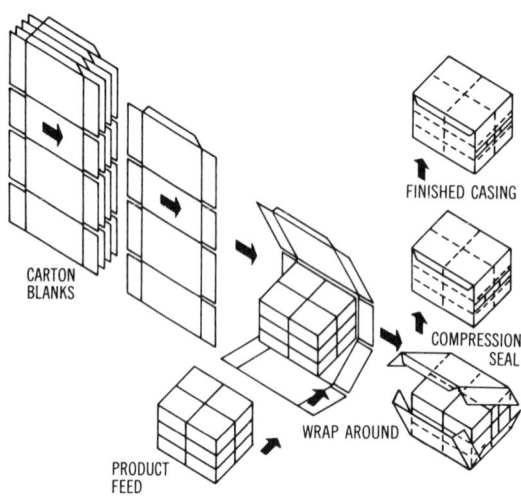

Fig. 13-1. Principles of *wrap-around* caseforming and loading. Courtesy FMC Corp., Packaging Machinery Division.

PACKAGING EQUIPMENT

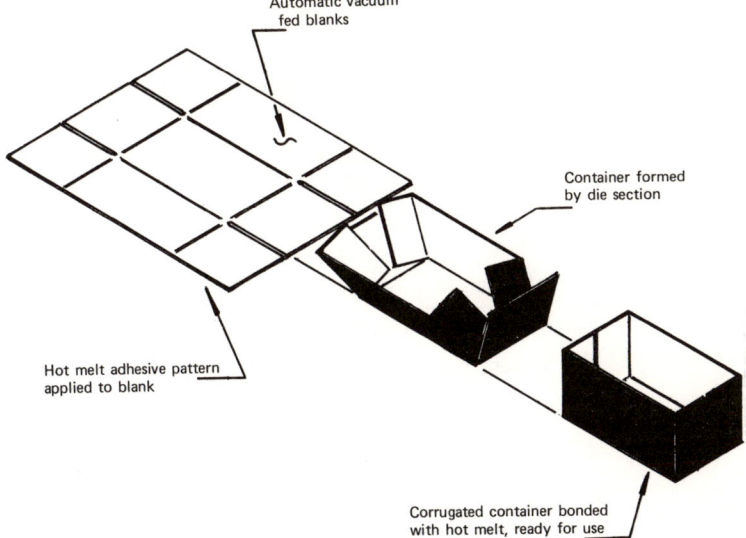

(a) Principle of automatic box forming without contents.

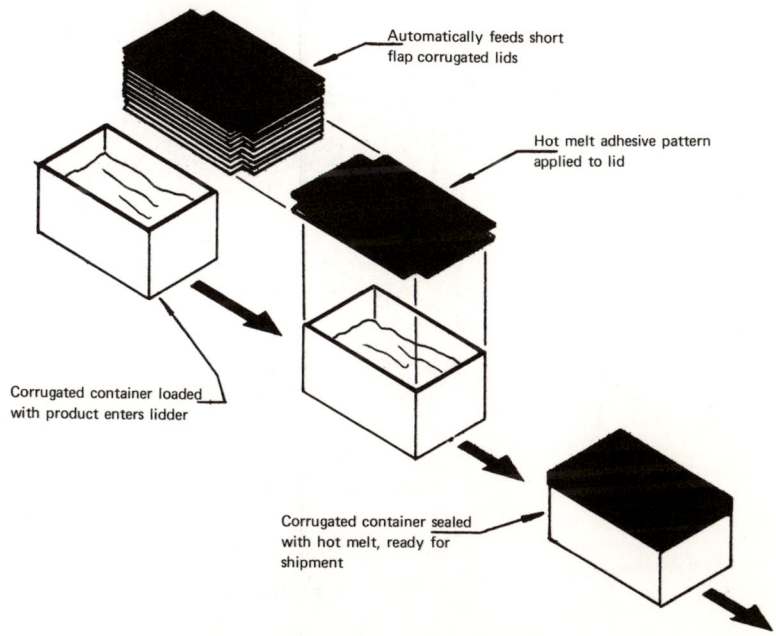

(b) Principle of lid forming assembly.

Fig. 13-2. Case forming without contents. Courtesy Sim-pak.

506   **DISTRIBUTION PACKAGING**

Other automatic machines are available to form a fibreboard container from a blank where no contents are involved (Fig. 13-2). This technique provides on-site assembly for a partial telescope style box immediately prior to filling. Complementary equipment is also available to form and apply the lid for these boxes.

The manual opening, forming and positioning of a box can be assisted on an air operated device as shown in Fig. 13-3.

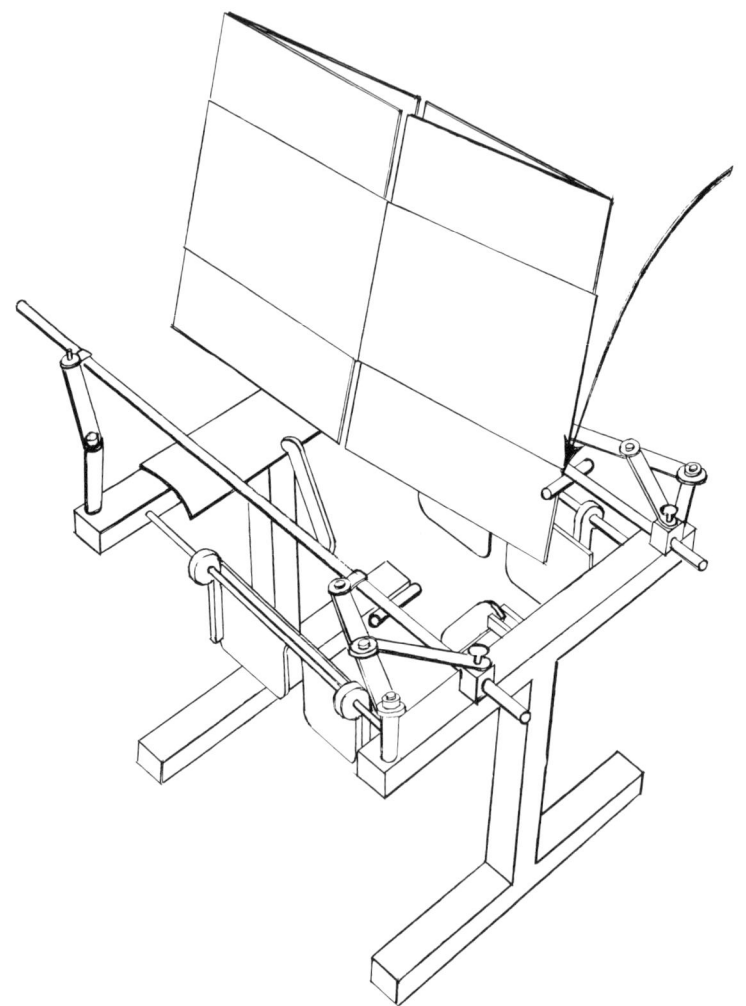

Fig. 13-3. Device to assist box opening and forming. Courtesy Pratt-Spector Corp.

PACKAGING EQUIPMENT

(*a*) Bottle loader

(*b*) Case loader

Fig. 13-4. Two types of automatic top loading machines.

Loaders and unloaders. Loading devices assist in the mechanical insertion of the product into a shipping container. The greatest degree of development in these devices is for the loading of consumer packages such as cans and bottles into shipping containers. High-speed packing lines require either fully or semi-automatic equipment of this type which is found in dairies, breweries, canneries, etc. Basically, loading is performed by two principles: (1) top loading and (2) end or side loading. In Fig. 13-4 two types of automatic top loading machines are shown. The principle of side loading, illustrating different content arrangement, is demonstrated in Fig. 13-5. A fully automatic

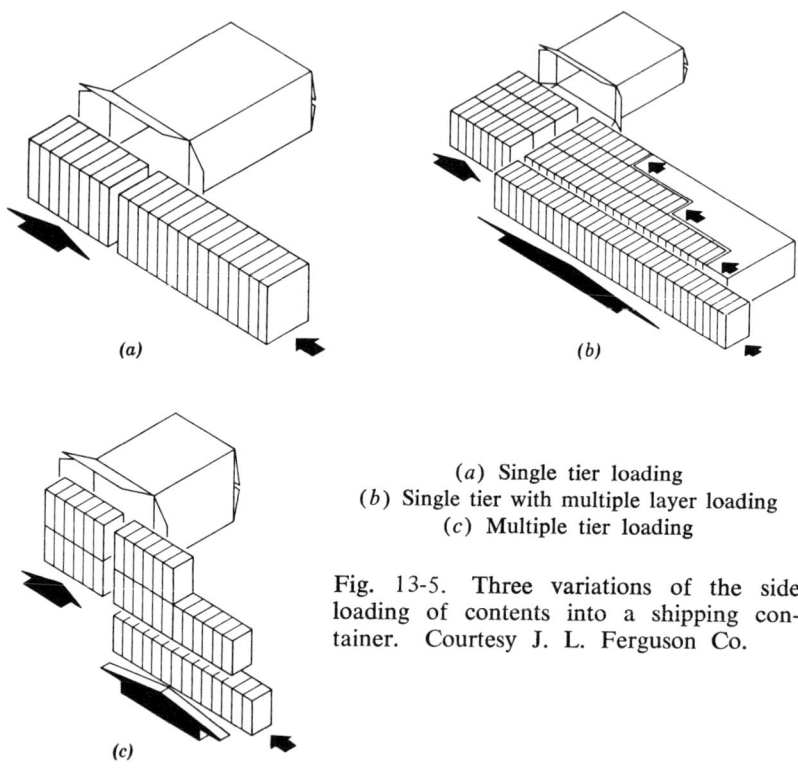

(a) Single tier loading
(b) Single tier with multiple layer loading
(c) Multiple tier loading

Fig. 13-5. Three variations of the side loading of contents into a shipping container. Courtesy J. L. Ferguson Co.

loading machine, in conjunction with box set-up and sealing, was shown in Chapter 9. The operation of a semi-automatic loader is illustrated in Fig. 13-6. Forming of the shipping container over the loading tube is manually performed by the operator. Arrangement of the contents and insertion into the container is mechanical. The container, with its bottom unsealed, is then mechanically lowered onto a belt conveyor which feeds it to a top and bottom case sealer. Similar loading equipment is available for the insertion of contents into flexible shipping containers (Fig. 13-7).

PACKAGING EQUIPMENT

Fig. 13-6. Semi-automatic case loader. Courtesy FMC Corp.

Fig. 13-7. Loading equipment for flexible containers.

(a) Courtesy ABC Packaging Machine Corp.

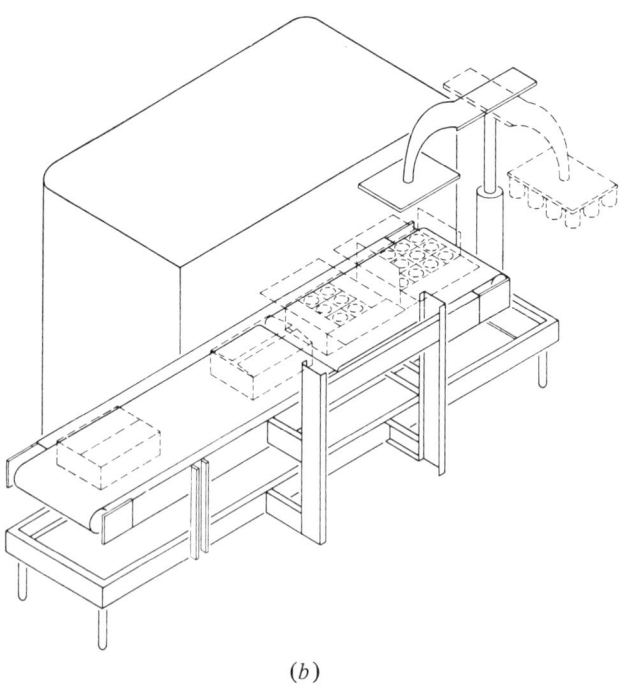

(b)

Fig. 13-8. Two types of case unloaders and unscramblers.

PACKAGING EQUIPMENT

When empty consumer packages such as bottles and cans are received in unsealed shipping containers, they may be automatically unloaded and unscrambled by specialized equipment of the types shown in Fig. 13-8. Fig. 13-8a shows an unloading system with the contents dropping onto the unscrambler unit as the unsealed cases are raised on an inclined conveyor. In Fig. 13-8b the contents are lifted by suction from the case and then placed on an unscrambler unit.

Weighers. Weighing of the product or the filled container is either performed in conjunction with filling or loading equipment or as a separate function. Illustrations of the former were included in conjunction with bulk packaging of shipping sacks. If the product is weighed prior to packaging the term *net weighing* is applicable. Weighing of the complete package, including the container and the product, is termed *gross weighing*.

In packaging, weighing serves a two-fold purpose: (1) to fill the package to a given weight with a minimum overage or underage and (2) to record the weight for freight charge determination. If the weight of individual units within the shipping container is controlled, repetitive weighing of the shipping container is not essential, and a predetermined standard for freight charge determination is applied.

Weighing devices range from manually operated scales on which the package rests while filling to fully automatic devices which are coupled to classifying, sorting, or rejection controls, or provide direct input to a computer. Automatic weighing devices of this type are generally referred to as

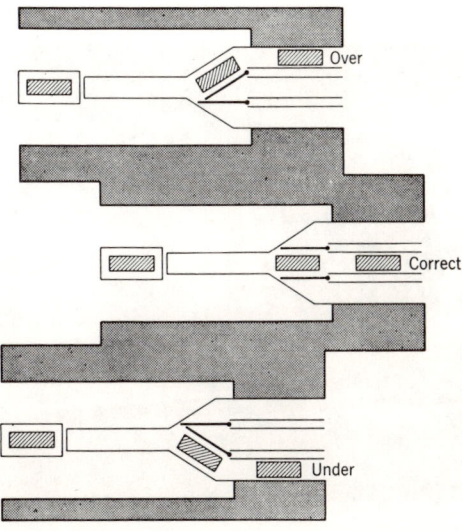

Fig. 13-9. Rejection system for folding cartons based on weight control.

512 **DISTRIBUTION PACKAGING**

check weighers. In Fig. 13-9 is a diagram of a rejection system for folding cartons based on weight.

For shipping containers a system that rejects only underweight units is described in Fig. 13-10. The controls operate on a photoelectric principle and are set for a predetermined weight with acceptable weight tolerances.

If weights of the finished package vary, a printing device to automatically record the weight of each package can be incorporated into the scale. At the same time, a label or shipping tab may be imprinted. This tab provides accuracy and a continuous record of packages passing over the scale (Fig. 13-11). A fully integrated system providing for both weighing and imprinting of the shipping container at speeds up to 30 boxes per minute is shown in Fig. 13-12.

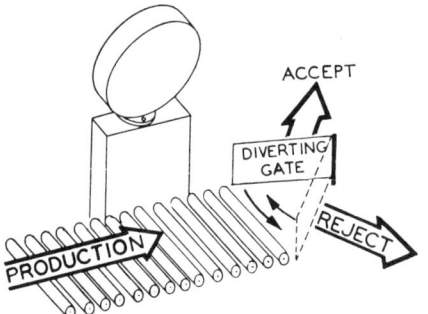

Fig. 13-10. Rejection system for underweight shipping containers. Courtesy Toledo Scale Division of Reliance Electric Co.

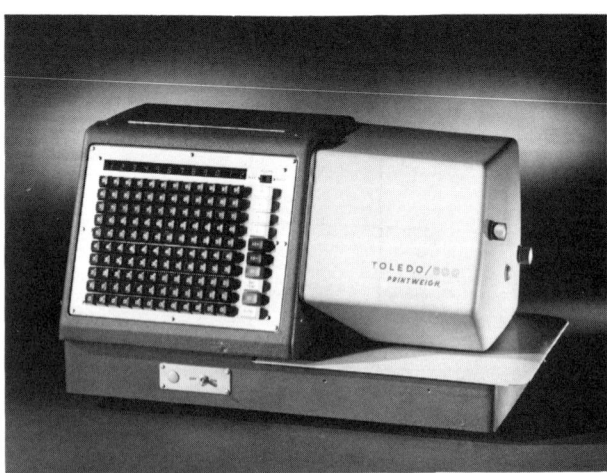

Fig. 13-11. Scale incorporating automatic printing device and recording unit. Courtesy Toledo Scale Division of Reliance Electric Co.

PACKAGING EQUIPMENT

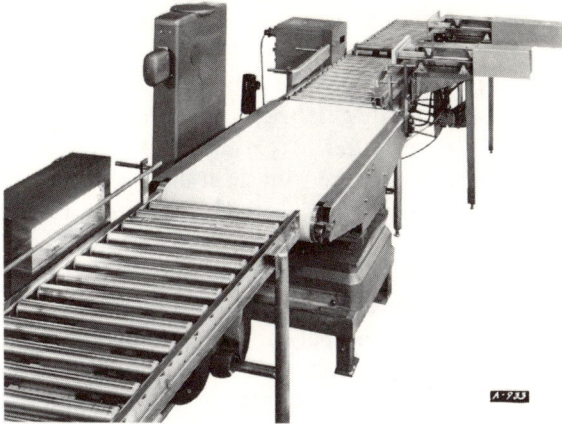

Fig. 13-12. Automatic weighing and imprinting system. Courtesy Toledo Scale Division of Reliance Electric Co.

Fig. 13-13. Typical bench-type scale set in gravity conveyor line. Courtesy Toledo Scale Division of Reliance Electric Co.

Bench, mobile, floor, and crane scales are available to be placed at any convenient location. A typical bench scale set in a gravity conveyor line in a shipping department is illustrated in Fig. 13-13.

Counting devices. Counting may be defined as the physical tally of units for control purposes. Counting is, of course, employed throughout the industrial process in many different operations. In distribution packaging counting is of primary importance with small units which are packed either into a unit container or directly into a shipping container. In many instances, weigh-counting is feasible. The advantage of weigh-counting is the ability to treat the product as a bulk material, eliminating the need to individually count each item. Predetermination of a standard quantity for a given weight with suitable weight tolerances applying, permits the use of most standard scales in the weigh-counting operation. When each individual item must be counted prior to loading into a package, two counting principles are applicable as follows:

1. Mechanical counting.
2. Electronic counting.

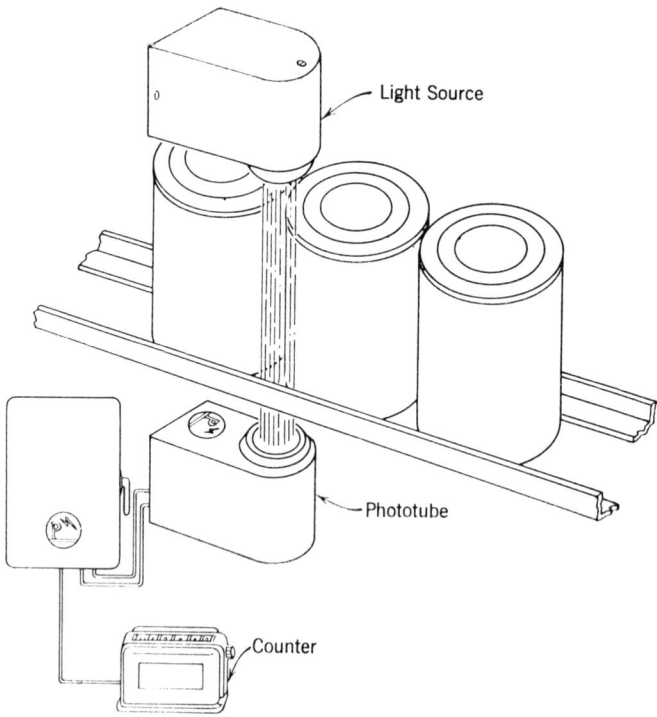

Fig. 13-14. Basic principle of electronic counting. Courtesy Photoswitch Division, Electronics Corp. of America.

PACKAGING EQUIPMENT

In mechanical counting the moving item trips a lever which in turn activates a counter. This method of counting is only applicable when the item is of adequate weight, is properly oriented in relation to the lever, and is not in any way damaged by the lever. Mechanical counters are therefore applied on conveyor lines for unit packages and shipping containers. Counters of this type are also incoporated in various types of packaging and marking equipment.

Electronic counting functions on the photoelectronic principle using a light source and a phototube to activate a counter when the light source is interrupted. Electronic counting can operate on high-speed lines since there are no limitations due to mechanical motion. In addition, selective counting by height, length, or contour is possible if a product mix of different items is in the same flow path. In Fig. 13-14, the basic principle of electronic counting is illustrated.

Counting mechanisms, as important control devices, are incorporated with scheduling, order selection, billing, and other clerical and record keeping functions, or are directly integrated with a computer.

Volume calculators. This recently developed equipment automatically computes the cube of a shipping container as it is transported on a conveyor. These devices measure all three dimensions of the package and then translate this information into a cube figure. This figure may be imprinted, freight rates can be determined and other functions performed.

Equipment of this type is used for the building of container loads for air cargo and marine shipments.

Auxiliary aids. In this section a brief survey is made of some of the auxiliary aids available to more efficiently carry out the distribution packaging function. No attempt is made to include material handling devices which contribute greatly to the efficiency of many packaging operations, since material handling, by itself, requires comprehensive treatment. Several excellent texts and handbooks devoted exclusively to this subject have been prepared.

Reference to auxiliary packaging aids has already been made in previous chapters. In this section, other devices, including packing benches, heat sealing equipment, filling devices for cushioning materials and cleaners are covered.

Packing benches. If packaging is performed primarily by manual methods, the layout of a packing station to minimize wasted motions and to reduce fatigue is of great importance. The orderly arrangement of the necessary containers, interior packing pieces, dispensers, and labels is often ignored. In Fig. 13-15, a well-arranged packing bench is shown. Where heavy crates are packaged, lowering of the crate to ground level for subsequent pick up by mechanical handling equipment is assisted by an hydraulic tilting table device which also contains space for some essential packing components (Fig. 13-16).

Fig. 13-15. Well-organized packing bench.

Fig. 13-16. Hydraulic tilting table.

PACKAGING EQUIPMENT

Various fixtures can be mounted to packing benches or the benches can be irregular shaped to minimize reaching by the packer. Such fixtures include loading funnels or chutes, heat sealers, and staplers. The positioning of the merchandise to be packed on conveyors, pallets, floor trucks, or other conveyances, in relation to the packing bench, as well as the location of a removal system, further influences the efficiency of the packaging operation. Double-tiered conveyors are an excellent method of conserving space. Floor chutes are also a method to remove packed merchandise in multi-story operations.

Cushioning materials storage and dispensing equipment. Increased use of free flowing cushioning materials such as expanded polystyrene has led to the development of various aids to facilitate efficient use of these materials. This includes various techniques of overhead storage and means of dispensing at a packing station. One of the simplest forms consists of a canvas storage hopper usually positioned at a conveyorized loading station. The packer squeezes a scissor release to dispense the cushioning material. A more sophisticated centralized airveying system is shown in Fig. 13-17. This eliminates the individual refilling of hoppers. Equipment has also been perfected which accomplishes the automatic filling of a container with such cushioning materials (Fig. 13-18). Any excess above the top score line is wiped off and is recirculated.

Fig. 13-17. Airveying system making effective use of overhead space at multiple packing station system. Courtesy Norel Plastic Corp.

Fig. 13-18. Automatic filling machine utilizing free flowing cushioning material. Courtesy Norel Plastic Corp.

Heat-sealing equipment. Heat sealing is defined as a method of unitizing two or more surfaces by fusion, either of the coatings or of the base materials, under controlled conditions of temperature, pressure, and time (dwell). Most heat sealing is performed on consumer packages, either separately or in conjunction with other packaging operations. For distribution packaging purposes, heat sealing is primarily performed where a barrier material is employed and a moisture vapor-proof seal is essential. In these applications, either portable or bench-type heat-sealing units are usually employed. Where high-speed heat sealing of identical products packed in bags or pouches is required, many of the available consumer-type continuous sealers are applicable. Bench type and portable units are designed either with or without adjustable heat controls and are furnished in rotary or pressure bar design.

In Fig. 13-19 a rotary sealer, specially designed for the sealing of reinforced papers, is depicted. A jaw-type sealer, for heavy duty service, adapted for sealing such materials as cellophane, glassine, and military papers is illustrated in Fig. 13-20.

PACKAGING EQUIPMENT

The following factors directly relate to the efficiency of a heat sealing operation:

1. Machine factors including clamp or dwell time, sealing bar or jaw temperature and pressure.

2. Resin factors, i.e., density, melt index and slip additives.

3. Film factors such as gauge, bag style (gusseted or not gusseted), heat treatment for printability, film making procedure (blow versus cast) and sealing direction (machine versus transverse).

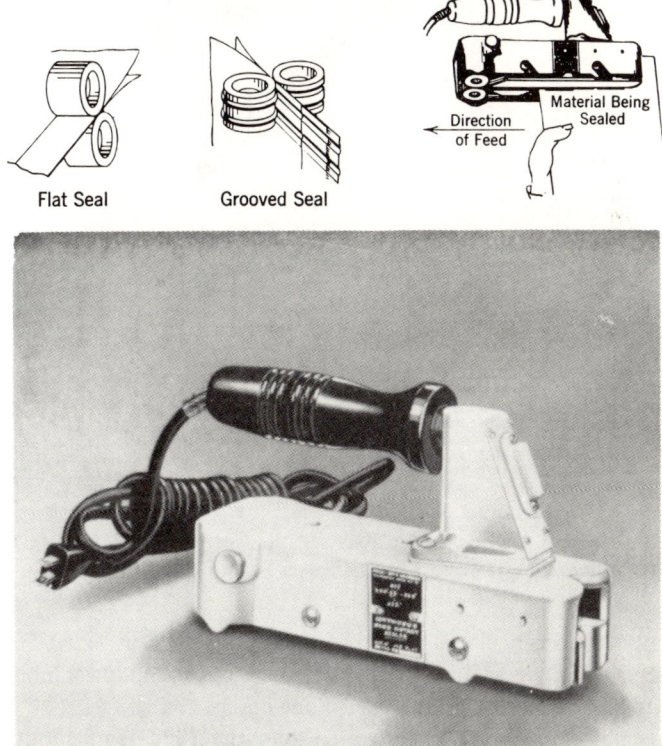

Fig. 13-19. Rotary hand sealer for kraft and laminated papers showing two kinds of sealing rollers. Courtesy Pack-Rite Machines, Toledo Scale, Div. of Reliance Electric Co.

No matter what type of polyethylene film, bag, or resin has to be heat-sealed, there is practically always a sufficiently wide operating range of clamp time, sealing temperature, and sealing pressure to achieve satisfactory seals.

Generally speaking, the widest permissible temperature ranges occur at short clamp times and low pressures; the widest permissible clamp time ranges occur at low temperatures and low pressures. However, since machine condi-

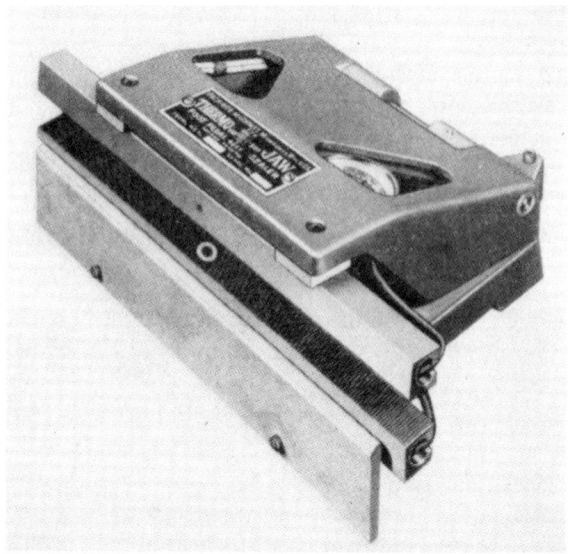

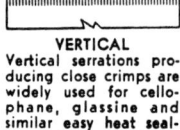
VERTICAL
Vertical serrations producing close crimps are widely used for cellophane, glassine and similar easy heat sealing, lighter weight materials.

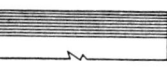
HORIZONTAL
Where a horizontal crimp is preferred, bars are furnished with sufficient serrations to make a seal with horizontal impressions.

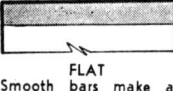
FLAT
Smooth bars make a flat seal or where extra heat penetration is required. Recommended for foil, kraft and scrim-backed laminates and overcap labels.

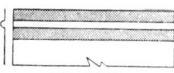
GROOVED
For a grooved seal, one bar has a groove the full length, the other bar has a matching bead — with a line of short vertical crimps each side of groove

Fig. 13-20. Jaw-type heat sealer and four variations of sealing impressions. Courtesy Pack-Rite Machines, Toledo Scale, Div. of Reliance Electric Co.

tions may occasionally fluctuate and get out of control, it is generally preferable to operate in a median region between the upper and lower limits, such as those roughly indicated by the dashed line in Fig. 13-21.

The graphical relationship between clamp time and temperature for a typical film with a 40 lb per sq in. pressure is reproduced (Fig. 13-21). Similar graphical relationships have been developed by the U.S. Industrial Chemicals Company in research on heat sealing.

Ultrasonic sealing equipment. Ultrasonic sealing is a mechanical rather than a thermal method of welding two films together; only a minute amount of heat is generated in the films to be sealed.

Mechanical impulses are exerted on the two layers of film by means of a metallic sealing tip, or welding head, which vibrates up and down at a high frequency. The frequency is just beyond the high-pitch end of the audible spectrum, 18,000 to 20,000 and more per second. Hence, the name ultrasonic.

PACKAGING EQUIPMENT

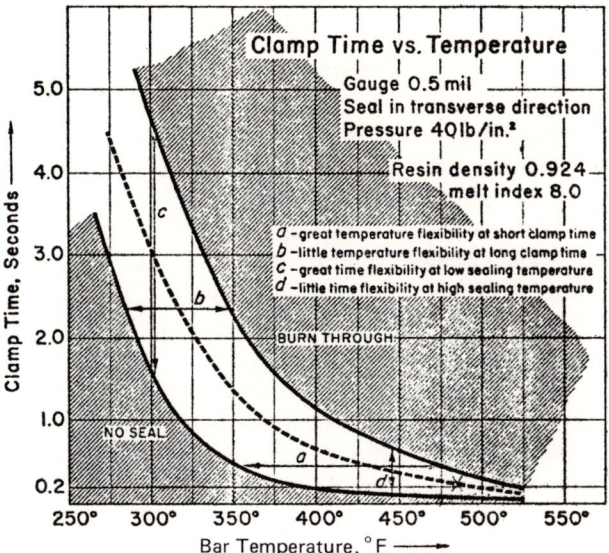

Fig. 13-21. Graphical relationship of clamp time and temperature for a polyethylene film. Courtesy U.S. Industrial Chemicals Co., Division of National Distillers and Chemical Corp.

This technique generally produces a good seal with usually little shrinkage, deformation, or deterioration of the film in and around the seal area. This is one reason why ultrasonic welding can be used to seal oriented films, even biaxially oriented polypropylene films, which are commonly considered non-sealable. Thick-gauge films are apparently easier to seal ultrasonically than very thin films of say, less than 1 mil (25 micron) gauge which tend to curl up due to shrinking or deorientation.

Ultrasonic film sealing is still in the developmental stage. The equipment has the advantages of being noiseless, easy and safe to operate and of staying cool. Good seals at sealing speeds (film feed rates) exceeding 100 ft/min (30 m/min) are said to have been obtained with commercial equipment using a 400-watt generator. The thinner the sealing tip, the faster the feed rate which still permits obtaining a strong seal.

Quite generally, ultrasonic sealing appears to be less suitable for flexible polyethylene film than for stiffer films such as polypropylene.

Cleaning devices. Prior to packaging, cleaning of the product or the container may be necessary. Product cleaning is required for subsequent corrosion protection and equipment and processes essentially are designed for the product itself. Such operations as degreasing and rust removal are practiced. Subsequent treatments of the cleaned product follow established corrosion prevention techniques including dipping, coating, and spraying.

With volatile corrosion inhibitors no such further treatment after cleaning may be required for some products.

Cleaning of containers is primarily necessary with re-usable containers. New containers are procured in such a manner that contamination in shipment is minimized. Exceptions (when cleaning and/or sterilization must be practiced whether the container is new or re-used), are in such industries as food, beverage, and ethical pharmaceutical. Specialized equipment including bottle washers, bottle sterilizers, etc. are used for this purpose.

Waste utilization machines. For the filling of voids or for cushioning of products within a shipping container, various waste products are used. This includes shredded newsprint, cellophane shavings, scrap fibreboard, etc. The waste can be procured in shredded form or prepared by the user prior to packaging. For flexible materials, various types of shredders are marketed. Today's high labor cost discourages the salvaging of waste products for further packaging use. However, this situation may change with the growing emphasis on recycling of materials and during material shortages.

SELECTION OF PACKAGING MACHINERY

The selection of a suitable packaging machine or auxiliary device is dependent upon: (1) technical requirements in terms of mechanically performing the packaging function either completely or partially and (2) the economic justification for the capital expenditure involved. The former relates to the performance of standard existing equipment or to the potentiality to modify such equipment or to design new equipment. The latter rates a technically feasible system in terms of its capital and operating costs and compares it with the existing cost of the operation in question.

In most cases, the selection of packaging equipment cannot be isolated from packaging material cost. Further, since equipment must be integrated with material handling devices, a systems approach, considering all technical and economic factors, must be pursued.

TABLE 13-2

ECONOMIC SUMMARY OF TWO ALTERNATE INTEGRATED PACKAGING SYSTEMS

Type of Carton Former Used in System	Labor	Savings Material	Total	Capital Investment	Return (Years)
Alternate 1: Automatic carton former	$34,000	$12,000	$48,000	$65,000	1.35
Alternate 2: Semi-automatic carton former	17,000	13,000	30,000	37,000	1.23

PACKAGING EQUIPMENT

In an analysis to modernize packaging for a food manufacturer, whose operations were essentially manual, a number of alternate systems were studied and the cost of container materials, packaging machinery and material handling equipment was rated. The alternates included represented the attainment of varying levels of mechanization. The system finally selected was semi-automatic and saved approximately $30,000 a year in materials and labor with a one-time capital investment of $37,000. A more mechanized version, requiring an investment of $65,000 would have saved $48,000 annually in materials and labor, but had a lower return on the investment. A summary of the costs, savings and return is listed in Table 13-2.

As an aid to rate packaging equipment, and to include all pertinent details in purchase orders, the Packaging Institute, USA has prepared a check list reproduced in Fig. 13-23.

CHECK LIST OF ITEMS TO BE INCLUDED ON PURCHASE ORDERS FOR PACKAGING EQUIPMENT

SECTION 1.—Physical details of the equipment.
1. Name and Model ☐
2. Right or Left Hand Operation ☐
3. Materials of Construction ☐
   - Regular ☐
   - Special for contact surfaces ☐
   - " " wearing surfaces ☐
   - " " pumps ☐
   - " " piping and valves ☐
   - " " gaskets ☐
   - " " etc. ☐
4. Electrical Requirements ☐
   - Voltage, etc. ☐
   - Underwriter's classification ☐
   - Motors—Manufacturer ☐
   - Capacity ☐
   - Starters—Manufacturer ☐
   - Capacity ☐
   - Location ☐
   - Switches—Manufacturer ☐
   - Capacity ☐
   - Location ☐
5. Lubrication Requirements ☐
   - Fittings ☐
   - At point of use ☐
   - Central location ☐
   - One shot system ☐
6. Safety Features ☐
   - State requirements ☐
   - Local " ☐
   - Plant " ☐

Fig. 13-23. Check list for purchasing of packaging equipment. Courtesy The Packaging Institute, USA.

7. Paint . . . . ☐
         Type . . . . ☐
         Colors . . . . ☐
   8. Special Items . . . . ☐
         Shrouding . . . . ☐
   9. Auxiliary Equipment & Miscellaneous Items . . . . ☐
         Variable speed drives . . . . ☐
               Speed range . . . . ☐
         Pumps . . . . ☐
               Pressure . . . . ☐
               Capacity . . . . ☐
               Type . . . . ☐
         Vacuum Pumps . . . . ☐
               Vacuum . . . . ☐
               Capacity . . . . ☐
               Type . . . . ☐
         Clutches . . . . ☐
         Heaters. . . . . ☐
         Counters . . . . ☐
         Indicators . . . . ☐
         Micro-switches. . . . . ☐
         Electronic controls . . . . ☐
         Timing dials . . . . ☐
         Tachometers. . . . . ☐
         Etc. . . . . ☐

SECTION 2.—Change Parts
   1. Quantity required. . . . . ☐
         Individual parts, or . . . . ☐
         Sets. . . . . ☐
   2. Marking . . . . ☐
   3. Interchangeability. . . . . ☐
   4. Time required to change over . . . . ☐
   5. On delivered equipment . . . . ☐
   6. Shipped separately . . . . ☐

SECTION 3.—Engineering Data and Details necessary for installation and interconnecting.
   1. Equipment. . . . . ☐
         Floor Plan . . . . ☐
         Elevation view. . . . . ☐
         Assembled weight. . . . . ☐
         Floor loading . . . . ☐
   2. Details of interconnection . . . . ☐
   3. Limiting factors affecting installation . . . . ☐
         Doorways and Halls. . . . . ☐
               Heights. . . . . ☐
               Widths . . . . ☐
         Elevator Capacities . . . . ☐
         Floor Loads . . . . ☐

PACKAGING EQUIPMENT 525

SECTION 4.–Operational Details and Operational Guarantee.
1. Description of job expected. . . . . ☐
2. Speed of operation . . . . ☐
3. Tolerances allowable . . . . ☐
4. Efficiency expected. . . . . ☐
    Defined as to speed . . . . ☐
    " " " period of time . . . . ☐
    " " " acceptable quality . . . . ☐
5. Packaging Materials to be used on equipment . . . . ☐
    Specifications . . . . ☐
    Samples . . . . ☐
6. Materials to be Packaged . . . . ☐
    Special precautions . . . . ☐
7. Operators . . . . ☐
    Number required . . . . ☐
    Skill . . . . ☐
    Sex . . . . ☐

SECTION 5.–Inspection and Test at Manufacturer's Plant using actual materials.

SECTION 6.–Delivery, Installation and Test Period.
1. Delivery Date . . . . ☐
2. Method of Shipment . . . . ☐
    Rail) – Skid) – knocked down . . . . ☐
    Truck) – Crate) – assembled . . . . ☐
3. Movement into position and uncrating . . . . ☐
4. Installation. . . . . ☐
    Supervision of manufacturer's engineer . . . . ☐
    Labor and materials by purchaser . . . . ☐
5. Start up . . . . ☐
    Supervision of manufacturer's engineer . . . . ☐
6. Instruction and demonstration of . . . . ☐
    Operation by manufacturer's engineer . . . . ☐
    To operators and mechanics . . . . ☐
7. Test Period . . . . ☐
    Time in which guarantee is to be met . . . . ☐
    Supervision of manufacturer's engineer . . . . ☐
    Cover change parts, or special agreement . . . . ☐

SECTION 7.–Manuals and Data.
1. Operating Manual . . . . ☐
    Operating Instructions . . . . ☐
    For specific model . . . . ☐
    Illustrated . . . . ☐
    Change over instructions . . . . ☐
2. Parts Manual . . . . ☐
    For specific model . . . . ☐
    Illustrated . . . . ☐
    Change parts included . . . . ☐

Parts stocked by manufacturer . . . . ☐
Assembly drawing. . . . . ☐
3. Wiring diagrams . . . . ☐
4. Lubrication Manual. . . . . ☐
5. Data for Timing Charts. . . . . ☐

SECTION 8.—Guarantee by Manufacturer against
Defective workmanship . . . . ☐
Defective materials . . . . ☐
Patent Claims . . . . ☐

SECTION 9.—Maintenance Spare Parts.
1. To be delivered with equipment. . . . . ☐
2. To be shipped separately. . . . . ☐

BIBLIOGRAPHY

Texts and Manuals

Ainsworth, J. H., *Paper—The Fifth Wonder,* Thomas Publishing Co., N.Y.
Barail, Louis C., *Packaging Engineering,* Reinhold Publishing Corp., N.Y., 1954.
Barlow, C. Wayne, *Corporate Packaging Management,* American Management Association, Inc., 1969.
Bettendorf, H. J., *A History of Paperboard and Paperboard Containers,* Board Products Publishing Co., Chicago, Ill.
Bigger, R. P., *The Manufacture of Fibre Cans,* Board Products Publishing Co., Chicago, Ill.
Bolz, H. A., and G. E. Hagemann, *Materials Handling Handbook,* The Ronald Press Co., N.Y., 1958.
British Standards Institution, *British Standard Glossary of Packaging Terms,* London, 1959.
Brody, Aaron L., *Flexible Packaging of Foods,* The Chemical Rubber Co., 1970.
Brown, Kenneth, *Package Design Engineering,* John Wiley & Sons, Inc., N.Y., 1959.
Bruce, Harry J., *Distribution and Transportation Handbook,* Cahners Publishing Co., Inc., 1971.
Bruce, Harry J., *How to Apply Statistics to Physical Distribution,* Distribution Age, Chilton Co., Philadelphia, 1967.
Calcin, John B., and George S. Whitman, *Modern Pulp and Paper Making,* Reinhold Publishing Corp., N.Y., 1957.
Cole, E. J., and M. Todd, *Pulp and Paper Mill Instrumentation,* Lockwood Trade Journal Co., Inc., N.Y., 1957.
Colton, Richard C., and Edmund S. Ward., *Practical Handbook of Industrial Traffic Management—Fifth Edition,* Washington, D.C., The Traffic Service Corporation, 1973.
Containerisation International Yearbook, The National Magazine Co., Ltd., London.
De Bruyne, N. A., and R. Houwick, *Adhesion and Adhesives,* Elsevier Press, Inc., Houston, Texas, 1951.
Delmonte, John, *Technology of Adhesives,* Reinhold Publishing Corp., N.Y.
Elliot, N. J., *Pricing and Estimating Methods,* Haywood Publishing Co., Chicago, Ill.
Fladager, Vernon L., *The Selling Power of Packaging,* McGraw-Hill Book Co., New York, 1956.
Freezing Preservation of Foods, The, in four volumes, The AVI Publishing Co., Inc., Westport, Conn., 1968.
Friedman, Walter F. and Jerome J. Kipnees, *Industrial Packaging,* John Wiley & Sons, N.Y., 1960.
Fundamentals of Packaging, Blackie & Son Ltd., London, 1962.
Glossary of Packaging Terms, Packaging Institute U.S.A., N.Y.
Glossary of Traffic Terms, A, The Traffic Service Corp., Washington, D.C.
Gordon, G. A., and E. E. Wheeler, *The Properties of Cushioning Materials For Use in Packaging* – The Printing, Packaging and Allied Trades Research Assoc., Surrey, England, 1961.
Griffin, Roger C., Jr., and Stanley Sacharow, *Principles of Package Development,* The AVI Publishing Co., Inc., Westport, Conn., 1972.

Hackamack, Lawrence C., *Making Equipment-Replacement Decisions,* American Management Association, Inc., N.Y., 1969.
Hanlon, Joseph F., *Handbook of Package Engineering,* McGraw-Hill Book Co., N.Y., 1971.
Harris, C. M., and C. E. Crede, *Shock and Vibration Handbook,* Vols. 1, 2 & 3, McGraw-Hill Book Co., N.Y., 1961.
Jenness, Lyle C., and John Lewis, *Industrial Lectures on Pulp and Paper Manufacture, University of Maine,* The Lockwood Trade Journal Co., Inc., N.Y., 1953.
Koehler, Arthur, *Properties and Uses of Wood,* McGraw-Hill Book Co., N.Y., 1924.
Leonard, Edmund A., *Introduction to the Economics of Packaging,* Edmund A. Leonard, N.Y., 1968.
Long, Robert P., *Package Printing,* Graphic Magazines, Garden City, N.Y., 1964.
Material Handling Engineering Handbook & Directory, Industrial Publishing Co., Cleveland, Ohio.
McGuire, E. Patrick, *Packaging and Paper Converting Adhesives,* Palmerton Publishing Co., N.Y., 1963.
Modern Packaging Encyclopedia/Packaging Planning Guide, McGraw-Hill, Inc.
Modern Plastics Encyclopedia, McGraw-Hill, Inc.
Packaging For The Small Parcel Environment, United Parcel Service of America, Inc., N.Y., 1971.
Paine, F. A., *Packaging Materials and Containers,* Blackie & Son Ltd., London, 1967.
Rath, Eric, *Container Systems,* John Wiley & Sons, Inc., N.Y., 1973.
Rennicke, Norbert G., *The Manufacture of Paperboard,* Board Products Publishing Co., Chicago, Ill.
Sacharow, Stanley and Roger C. Griffin, Jr., *Food Packaging,* The AVI Publishing Co., Inc., Westport, Conn., 1970.
Stephenson, Newell J., *Pulp and Paper Manufacture,* McGraw-Hill Book Co., N.Y.
    Vol. 1 *Preparation and Treatment of Wood Pulp,* 1950.
    Vol. 2 *Preparation of Stock for Paper Making,* 1951.
    Vol. 3 *Manufacture and Testing of Paper and Board,* 1952.
    Vol. 4 *Auxiliary Paper Mills,* 1955.
Stern, Walter, *The Package Engineering Handbook,* Board Products Publishing Co., Chicago, Ill., 1954.
Textbook of Corrugated Box Production, A, Haywood Publishing Co., Chicago, Ill.
Vail, James G., *Soluble Silicates* (Vol. 2), Reinhold Publishing Corp., N.Y., 1952.
Werner, A. W., *The Manufacture of Fibre Boards,* Board Products Publishing Co., Chicago, Ill., 1954.

Indexes, Lists, Etc.

American Society for Testing Materials, *Index to Standards.*
American Standards Association, *Catalog of American Standards.*
Chemical Specialties Manufacturers Association, *Agencies and Regulations of Interest to the Aerosol Industry,* N.Y., 1962.
Packaging Abstracts, Packaging Division of the Printing, Packaging, and Allied Trades Research Association, Leatherhead, Surrey, Great Britain.
Packaging Institute, USA, *List of Publications.*
Sources of Information on Containers and Packaging, Superintendent of Documents, U.S. Government Printing Office.
Technical Association of the Pulp and Paper Industry, *Bibliography of Papermaking and U.S. Patents.*

U.S. Clearinghouse for Federal Scientific and Technical Information, *Selective Bibliography of Government Research Reports and Translations, Containers, Packaging, Storage, Etc.*, Springfield, Va., 1965.

U.S. Department of Agriculture, Forest Service, Forest Products Laboratory, *List of Publications on Box and Crate Construction and Packaging Data*, 65-038, January 1966.

World Container Shipping Guide, Sea and Air, Container News, 1973, Vol. 8, Nos. 1 & 2.

Carrier Agency and Government Publications

Annual Survey of Manufacturers, U.S. Department of Commerce, Social and Economic Statistics Administration, Bureau of the Census.

Association of American Railroads, Freight Loading and Container Section, *Container Bulletins.*

Breakage and Damage–In Grocery Warehouses and Retail Food Stores–Transportation and Facilities Research Division, Agricultural Marketing Service, U.S. Department of Agriculture, Marketing Research Report No. 652.

Department of Defense, Military Standardization Handbook, *Package Cushioning Design,* MIL-HDBK-304, 1964.

Department of Transportation – *CFR-49 Surface Transportation*
CFR-46 Water Transportation
CFR-14 Air Transportation

Federal Aviation Regulations.

Gigliotti, M. E., *Design Criteria For Plastic Package Cushioning Materials,* Plastics Technical Evaluation Center, Picatinny Arsenal, Dover, N.J., 1962.

Green, Frank W., *New York Export Packaging Survey,* Maritime Association of the Port of New York.

Hasset, Harry E., *Motor Carrier Packaging and Inspection,* The Motor Truck Association of Southern California.

Hazardous Materials Regulations of the Department of Transportation (R. M. Graziano's Tariff).

Henny, C. and F. Leslie, *An Approach to the Solution of Shock and Vibration Isolation Problems As Applied to Package Cushioning Materials.* Shock, Vibration and Associated Environments Bulletin No. 30, Part II. Office of the Secretary of Defense, Research and Engineering, Washington, D.C., 1962.

International Air Transit Association (IATA).

Interstate Commerce Commission Regulations.

Maritime Consultive Organization Regulations (IMCO).

National Furniture Packing Specifications, National Furniture Traffic Conference, Inc.

National Motor Freight Classification.

Orba, A., *Progress in Package Cushioning Design With Reference to the Measurement of Cushion Performance.* Proceedings of Symposium on Dynamics of Package Cushioning. Royal Radar Establishment, Ministry of Aviation, Malvern, England, 1960.

Postal Manual.

Rock Island Arsenal Laboratory, Rock Island, Ill., *Use of Volatile Corrosion Inhibitors With Ferrous and Nonferrous Metal Finishes,* by R. E. Johnson, 1963.

Stern, R. K., *Trends in the Isolation of Packaged Items.* Shock, Vibration and Associated Environments Bulletin No. 30, Part II. Office of the Secretary of Defense, Research and Engineering, Washington, D.C., 1962.

Uniform Freight Classification–Ratings, Rules and Regulations.

United Parcel Service Guide.

U.S. Business & Defense Service Administration, *Sources of Information on Containers and Packaging*, by James M. Devlin, U.S. Government Printing Office, Washington, D.C., 1965.
U.S. Business and Defense Services Administration, *Western Europe Standardizing Packaging Dimensions.*
U.S. Department of Commerce, Domestic and International Business Administration, *U.S. Industrial Outlook, Annual*, U.S. Government Printing Office, Washington, D.C.
U.S. National Bureau of, Standards. *Commercial Standards and Simplified Practice Recommendations.*

Professional Organization Publications and Selected Articles

American Management Association, *Management Bulletins and Reports.*
American Material Handling Society, *Packaging-Handling Coordination (PHC): A Guidance Handbook*, Cleveland, Ohio, 1962.
American Society for Testing Materials, *Paper and Paperboard—Characteristics, Nomenclature, and Significance of Tests. Book of ASTM Standards, Special Technical Publications.*
"Beefed-Up Box Stitchers Boost Output, Cut Maintenance", *Package Engineering,* Nov. 1973, Vol. 18, No. 11, pp. 53-54.
Botsford, A. C., *Air Cargo Shock Damage Experience in Handling and Shipping,* Society of Automotive Engineers, #494, April 1955.
Boyer, Paul M., "How to Cut A Package's Costs Without Hurting Its Performance", *Package Engineering,* June 1973, Vol. 18, No. 6, pp. 58-59.
Brody, Aaron L., "Systems Integration of Packaging", *Modern Packaging,* May 1968, pp. 159-162.
Brooks, Durward L., "Essentials of Export Packaging and Marking", *Transportation & Distribution Management,* June 1973, Vol. 13, No. 6, pp. 42-47.
"Build Smooth Flow Into Line For Hard-To-Pack Products", *Package Engineering,* Dec. 1972, Vol. 17, No. 14, pp. 50-52.
Dixon, James M., "Taking A Long Step Toward Damage-Free Transportation", *Distribution Worldwide,* June 1972, Vol. 71, No. 6, pp. 29.
Einbinder, David, "On-Demand Supply System Feeds Lines Faster", *Package Engineering,* Oct. 1973, Vol. 18, No. 10, pp. 62-64.
Einbinder, David, "Packaging Labor Costs: The Need For Accurate Accounting Methods", *Package Development,* Sept./Oct. 1974, pp. 72-76.
Forsyth, F. E. and G. G. Maltenfort, "Trouble-Free Vacuum Handling of Cases Rests on The Air Flow", *Package Engineering,* Feb. 1973, Vol. 18, No. 2, pp. 50-53.
Friedman, Walter F., "The Need For A Broadened Focus in Package Design", *Package Development,* May/June 1975.
Friedman, Walter F., "The Role of Packaging in Physical Distribution", *Transportation & Distribution Management,* Feb. 1968, Vol. 8, No. 2, pp. 34-39.
Friedman, Walter F., "Where Does Packaging Stop and Physical Distribution Begin", NPW-7513, *The Packaging Institute, U.S.A.,* 1975.
Goldberg, Robert I., "The Packaging Systems Approach, Its Effect On Creative Development", *Package Development,* Nov./Dec. 1975, pp. 17-20.
Guins, Sergei G., "Pick Your Corrugated Board By Tests Instead of Guesses", *Package Engineering,* May 1973, Vol. 18, No. 5, pp. 58-60.
Harckham, Arthur W., "An Analytical Approach to Organizing, Positioning the Packaging Function", *Package Development,* July/Aug. 1975.

BIBLIOGRAPHY 531

Henry, Alan D., "Eleven Ways to Reduce Packaging Costs", *Army Logistician*, Jan-Feb. 1972, Vol. 4, No. 1, pp. 8-11, 38-39.
"How Fares The Shrink Shipper?", *Modern Packaging*, Sept. 1968, pp. 118-121.
"In-Line Tape Sealing Speeds Up Case Closing", *Package Engineering*, Dec. 1973, Vol. 18, No. 12, pp. 53-55.
Institute of Paper Chemistry, *Study of Paper Board Quality as Related to Fibre Box Performance*, 1955.
"It's Here: The Shrink-Packed Pallet Load.", *Modern Materials Handling*, Vol. 22, No. 8, Aug. 1967, pp. 44-45.
Kill, George N., "Mullen Test Clamping Pressure: An Old Controversy Revisited", *Package Development*, Sept./Oct. 1975, pp. 20-22.
Lansdale, David B., "Potential Problem Analysis: A Systematic Approach", *Package Development*, March/April 1974, pp. 17-22.
Leffler, Walter H., "Evaluating Existing Machinery Systems, Justifying the New, Part I & Part II", *Package Development*, Nov./Dec. 1973, pp. 33-37, Jan./Feb. 1974, pp. 23-30.
Lesser, Lawrence M., "Protective Packaging: What Shippers Are Doing", *Traffic Management*, April 1972, Vol. 11, No. 4, pp. 31-36.
Lewis, James E., "A System For Effective Packaging Specifications", *Package Development*, Mar./Apr. 1974, pp. 23-29.
Markle, Kenneth E., "New Package, Cushion Pay-Off: More Sales, Happier Customers", *Package Engineering*, Mar. 1973, Vol. 18, No. 3, pp. 74-77.
"Mechanized Case Packing: Now For Almost Everyone!", *Modern Materials Handling*, Sept. 1973, Vol. 28, No. 9, pp. 68-75.
"Military Ponders Its Packaging, The", *Modern Packaging*, July 1968, pp. 88-92.
National Safe Transit Committee, Inc., *Test Procedures*.
"No-Contact Case Printing Comes To Packaging Life", *Package Engineering*, Mar. 1973, Vol. 18, No. 3, pp. 62-64.
Packaging Institute, U.S.A., *Technical Reports, Glossary of Packaging Terms, Petroleum Packaging Notebook, Source Book for Closures, Source Book for Fundamental Resins for Packaging Adhesion to Board Manual*.
"Packaging's Role in Physical Distribution", American Management Association, *Management Bulletin #77*, 1966.
Pizzirusso, Joseph, "Develop New Foam Cushioning To Withstand Repeated Impacts", *Package Engineering*, July 1973, Vol. 18, No. 7, pp. 54-56.
"Reusable Pail Scores Double Pay-Off: Shipping Costs Drop, Sales Rise", *Package Engineering*, Nov. 1973, Vol. 18, No. 11, pp. 54-55.
Schreiber, F. W., "How Package Development Can Work Most Effectively With Marketing", *Package Development*, Mar./Apr. 1975, pp. 23-24.
Seely, Robert E., "Establishing An Incoming Quality Control Program For Packaging", *Package Development*, Nov./Dec. 1974, pp. 8-16.
Seely, Robert E., "A Procedure For Estimating The Quality Of A Production Lot", *Package Development*, Nov./Dec. 1975, pp. 14-16.
Stern, E. George, *The Nail, An Indispensable Fastener*, ASME #51-S-18.
"Structural Redesign Cuts Cost 50 Per Cent", *Package Engineering*, Dec. 1972, Vol. 17, pp. 56-57.
TAPPI, *Fabrication Manual for Corrugated Box Plants, Pulp and Paper Manufacture: Bibliography and United States Patents, Annual*.
TAPPI *Standards and Provisional Methods*.
Technical Association of the Pulp and Paper Industry, *TAPPI Monograph Series, Special Association Publications*.

"Testing Packages Under 'Safe Transit' Procedure Wins Recognition From Carriers", *Traffic World*, Vol. 129, No. 12, March 25, 1967, pp. 44-46.

"Thinking Steps in a Program to Standardize Shipping Containers", *Package Development*, July/August 1974, pp. 30-34.

"Tight Loads For Mixed-Up Cases Enhance Profits Three Ways", *Package Engineering*, Apr. 1973, Vol. 18, No. 4, pp. 68-69.

"Total Systems Approach to Corrugated Container Design, A", *Package Development*, May/June 1974.

"Unit-Load Shrink-Wrapping Comes of Age", *Modern Materials Handling*, May 1973, Vol. 28, No. 5, pp. 39-54.

Van Crane Brock, Allen, "Pre-Shipment Testing: A Potent Damage-Prevention Weapon", *Traffic Management*, April 1968, pp. 51-55.

Weber, R. Frank, "Applying the Systems Approach to Packaging", *Modern Materials Handling*, Aug. 1967, Vol. 22, No. 8, pp. 61-63.

"What? Shrink-Wrap Loads at $34°$?", *Package Engineering*, Aug. 1973, Vol. 18, No. 8, pp. 54-55.

Williams, Stewart E., "Recognizing, Analyzing and Solving Corrugated Box Problems", *Package Development*, Sept./Oct. 1974, pp. 63-71.

Commercial and Trade Association Publications

Aluminum Co. of America, *ALCOA Aluminum—Its Properties and Uses*.

American National Standard — *Specifications for Metal Drums and Pails*, American National Standards Institute, Inc.

American Paper Institute, *Combination Paperboard Standards*, Paperboard Group, N.Y. 1969.

Associated Cooperage Industries of America, Inc., *The Wooden Barrel Manual*.
  Grade Rules and Specifications (Slack and Tight Barrels).

Automatic Identification Manufacturers ... Present the Emerging Impact of Automatic Identification Systems on Material Handling, The Material Handling Institute.

Fibre Box Association, *Transportation & Packing Survey, 1952*.
  Showcase 70's.
  Statistical History of the Fibre Box Industry.
  Fibre Box Handbook, 100 Years U.S.A.

Glass Container Manufacturers Institute, Inc., *The Glass Container Industry*.
  The History of Glass Containers.

International Union of Marine Insurance, *Cargo Loss Prevention Recommendations*.

Insurance Co. of North America Companies, *Ports of the World*.

Manual of Container Linings, Container Division, Jones & Laughlin Steel Corp.

National Paper Box Association, *The Rigid Paper Box (Set-up)*.

Package Machinery Directory, Packaging Machinery Manufacturers Institute, Washington, D.C.

Package Research Laboratory, *Wirebound Textbook*.
  Design Facts.

Packaging in Perspective — Report to the AD HOC Committee on Packaging, Jan. 1974, Arthur D. Little Inc.

Paper Shipping Sack Manufacturers Association, *General Manual Number 5*.

Paperboard Packaging Council, *The Folding Carton*.

The Port of New York Authority, *Guide to Air Shipping Via Port of New York*.

BIBLIOGRAPHY

Pressure-Sensitive Tape Council, Specifications and Technical Committee, *Test Methods for Pressure Sensitive Tapes.*
Sutermeister, Edwin, *The Story of Papermaking,* S. D. Warren Co., Boston, 1954.
U.S.I. Film Products, Division of National Distillers and Chemical Corp., *Heat Sealing Polyolefin Films.*
Wirebound Manufacturers Association, Inc., *Wirebound Specification Data.*

INDEX

Abrasion, 90, 115, 136, 137
Absorption, boxboard, 136-137
Acetate tapes, film and fibre, 380
Addressograph, 480
Adhesion, boxboard, 136
  quality corrugated fibreboard, 72, 102
Adhesives, closure, 362-372, 444-446
Adhesives, corrugated fibreboard, 71, 72, 108, 109
  pressure sensitive, 115, 373, 374, 403
Adhesive, easy-opening, 363, 452, 453, 455, 466
  gumming, 375
Adhesive patterns, 372
Adhesive setting time, 362-363, 367, 368, 370
Aluminum foil, 289, 340-342
All bound box, 253-254
American Management Association, 29
American Paper Institute, 126, 127, 128, 132, 135
American Society for Testing Materials, 27
American Standards Association, 29
Animal adhesive, 375
Armament Development Test Center, 29
Asphalt laminant, 176, 289, 324, 325
Association of American Railroads, 29
Auger packing, 186, 188, 306
Automation, 59
Auxiliary aids, 515

Bag flattener, 205, 206
Bagging, 320, 321, 325
Bags, 173
Bag tape sealer, 383
Bail, 280, 304
Baler pack, 183, 186
Banding, rigid boxes, 141, 142
Barbed nail, 216
Barge, 3
Barrel, 280
  wood slack, 281, 282
  wood tight, 281, 282
Barrels, metal, statistics, 15
Barriers, 32
Barrier materials, 313, 314
Base, nailed wooden crate, 228, 229
  wirebound crate, 257, 259, 260, 263
Battens, nailed wood, 228, 231
  wirebound, 254, 255
Belsinger style box, 418
Bending board, 127, 132, 133
Binding wires, 253, 254, 260
Binary coded digit, 475-478
Bleached manila lined board, 128
Blocking and bracing, 90, 230, 240, 257
Bogus corrugating medium, 71
Bolts, 228, 230, 231, 360, 361
Bottle washing, 522
Boxboard definitions, 128-135
  physical properties, 136-137
  specifications, 162
Boxes, cleated, 266-277
  corrugated fibreboard, 77-90
  nailed wooden, 209-251
  rigid, 137-147
  wirebound, 252-257
Box nails, 216, 217

535

Box shook, 210, 245, 246
Box shops, 5
Box wrap, specifications, 142
Bright nail, 216
Brightness, boxboard, 127, 136
Brightwood, 150
Brush and roller stencil ink applicators, 488
Bulk sack, 183
Bullseye, 477
Bundle, boxboard, 129
Bundling, 33, 412-447, 502
  materials and equipment, 417-441
  selection factors, 441-450
Bureau of Explosives, 29
Burlap, 176-177
Bursting strength, boxboard, 101, 136, 162, 163
  fibreboard, 71, 74, 101, 102
Butt joint, 233
By-products, 58

Cables, 417
Cady tester, 71, 136
Caliper, boxboard, 130, 131, 132-136, 144, 161-163
Can, definition, 280
  metal, statistics, 14
Cane fibreboard, 343, 344
Cane fibres, 58, 347
Canister, 280
Canning, 31
Carboy, 282
Carloading, 42, 328, 420
Carrier loading practice, 33
Carrier regulations, 74, 108, 399-402, 445, 448, 449, 450, 466, 499-500
Carrier utilization, 33
Cartoning equipment, 167
Carton wrapping, 504-506
Case forming, 504-505
Case loaders, 503, 504, 508-509
Case opener, 369, 371
Case removal mechanism, 371, 372
Case sealer, 364-371, 461, 503
Case unloaders, 510, 511
Cask, 280
Cautionary markings, 469, 471, 498, 500
Cellophane, 328-329
Cellophane shavings, 522
Cellophane tape, 380
Cellulose acetate, 290, 329, 330, 333
Cellulose wadding, 343, 344
Cement coated nail, 216, 218

Center special overlap slotted container, 78, 80, 113
Center special slotted container, 78, 79, 113
Chain stitch, 192, 452, 454
Checking, wood, 214
Chimes, 301
Chip, bleached manila lined, 127
  boxboard, 127, 133
  corrugating medium, 71
  news vat-lined, 126
  white vat-lined, 126
Circuate wire, 388
Clay coated boxboard, 128
Cleaning, 522
Cleaning devices, 521-522
Clearance, 90
Cleated containers, 251, 266-276
  styles, 267-276
Cleats, wood, 226, 227, 231, 233, 234, 235, 237, 241, 242, 251, 254, 256, 257, 261, 266-269
Clinches, 390-395
Closed-cell plastics, 352-355
Closure, adhesives, 362-373
  definition, 360
  fibre cans, 283, 285
  fibre drums, 290, 291, 387
  looped wire, 253
  multi-wall sack, 19, 197
  selection factors, 365-373
  tape, 373-386, 444-445
  water resistant, 375
Closures, fibreboard boxes, 97, 362-405
Cloth tape, 378, 380-381
Coated paper, 322
Coatings, fibreboard, 115
Code formats, 474-478
Coding, 468-470, 472-479
  selection factors, 479
Coding identification, 467, 472
  appearance factors, 496, 497
  economic factors, 497-499
  regulations, 499-500
Cold, influence, 37
Collars, corrugated fibreboard, 90, 92, 93
Color selection, 31
Combiner, fibreboard, 461
Common nail, 216
Compression, 37, 40, 44-45, 104, 117, 363, 364, 453, 460
  corrugated fibreboard, 402, 403
Compression unit, 363-366, 369, 371

INDEX 537

Conical spring suspension, 354
Configuration considerations, 31
Consulting services, 16
Consumer packaging, definition, 31
Containers, corrugated and solid fibreboard, 77-90
  statistics, 64, 65
  cylindrical, 277-312
    definitions, 279-282
  early European, 3
  fibre cylindrical, 283-294
  glass, 4, 283
    statistics, 13
  metal, 4
    statistics, 14-15
  packaging, statistics, 10
  paper and paperboard, 4, 125-172
    statistics, 15, 124
  plastic cylindrical, 294-296
  primitive types, 1
  reusable, 54
  rubber inflatable, 34
  steel cylindrical, 296-305
  textile, 4, 175-180
    statistics, 15
  wooden, 4, 209-241, 360, 361, 442, 443, 446, 447
  wood cylindrical, 282-283
Containerization, 33
Contamination, 37, 40, 50
Contract packagers, 5
Conveyor imprinter, 479-482
Convolute winding, 287-288
Cooler nail, 216, 217
Cordage, 417, 432, 433
Corner constructions, wooden containers, 237-240
Corner protectors, fibreboard, 91-93
Corrosion, 421, 521-522
Corrugated blocks, 91, 96
Corrugated fibreboard, 69-122, 261, 266, 345-347
  components, 71, 74
  cushioning, 67, 68, 90, 103-104, 115, 117, 345, 346
  double faced, 66, 67, 74, 75, 103, 104
  double wall, 66, 67, 74, 75, 103, 104
  interior packing pieces, 90-97
  single faced, 66, 67, 104, 345, 346
  single wall, 66, 67
  triple wall, 66, 67
  unlined, 63, 66
Corrugated metal fasteners, 216, 237, 248
Corrugating medium, 71, 76, 103, 116

Corrugators, 5
Cost estimating, fibreboard, 114
Cotton, 4, 176, 177, 323, 417
Count, boxboard, 129
Counting devices, 514-515
Covering agents, 470
Covering materials, rigid boxes, 141, 142
Crates, wirebound, 257-260
Crates, wooden, 209-251, 228-230
Creped back paper tape, 380
Creped paper, 176
Crimp cover pail, 305
Crimping, strapping, 426
Cross grain wood, 214
Crushing, 40
Cryovac, 334
Cupping, wood, 213, 215
Curled hair, 347, 348
Cushioning, 58
  corrugated fibreboard, 68, 69, 90, 104, 115, 117, 345, 346
Cushioning devices, 355, 356
Cushioning materials, 342-356
  definition, 314
Cushioning materials storage and dispensing equipment, 517, 518
Cylinder kraft, 71, 73, 314-317
Cylinder paper machines, 64, 125
Cylindrical container imprinting, 468, 472, 482
Cylindrical containers, equipment, 309-312

Damage causes, 51, 359
Damage level, 38
Dangerous products, 33, 36, 50
Data processing printer, stenciling, 489
Degreasing, 521
Defects in wood, 214-215
Department of Commerce, 29
Department of Defense, 29
Department of Transportation, 29, 500
Design, fibreboard, 97-104
  folding cartons, 161-166
  rigid boxes, 142-147 147
  wirebound boxes and crates, 260-261, 263
  wooden boxes and crates, 231-243
    types of loads, 231-233
Diagonal fold style tray, 151, 171
Diagonals, wood, 237, 240, 263, 264
Dipping, 521
Dispenser, labels, 483-485
Dispensers, tape, 375-386, 419-420, 444, 446

Display features, 89, 138
Distortion, 37, 339
Distribution hazards, 37-52
Distribution packaging, definition, 31
  dynamics, 57-59
  economics, 52-57
  fabrics, 323-328
  fundamentals, 33-36
  paper, 313-328
    properties, 314-322
  scope, 32, 33, 57-59
Double cover box, 81
Double-lined slide box, 84
Double slide box, 85, 86
Drum, definition, 280
  fibre, 285-294
  steel, 297-301
    statistics, 15
  light gauge openings and covers, 302-304
Dual purpose boxes, 89, 90
Duplex tape, 376
Dust flaps, folding cartons, 152

Earthenware, 2, 3, 279, 282, 283
Easy-opening, 21, 32, 451-466
  adhesives, 363, 452, 453, 455, 456
  application equipment, 460-462
  fibreboard box, 452-460
    manufacturer's joint, 456
    peripheral box, 457-460
  sealing tape, 451, 452, 457
  selection factors, 462-463
  strength considerations, 465-466
Economics, bundling, 447-448
  closures, 403, 405
  considerations, 31
  easy opening, 463-465
  fibreboard boxes, 111-114
  fiber box user fabrication, 120-122
  folding cartons, 161-166
  manufacturer's joints, 110, 111, 114
  marking, labeling, and coding, 497-499
  nailed wooden containers, 243, 244
  packaging machinery, 522-523
  reinforcing, 447-448
  rigid boxes, 144-147
  rigid box fabrication, 144-147
  shipping sacks, 197, 206
  wirebound containers, 261, 262
Edge slats, 263
Electronic counting, 514-515
Electronic data processing, 467
Embossing, 472
Empty-box sealing, 370, 371

Ended style rigid box, 138, 140
End-opening box, 111, 369-371, 455
Engineering Research and Development Laboratories, 29
Environmental hazards, 37-40, 105
Equipment, cylindrical containers, 309-312
  counting, 502-503, 514-515
  folding carton packaging, 166-169
  heat sealing, 518-520
  imprinting, 479-482
  labeling, 483-485
  loading and unloading, 507-511
  packaging engineer, 26, 27
  selection factors, 56, 365, 367
  shipping, sack filling and weighing, 186-191, 193
  stenciling, 486-490
  strapping, 421-431
  ultrasonic sealing, 520, 521
  waste utilization, 522
  weighing, 511-514
  wrapping, 504-506
"Evans" container, 276
Excelsior, 97, 343, 350, 355
Expendable containers, 55
Explosives, 50
Extruders, 5

Face boards, 261
Facings, corrugated fibreboard, 71, 74-77, 101
Fasteners, wood, 216-226, 243, 265, 360-362
Feed-wheel stretcher, 424
Felt, 243, 348
Fibre cans, 283-285
  containers, cylindrical, 283
Fibre drums, 285-294
  barriers, 289
  closures, 290
  convolute winding, 288
  linings and coatings, 290
  specifications and regulations, 291
  top and bottom, 286
  types, 289
Fibreboard, history, 63
  grades, 73-77, 101, 105
  manufacture, 64
  physical properties, 97-104
  special treatments and constructions, 114, 119
Fibreboard boxes, properties, 104-111
  statistics, 12
  styles, 77-90
  user fabrication, 119-122

INDEX

Fibreboard footage, formulas, 111-114
Filler board, 73
Filling, 186-191, 307, 309-312, 503
Film factors, 519
Films, packaging, 328-337
  properties, 330-332
  transparent, statistics, 16
Finish boxboard, 131
  surface, facings, 71
Five panel folder, 82, 83, 418
Fixed-beam readers or scanners, 492-493
Flap contact area, 372
Flap gap, 373
Flat-back paper tape, 380
Flat crush resistance, 103, 104
Flat metal strapping, 420-425, 446-447
"Flat-pack closer", 265
Flutes, corrugated fibreboard, 68-70, 101-105, 345
Foamed latex, 348, 349
Foam, plastic, paper laminated, 117, 119
Foam rubber, 348, 349
Foil, laminations, 340-342
  metal, 173, 289, 337, 342
Folding carton, board footage, 163
  definition, 149
  lining, 158, 165, 170, 171
  manufacture, 149
  statistics, 114
  styles, 150-161
Folding cartons, 123-137, 149-172
  history, 123-125
Forest Products Laboratory, 29, 210
Fourdrinier kraft, 71, 73, 182, 271
Fourdrinier paper machines, 64, 126, 315
"Framerap", 88
"Frearson" screw head, 225
Friction covers, drums, 290, 302, 452
Fruit and vegetable nailed wooden containers, 240, 241
Full-overlap slotted containers, 78, 79, 112, 113, 418
Fungi, 241, 348

Gas transmission, 329
Gauge lists, boxboard, 131-135
Glass, 2, 3, 345
  containers, 4, 283
    statistics, 13
  fibres, 323
Glassine, 170, 174, 290, 316, 318, 320
Glassmaking, history, 2, 3
Glue applicator, labels, 483
Glued closures, 362-373

Glued joints, 107
Glue patterns, 372-373
Governmental agencies, 29, 499-500
Grain, wood, 214, 220, 237, 238
Gravity feed filling, 186-188, 307
Grease proof paper, 314-318, 321
Grease resistance, 314, 317, 321
Gross weigher, 188-191, 511
"Guard bars", 478
Gummed labels, 483
Gummed tape, 373-378, 402, 403, 404
Gum reactivation, 375

Half-slotted containers, 80, 115
Hammer staples, 397, 398
Hand-held readers or scanners, 492, 496
Hand holes, fibreboard boxes, 116, 455, 456
Hand stencil applicator, 489
Handling considerations, 32, 52-53
"Hardened steel nails", 219
Hard woods, 211-212, 261
Hawser, 417
Hazards, early trade, 2
  environmental, 37-42, 105
  miscellaneous, 37, 50
  physical, 37, 40-52
  transit, 32, 402
Heat, influence, 37, 109
Heat seal coated labels, 485
Heat sealing, 319, 329, 342, 518-520
"Hinged-corner" crate, 276
Hinged cover rigid box, 139, 147
Honeycomb paper structures, 117, 118, 348
Hood crate, 257, 259
Humping and switching, 42, 51
Hydraulic tilting table, 515, 516

Identification considerations, 32, 467-470
Impacts, 37, 41-44, 111, 442, 445
Impeller-type filling, 307, 310
Impeller-type valve packer, 187, 188
Impregnations, 115, 320
Imprinting, 376, 472, 479-482
Infestation, 37, 50, 177, 329, 348
Ink, stencil, 486
Ink penetration, boxboard, 136
Inorganic fibres, 58, 323
Insecticides, 50
Insect repellants, 50
Insulation, 90, 117
Interior packing pieces, fibreboard, 90-97

International Air Transit Association, 500
International Material Management Society, 28
Invoice preparation, 470, 487, 489

Jaw-type heat sealer, 518
Joints, lumber, 221, 233, 236, 237
Jute, 417
  fibre, 176
  paper, 315
  paperboard, 71, 73

Kegs, metal, statistics, 15
  wooden, 57, 280
Kit, 280
Knots, wood, 214, 215
Kraft, shipping sack, 180-186
Kraft paper, 314-317
Kraft paperboard, 71

Labeling, 31, 467-500
  selection factors, 496-497
Labels, 469-472
  computer-generated, 470
  machine readable, 470
  pressure sensitive, 470
Lag screws, 361
Laminated fabrics, 323
Laminated films, 336
Laminated papers, 323-325
Laminations, foil, 340-342
  shipping sacks, 180, 181
  textile, 175-176
  types and properties, 338-341
Laminators, 5
Latex laminant, 324, 325
Lead foil, 337
Lever tape dispensers, 375, 376, 382-384
Lidding, wooden containers, 241, 248-250
Light gauge drums and pails, cover features, 302-307
Linderman joint, 233, 236, 237
Linen rags, 4
Liners, corrugated fibreboard, 90, 92, 93
  wirebound boxes, 254, 255
Linings, fibre cans, 283, 284
  fibre drums, 287, 289
  steel containers, 307, 308
Linseed oil, 486
Loaders and unloaders, 507-511
Load types, wooden containers, 231, 232, 233
Lock corner folding carton, 153, 154
"Lock Corner" wooden box, 228

Lock-end folding carton, 156, 157, 159, 163, 171
Locking rings, 291, 303
Looped wire closures, 252, 253
Low tensile adhesive, 455
Lug cover, 304
Lumber defects, 214-215
Lumber, resawn, 252, 254
  rotary cut, 252, 254

Macerated paper, 343
Machine clamp and dwell time, 519
Machine-readable label, 470
Manila boxboard, 128
Manufacturer's joints, fibreboard, 97, 105, 107-111, 122, 456, 458, 459
Marketing considerations, 496-497
Marking, 467-500
  selection factors, 496-499
Maritime Consultive Organization Regulations, 500
Material fatigue, 44, 105
Materials, cost, 53-55
Materials, packaging engineer, 25, 26
Matted fibre structures, 343-348
Mechanical counting, 514-515
Mechanization, packaging, 59
  wirebound box and crate assembly, 264-266
Metal stayed set-up boxes, 146, 147
Microorganisms, 50
"Mimeograph", 487
Miscellaneous wood defects, 215
Mitered cleats, 256, 261
Moisture, influence on stacking strength, 44-45
Moisture content, wood, 213, 214
Mold, 110, 115
Mortise cleats, 256, 261
Moving-beam readers or scanners, 492-495
Mullen tester, 74, 101, 102, 136
Multi-corrugating material, 117, 118, 345
Multigraph, 480
Multilith, 480
Multiple head stitcher, 391
Mylar, 334, 335, 336

Nailed wooden boxes, 57, 209-251
  history of, 209
  manufacturing techniques, 209-210
Nail holding, 216-221
Nailing errors, 222
Nailing machine, 247, 248
Nail-less and nail-on strapping, 420

INDEX 541

Nails, 216-221, 247, 360
National Motor Freight Classification, 74, 400
National Safe Transit Association, 29, 38-43
Natural fibres, 58, 323
Naval Logistics Engineering Group, 23
Navy Materials Handling Laboratory, 29
Net weigher, 188, 189
News, filled, 126
  white patent coated, 127
Non-bending boxboard, 126, 127, 132
Non-cellular materials, 343-348
Non-skid fibre boxes, 115
Notching, strapping, 426
Nylon, 323, 330, 374, 417, 431, 433, 445

One-piece folder, 82, 83
Open-cell plastics, 348-349, 350-351
Open crate, 228, 230
Open-mouth sack, 180, 182, 183, 185-186, 193, 201, 202
Optical scanning systems, 495
Osnaburg cloth, 176, 177
Overlap slotted container, 78, 79, 418
Overwrap, 315, 328

Package engineering, definition, 17
  education, 25, 30, 31
  function, 18, 22-24, 25
  organization, 21, 24, 25, 26
Packaging, definition, 21-22
  effect of transportation, 3
  equipment classification, 502-504
  machinery, 59, 119-122, 166-172, 309-312, 394-398, 467-474, 503-522
    selection, 522
    specifications, 523-526
  materials, selection, 58-59
  value and growth, 8-9
  service organizations, 5
  value, 6-7
Packaging committees, 20-22
Packaging coordinator, 24
Packaging Engineer, 25
Packaging foils, 337, 340, 341
Packaging films, 328-337
Packaging industry, evolution, 2
  statistics, 4-15
Packaging Institute, 28
Packaging papers, 314-328
  coated papers, 322
  glassine, 320
  laminated papers, 323

plain greaseproof, 321
pouch paper, 322
reinforced papers, 323-328
vegetable parchment, 322
waxed papers, 322
Packing benches, 515-517
Packing, definition, 52
Pads, corrugated fibreboard, 90, 92
Pails, definition, 280
  metal, statistics, 15
  steel, 304-307
Pallet boxes, 54, 241-242
Pallet patterns, 365, 469
"Panel-Lox-Box", 274
Paper, extensible, 117, 181, 207
  history of manufacturing, 3
Paper and paperboard statistics, 124
Paperboard, 125-137, 149, 347
Paper-overlaid veneer, 209, 266, 271-274
Papier mâché, 346, 347
Papers, coarse, 4, 314-315
Paraffin, 115
Parchment, 170, 290, 316, 317
  vegetable, 322
Partitions, fibreboard slotted, 90, 91, 94, 95
Perforating, 453, 454-455, 457-460
Perishables, spoilage, 2, 240-241
"Phillip" screw head, 225
Pilferage, 37, 50, 465, 468, 490
Pin feed labels, 485
Pinch type sack, 182, 185, 186
Pitch, code pattern, 494
Plain back labels, 483
Plastic cylindrical containers, 294-296
Plastics, flexible, 58, 352-355
  rigid, 58, 348-349, 350-351
Plastic strapping, 431-441
Pliofilm, 290, 329, 331, 333
"Ply-Fold" box, 275, 276
Plywood, 209, 254, 266, 269, 270, 272
Plywood boxes, cleated, 267-270
Plywood drums, 282, 283
Polyester, 331, 334, 335
  film, 336
  strapping, 431-435
Polyethylene, 115, 175-176, 289, 290, 331, 333-334
  container, 294-296
  foam, 351, 353
  molded film, 348-350
Polypropylene foam, 351, 353
  strapping, 431, 432, 434

542 DISTRIBUTION PACKAGING

Polystyrene, 332, 334
  expanded, 345, 346, 351, 353-355
Polyurethane, flexible, 348
  foam-in-place, 348, 349, 351
Polyvinyl alcohol, 332
Polyvinyl Chloride film, 332, 334-336
Polyvinylidene-chloride film, 333-335
Positioning, 90
Post Office Department, 23
Post-type stapler, 394
Post-type stitcher, 391, 393
Pouch forming, 518-519
Pouch paper, 322
Power strapping tools, 428-431
Pressure, influence, 38
Pressure-sensitive labels, 470, 483-485
Pressure-sensitive tapes, 378-386, 484
  types, 380-381, 490-491
  dispensers, 381-386, 491
Printer-slotter, 462
Printing, 468, 469, 479-482
  attachments, 481, 492
  patterns, 31, 105, 454, 462
  positioning errors, 465
  quality, 115, 496, 497
  weight control, 412
Product identification, 19, 31, 467-469
  appearance factors, 496, 497
  economic factors, 497-499
  regulations, 499-500
Production considerations, 31
Professional societies, packaging, 27-30
Protection requirements, 31, 36-52, 111, 112, 442-445
Pull and tear tape dispenser, 375, 382
Pulp, molded, 119, 347
  wood, 4, 347
Pulp additives, 115
Puncture, 37, 102, 139, 279, 323, 460
  corrugated fibreboard, 69, 71, 102, 103

Quality control, 19, 105

Rack gear stretcher, 424
Radioactive materials, 50
Radio pads, fibreboard, 91, 92
Railroad cars, 3, 7, 326
Rat trap, fibreboard, 91, 92
Rayon, 323, 374, 417, 445
Reflectances, scanning, 474
Reflection, scanning, 474
Regional shipping damages, 40
Regular number, boxboard, 129

Regular slotted containers, 77, 78, 112, 113, 391, 399, 418, 454-455
Regulatory considerations, 31
Reinforced papers, 323-328
Reinforcing, 32, 408-447
  materials and equipment, 417-441
  selection factors, 441-450
Reinforcement, fibreboard boxes, 97, 116-117, 417
  shipping sacks, 206, 207
  wooden containers, 231, 240, 408, 435, 442, 447
Rejection system, weighing, 511-512
Release agent, 115
Research organizations, 5, 10, 17
Resin density and melt index, 519
Retention, 90, 402
Retractable anvil-type stapler, 394-397
Returnable containers, 297, 299
Reusable stencil, 486
Reusability, 54, 55, 58, 240, 241-243, 253, 374, 452
Rigid box, definition, 137
  history, 125
  manufacture, 137-138
  statistics, 124
  styles, 138-142
Rip cord, 452, 464
Rock fastener, 252, 253, 265, 266, 452
Rodenticides, 50
Rodent repellants, 50, 115
Rolling hoops, 301, 302
Rope, 417
Rotary heat sealer, 518
Rotational wrap, 439, 440
Round wire strapping, 420, 421, 423, 426, 427-431
Rubber, 348
Rubber container, inflatable, 34
Rubber-hydrochloride film, 329, 331, 333, 334, 335

Sack, definition, 173-175
Sacking, 336
Saran, 331, 333-335
Satchel bottom sack, pasted, 180
Sawdust, 343, 350, 355
Scales, 511-514
Scannable codes, 473-478
Scanning, 492-496
Screw cap, pails, 305, 307
Screws, wooden boxes, 223-226, 360-361
"Screwtite" nails, 219

INDEX 543

Scuffing, 115, 322, 483
Seal-end folding cartons, 155-157, 170, 171
Sealer, strapping, 423, 424
Sealers, end-opening boxes, 370
Sealing, 31, 32, 360-405
Sealing, adhesives, 363-386
  manual, 363
  fully-automatic, 364-371
  semi-automatic, 364
Seal-less strapping tools, 427, 428
Seals, strapping, 426, 427, 428
  wrapping, 519
Seaming, steel drums, 298, 300, 301
Seaming machine, fibre cans, 284
Seams, shipping sacks, 177-180
Selective counting, 515
"Self-drilling" wood screw, 224
Semi-chemical board, 71
Separation, 90
Sewed open-corner sack, 182, 183
Sewing, 177-180, 185, 191-197
Sewing machines, 193-194
Shape, influence, 105, 111-114, 228, 262, 277, 279, 370
Shear, 50
Sheathed crates, 228-230
Ship lap joint, 233, 237
Shipping identification, 467, 469-471
  appearance factors, 496, 497
  economic factors, 497-499
  regulations, 499-500
Shipping sacks, 173-207
  history, 174-177
  multiwall paper, 57, 180-207
    constructions, 181, 182
  selection, 205-206
  specifications, 198-203
  styles, 182-186
  textile and laminated, 175-180
Shock mounts, 231, 348, 349, 353-356
Shock recording device, 42
Shock resistance, wood, 211, 212
Shook, box, 210, 246, 247
Short case sealer, 365, 366
Shredders, 522
Shrink films, 337, 436, 437, 440
  selector chart, 334-335
Siftproof folding carton, 170
Single-lined slide box, 84, 85
Single manila-lined board, 128
Single trip containers, 297, 298
Single-use stencil, 486
Sinker nail, 216, 217
Sisal, 323, 374, 417

Size, influence, 58, 229, 230, 260, 261
Skew, code pattern, 494
Slide style rigid box, 138, 139
Slitting, 457-461
Slotting, 454-455
Society of Packaging and Handling Engineers, 28
Sodium silicate adhesive, 72, 362
"Soft nails", 219
Soft woods, 211-212, 261
Solid bleached sulfate paper board, 128
Solid fibreboard, 64, 73, 252, 266, 270
Spiral stretch wrap, 440-441
Spiral winding, 288, 289
Spiral wrapping, 440-441
Split drum tensioning tool, 424-426
Splitting resistance wood, 211, 212
Sponge rubber, 348
Spring pads, fibreboards, 91-93
Springs, 345, 355
Stacking, 44, 105, 117, 241, 270, 465
Stainless steel strapping, 421
Stamping, 472
Staples, wirebound boxes, 261, 386
  wooden boxes, 221, 223, 247, 386
Staple types, 388-391
Stapling, 387, 394-399, 403
Stapling equipment, 394-398
  anvil stapler, 394
  hammer stapler, 397, 398
Starch adhesive, 72, 73
Stays, rigid boxes, 141, 142, 146, 147
Steamship hold, 3
Steel drums, 297-301
Steel strapping, 420-425
  properties, 421-423
Stencil board, 486
Stenciling, 469
  methods, 486-490
Stepped end bag, 207
"Stiff-stock nails", 219
Sterilization, 522
Stitch closure, 387-393
Stitch defects, 392, 393
Stitch types, 390-391
Stitch yield, 388-389, 403
Stitched joints, 108
Stitching equipment, 390-393
Stitching head, 391
Stitching wire, 388-390, 400
Stock control, 468
"Straparound" box, 273
Strap finishes, 420-421

544   DISTRIBUTION PACKAGING

Strapping, 33, 107, 231, 242, 420-434
   equipment, 421-431
Strap yield, 442
Straw, 2, 4, 343
Stresses, dynamic, 40, 44, 50
   static, 40
Stretcher, strapping, 423-424
String, 417
Strip covering, rigid boxes, 141, 142
Strip staying, rigid boxes, 142-143
Subsurface filling, 310
Strippable reinforced gummed tape, 456, 457
Surface design considerations, 31
Suspension, 90, 345, 355
Sunburst, 459
Supplier-printed label, 497
Synthetic fabrics, 323
Synthetic fibres, 58, 323, 325, 381
Synthetic resins, 362

Tacker Stapler, 397-399
Tag board, 486
Tagging, 469, 479, 480, 485, 486
Tape, easy-opening, 452-454
   filament, pressure sensitive, 381, 419, 444-446
   gummed cloth, 107, 108, 374, 380
   gummed filament, 374, 419, 444
   gummed paper, 107, 108, 374, 403-405, 491
   gummed reinforced paper, 107, 108, 374, 403-405
   heat-activated, 367, 368, 370, 385, 485
   pressure-sensitive, 367, 373-384, 402-403, 419, 445, 491
   printed, 479
Tape closure, 369-386, 402-403, 490
   pilferage deterrent, 490
Tape, cloth, pressure sensitive, 378
Tape dispensers 375-377, 381-384, 419, 444, 446, 491
   application equipment, 491-492
   attachment, 491
Tape joints, 107-108
Tapes, 373-386
   dispenser's, 375, 376, 491
   gummed, 374, 375, 491
   pressure sensitive, 378-386, 490-491
   sealers, 377
Tare weight, 52, 244, 267, 300
Tariff, R.M. Graziano, 500

Tear, 40, 50
   boxboard, 136, 137, 163
   corrugated fibreboard, 69
   shipping sack kraft, 181, 182
Tear tape, 454
Technical Association of the Pulp and Paper Industry, 28
Telescope box, fibreboard, 387, 407, 418, 445
Telescope rigid box, 139, 143, 147
Telescope style folding carton, 153
Tensile strength, boxboard, 136, 137
   shipping sack kraft, 181, 182
   tape, 381
Tension, 50
   mount, 356
Tensioner, strapping, 427
Testing, boxboard, 136, 137
   combined corrugated, fibreboard, 101-104
   fibreboard boxes, 44, 45, 104, 105, 109, 110
   machine manufacturers, 523
   manufacturer's joints, 110
Testing and research organizations, 16
Tests, drop, 203
   sack closure, 194-197
   sacks, 203
Textile and textile laminated sacks, 175-180
Textile bags, 460
Textiles, 3, 176-177
Three-way corner, 338-340
Three-piece folder, 83
Tierce, 280
Tie-stitched joint, 107
Tightener, strapping, 423
Tight-head drum, 297-299, 304
Tight-head pail, 304, 305
Tight wrapping, rigid boxes, 141, 142
Tilt, code pattern, 494
Tin-foil, 337
Tissue paper, 115, 320
Tongue and groove, 233, 236, 237, 256, 261
Tops, wirebound crates, 255, 259, 260
Torsion, 50
Torsion bar suspension, 356
Tote boxes, 241-243
Trade associations, 17
Trays, corrugated fibreboard, 90, 92, 93
Tray-type folding cartons, 150-154, 171, 172
Trimming, rigid boxes, 141, 142

INDEX 545

Triple slide box, 85, 86
Transportation and packing survey, 46-52
Tropical environment, influence, 38
Tub, 280
Tube-type folding cartons, 154-161, 170-172
Tuck-end folding cartons, 156-160, 171
Twine, 192, 360, 417
"Twinfast" wood screw, 224
Twist wire closure, 252, 253, 257
Two-piece folder, 82, 83
Tying machine, 417
Tying, sack, 191, 192
Typewritten stencil, 488

Ultrasonic sealing equipment, 520-521
Unicellular rubber, 348
Uniform Freight Classification, 74, 181, 182, 198, 291-294, 400, 454, 458, 500
United Nations Label System, 500
United Parcel Service Guide, 500
U.S. Army Laboratories, 29
U.S. Department of Agriculture, 29
U.S. Postal Regulations, 500
Unit load, 32, 407
Unitizing, 2, 407, 412-416, 504
Unitizing with films, 436-441
　rotational wrap, 439-441
　shrink films, 436, 437
　stretch films, 436, 440-441
Universal Product Code, 468, 470, 475, 478
Unscrambler, 510, 511
Unsheathed crates, 228-230
Upright crate, 257, 265, 266
Urea formaldehyde resin, 73
Utility considerations, 31

Vacuum filling, 310
Valve packer, 186-188
Valve-type sack, 180, 182-185, 193, 203, 204
V-diagonals, 240, 263
Veneer, 214, 251, 267-273, 283
Vent holes, fibreboard boxes, 116
Vents, pail, 306, 307
Vibration, 37, 41
Vibratory feeder, 190, 191
Vinyl chloride, 336
Vinyl films, 332-336
Vinylidene chloride, 333
Volatile corrosion inhibitor, 522

Volume calculators, 515
Volumetric filling, 310

Walking gripper, 434
Waste products, 58, 315
Water, influence, 38
Water vapor, influence, 38, 195
　transmission, 314, 322
"Watkins" box, 275, 276
Waxed paper, 115, 170, 314, 318, 319, 322
Weatherproof fibreboards, 72, 115-116, 270, 442
Web-wrapping, 438, 440
Weighers, 511-514
Weighing, shipping sacks, 186-190, 193
Weight controls, 511-512
Weight, influence, 229-230
Wet strength paper, 72, 176, 177, 181, 182, 314, 315, 317, 320, 322
White patent coated board, 128
Windows, box, 138
Wirebound box, styles, 252-257
Wirebound box and crate assembly, 264-266
Wirebound containers, 251-276
Wirebound crates, 257-260
Wood, board footage, 243, 244
Wood characteristics, 210-211
Wooden Boxes, 209-251, 408
　selection guides, 234, 235
　styles, 226-228
Wooden container, fabrication and assembly, 244-250
Wooden crates, 209-251
　styles, 228-230
Wood groupings, 211-212, 220, 260, 261, 443
Wood pulp, 347
Wood strength, 211-215
Wood thickness, 220, 233, 236, 237
Wood-working machinery, 245-250
Wool, 348
Wrap-around crates, 257-259
Wrapping, 171, 314, 315, 504-506
Wrapping machines, 504-506
Wright-Air Development Center, 29

X-diagonal, 239, 240, 263

Zinc-coated strapping, 421
Zone, shock recording, 43

TS
195
.F74
1977

91638